Mitochondrial Dysfunction in Aging and Diseases of Aging

Mitochondrial Dysfunction in Aging and Diseases of Aging

Special Issue Editor
Richard H. Haas

MDPI • Basel • Beijing • Wuhan • Barcelona • Belgrade

Special Issue Editor
Richard H. Haas
UC San Diego School of Medicine
USA

Editorial Office
MDPI
St. Alban-Anlage 66
4052 Basel, Switzerland

This is a reprint of articles from the Special Issue published online in the open access journal *Biology* (ISSN 2079-7737) in 2019 (available at: https://www.mdpi.com/journal/biology/special_issues/Mitochondria).

For citation purposes, cite each article independently as indicated on the article page online and as indicated below:

LastName, A.A.; LastName, B.B.; LastName, C.C. Article Title. *Journal Name* **Year**, *Article Number*, Page Range.

ISBN 978-3-03921-327-6 (Pbk)
ISBN 978-3-03921-328-3 (PDF)

Cover image courtesy of Scott McAvoy.

© 2019 by the authors. Articles in this book are Open Access and distributed under the Creative Commons Attribution (CC BY) license, which allows users to download, copy and build upon published articles, as long as the author and publisher are properly credited, which ensures maximum dissemination and a wider impact of our publications.

The book as a whole is distributed by MDPI under the terms and conditions of the Creative Commons license CC BY-NC-ND.

Contents

About the Special Issue Editor vii

Preface to "Mitochondrial Dysfunction in Aging and Diseases of Aging" ix

Richard H. Haas
Mitochondrial Dysfunction in Aging and Diseases of Aging
Reprinted from: Biology **2019**, *8*, 48, doi:10.3390/biology8020048 1

Peter J. McGuire
Mitochondrial Dysfunction and the Aging Immune System
Reprinted from: Biology **2019**, *8*, 26, doi:10.3390/biology8020026 6

Robert K. Naviaux
Incomplete Healing as a Cause of Aging: The Role of Mitochondria and the Cell Danger Response
Reprinted from: Biology **2019**, *8*, 27, doi:10.3390/biology8020027 16

Isabella Peixoto de Barcelos and Richard H. Haas
CoQ10 and Aging
Reprinted from: Biology **2019**, *8*, 28, doi:10.3390/biology8020028 34

Roisin Stout and Mark Birch-Machin
Mitochondria's Role in Skin Ageing
Reprinted from: Biology **2019**, *8*, 29, doi:10.3390/biology8020029 56

Sunil J. Wimalawansa
Vitamin D Deficiency: Effects on Oxidative Stress, Epigenetics, Gene Regulation, and Aging
Reprinted from: Biology **2019**, *8*, 30, doi:10.3390/biology8020030 68

Janis T. Eells
Mitochondrial Dysfunction in the Aging Retina
Reprinted from: Biology **2019**, *8*, 31, doi:10.3390/biology8020031 83

Yvonne Will, Jefry E. Shields and Kendall B. Wallace
Drug-Induced Mitochondrial Toxicity in the Geriatric Population: Challenges and Future Directions
Reprinted from: Biology **2019**, *8*, 32, doi:10.3390/biology8020032 92

Magdalene K Montgomery
Mitochondrial Dysfunction and Diabetes: Is Mitochondrial Transfer a Friend or Foe?
Reprinted from: Biology **2019**, *8*, 33, doi:10.3390/biology8020033 106

Jason Duran, Armando Martinez and Eric Adler
Cardiovascular Manifestations of Mitochondrial Disease
Reprinted from: Biology **2019**, *8*, 34, doi:10.3390/biology8020034 123

Nima B. Fakouri, Thomas Lau Hansen, Claus Desler, Sharath Anugula and Lene Juel Rasmussen
From Powerhouse to Perpetrator—Mitochondria in Health and Disease
Reprinted from: Biology **2019**, *8*, 35, doi:10.3390/biology8020035 150

Veronica Granatiero and Giovanni Manfredi
Mitochondrial Transport and Turnover in the Pathogenesis of Amyotrophic Lateral Sclerosis
Reprinted from: *Biology* **2019**, *8*, 36, doi:10.3390/biology8020036 . . . . **165**

Isabella Peixoto de Barcelos, Regina M. Troxell and Jennifer S. Graves
Mitochondrial Dysfunction and Multiple Sclerosis
Reprinted from: *Biology* **2019**, *8*, 37, doi:10.3390/biology8020037 . . . . **179**

Chun Chen, Doug M. Turnbull and Amy K. Reeve
Mitochondrial Dysfunction in Parkinson's Disease—Cause or Consequence?
Reprinted from: *Biology* **2019**, *8*, 38, doi:10.3390/biology8020038 . . . . **195**

Ian Weidling and Russell H. Swerdlow
Mitochondrial Dysfunction and Stress Responses in Alzheimer's Disease
Reprinted from: *Biology* **2019**, *8*, 39, doi:10.3390/biology8020039 . . . . **221**

Mats I Nilsson and Mark A Tarnopolsky
Mitochondria and Aging—The Role of Exercise as a Countermeasure
Reprinted from: *Biology* **2019**, *8*, 40, doi:10.3390/biology8020040 . . . . **239**

About the Special Issue Editor

Richard Haas M.A., M.B., B.Chir., M.R.C.P. is a Professor of Neurosciences and Pediatrics at the University of California San Diego. He is Director of the Neurometabolic Clinics at Rady Children's Hospital San Diego and the UCSD Adult Neurometabolic Clinic. He directs the UCSD Mitochondrial Disease Laboratory and is Co-director of the UCSD Mitochondrial and Metabolic Disease Center. He directs the fellowship training program of the North American Mitochondrial Disease Consortium. Dr. Haas received his undergraduate training at Cambridge University and studied clinical medicine at University College Hospital in London, Great Ormond Street Hospital and the Royal Postgraduate Medical School Hammersmith. He trained in metabolism and pediatric neurology as NIH Fellow in Biochemistry of Mental Retardation at the University of Colorado, Denver. His clinical and research focus for 40 years has been on mitochondrial disease and in 1998 he was Founding President of the Mitochondrial Medicine Society. In 2016 he was awarded the Vanguard award by the United Mitochondrial Disease Foundation and he serves on the Scientific and Medical Advisory Board.

Preface to "Mitochondrial Dysfunction in Aging and Diseases of Aging"

Mitochondria are increasingly recognized as important players in the aging process. Most aging-related diseases, particularly neurodegenerative diseases, involve the mitochondria. A PubMed search for mitochondria and aging produced 348 articles from the first seven months of 2018. This is not surprising considering that mitochondria are not only involved in energy production through oxidative phosphorylation, but also play an important role in intracellular homeostasis, calcium balance, and the metabolism and interconversion of dietary substrates—fats, proteins, and carbohydrates in the fed and fasting states. They signal their metabolic state to the nucleus, to other cells and in response to stress. Mitochondria have their own protein synthetic apparatus and are able to replicate the pathways readily disrupted in disease and aging. They are constantly involved in fusion and fission, the balance of which is essential for cell health. These organelles participate in apoptosis, make most of the cell's free radicals, and are crucial for innate immunity. Mitochondrial DNA has an estimated 10-fold greater mutation rate than nuclear DNA, and less repair capacity, important factors in aging and cancer. Mitochondria are impacted by environmental factors and toxins, and different mtDNA haplogroups that originally adapted to different geographic origins are important contributors to disease. As mitochondria play a critical metabolic role in all organ systems, they are particularly impacted by disease and contribute to the aging process itself. The invited review articles in this supplement cover most of the common diseases of aging. This edition of *Biology* aims to review the current state of knowledge about the role of mitochondria in the aging process. The international group of contributing authors includes many leading experts in their fields.

Richard H. Haas
Special Issue Editor

Editorial

Mitochondrial Dysfunction in Aging and Diseases of Aging

Richard H. Haas [1,2]

1. Department of Neurosciences, University of California San Diego, San Diego, CA 92093-0935, USA; rhaas@ucsd.edu
2. Department of Pediatrics, University of California San Diego, San Diego, CA 92093-0935, USA

Received: 10 June 2019; Accepted: 11 June 2019; Published: 17 June 2019

Mitochondria have been increasingly recognized as the important players in the aging process. Most aging-related diseases, particularly, neurodegenerative diseases, have mitochondrial involvement. A PubMed search for mitochondria and aging lists 704 articles in 2018. This is not surprising as mitochondria are involved not only in energy production through oxidative phosphorylation but also play an important role in intracellular homeostasis, calcium balance, and the metabolism and interconversion of our dietary substrates, fats, proteins, and carbohydrates, in the fed and fasting states. They have an important role in signaling their metabolic state to the nucleus and to other cells in response to stress. Mitochondria have their own protein synthetic apparatus and replicate themselves, pathways readily disrupted in disease and aging. They are constantly involved in fusion and fission, the balance of which is essential for cell health. These organelles participate in apoptosis, they make most of the cell's free radicals, and they are crucially important for innate immunity. Mitochondrial DNA has an estimated 10-fold greater mutation rate than nuclear DNA and less repair capacity, and this plays an important role in aging and cancer. Mitochondria are impacted by environmental factors and toxins, and different mtDNA haplogroups originally adapted to geographically different origins make an important background contribution to disease. As mitochondria play a critical metabolic role in all organ systems, they are particularly impacted by disease and contribute to the aging process itself. The invited review articles in this special supplement cover most of the common diseases of aging. Enthusiasm for this supplement in *Biology* was driven by the opportunity to review the current state of knowledge about the role of mitochondria in the aging process. The international group of contributing authors includes many of the leaders in their fields.

This special issue starts with the discussion on the role of mitochondria in a number of critical organ systems; starting with the immune system, the cell danger response and healing, skin aging, the role of coenzyme Q and vitamin D, mitochondria and the retina, and drug toxicity in the geriatric population. The focus then moves to specific diseases of aging the role of mitochondria in diabetes, cancer, cardiovascular disease, and neurodegenerative disorders;amyotrophic lateral sclerosis (ALS), multiple sclerosis, Parkinson's disease, and Alzheimer's disease. Finally, an important paper focusing on muscle and aging points out the important therapeutic role of exercise.

Increasingly the role of mitochondria in innate immunity has been studied but, as noted by Peter McGuire [1], the importance of mitochondrial dysfunction in aging and immunity is less discussed. He provides an overview of three main effects of aging on this system, inflammation with aging, susceptibility to viral infections, and declining T-cell function. He points out the role of mitochondrial damage associated molecular patterns (mtDAMPs), which when released from mitochondria as a consequence of stress, apoptosis, or necrosis, trigger caspase-1 activation with the release of pro-inflammatory cytokines. He discusses the increased susceptibility of older adults to viral infections, and the role that mitochondria play in innate immune signaling against viruses and the production of protective type I interferons. Finally, he discusses the hypothesis that T-cell dysfunction in aging is due to a decline in mitochondrial function.

Robert Naviaux [2] discusses a holistic new model of incomplete healing and its role in aging. He explores the role of mitochondria in the healing process and the effects of aging. He points out that healing involves the cell danger response and that metabolic cross-talk between mitochondria and the nucleus, between neighboring and distant cells via signaling metabokines regulates the completeness of healing. He discusses the causes of cellular stress and the role of mitochondria in the cell danger response, pointing out the critical roles of purinergic and sphingolipid signaling pathways. Finally, this paper discusses that cellular arrest in the various phases of the cell danger response leads to chronic inflammatory and pain syndromes, to susceptibility to bacterial and viral infections, to a variety of aging-related diseases as well as autoimmune disorders, and to neurodegenerative diseases.

Isabella Peixoto de Barcelos and Richard H. Haas [3] review, from a translational perspective, the data regarding the association of CoQ_{10} and aging. They discuss the changes in coenzyme Q levels during the aging process and its putative contribution to aging diseases through a wide variety of metabolic roles. CoQ_{10} functions in membranes throughout the cell where antioxidant and signaling roles predominate. They explore the growing evidence that oxidative stress is a major component of cellular senescence, a multifactorial process that involves DNA, protein, and lipid damage and activation of signaling pathways associated with aging. They discuss the state of the evidence that CoQ_{10} supplementation may be helpful for diseases of aging and aging itself.

Roisin Stout and Mark Birch-Machin [4] review the increasing evidence that mitochondrial dysfunction and oxidative stress contribute to skin aging. They discuss the important mitochondrial role and energy production in cell signaling, wound healing, pigmentation, vasculature homeostasis, and hair growth, as well as defense against infection. They explore the free radical theory of aging in the skin and point out that mtDNA deletions are increased in aged UV exposed epidermis. They review the role of calorie restriction on skin models of aging and the role of mitochondria in the pigmentary changes of aging. Finally, they discuss the role of mitochondria in photoaging, effects of pollution, stress-induced skin wrinkle formation, and hair loss and greying.

Sunil J. Wimalawansa [5] explores the role of vitamin D and its metabolites in the aging process. He discusses newly recognized functions of vitamin D as controllers of systemic inflammation, oxidative stress, and mitochondrial respiratory function. He reviews the role of the active metabolite 1,25(OH)2D as a gene regulator, inhibiting NF-κB expression, which is thought to play a role in many aging-related disorders, including inflammation and cancer. He explores the evidence that the vitamin D modulates mitochondrial function and functions as a powerful antioxidant. He points out that hypovitaminosis D increases the incidence and severity of several age-related common diseases and metabolic disorders that are linked to oxidative stress, including obesity, insulin resistance, type 2 diabetes, hypertension, pregnancy complications, memory disorders, osteoporosis, autoimmune diseases, certain cancers, and systemic inflammatory diseases. Finally, the importance of worldwide vitamin D supplementation is emphasized.

Janis T. Eells [6] reviews the role of mitochondria in one of the body's most bioenergetic organs, the retina. The eye is exposed to visible light and has extensive antioxidant protective mechanisms. Aging retinal pigment epithelial cells have impaired mitochondrial function with increased reactive oxygen species production. She notes that aging-related mitochondrial dysfunction causes increased oxidative injury, which, coupled with impaired repair mechanisms, results in retinal dysfunction and retinal cell loss, leading to visual impairment. Janis T. Eells discusses the most common cause of age-related blindness in developed countries, that is, age-related macular degeneration. She discusses the relationship between mitochondrial function and complement factor H, mutations of which are a putative risk factor for age-related macular degeneration. Finally, she discusses the role of mitochondrial dysfunction in diabetic retinopathy and glaucoma.

Yvonne Will, Jefry E. Shields and Kendall B. Wallace [7] bring together extensive academic and pharmaceutical research experience to discuss the role of mitochondria in drug toxicity in the elderly. This is an under-explored topic despite its great importance. They note that 'Drug-induced mitochondrial toxicity has been described for many different drug classes and can lead to liver, muscle, kidney, and central nervous system injury and, in rare cases, to death'. They discuss the progressive loss of mitochondrial function with age with decreased mitophagy as a cellular surveillance system failure. They point out that drug mitochondrial toxicity has often not been identified in pre-clinical studies because young healthy animals are used rather than aged animals with more susceptible mitochondria. For the many drug classes, the particular drugs with most severe mitochondrial toxic effects can be revealed by in vitro systems where cells are forced to use mitochondrial respiration. Polypharmacy in the elderly, coupled with access to over the counter drugs, which inhibit mitochondrial function, compounds the problem of mitochondrial toxicity. A very useful table of commonly used drugs and their mitochondrial effects has been provided. Finally, the important effects of lifestyle and diet on susceptibility to mitochondrial toxicity is discussed.

Magdalene K Montgomery [8] provides a valuable review of the role of mitochondria in obesity and type 2 diabetes. A discussion of the role of mitochondria in diabetic organ damage focuses on diabetic cardiomyopathy. A review of the role of mitochondria in insulin resistance finds that while most of the antidiabetic drugs increase mitochondrial biogenesis with an improvement in insulin efficacy, particularly, those that modulate peroxisome proliferator-activated receptors, there are mitochondrial toxic effects caused by many antidiabetic drugs. Finally, she explores the beneficial or detrimental role of intercellular exchange of mitochondria, mitochondrial DNA, and mitochondrial fragments through the exchange of exosomes and through nanotubes.

Jason Duran, Armando Martinez and Eric Adler [9] discuss cardiovascular manifestations of mitochondrial disease and the role of mitochondria in myocardial ischemia and diabetic cardiomyopathy. Cardiomyocytes are among the most energy dependent cells in the body. They review the mitochondrial changes with aging and the effects on the heart. They then discuss the cardiac and cardiovascular manifestations of canonical mitochondrial syndromes. In these disorders, hypertrophic and dilated cardiomyopathies are commonplace, and a variety of cardiac conduction defects can be life-threatening. A discussion of mitochondrial dysfunction in cardiac ischemia details the role of mitochondrial dysfunction and oxidative stress in reperfusion injury, which increases the size of myocardial infarction through necrosis and mitochondrially mediated apoptosis. Finally, the mitochondrial role in diabetic cardiomyopathy is discussed.

The next article by Nima B. Fakouri, Thomas Lau Hansen, Claus Desler, Sharath Anugula and Lene Juel Rasmussen [10] addresses the interactions between mitochondria and cancer. The authors focus their review on genomic instability, dysregulation of cellular energetics, and mitochondrial function. They note that DNA damage, secondary to oxidative stress produced by reactive oxygen and nitrogen species, activates the DNA damage response (DDR), which is an energy-dependent process. They then elegantly discuss how the DDR can both activate and impair mitochondrial function, the latter through poly (ADP-ribose) polymerase enzyme hyper-activation. They next discuss the role of mitochondria-nuclear signaling in aging and cancer, along with the role of ROS. Mitochondria have an epigenetic role, and, in cancer, mtDNA mutations are associated with a poor prognosis—and they note a correlation between mitochondrial respiration, cytosolic dNTP pools, and chromosomal instability. In summary, the connection between mitochondria and cancer is complex but, as the authors note, 'the hallmarks of cancer include genomic instability, dysregulation of cellular energetics, and mitochondrial dysfunction, which also are common pathways important for cellular aging'.

Moving on to neurodegenerative diseases, the review, by Veronica Granatiero and Giovanni Manfredi [11], discusses the role of mitochondrial dysfunction in the devastating disorder amyotrophic lateral sclerosis (ALS). They note that neurons are very energy dependent, and mitochondrial transport and turnover is critical for neuronal and axonal health. In both genetic ALS and 90% of sporadic cases, mitochondrial dysfunction is manifest by changes in fusion, fission, and transport. These protein 'motors' are ATP dependent. They note that in ALS, microtubular kinesin anterograde mitochondrial transport and dynein retrograde transport are both disrupted. The ATP/ADP ratio is an important component of the signaling system for mitochondrial transport. The hypothesis that protein aggregates in ALS bind to and damage mitochondria causing dysfunction is discussed. Gene mutations linked to ALS are involved in mitochondrial quality control. The authors note that a recent study shows that the actin cytoskeleton plays a part in isolating damaged mitochondria from the rest of the network. Mitophagy is increased, and Parkin levels are decreased, in animal models and human ALS, suggesting that the decline in Parkin protein is related to mitophagy. Whether mitochondrial dysfunction in ALS is a cause or an effect remains unclear, but mitochondrial fusion, fission, and transport have an important role in this disease.

Isabella Peixoto de Barcelos, Regina M. Troxell and Jennifer S. Graves [12] discuss the role of mitochondria in multiple sclerosis (MS). Decreases in oxidative phosphorylation and mitochondrial transport have been documented. MS is an inflammatory disorder with acute and chronic phases, involving both mitochondrial innate immunity and chronic inflammation followed by neurodegeneration. They note that mitochondrial dysfunction occurs as a consequence of inflammation and oxidative stress with ROS production. Nucleic acid, protein, and lipid damage ensue. A fall in ATP production ultimately triggers mitochondrially mediated apoptosis. Mitochondrial stress impairs oligodendrocyte function. Studies in human cortex confirm increased levels of mtDNA mutations, mtDNA depletion, and impairment in complex I and III enzyme activity. Mitochondrial changes in animal models are reviewed. Experimental autoimmune encephalomyelitis, an animal model of MS, shows mitochondrial swelling and dysfunction with some evidence for antioxidant rescue. The authors note that some mtDNA diseases due to point mutations, such as Leber's Hereditary Optic Neuropathy, can have an MS phenotype, and MS-like demyelinating disease has been reported in a variety of mitochondrial-nuclear gene defects, including PolG and OPA1. Large population studies have failed to confirm an increased incidence of mtDNA mutations in the MS population, although the JT haplogroup does have an increased MS risk. Finally, the authors discuss the role of mitochondrial therapies in MS.

Chun Chen, Doug M. Turnbull and Amy K. Reeve [13] provide a current review of the mitochondrial role in Parkinson's disease (PD). Evidence for mitochondrial involvement stretches back 40 years. They discuss mitochondrial pathways involved in PD and the rapid growth of knowledge resulting from genomic sequencing of familial cases. Evidence for complex I deficiency in animal models and human disease is discussed. High levels of mtDNA clonal deletions are found in substantia nigra neurons in normal aging and PD, with ROS as a likely contributing factor. This aging-related mtDNA pathology in substantia nigra likely contributes to susceptibility to PD, and alpha-synuclein can increase mitochondrial membrane permeability, ROS production, and cell death. The authors discuss recent data on cytosolic calcium oscillations with the import of calcium into mitochondria, which likely plays a role in cell death. Mutations in two genes *VSP35* and *CHCHD2*, which impair the mitochondrial function, are responsible for autosomal dominant PD. These and the roles of mutations in PINK1 and Parkin in mitophagy are discussed. Finally, we are treated to a detailed discussion of mitochondrial turnover and dynamics and the putative role of protein aggregation in PD. The authors conclude that mitochondrial complex I deficiency plays a key role in PD, and the likely causes are elegantly discussed.

Ian Weidling and Russell H. Swerdlow [14] review the evidence for a mitochondrial role in Alzheimer's disease (AD). Starting with longstanding evidence of brain hypometabolism, they discuss findings of mitochondrial electron transport enzyme deficiencies and, in particular, cytochrome oxidase (COX) deficiency in Alzheimer's brain. Recent studies confirm mitochondrial structural abnormalities in Alzheimer's brain, likely the result of a fusion-fission malfunction, which may be related to amyloid beta accumulation. An interesting reciprocal relationship between mtDNA deletions and aging noted in AD brain is discussed, and the evidence for increased oxidative injury in AD brain compared to age-matched controls is reviewed. Cybrids created from AD mtDNA show similar changes to the brain with COX deficiency. A mitochondrial interaction with Alzheimer's protein aggregates is discussed. Mitochondrial toxins can induce the formation of AD-like Tau alterations, and, in mouse models, mitochondrial dysfunction precedes amyloid plaque accumulation. The authors note that it is unclear whether mitochondrial changes are the cause or effect of abnormal protein accumulation in AD. Amyloid beta inhibits COX activity, and Tau accumulation disrupts mitochondrial transport. ApoE4 overexpression impairs electron transport complexes. The role of mitochondria in clearing protein aggregates is reviewed. Finally, the multiple effects of mitochondrial dysfunction and the Integrated Stress Response (activated in AD) on gene expression are discussed. The authors conclude that given the multiple lines of evidence of mitochondrial dysfunction in AD—this is a reasonable therapeutic target.

This special edition on mitochondrial dysfunction in aging concludes with a discussion of treatment. Sarcopenia is an inevitable consequence of aging. Mats I Nilsson and Mark A Tarnopolsky [15] detail the role of exercise as a treatment for mitochondrial aging. An overview of mitochondrial evolution leads on to a discussion of the homeostatic role of mitochondria and their role in defense against the three main aging changes, oxidative injury, protein aggregation, and inflammation. An integrated system's hypothesis of aging is developed with mitochondria playing a central role. The role of mitochondrial ROS is discussed, and the authors point out that oxidative phosphorylation decline is an aging phenomenon in all species along with a variety of structural and functional mitochondrial changes. Age-related protein aggregation and lipofuscin accumulation result in an accumulation of cellular debris impervious to lysosomal and proteasomal degradation. The mitochondrial role in the chronic inflammation of aging is explored. The authors then turn to the role of exercise in correcting the cellular decline of aging. A convincing case is made for the benefit of exercise, slowing aging-related changes, including reducing intracellular danger signals, rejuvenating mitochondria, assisting with intracellular garbage clearance, and decreasing aging-related inflammation.

Conflicts of Interest: The authors declare no conflict of interest.

References

1. McGuire, P.J. Mitochondrial dysfunction and the aging immune system. *Biology* **2019**, *8*, 26. [CrossRef] [PubMed]
2. Naviaux, R.K. Incomplete healing as a cause of aging: The role of mitochondria and the cell danger response. *Biology* **2019**, *8*, 27. [CrossRef] [PubMed]
3. Barcelos, I.P.d.; Haas, R.H. Coq10 and aging. *Biology* **2019**, *8*, 28. [CrossRef] [PubMed]
4. Stout, R.; Birch-Machin, M. Mitochondria's role in skin ageing. *Biology* **2019**, *8*, 29. [CrossRef] [PubMed]
5. Wimalawansa, S.J. Vitamin d deficiency: Effects on oxidative stress, epigenetics, gene regulation, and aging. *Biology* **2019**, *8*, 30. [CrossRef] [PubMed]
6. Eells, J.T. Mitochondrial dysfunction in the aging retina. *Biology* **2019**, *8*, 31. [CrossRef] [PubMed]
7. Will, Y.; Shields, J.E.; Wallace, K.B. Drug-induced mitochondrial toxicity in the geriatric population: Challenges and future directions. *Biology* **2019**, *8*, 32. [CrossRef] [PubMed]
8. Montgomery, M.K. Mitochondrial dysfunction and diabetes: Is mitochondrial transfer a friend or foe? *Biology* **2019**, *8*, 33. [CrossRef] [PubMed]
9. Duran, J.; Martinez, A.; Adler, E. Cardiovascular manifestations of mitochondrial disease. *Biology* **2019**, *8*, 34. [CrossRef] [PubMed]
10. Fakouri, N.B.; Hansen, T.L.; Desler, C.; Anugula, S.; Rasmussen, L.J. From powerhouse to perpetrator—Mitochondria in health and disease. *Biology* **2019**, *8*, 35. [CrossRef] [PubMed]
11. Granatiero, V.; Manfredi, G. Mitochondrial transport and turnover in the pathogenesis of amyotrophic lateral sclerosis. *Biology* **2019**, *8*, 36. [CrossRef] [PubMed]
12. Barcelos, I.P.d.; Troxell, R.M.; Graves, J.S. Mitochondrial dysfunction and multiple sclerosis. *Biology* **2019**, *8*, 37. [CrossRef] [PubMed]
13. Chen, C.; Turnbull, D.M.; Reeve, A.K. Mitochondrial dysfunction in parkinson's disease—Cause or consequence? *Biology* **2019**, *8*, 38. [CrossRef] [PubMed]
14. Weidling, I.; Swerdlow, R.H. Mitochondrial dysfunction and stress responses in alzheimer's disease. *Biology* **2019**, *8*, 39. [CrossRef] [PubMed]
15. Nilsson, M.I.; Tarnopolsky, M.A. Mitochondria and aging—The role of exercise as a countermeasure. *Biology* **2019**, *8*, 40. [CrossRef] [PubMed]

© 2019 by the author. Licensee MDPI, Basel, Switzerland. This article is an open access article distributed under the terms and conditions of the Creative Commons Attribution (CC BY) license (http://creativecommons.org/licenses/by/4.0/).

Review

Mitochondrial Dysfunction and the Aging Immune System

Peter J. McGuire

Metabolism, Infection and Immunity Section, National Human Genome Research Institute, National Institutes of Health, Bethesda, MD 20892, USA; peter.mcguire@nih.gov; Tel.: +1-301-451-7716

Received: 9 November 2018; Accepted: 16 January 2019; Published: 11 May 2019

Abstract: Mitochondria are ancient organelles that have co-evolved with their cellular hosts, developing a mutually beneficial arrangement. In addition to making energy, mitochondria are multifaceted, being involved in heat production, calcium storage, apoptosis, cell signaling, biosynthesis, and aging. Many of these mitochondrial functions decline with age, and are the basis for many diseases of aging. Despite the vast amount of research dedicated to this subject, the relationship between aging mitochondria and immune function is largely absent from the literature. In this review, three main issues facing the aging immune system are discussed: (1) inflamm-aging; (2) susceptibility to infection and (3) declining T-cell function. These issues are re-evaluated using the lens of mitochondrial dysfunction with aging. With the recent expansion of numerous profiling technologies, there has been a resurgence of interest in the role of metabolism in immunity, with mitochondria taking center stage. Building upon this recent accumulation of knowledge in immunometabolism, this review will advance the hypothesis that the decline in immunity and associated pathologies are partially related to the natural progression of mitochondrial dysfunction with aging.

Keywords: aging; mitochondria; inflammation; innate immunity; adaptive immunity; immunosenescence

1. Introduction

The ancestry of the mitochondrion originated ~2.5 billion years ago within the bacterial phylum α-Proteobacteria, during the rise of eukaryotes [1]. The endosymbiotic theory, advanced with microbial evidence by Dr. Lynn Margulis in the 1960s, proposed that one prokaryote engulfed another resulting in a quid pro quo arrangement and survival advantage [2]. The ability of mitochondria to convert organic molecules from the environment to energy led to the persistence of this pact.

Since most cells contain mitochondria, the clinical effects of mitochondrial dysfunction are potentially multisystemic, and involve organs with large energy requirements [3]. In addition to making energy, the basis of life, mitochondria are also involved in heat production, calcium storage, apoptosis, cell signaling, biosynthesis, and aging—all important for cell survival and function [4–7]. A decline in mitochondrial function and oxidant production has been connected to normal aging and with the development of a variety of diseases of aging. These topics are explored more thoroughly in other articles in this special edition. While the human immune system undergoes dramatic changes during aging, eventually progressing to immunosenescence [8], the role of mitochondrial dysfunction in this process remains largely absent in the literature. Consequently, the purpose of this review is to highlight three important issues in the aging immune system: (1) inflammation with aging; (2) susceptibility to viral infections; (3) impaired T-cell immunity. These clinical phenotypes will be related to our current knowledge on the role of the mitochondria in immune function. As the associations discussed are largely speculative, it is hoped that this review will serve as a stimulus for further investigation into these issues.

2. Is There a Mitochondrial Etiology for "Inflamm-Aging"?

The term "inflamm-aging" (IA) refers to a low-grade, chronic inflammatory state that can be found in the elderly [9]. IA increases morbidity and mortality in older adults, and nearly all diseases of aging share an inflammatory pathogenesis including Alzheimer's disease, atherosclerosis, heart disease, type II diabetes, and cancer [9]. Nevertheless, the precise etiology of IA and its causal role in contributing to adverse health outcomes remain largely unknown.

The ability of the innate system to respond to a wide variety of pathogens lies in germline-encoded receptors, whose recognition is based on repetitive molecular signatures. These pattern recognition receptors (PRRs) are present on the cell surface and intracellular compartments. Toll-like receptors (TLRs), retinoic acid-inducible gene I-like receptors (RLRs), nucleotide oligomerization domain-like receptors (NLRs) and cytosolic DNA sensors (cGAS and STING) are prime examples [10]. Ligands for these receptor systems comprise pathogen associated molecular patterns (PAMPs) and damage associated molecular patterns (DAMPs) [11]. PAMPs are derived from components of microorganisms and are recognized by innate immune cells bearing PRRs. In contrast to PAMPs, DAMPs are endogenous "danger signals" that are released by cells during stress, apoptosis or necrosis. DAMPs can arise from a variety of components normally sequestered to the mitochondria, when upon release, induce inflammation via recognition by the same PRRs that recognize PAMPs [12,13]. Events downstream of PRR engagement include caspase-1 activation with the release of pro-inflammatory cytokines [14]. Examples of mitochondrial DAMPs (mtDAMPs) include cardiolipin, n-formyl peptides (e.g., fMet), mitochondrial transcription factor A (TFAM), adenosine triphosphate (ATP), reactive oxygen species (mtROS), and mitochondrial DNA (mtDNA) (Figure 1). From an evolutionary standpoint, select mitochondrial products produce inflammation due to their prokaryotic origins: e.g., cardiolipin (TLR), fMet (formyl peptide receptor 1, FPR1), and mtDNA (TLR, NLR, cGAS) [15–22]. However, mtDAMPs are not just limited to bacterial mimics. TFAM, a nuclear gene and key regulator of mtDNA transcription and replication, activates immune cells via receptors for advanced glycation end products (RAGE) and TLR9 [23,24]. Products of oxidative phosphorylation (OXPHOS) can also stimulate innate immune cells. Released from apoptotic or necrotic cells, ATP binds to purigenic receptors initiating inflammation [25], while mtROS modifies core immune signaling pathways involving hypoxia inducible factor 1 alpha (HIF1α) and nuclear factor kappa light chain enhancer of activated B-cells (NFkB) [26,27].

mtDAMPs contribute to a host of inflammatory diseases, including sepsis, systemic inflammatory response syndrome (SIRS), ischemic reperfusion injury, and aging [28]. One of the consequences of failing mitochondria due to aging, beyond mtROS, is the release of mtDNA. Plasma levels of mtDNA increase gradually after the fifth decade of life, correlating with elevated levels of pro-inflammatory cytokines (i.e., TNF-α, IL-6, RANTES, and IL-1ra) [29]. These data indicate that mtDNA may promote the production of pro-inflammatory cytokines in aging. Because cell stress, senescence and death are a part of the pathophysiology of aging [30], designing new therapeutic strategies against circulating mtDNA, or other mtDAMPs, or their cognate receptors (e.g., TLRs or FPR1) may be a viable strategy to approaching IA and its associated conditions.

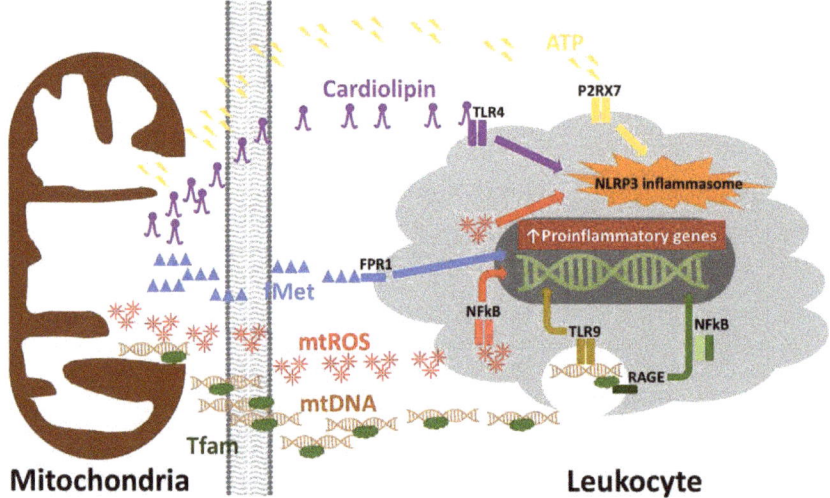

Figure 1. Mitochondrial damage associated molecular patterns (DAMPs). DAMPs derived from mitochondrial components may be released during cellular injury, apoptosis or necrosis. Once these mitochondrial components are released into the extracellular space, they can lead to the activation of innate and adaptive immune cells. The recognition of mitochondrial DAMPs involves toll-like receptors (TLR), formyl peptide receptors (FPR) and purigenic receptors (P2RX7). By binding their cognate ligands or by direct interaction (i.e., reactive oxygen species, ROS), intracellular signaling pathways such as NFkB and the NLRP3 inflammasome become activated resulting in a proinflammatory response. TLR4 = toll-like receptor 4, TLR9 = toll-like receptor 9, P2RX7 = purigenic receptor, FPR1 = formyl peptide receptor 1, NLRP3 = NLR Family Pyrin Domain Containing 3, fMet = N-formylmethionine, mtROS = mitochondrial reactive oxygen species, mtDNA = mitochondrial DNA, Tfam = transcription factor A, mitochondrial, RAGE = receptors for advanced glycation end-products, NFkB = nuclear factor kappa-light-chain-enhancer of activated B cells.

3. Is Increased Susceptibility to Viral Infections Related to Depressed Mitochondrial Anti-Viral Signaling Pathways?

In general, older adults are more susceptible to a variety of viral infections, especially respiratory viral infections, resulting in high morbidity and mortality. For example, adults over the age of 65 exhibit a vulnerability to influenza A virus (IAV), and account for ≥90% of IAV-related deaths annually [31,32]. Type I interferons (e.g., IFN-α and IFN-β) are essential cytokines involved in the host antiviral response. Secreted by numerous cell types such as lymphocytes, monocytes, macrophages, dendritic cells, fibroblasts, endothelial cells, osteoblasts and others, type I interferons: (1) limit viral spread by inducing antiviral states in infected and neighboring cells; (2) stimulate antigen presentation and natural killer cell function; and (3) promote antigen-specific T and B cell responses and immunological memory. Interestingly, mitochondria play a major part in innate immune signaling against viruses and the production of type I interferons and will be discussed further.

RLRs (e.g., RIG-I and MDA5) are cytosolic receptors that recognize viral RNA. Consequent to binding viral RNA, RIG-I and MDA5 mobilize the mitochondrial antiviral signaling protein (MAVS) [33,34]. MAVS is a 56 kDa protein which contains an N-terminal caspase recruitment domain (CARD), a proline-rich region and a C-terminal transmembrane domain. Anchored on the outer membrane of the mitochondria, peroxisomes and mitochondrial associated membranes (e.g., endoplasmic reticulum), MAVS assembles into prion-like aggregates following RIG-I or MDA5 binding (Figure 2). MAVS aggregates serve as a scaffold to recruit various TNF receptor associated factors (TRAFs), resulting in phosphorylation and nuclear translocation of interferon regulatory factors

(IRFs) [35]. Downstream of MAVS, IRF3, IRF5 and IRF7 bind to their cognate promoters, leading to the production of type I interferons [36]. The localization of MAVS to the outer mitochondrial membrane is not coincidental. MAVS activity has been found to be dependent upon intact mitochondrial membrane potential, and by extension OXPHOS function [37].

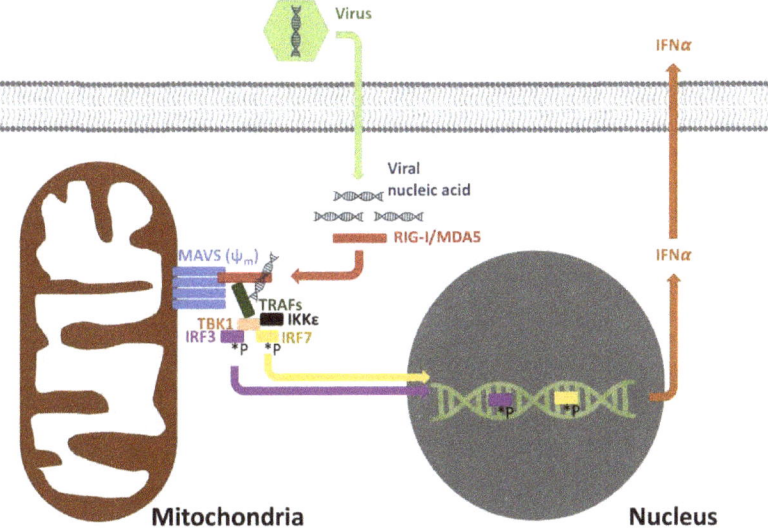

Figure 2. Mitochondrial antiviral response. The recognition of viral nucleic acids involves mitochondria and intact membrane potential (ψ_m). RIG-I and MDA5 recognize cytoplasmic viral nucleic acids, leading to the oligomerization of MAVS. MAVS then sets in motion a signaling pathway that eventually leads to the phosphorylation (*P) of IRF3/7 with subsequent induction of IFNα to offer antiviral cellular protection. RIG-I = retinoic acid inducible gene I, MDA5 = melanoma differentiation-associated protein 5, MAVS = mitochondrial antiviral-signaling protein, TRAFs = TNF receptor associated factors, TBK1 = TANK binding kinase 1, IKKε = inhibitor of nuclear factor kappa-B kinase subunit epsilon, IRF 3 = interferon response factor 3, IRF 7 = interferon response factor 7, INFα = interferon alpha.

To date, studies addressing MAVS function during aging and its relationship to waning antiviral immunity are lacking. Decreased mitochondrial membrane potential, mitochondrial dysfunction and declining mitophagy occur in a variety of aging cell types [38,39], raising the question of whether MAVS dysfunction can occur due to mitochondrial failure with aging. Mitochondrial respiratory capacity is impaired in aging monocytes [40] as is phosphorylation of IRF3 and IRF7, suggesting a link with MAVS [41]. As a result, type I IFN synthesis is significantly lower in dendritic cells and monocytes from aging individuals [42,43]. In addition to a decline in mitochondrial respiration, oxidative stress, another consequence of aging, may also be involved in this process [43].

4. Is Impaired T-Cell Immunity in Aging Related to a Decline in Mitochondrial Function?

Aging-related decline in immune function (i.e., immunosenescence) renders older individuals more vulnerable to infectious diseases and cancer, resulting in increased morbidity and mortality. Besides increased susceptibility to infection, vaccine efficacy is significantly reduced in the elderly, limiting the utility of prophylaxis [44,45]. Undeniably, profound changes in T-cell function are evident in older individuals, and these changes may be related to a decline in mitochondrial function.

T-cells play a central role in the coordination of adaptive immune responses and cell-mediated immunity. The ability of T-cells to fulfill this role is dependent upon rapid cellular proliferation and differentiation. In response to infection, T-cells proliferate every 4–6 h, generating >10^{12} cells in one week [46,47]. This is accompanied by an increase in size, DNA remodeling, up-regulation of transcription factors and effector molecules, and increased expression of surface proteins [48,49], thus necessitating a large metabolic demand. To accomplish this task, metabolic fuels including fats, sugars and amino acids are actively transported across the cell membrane to feed the increase in energetic demands [50,51]. Along with this increased transport, T-cells undergo metabolic reprogramming during their transition from a naïve state to activated and differentiated cell types (e.g., effector, regulatory and memory cells).

The diverse roles played by mitochondria in T-cell activation emphasizes the potential mechanisms by which aging-related mitochondrial decline may contribute to immune dysfunction. Following stimulation of the T-cell receptor, T-cells undergo substantial changes in intermediary metabolism including an increase in glycolysis and OXPHOS [52–57]. In the presence of oxygen, pyruvate produced via glycolysis is fully oxidized in the mitochondria for energy in many cell types [58,59]. In T-cells, a significant proportion of glucose is not oxidized, but rather fermented to lactate from pyruvate via lactate dehydrogenase. This is done despite the presence of oxygen, and is termed aerobic glycolysis or Warburg metabolism [50,56,60]. Although Warburg metabolism is viewed as energetically inefficient, the rate of glycolysis is 10–100 times faster than glucose oxidation by the mitochondria, yielding equivalent amounts of ATP [61]. The additional payoff of Warburg metabolism lies in pathways that are branch points off of glycolysis (e.g., pentose phosphate pathway) which yield reducing equivalents for biosynthesis and nucleotides. Despite this adoption of the Warburg phenotype, OXPHOS is still required for T-cell activation [57]. ATP derived from the mitochondrial respiration promotes enhanced glycolysis as well as the initiation of proliferation in activated T-cells [62]. While pyruvate is mostly diverted to lactate rather than acetyl-CoA via pyruvate dehydrogenase, TCA function and the generation of reducing equivalents in highly proliferating cells is still maintained through anapleurosis: glutamine is converted to α-ketoglutarate via glutaminolysis [63,64]. Bioenergetic studies of aging tissues are consistent with a progressive decline in mitochondrial respiratory function due to a decrease in respiratory complex activity, mitochondrial membrane potential, and impaired mitophagy [39,65]. As a result, impaired OXPHOS results in reduced ATP production, thus potentially limiting glycolysis, biosynthesis and the attainment of biomass during T-cell activation and proliferation.

Besides engaging in bioenergetics, mitochondria also function in T-cell activation by modulating secondary messengers including calcium (Ca^{2+}) and reactive oxygen species (ROS). In activated T-cells, mitochondria localize to the immune synapse, and where they regulate Ca^{2+} flux [5,6]. In response to this calcium flux, ROS production via complex III of the respiratory chain is amplified, leading to nuclear factor of activated T-cells (NFAT) activation and subsequent interleukin-2 (IL-2) production [66]. Aged T-cells show reduced Ca^{2+} signaling, which could be partly due to deficits in Ca^{2+} regulation found in mitochondria of aged cells [67,68], theoretically yielding perturbations at the immune synapse causing diminished T-cell signaling and activation.

Depending on the cytokine milieu, helper T-cells (Th), marked by the surface expression of CD4, differentiate into various effector subsets comprising T-helper 1 (Th1), T-helper 2 (Th2), T-helper 17 (Th17), regulatory T-cells (Treg). Each of these T-cell subsets are unique in their responsibilities and are identified by their cytokine signatures. Accompanying these functional distinctions are differences in metabolic reprogramming (Figure 3). For example, for T-cells subsets involved in inflammation (e.g., Th1 and Th17), the Warburg metabolism instituted at T-cell activation persists [69]. Despite this primary use of glycolysis, intact OXPHOS is still necessary for their function [57]. The effects of mitochondrial dysfunction may be more readily seen in regulatory (Treg) and memory (Tmem) T-cells. Tregs, which serve to modulate the immune system and maintain tolerance, revert back to OXPHOS as their main pathway for generating energy upon differentiation [69]. Tmem follow a similar metabolic path. Tmem are critical for adaptive immune responses characterized by robust responses to secondary immune challenges. Unlike effector T-cells, Tmem do not undergo extensive proliferation and produce little or no cytokines. As such, the metabolic profile of Tmem are essentially catabolic, relying on OXPHOS and fatty acid oxidation [70,71]. Therefore, it is not surprising to find that CD8+ cytotoxic memory T-cells have high respiratory capacity and increased mitochondrial mass, which allows them to rapidly reactivate upon re-exposure to their cognate antigens [62,72]. Given the age-related decline in mitochondrial function as described above, T-cell subsets which are critical for immunosurveillance and the clearance of invading pathogens could be functionally impaired and may partially explain the vulnerability to infection and cancer with aging [57]. Emerging data also suggest that aging significantly affects Treg frequencies, subsets and function [73], potentially leading to the increased incidence of autoimmunity, oftentimes seen with aging [74]. As noted above, Tmem also rely heavily on OXPHOS. Therefore, aging-related deficiencies in Tmem may also be traced to declining OXPHOS, manifesting as impaired immune memory to novel antigens and suboptimal boosts to existing memory [75].

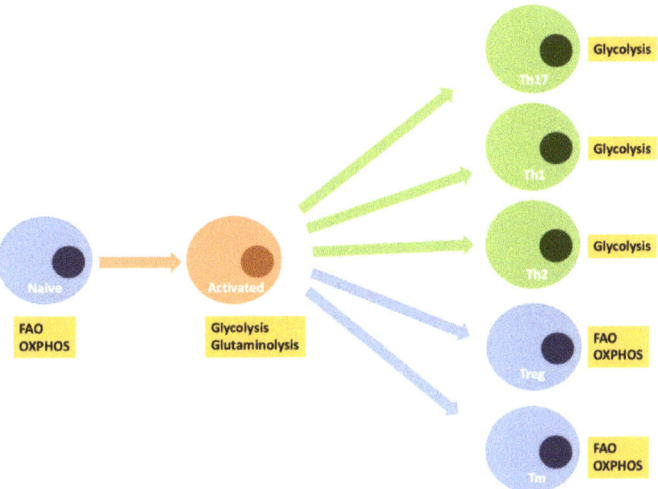

Figure 3. T-cell activation and differentiation involved metabolic reprogramming. At rest, naïve T-cells primarily use OXPHOS to derive their energy. Following activation, T-cells switch to Warburg metabolism and glutaminolysis to support their proliferative needs. Differentiation into T-helper subsets can involve either the maintenance of the Warburg phenotype (i.e., Th17, Th1, Th2), or the reversion to OXPHOS with FAO (i.e., Treg, Tm) as an important fuel. FAO = mitochondrial fatty acid oxidation, OXPHOS = oxidative phosphorylation, Th17 = T-helper cell 17, Th1 = T-helper cell 1, Th2 = T-helper cell 2, Treg = regulatory T-cells, Tm = memory T-cells.

5. Conclusions

Virtually every country in the world is experiencing the challenges associated with accelerated growth in the aging population. With this graying of the population comes an increased incidence in diseases of aging, many of which have an immune component. As a result, understanding the pathophysiology of diseases of aging is now more important than ever. In this review, three main immune issues prevalent in the aging population were addressed: (1) inflamm-aging; (2) increased vulnerability to infection; and (3) declining T-cell immunity. The role of the mitochondria in inflammation and immunity, combined with the knowledge of a decline in mitochondrial function with aging, has been synthesized in this review in an effort to partially explain the immune phenotype associated with aging. However, further examination of this relationship is needed. As the methods of inquiry into mitochondrial biology continue to expand, so will investigations into the relationship between this ancient organelle and immunity in the aging population.

Funding: This work was supported by the intramural research program of the National Institutes of Health (HG200381-03).

Conflicts of Interest: The author declares no conflict of interest.

References

1. Martin, W.F.; Garg, S.; Zimorski, V. Endosymbiotic theories for eukaryote origin. *Philos. Trans. R. Soc. Lond. B Biol. Sci.* **2015**, *370*, 20140330. [CrossRef]
2. Gray, M.W. Mitochondrial evolution. *Cold Spring Harb. Perspect. Biol.* **2012**, *4*, a011403. [CrossRef] [PubMed]
3. Vafai, S.B.; Mootha, V.K. Mitochondrial disorders as windows into an ancient organelle. *Nature* **2012**, *491*, 374–383. [CrossRef]
4. Antico Arciuch, V.G.; Elguero, M.E.; Poderoso, J.J.; Carreras, M.C. Mitochondrial regulation of cell cycle and proliferation. *Antioxid. Redox Signal.* **2012**, *16*, 1150–1180. [CrossRef] [PubMed]
5. Quintana, A.; Schwindling, C.; Wenning, A.S.; Becherer, U.; Rettig, J.; Schwarz, E.C.; Hoth, M. T cell activation requires mitochondrial translocation to the immunological synapse. *Proc. Natl. Acad. Sci. USA* **2007**, *104*, 14418–14423. [CrossRef]
6. Schwindling, C.; Quintana, A.; Krause, E.; Hoth, M. Mitochondria positioning controls local calcium influx in T cells. *J. Immunol.* **2010**, *184*, 184–190. [CrossRef]
7. Sena, L.A.; Chandel, N.S. Physiological roles of mitochondrial reactive oxygen species. *Mol. Cell* **2012**, *48*, 158–167. [CrossRef] [PubMed]
8. Weyand, C.M.; Goronzy, J.J. Aging of the Immune System. Mechanisms and Therapeutic Targets. *Ann. Am. Thorac. Soc.* **2016**, *13* (Suppl. 5), S422–S428. [CrossRef]
9. Xia, S.; Zhang, X.; Zheng, S.; Khanabdali, R.; Kalionis, B.; Wu, J.; Wan, W.; Tai, X. An Update on Inflamm-Aging: Mechanisms, Prevention, and Treatment. *J. Immunol. Res.* **2016**, *2016*, 8426874. [CrossRef]
10. Thompson, M.R.; Kaminski, J.J.; Kurt-Jones, E.A.; Fitzgerald, K.A. Pattern recognition receptors and the innate immune response to viral infection. *Viruses* **2011**, *3*, 920–940. [CrossRef]
11. Tang, D.; Kang, R.; Coyne, C.B.; Zeh, H.J.; Lotze, M.T. PAMPs and DAMPs: Signal 0s that spur autophagy and immunity. *Immunol. Rev.* **2012**, *249*, 158–175. [CrossRef]
12. Janeway, C.A., Jr.; Medzhitov, R. Innate immune recognition. *Annu. Rev. Immunol.* **2002**, *20*, 197–216. [CrossRef]
13. Medzhitov, R. Innate immunity: Quo vadis? *Nat. Immunol.* **2010**, *11*, 551–553. [CrossRef]
14. Martinon, F.; Burns, K.; Tschopp, J. The inflammasome: A molecular platform triggering activation of inflammatory caspases and processing of proIL-beta. *Mol. Cell* **2002**, *10*, 417–426. [CrossRef]
15. Wenceslau, C.F.; McCarthy, C.G.; Goulopoulou, S.; Szasz, T.; NeSmith, E.G.; Webb, R.C. Mitochondrial-derived N-formyl peptides: Novel links between trauma, vascular collapse and sepsis. *Med. Hypotheses* **2013**, *81*, 532–535. [CrossRef]
16. Collins, L.V.; Hajizadeh, S.; Holme, E.; Jonsson, I.M.; Tarkowski, A. Endogenously oxidized mitochondrial DNA induces in vivo and in vitro inflammatory responses. *J. Leukoc. Biol.* **2004**, *75*, 995–1000. [CrossRef]
17. Hauser, C.J.; Sursal, T.; Rodriguez, E.K.; Appleton, P.T.; Zhang, Q.; Itagaki, K. Mitochondrial damage associated molecular patterns from femoral reamings activate neutrophils through formyl peptide receptors and P44/42 MAP kinase. *J. Orthop. Trauma* **2010**, *24*, 534–538. [CrossRef]
18. Zhang, Q.; Raoof, M.; Chen, Y.; Sumi, Y.; Sursal, T.; Junger, W.; Brohi, K.; Itagaki, K.; Hauser, C.J. Circulating mitochondrial DAMPs cause inflammatory responses to injury. *Nature* **2010**, *464*, 104–107. [CrossRef]
19. Zhang, Q.; Itagaki, K.; Hauser, C.J. Mitochondrial DNA Is Released by Shock and Activates Neutrophils Via P38 Map Kinase. *Shock* **2010**, *34*, 55–59. [CrossRef]
20. Iyer, S.S.; He, Q.; Janczy, J.R.; Elliott, E.I.; Zhong, Z.; Olivier, A.K.; Sadler, J.J.; Knepper-Adrian, V.; Han, R.; Qiao, L.; et al. Mitochondrial cardiolipin is required for Nlrp3 inflammasome activation. *Immunity* **2013**, *39*, 311–323. [CrossRef]
21. Toksoy, A.; Sennefelder, H.; Adam, C.; Hofmann, S.; Trautmann, A.; Goebeler, M.; Schmidt, M. Potent NLRP3 Inflammasome Activation by the HIV Reverse Transcriptase Inhibitor Abacavir. *J. Biol. Chem.* **2017**, *292*, 2805–2814. [CrossRef]

22. Rossen, R.D.; Michael, L.H.; Hawkins, H.K.; Youker, K.; Dreyer, W.J.; Baughn, R.E.; Entman, M.L. Cardiolipin-Protein Complexes and Initiation of Complement Activation after Coronary-Artery Occlusion. *Circ. Res.* **1994**, *75*, 546–555. [CrossRef]
23. Chaung, W.W.; Wu, R.; Ji, Y.; Dong, W.; Wang, P. Mitochondrial transcription factor A is a proinflammatory mediator in hemorrhagic shock. *Int. J. Mol. Med.* **2012**, *30*, 199–203.
24. Julian, M.W.; Shao, G.; Vangundy, Z.C.; Papenfuss, T.L.; Crouser, E.D. Mitochondrial transcription factor A, an endogenous danger signal, promotes TNFalpha release via RAGE- and TLR9-responsive plasmacytoid dendritic cells. *PLoS ONE* **2013**, *8*, e72354. [CrossRef]
25. Iyer, S.S.; Pulskens, W.P.; Sadler, J.J.; Butter, L.M.; Teske, G.J.; Ulland, T.K.; Eisenbarth, S.C.; Florquin, S.; Flavell, R.A.; Leemans, J.C.; et al. Necrotic cells trigger a sterile inflammatory response through the Nlrp3 inflammasome. *Proc. Natl. Acad. Sci. USA* **2009**, *106*, 20388–20393. [CrossRef]
26. Brunelle, J.K.; Bell, E.L.; Quesada, N.M.; Vercauteren, K.; Tiranti, V.; Zeviani, M.; Scarpulla, R.C.; Chandel, N.S. Oxygen sensing requires mitochondrial ROS but not oxidative phosphorylation. *Cell Metab.* **2005**, *1*, 409–414. [CrossRef]
27. Chandel, N.S.; Trzyna, W.C.; McClintock, D.S.; Schumacker, P.T. Role of oxidants in NF-kappa B activation and TNF-alpha gene transcription induced by hypoxia and endotoxin. *J. Immunol.* **2000**, *165*, 1013–1021. [CrossRef]
28. Nakahira, K.; Hisata, S.; Choi, A.M. The Roles of Mitochondrial Damage-Associated Molecular Patterns in Diseases. *Antioxid. Redox Signal.* **2015**, *23*, 1329–1350. [CrossRef]
29. Pinti, M.; Cevenini, E.; Nasi, M.; De Biasi, S.; Salvioli, S.; Monti, D.; Benatti, S.; Gibellini, L.; Cotichini, R.; Stazi, M.A.; et al. Circulating mitochondrial DNA increases with age and is a familiar trait: Implications for "inflamm-aging". *Eur. J. Immunol.* **2014**, *44*, 1552–1562. [CrossRef]
30. Childs, B.G.; Durik, M.; Baker, D.J.; van Deursen, J.M. Cellular senescence in aging and age-related disease: From mechanisms to therapy. *Nat. Med.* **2015**, *21*, 1424–1435. [CrossRef]
31. Thompson, W.W.; Shay, D.K.; Weintraub, E.; Brammer, L.; Cox, N.; Anderson, L.J.; Fukuda, K. Mortality associated with influenza and respiratory syncytial virus in the United States. *JAMA* **2003**, *289*, 179–186. [CrossRef]
32. Simonsen, L.; Clarke, M.J.; Schonberger, L.B.; Arden, N.H.; Cox, N.J.; Fukuda, K. Pandemic versus epidemic influenza mortality: A pattern of changing age distribution. *J. Infect. Dis.* **1998**, *178*, 53–60. [CrossRef]
33. Kumar, H.; Kawai, T.; Kato, H.; Sato, S.; Takahashi, K.; Coban, C.; Yamamoto, M.; Uematsu, S.; Ishii, K.J.; Takeuchi, O.; et al. Essential role of IPS-1 in innate immune responses against RNA viruses. *J. Exp. Med.* **2006**, *203*, 1795–1803. [CrossRef]
34. Lin, R.T.; Lacoste, J.; Nakhaei, P.; Sun, Q.; Yang, L.; Paz, S.; Wilkinson, P.; Julkunen, I.; Vitour, D.; Meurs, E.; et al. Dissociation of a MAVS/IPS-1/VISA/Cardif-1KK epsilon molecular complex from the mitochondrial outer membrane by hepatitis C virus NS3-4A proteolytic cleavage. *J. Virol.* **2006**, *80*, 6072–6083. [CrossRef]
35. Liu, B.Y.; Gao, C.J. Regulation of MAVS activation through post-translational modifications. *Curr. Opin. Immunol.* **2018**, *50*, 75–81. [CrossRef]
36. Lazear, H.M.; Lancaster, A.; Wilkins, C.; Suthar, M.S.; Huang, A.; Vick, S.C.; Clepper, L.; Thackray, L.; Brassil, M.M.; Virgin, H.W.; et al. IRF-3, IRF-5, and IRF-7 coordinately regulate the type I IFN response in myeloid dendritic cells downstream of MAVS signaling. *PLoS Pathog.* **2013**, *9*, e1003118. [CrossRef]
37. Koshiba, T.; Yasukawa, K.; Yanagi, Y.; Kawabata, S. Mitochondrial Membrane Potential Is Required for MAVS-Mediated Antiviral Signaling. *Sci. Signal.* **2011**, *4*, ra7. [CrossRef]
38. Sugrue, M.M.; Tatton, W.G. Mitochondrial membrane potential in aging cells. *Neurosignals* **2001**, *10*, 176–188. [CrossRef]
39. Diot, A.; Morten, K.; Poulton, J. Mitophagy plays a central role in mitochondrial ageing. *Mamm. Genome* **2016**, *27*, 381–395. [CrossRef]
40. Pence, B.D.; Yarbro, J.R. Aging impairs mitochondrial respiratory capacity in classical monocytes. *Exp. Gerontol.* **2018**, *108*, 112–117. [CrossRef]
41. Molony, R.D.; Nguyen, J.T.; Kong, Y.; Montgomery, R.R.; Shaw, A.C.; Iwasaki, A. Aging impairs both primary and secondary RIG-I signaling for interferon induction in human monocytes. *Sci. Signal.* **2017**, *10*, eaan2392. [CrossRef]

42. Qian, F.; Wang, X.; Zhang, L.; Lin, A.; Zhao, H.; Fikrig, E.; Montgomery, R.R. Impaired interferon signaling in dendritic cells from older donors infected in vitro with West Nile virus. *J. Infect. Dis.* **2011**, *203*, 1415–1424. [CrossRef]
43. Stout-Delgado, H.W.; Yang, X.; Walker, W.E.; Tesar, B.M.; Goldstein, D.R. Aging impairs IFN regulatory factor 7 up-regulation in plasmacytoid dendritic cells during TLR9 activation. *J. Immunol.* **2008**, *181*, 6747–6756. [CrossRef]
44. Burns, E.A.; Lum, L.G.; L'Hommedieu, G.; Goodwin, J.S. Specific humoral immunity in the elderly: In vivo and in vitro response to vaccination. *J. Gerontol.* **1993**, *48*, B231–B236. [CrossRef]
45. Musher, D.M.; Chapman, A.J.; Goree, A.; Jonsson, S.; Briles, D.; Baughn, R.E. Natural and vaccine-related immunity to Streptococcus pneumoniae. *J. Infect. Dis.* **1986**, *154*, 245–256. [CrossRef]
46. Badovinac, V.P.; Haring, J.S.; Harty, J.T. Initial T cell receptor transgenic cell precursor frequency dictates critical aspects of the CD8(+) T cell response to infection. *Immunity* **2007**, *26*, 827–841. [CrossRef]
47. Yoon, H.; Kim, T.S.; Braciale, T.J. The Cell Cycle Time of CD8(+) T Cells Responding In Vivo Is Controlled by the Type of Antigenic Stimulus. *PLoS ONE* **2010**, *5*, e15423. [CrossRef]
48. Gray, S.M.; Kaech, S.M.; Staron, M.M. The interface between transcriptional and epigenetic control of effector and memory CD8(+) T-cell differentiation. *Immunol. Rev.* **2014**, *261*, 157–168. [CrossRef]
49. Zhang, N.; Bevan, M.J. CD8(+) T cells: Foot soldiers of the immune system. *Immunity* **2011**, *35*, 161–168. [CrossRef]
50. Macintyre, A.N.; Gerriets, V.A.; Nichols, A.G.; Michalek, R.D.; Rudolph, M.C.; Deoliveira, D.; Anderson, S.M.; Abel, E.D.; Chen, B.J.; Hale, L.P.; et al. The Glucose Transporter Glut1 Is Selectively Essential for CD4 T Cell Activation and Effector Function. *Cell Metab.* **2014**, *20*, 61–72. [CrossRef]
51. Sinclair, L.V.; Rolf, J.; Emslie, E.; Shi, Y.B.; Taylor, P.M.; Cantrell, D.A. Control of amino-acid transport by antigen receptors coordinates the metabolic reprogramming essential for T cell differentiation. *Nat. Immunol.* **2013**, *14*, 500–508. [CrossRef]
52. Frauwirth, K.A.; Riley, J.L.; Harris, M.H.; Parry, R.V.; Rathmell, J.C.; Plas, D.R.; Elstrom, R.L.; June, C.H.; Thompson, C.B. The CD28 signaling pathway regulates glucose metabolism. *Immunity* **2002**, *16*, 769–777. [CrossRef]
53. Marjanovic, S.; Eriksson, I.; Nelson, B.D. Expression of a new set of glycolytic isozymes in activated human peripheral lymphocytes. *Biochim. Biophys. Acta* **1990**, *1087*, 1–6. [CrossRef]
54. Marjanovic, S.; Wielburski, A.; Nelson, B.D. Effect of phorbol myristate acetate and concanavalin A on the glycolytic enzymes of human peripheral lymphocytes. *Biochim. Biophys. Acta* **1988**, *970*, 1–6. [CrossRef]
55. O'Rourke, A.M.; Rider, C.C. Glucose, glutamine and ketone body utilisation by resting and concanavalin A activated rat splenic lymphocytes. *Biochim. Biophys. Acta* **1989**, *1010*, 342–345. [CrossRef]
56. Wang, R.N.; Dillon, C.P.; Shi, L.Z.; Milasta, S.; Carter, R.; Finkelstein, D.; McCormick, L.L.; Fitzgerald, P.; Chi, H.B.; Munger, J.; et al. The Transcription Factor Myc Controls Metabolic Reprogramming upon T Lymphocyte Activation. *Immunity* **2011**, *35*, 871–882. [CrossRef]
57. Tarasenko, T.N.; Pacheco, S.E.; Koenig, M.K.; Gomez-Rodriguez, J.; Kapnick, S.M.; Diaz, F.; Zerfas, P.M.; Barca, E.; Sudderth, J.; DeBerardinis, R.J.; et al. Cytochrome c Oxidase Activity Is a Metabolic Checkpoint that Regulates Cell Fate Decisions During T Cell Activation and Differentiation. *Cell Metab.* **2017**, *25*, 1268 e1257. [CrossRef]
58. Jones, R.G.; Thompson, C.B. Revving the engine: Signal transduction fuels T cell activation. *Immunity* **2007**, *27*, 173–178. [CrossRef]
59. Krebs, H.A. The tricarboxylic acid cycle. *Harvey Lect.* **1948**, *Series 44*, 165–199.
60. Macintyre, A.N.; Rathmell, J.C. Activated lymphocytes as a metabolic model for carcinogenesis. *Cancer Metab.* **2013**, *1*, 5. [CrossRef]
61. Shestov, A.A.; Liu, X.J.; Ser, Z.; Cluntun, A.A.; Hung, Y.P.; Huang, L.; Kim, D.; Le, A.; Yellen, G.; Albeck, J.G.; et al. Quantitative determinants of aerobic glycolysis identify flux through the enzyme GAPDH as a limiting step. *eLife* **2014**, *3*, e03342. [CrossRef]
62. Van der Windt, G.J.; O'Sullivan, D.; Everts, B.; Huang, S.C.; Buck, M.D.; Curtis, J.D.; Chang, C.H.; Smith, A.M.; Ai, T.; Faubert, B.; et al. CD8 memory T cells have a bioenergetic advantage that underlies their rapid recall ability. *Proc. Natl. Acad. Sci. USA* **2013**, *110*, 14336–14341. [CrossRef]

63. DeBerardinis, R.J.; Mancuso, A.; Daikhin, E.; Nissim, I.; Yudkoff, M.; Wehrli, S.; Thompson, C.B. Beyond aerobic glycolysis: Transformed cells can engage in glutamine metabolism that exceeds the requirement for protein and nucleotide synthesis. *Proc. Natl. Acad. Sci. USA* **2007**, *104*, 19345–19350. [CrossRef]
64. Csibi, A.; Fendt, S.M.; Li, C.; Poulogiannis, G.; Choo, A.Y.; Chapski, D.J.; Jeong, S.M.; Dempsey, J.M.; Parkhitko, A.; Morrison, T.; et al. The mTORC1 pathway stimulates glutamine metabolism and cell proliferation by repressing SIRT4. *Cell* **2013**, *153*, 840–854. [CrossRef]
65. Lee, H.C.; Wei, Y.H. Mitochondria and aging. *Adv. Exp. Med. Biol.* **2012**, *942*, 311–327.
66. Sena, L.A.; Li, S.; Jairaman, A.; Prakriya, M.; Ezponda, T.; Hildeman, D.A.; Wang, C.R.; Schumacker, P.T.; Licht, J.D.; Perlman, H.; et al. Mitochondria are required for antigen-specific T cell activation through reactive oxygen species signaling. *Immunity* **2013**, *38*, 225–236. [CrossRef]
67. Mather, M.W.; Rottenberg, H. The inhibition of calcium signaling in T lymphocytes from old mice results from enhanced activation of the mitochondrial permeability transition pore. *Mech. Ageing Dev.* **2002**, *123*, 707–724. [CrossRef]
68. Murchison, D.; Zawieja, D.C.; Griffith, W.H. Reduced mitochondrial buffering of voltage-gated calcium influx in aged rat basal forebrain neurons. *Cell Calcium* **2004**, *36*, 61–75. [CrossRef]
69. Buck, M.D.; O'Sullivan, D.; Klein Geltink, R.I.; Curtis, J.D.; Chang, C.H.; Sanin, D.E.; Qiu, J.; Kretz, O.; Braas, D.; van der Windt, G.J.; et al. Mitochondrial Dynamics Controls T Cell Fate through Metabolic Programming. *Cell* **2016**, *166*, 63–76. [CrossRef]
70. Fox, C.J.; Hammerman, P.S.; Thompson, C.B. Fuel feeds function: Energy metabolism and the T-cell response. *Nat. Rev. Immunol.* **2005**, *5*, 844–852. [CrossRef]
71. Yusuf, I.; Fruman, D.A. Regulation of quiescence in lymphocytes. *Trends Immunol.* **2003**, *24*, 380–386. [CrossRef]
72. Van der Windt, G.J.; Everts, B.; Chang, C.H.; Curtis, J.D.; Freitas, T.C.; Amiel, E.; Pearce, E.J.; Pearce, E.L. Mitochondrial respiratory capacity is a critical regulator of CD8+ T cell memory development. *Immunity* **2012**, *36*, 68–78. [CrossRef]
73. Jagger, A.; Shimojima, Y.; Goronzy, J.J.; Weyand, C.M. Regulatory T cells and the immune aging process: A mini-review. *Gerontology* **2014**, *60*, 130–137. [CrossRef]
74. Vadasz, Z.; Haj, T.; Kessel, A.; Toubi, E. Age-related autoimmunity. *BMC Med.* **2013**, *11*, 94. [CrossRef]
75. Goronzy, J.J.; Weyand, C.M. Successful and Maladaptive T Cell Aging. *Immunity* **2017**, *46*, 364–378. [CrossRef]

© 2019 by the author. Licensee MDPI, Basel, Switzerland. This article is an open access article distributed under the terms and conditions of the Creative Commons Attribution (CC BY) license (http://creativecommons.org/licenses/by/4.0/).

Review

Incomplete Healing as a Cause of Aging: The Role of Mitochondria and the Cell Danger Response

Robert K. Naviaux

The Mitochondrial and Metabolic Disease Center, Departments of Medicine, Pediatrics, Pathology, University of California, San Diego School of Medicine, San Diego, CA 92103, USA; Naviaux@ucsd.edu; Tel.: +1-619-993-2904

Received: 9 January 2019; Accepted: 20 February 2019; Published: 11 May 2019

Abstract: The rate of biological aging varies cyclically and episodically in response to changing environmental conditions and the developmentally-controlled biological systems that sense and respond to those changes. Mitochondria and metabolism are fundamental regulators, and the cell is the fundamental unit of aging. However, aging occurs at all anatomical levels. At levels above the cell, aging in different tissues is qualitatively, quantitatively, and chronologically distinct. For example, the heart can age faster and differently than the kidney and vice versa. Two multicellular features of aging that are universal are: (1) a decrease in physiologic reserve capacity, and (2) a decline in the functional communication between cells and organ systems, leading to death. Decreases in reserve capacity and communication impose kinetic limits on the rate of healing after new injuries, resulting in dyssynchronous and incomplete healing. Exercise mitigates against these losses, but recovery times continue to increase with age. Reinjury before complete healing results in the stacking of incomplete cycles of healing. Developmentally delayed and arrested cells accumulate in the three stages of the cell danger response (CDR1, 2, and 3) that make up the healing cycle. Cells stuck in the CDR create physical and metabolic separation—buffer zones of reduced communication—between previously adjoining, synergistic, and metabolically interdependent cells. Mis-repairs and senescent cells accumulate, and repeated iterations of incomplete cycles of healing lead to progressively dysfunctional cellular mosaics in aging tissues. Metabolic cross-talk between mitochondria and the nucleus, and between neighboring and distant cells via signaling molecules called metabokines regulates the completeness of healing. Purinergic signaling and sphingolipids play key roles in this process. When viewed against the backdrop of the molecular features of the healing cycle, the incomplete healing model provides a new framework for understanding the hallmarks of aging and generates a number of testable hypotheses for new treatments.

Keywords: cell danger response; healing cycle; mitochondria; purinergic signaling; metabokines; sphingolipids; integrated cell stress response; de-emergence; crabtree effect; pasteur effect

1. Introduction

Some of the oldest [1], and the most recent [2,3] scientific publications on the biology of aging have focused on nutrition and metabolism as prime drivers. Mitochondria are located at the hub of the wheel of cellular metabolism. The mitochondrial proteome is transcriptionally and post-transcriptionally regulated according to tissue-specific needs [4], consists of about 1300 proteins [5], responds to injury [6], food quality [7], exercise [8], environmental pollution [9], and coordinates the cell danger response (CDR) [10]. The CDR is a term that was coined in 2014 [10], but includes elements of inflammation and healing that have been studied since before the time of Hippocrates (c. 460–370 BCE) [11]. The CDR is an evolutionarily conserved, multi-system response of multicellular organisms that is used to manage and heal from threat or injury. The CDR is a graded response that consists of nested layers that range from the molecular control of electron utilization and cellular oxygen consumption, through changes in the microbiome, mast cells and immune system, to the autonomic nervous system, enteric nervous system, and neuroendocrine circuits that are needed for whole-body integration of the response. When analyzed at the molecular and single-cell level, a widely-studied component of the CDR is known as the integrated cell stress response (ICSR) [12–14].

2. Defining Cellular Stress

In this paper, the word "stress" has a specific scientific meaning. Stress is any force, condition, chemical, pathogen or other stimulus that acts to perturb cellular function, requiring the expenditure of energy and resources to return the cell to its pre-stimulus or to a new steady state. Remarkably, psychological stresses, particularly early life stresses (ELS), regulate some of the same metabolic and gene expression networks used to defend the cell from microbial pathogens, physical injury, poisoning, and adversity of many other types [15–17]. In the case of major depressive disorder, innate immune/pro-inflammatory gene expression is increased, while adaptive immune/anti-inflammatory/pro-resolving gene expression is decreased [18]. Greater stresses stimulate proportionately more nucleotide and metabolite release through cell membrane channels [19], leading to greater metabokine and purinergic signaling [20,21]. Learning and development emerge from both conscious stresses and subconscious chemical stresses encountered throughout life. Stresses lead to metabolic memories that help cells and tissues improve future responses to previously encountered conditions [22]. Transient increases in the mitochondrial and cellular sources of reactive oxygen species (ROS) such as superoxide and hydrogen peroxide, reactive nitrogen species (RNS) such as nitric oxide (NO) and peroxynitrite (ONO_2^-), reactive aldehydes (RAs), and dissolved oxygen itself, help to regulate cellular redox. Redox, in turn, regulates the efflux of metabolites from the cell via membrane channels and transporters that contain redox-responsive cysteine disulfide residues [23]. Pulses of dissolved oxygen occur with a carbohydrate-rich meal because of the transient inhibitory effect of glucose on mitochondria produced by the Crabtree effect [24]. ROS, insulin, and IL1β are also stimulated naturally by every meal and can vary in magnitude according to the nutrient content of the meal [25]. As the dissolved oxygen concentration rises within the cell, glycolysis becomes inhibited by the Pasteur effect, creating a natural brake on an unchecked inhibition of mitochondrial oxphos from the meal-associated glucose and the Crabtree effect. However, glucose, oxygen, ROS, RNS, and RAs are only a part of the multi-faceted metabolic signaling network that is used to control cellular reactivity, epigenetic marking, and gene expression. Over 100 chemosensory G-protein coupled receptors respond to metabokines and peptides released after stress and injury and regulate the healing cycle [11].

3. The Healing Cycle

The mitochondrial responses that initiate and maintain the CDR are used to control cellular bioenergetics, oxygen utilization, redox signaling, and metabolism needed for healing [6] and regeneration [26] after injury or threat. The stages of the CDR are illustrated in Figure 1. Wakeful activity with nutrient intake, followed by restorative sleep are essential parts of health. These activities stimulate an integrated mix of three metabolic states that are controlled locally by metabolic signaling, and systemically by the central nervous system (CNS). Healthy whole-body function requires the coordination and use of cell-specific (1) glycolysis, (2) aerobic glycolysis, and (3) oxidative phosphorylation for energy and metabolism (Figure 1). Chemical activity associated with nutrient intake, brain activity, and basal metabolism, and added physical activity from natural child play or adult exercise lead to the graded release of extracellular ATP and related nucleotides, and to glutamate release [27,28] through stress-gated P2X7-pannexin and other channels in the cell membrane [19]. Once outside the cell, ATP and related nucleotides participate in purinergic signaling. Receptor binding to ATP and ADP leads to IP3-gated intracellular calcium release [29] and contextual changes in gene expression through autocrine and paracrine signaling pathways [30]. After extracellular metabolism by CD73 and CD39, ATP is converted to ADP, AMP, and to adenosine that acts to inhibit excess chemical stimulation and plays a key role in initiating and maintaining sleep by binding to P1/Adenosine /ADORA A2AR and A1R receptors [31]. This is illustrated as the restorative sleep cycle in Figure 1.

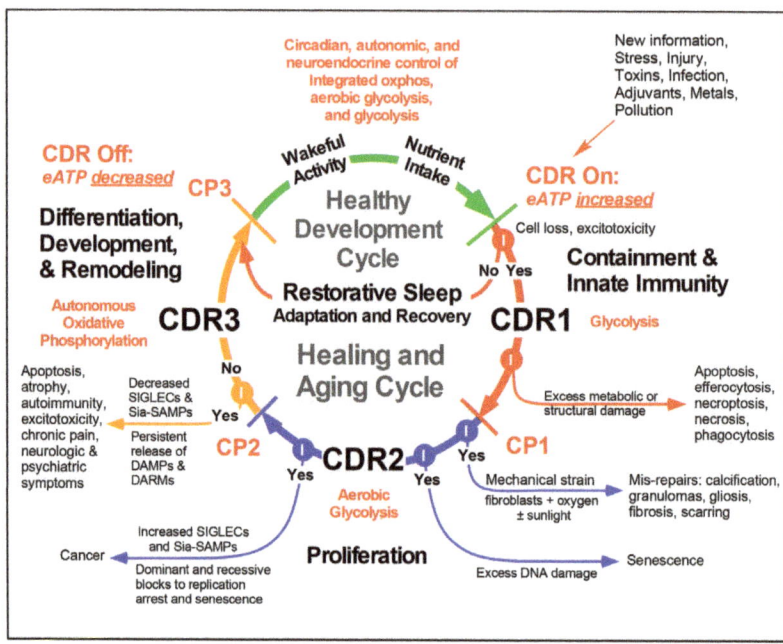

Figure 1. The metabolic features of the health and healing cycles. Abbreviations: CDR—cell danger response, eATP—extracellular ATP, CP1-3—checkpoints 1, 2, and 3, DAMP—damage-associated molecular pattern, DARM—damage-associated reactive metabolites, SIGLEC—sialic acid binding immunoglobulin-type lectin, e.g., CD33-related SIGLECs (CD33r-SIGLECs), Sia-SAMP—sialoglycan self-associated molecular pattern.

When the stress is of sufficient magnitude to cause cell death, more ATP is released and acts as a pro-inflammatory damage-associated molecular pattern (DAMP). Increased extracellular ATP triggers entry into the CDR1 stage of the healing cycle (Figure 1). Once the CDR is triggered, three different metabolic stages must be activated in sequence to heal. Healing cannot occur without activation of this metabolically-controlled cycle. CDR1 is characterized by the upregulation of anaerobic glycolysis and a reallocation of cellular resources for defense, damage containment, innate immunity, and repair at the expense of normal differentiated tissue function. Gap junctions between cells are decreased or lost as the tissue structure is disrupted by injury, and cell-autonomous functions become primary and metabolic cooperation between neighboring cells is decreased or suspended. Platelets and neutrophils are recruited to sites of injury or infection. CDR2 uses aerobic glycolysis to support stem cell recruitment and cell division needed for biomass replacement of cells lost in CDR1. If cell loss is not replaced, then age-related atrophy and sarcopenia occur. If excessive DNA damage is sustained by a cell in CDR2, replicative senescence occurs [32]. If the blocks to senescence are broken, then cancer can occur. If oxygen levels are high and significant mechanical strain is present, tissue fibrosis or wound scarring is stimulated (Figure 1). Fibrosis is one of several different types of mis-repairs that contribute to the symptoms of aging [33–35] (Figures 1–3).

In CDR3, cell-autonomous, aerobic metabolism by mitochondrial oxidative phosphorylation is gradually restored as new cells born in CDR2 take up residence in the recovering tissue and establish new tissue-specific contacts and gap junctions needed to extinguish the gene expression programs used for defense and growth in CDR1 and CDR2, and to restore normal differentiated cell, tissue, and organ function. If damage-associated molecular pattern (DAMP) and damage-associated reactive metabolite (DARM) release persists [36] or if ROS production is extinguished prematurely [37], autoimmunity can develop [38]. The adaptive immune response is regulated in CDR3. Persistent DAMP and DARM release can stimulate excitotoxicity in cells still in the CDR1 and CDR2 stages within a dyssynchronous mosaic of healing cells. With the completion of CDR3, the concentration of metabokines and DAMPS in the extracellular space is actively reduced to levels compatible with restoration of cell specialization and reintegration of cells back into a metabolically optimized cellular network. Re-integration and re-specialization are necessary for cells that have recently exited the CDR in order to re-establish their responsiveness to circadian patterns of wakeful activity and restorative sleep (Figure 1).

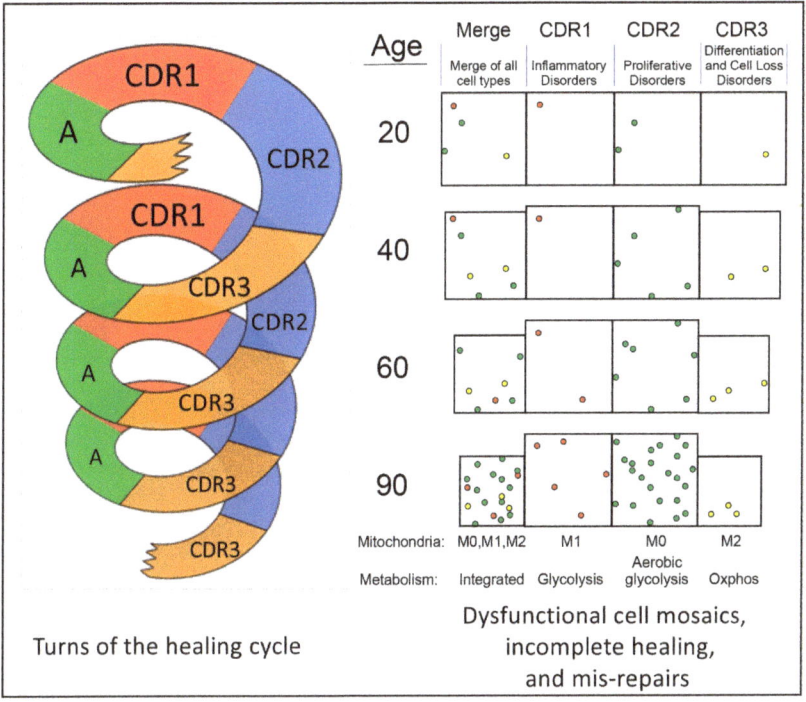

Figure 2. Repeated cycles of incomplete healing lead to aging and age-related disease. The spiral represents sequential turns of the healing cycle throughout life. Colored cells in the boxes on the right represent cells that have been delayed or arrested in a stage of the healing cycle. The decreased size of some boxes represents the loss in tissue volume from cell loss and atrophy. In this example, most arrested or delayed cells in the merge on the right after 60 and 90 years are in CDR2 (green). This will create an increased risk of proliferative disorders such as diabetes, heart disease, and cancer. Color code: CDR1 cells—red; CDR2 cells—green; CDR3 cells—yellow. Abbreviations: A—wakeful activity and nutrient intake. M1—mitochondria adapted for cell defense, reactive oxygen, nitrogen, and aldehyde production; M0—mitochondria adapted for cell growth and Warburg metabolism; M2—mitochondria adapted for oxidative phosphorylation (oxphos).

The sequence and stages of the healing cycle are highly conserved and tightly choreographed. Once pathological stress or cell death occurs, the same stages of the healing cycle are activated and restore normal function after any recoverable injury (Figure 2). Inevitably, some cells fail to complete the healing cycle and are left behind with each turn of the cycle. These arrested or delayed cells are unable to return to a normal metabolism and the gene expression pattern that is needed for peak differentiated cell and organ function (Figure 2). As non-specialized cells and mis-repairs such as fibrosis accumulate in incomplete stages of the healing cycle, the peak performance of tissues is degraded over time, and the risk for age-related disease is increased.

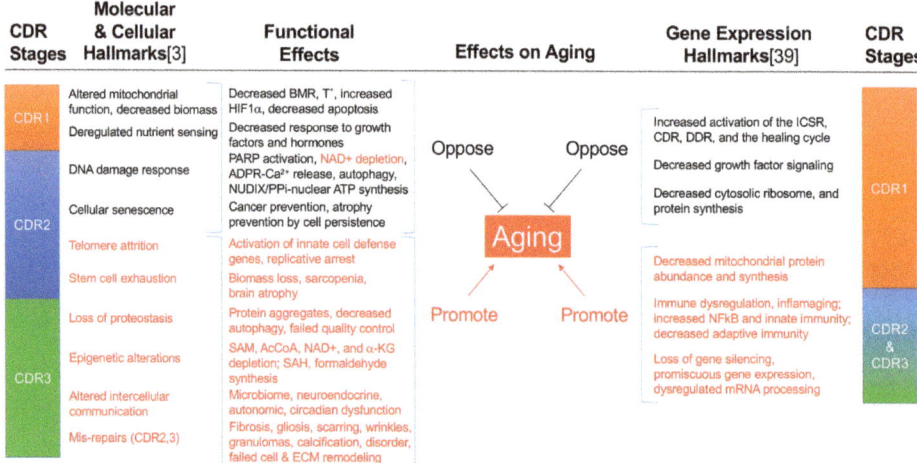

Figure 3. The hallmarks of aging as natural products of incomplete healing and the cell danger response [3,39]. Black font: hallmarks that oppose aging. Red font: hallmarks that promote aging. Abbreviations: CDR—cell danger response, ICSR—integrated cell stress response, DDR—DNA damage response, BMR—basal metabolic rate, T°—basal body temperature, HIF1α—hypoxia inducible factor 1α, PARP—poly ADP ribose polymerase, NAD+—nicotinamide adenine dinucleotide, ADPR—adenosine diphosphate ribose, SAM—S-adenosyl methionine, AcCoA—acetyl CoA, α-KG—alpha ketoglutarate, SAH—S-adenosyl homocysteine, PPi—pyrophosphate, NUDIX—nucleoside diphosphate X hydrolases, e.g., NUDT5, ECM—extracellular matrix.

4. Three Functionally-Polarized Forms of Mitochondria Are Used by the CDR

For over 60 years, mitochondria have been thought of as "damaged" or "dysfunctional" if they shift from energy production by oxidative phosphorylation to ROS production or shift from burning carbon skeletons to CO_2 and water to synthesizing new carbon skeletons for export as building blocks needed for cell growth. Yet mitochondria shift regularly and necessarily between these states throughout life. A simplified way of understanding the alternative differentiation states of mitochondria is to see the organelles as the metabolic gate-keepers of the electrons harvested by breaking down the carbon–carbon bonds in food. In this scheme, electrons can be used in mitochondria: (1) to fully reduce oxygen (O_2) to water (H_2O) while capturing the released chemical energy to make ATP by oxidative phosphorylation in the third stage of the cell danger response, CDR3 and in health, (2) to partially reduce oxygen to superoxide radicals (O_2^-) and hydrogen peroxide (H_2O_2) while relying on anaerobic glycolysis for ATP synthesis in CDR1 and cell defense, or (3) to use the electrons to synthesize new carbon–carbon bonds to produce and export building blocks such as citrate for cell membrane lipid synthesis or orotic acid for pyrimidine synthesis, while still consuming oxygen to make water and making ATP by aerobic glycolysis in CDR2 and cell growth (Figure 1).

A new nomenclature was introduced in 2017 to describe these three, functionally-polarized forms of mitochondria of the CDR [11,29]. All three "species" of mitochondria co-exist in different proportions in cells throughout development and aging in all multicellular organisms. The transition between states is determined by the interaction between nutrition and metabolism, the developmental and chronological age of the organism, the recent state of cell danger signaling and healing, and ambient environmental conditions. The naming of the differentiation states of mitochondria was based on the recognition that the pro-inflammatory state of macrophages known as M1 macrophages corresponded with the ROS producing capacity of their mitochondria and energy production by glycolysis [40]. The anti-inflammatory/pro-resolving M2 macrophages were found to contain mitochondria adapted for oxidative phosphorylation (oxphos). M0 macrophages are not yet committed to a fully M1 or M2 phenotype and contain mitochondria that are intermediate between M1 and M2. In the healing cycle, three different mitochondrial differentiation states are used to meet the specialized needs of each stage of the CDR. M1 mitochondria are used in CDR1. M0 mitochondria are used in CDR2, and M2 mitochondria are used in CDR3 (Figures 1 and 2, Table 1).

In vitro protocols have recently been developed that distinguish between M1, M0, and M2 macrophages experimentally based on mitochondrial phenotypes [41]. M1 mitochondria consume small amounts of oxygen, have a low spare respiratory capacity (SRC, or physiologic reserve capacity), can use fatty acids for ROS production and NLRP3 assembly [42], are not dependent on glutamine, and the cells containing M1 mitochondria produce large amounts of lactic acid. In contrast, M2 mitochondria consume more oxygen at baseline, can use both glucose oxidized to pyruvate and fatty acids for oxphos, have a higher SRC, are dependent on glutamine for the fully differentiated phenotype, and the cells produce small amounts of acid. M0, uncommitted or multipotential mitochondria have a low basal oxygen consumption similar to M1, shunt some glucose down the pentose phosphate pathway (PPP) for NADPH and building block production, have an SRC that is intermediate between M2 and M1 organelles, and produce small amounts of extracellular acid, similar to M2 [41] (Table 1).

Table 1. Phenotypic characteristics of animal cell mitochondria.

No.	Trait	Mitochondrial Phenotype [40,41,43]		
		M0	M1	M2
1	Cellular energy metabolism	Aerobic glycolysis	Glycolysis	Oxidative phosphorylation
2	Mitochondrial DNA copy number	Intermediate	Low	High
3	Predominant morphology	Intermediate	Punctate	Filamentous
4	Cell replicative potential	High (Warburg)	Intermediate	Low
5	Cell multilineage regenerative potential	High	Low	Low
6	Cell differentiation potential	Low	Intermediate	High
7	Cell cancer potential	High	Intermediate	Low
8	Inflammatory potential	Intermediate	High	Low
9	Cell susceptibility to killing by apoptosis	Low	Intermediate	High
10	Inducible organellar quality control	Intermediate	Low	High
11	Baseline oxygen consumption	Low	Low	High
12	Stressed (uncoupled) oxygen consumption above baseline (spare respiratory capacity)	Intermediate	Low	High
13	ROS production	Intermediate	High	Low
14	NLRP3 inflammasome assembly	Low	High	Low
15	Lactate release from cells	Intermediate	High	Low
16	Pentose phosphate pathway (PPP)	High—NADPH for biosynthesis and cell growth	Intermediate—NADPH for NOX	Intermediate—NADPH for redox

Table 1. Cont.

No.	Trait	Mitochondrial Phenotype [40,41,43]		
		M0	M1	M2
17	Use of fatty acid oxidation (FAO)	Fatty acid synthesis for growth > FAO	For ROS and NLRP3 activation	For oxphos
18	Use of glucose	Glycolysis and PPP	Glycolysis and lactate release	PPP and pyruvate for oxphos
19	Use of glutamine	High: citrate for ATP citrate lyase and Acetyl-CoA	Low	High: oxphos via alpha-ketoglutarate
20	Stage of greatest use in the healing cycle and cell danger response	CDR2	CDR1	CDR3

M0, M1, and M2 mitochondrial phenotypes also occur in solid tissues and can be recognized in part by morphological criteria. Mitochondria can be described informally as being distributed along a spaghetti (filamentous) and meatball (punctate) gradient. M2 mitochondria are filamentous and interconnected, and predominate in post-mitotic or slowly regenerating tissues. M1 mitochondria are punctate, and M0 mitochondria are intermediate in form, with both short filaments and punctate organelles reminiscent of the coccobacillary forms of their proteobacterial ancestors [44]. Mitochondrial fusion–fission dynamics naturally regulate the balance between fused and elongated mitochondria, and fragmented/fissioned mitochondria according the growth and metabolic characteristics of the cell [45]. M0, uncommitted, or stem-like mitochondria predominate in cells that survive after exposure to chemotherapeutic agents or toxins [46]. M2 mitochondria under these conditions are depleted by DRP1-dependent fission/fragmentation, and conversion to M1 organelles prior to cell removal by apoptosis. Similar transitions in mitochondrial structure and function are readily demonstrated in brain astroglia before and after treatment with pro-inflammatory triggers such as lipopolysaccharide (LPS) and interferon gamma (IFN-γ) [43].

The total mitochondrial biomass is decreased naturally after triggering the cell danger response (CDR) with toxins [46], physical injury [6], infection, or any trigger that activates the healing cycle [11]. Mitochondrial biomass can also be reduced experimentally by the dominant-negative expression of the D1135A allele of the mitochondrial polymerase gamma (POLG1) to deplete mitochondrial DNA copy numbers by about 50% [47]. This mouse model mimics the mitochondrial defects that are a well-established hallmark of aging [48]. The phenotypes of aging found in this model included increased NFkB and matrix metalloproteinase 9 (MMP9) expression, increased inflammation, age spots, wrinkled skin, and premature hair loss. These transcriptional and anatomical signs of aging were reversed within 1 month of treatment in mice by unblocking mtDNA synthesis and restoring normal mtDNA copy numbers [47].

5. The Importance of Nucleotides and Purinergic Signaling

Of an estimated 1300 mitochondrial protein coding genes, 1158 are catalogued in MitoCarta v2.0 [5]. Only 13 mitochondrial proteins are encoded by mitochondrial DNA (mtDNA). The remaining 1145 are encoded by nuclear genes that are subject to tissue-specific gene expression programs. At least 789 mitochondrial proteins (68% of 1158) are enzymes with catalytic functions that have been assigned enzyme commission (EC) numbers, or encode subunits of multi-protein enzyme complexes such as those in respiratory chain complexes I, II, III, IV, and V, or are transporters, or kinases that use ATP (Table S1). At least 433 (55% of 789) of these proteins are regulated by the availability of purine and pyrimidine nucleotides such as ATP, GTP, UTP, NAD(P)+, or FAD either as substrates, or as allosteric regulators (Table S1). No other single class of molecules regulates more of the mitochondrial proteome than the nucleotides. In addition, an unknown fraction of nuclear mitochondrial genes is regulated transcriptionally and post-transcriptionally by puringergic signaling via the 12 G-protein coupled receptors (GPCRs; 4 P1R and 8 P2YR) and seven ionotropic receptors (P2XR) that are widely distributed in all tissues [49]. Some purinergic receptors are also expressed in intracellular compartments such as mitochondria, lysosomes, and the nucleus [50]. For example, P2X6 receptors are translocated to the nucleus in an age-dependent manner, interact with mRNA splicing factor 3A1, decrease mRNA processing, and contribute to aging [51].

6. The Importance of Nutrition in Healing and Aging

Different stages in child development, human aging, and athletic performance have different specific nutrients and calories that produce the best outcomes at each stage [52–54]. Based on measured differences in mitochondrial substrate preferences between M1, M0, and M2 organelles (Table 1), it is hypothesized that each stage of the healing cycle—CDR 1, 2, and 3 in Figure 1, and even certain chronic illnesses stuck in one of the three stages of the CDR [11]—will also have different stage-specific nutritional needs for optimal outcomes. Both chemical mass action from substrate supply and metabolite signaling via metabokines [11] are likely to be important mechanistically. Some of the broadest clues that connect nutrition with aging, mitochondria, and healing come from studies of caloric restriction [3]. Natural aging results in a reduction in the rate of turnover of mitochondria (mitochondrial biogenesis, or "mitochondriogenesis"), and a decrease in mitochondrial protein synthesis, standing mitochondrial biomass, respiratory reserve capacity, and oxphos function [55] (Figure 3). Caloric restriction, on the other hand, stimulates mitochondrial turnover through pathways associated with AMPK-stimulated autophagy and mitophagy [56]. This leads to improved mitochondrial quality control, oxphos function, and reserve capacity, but does not increase overall mitochondrial biomass measured as mtDNA copy number [57]. Caloric restriction also leads to a decrease in circulating thyroid hormone, decreased resting energy expenditure (REE), and a decrease in circulating anabolic hormones such as IGFI, insulin, human growth hormone, and testosterone [58]. From an evolutionary point of view, this hypometabolic response allows fewer calories to be consumed while opposing weight loss during periods of seasonal hardship.

Caloric restriction is a classical trigger of a reversible, stress-resistant, non-reproductive stage in the nematode *Caenorhabditis elegans* called dauer. Dauer has been the source of discovery of many longevity genes and has stimulated a productive discussion of the difference between lifespan and healthspan extension [59]. Dauer permits animals to live for up to 4 months under harsh conditions instead of their normal lifespan of 2 weeks. However, longevity in dauer comes at the cost of much reduced function and many changes associated with altered sensory anatomy [60], repetitive behaviors [61] and metabolism [62]. When calories are restored to animals in dauer, they exit the stage and re-enter their normal life-cycle, picking up where they left off, as if little or no biological aging had occurred during the time they spent in dauer [63].

More specific changes in nutrient supply also play an important role in aging and healing. Dietary supplementation with branch chain amino acids at a level of 1.5 g/kg/day, corresponding to about 1% of daily calorie intake in mice, has been shown to extend lifespan by about 10% [64]. This effect required normal production of nitric oxide (NO) by endothelial nitric oxide synthase (eNOS) [64]. The selective addition of a stoichiometric mix of essential amino acids (EAAs) including branch chain amino acids, also has anti-cancer effects by promoting growth inhibition and apoptosis in transformed but not in normal cells [65]. EAAs also promoted wound healing by moderating inflammation in the early stages, and maintaining TGFβ needed for tissue remodeling in the later stages of healing [66]. This later effect is in what would be called CDR3 in the model presented here (Figure 1). Interestingly, TGF β (DAF-7) signaling also plays a key role in recovery from dauer and the re-establishment of normal development in *C. elegans* [67].

7. Progressively Dysfunctional Cellular Mosaics

The number of nucleated cells in a 70-kg adult male is about 5×10^{12}. Non-nucleated red blood cells are 5-times more abundant (2.5×10^{13}), but make up just 6.5% of the total mass of an adult [68]. The total number of human cells in an adult is 3.0×10^{13}. There are also 3.8×10^{13} bacterial cells in a typical adult body [68]. Cells die every day and must be replaced. It is estimated that an adult turns over about 1×10^{10} (10 billion; about 10 grams of) cells/day by apoptosis [69]. This is equivalent to about 1 in 500 nucleated cells/day removed by apoptosis. The distribution of cell turnover is highly heterogeneous. Most spontaneous, or physiologic cell death with replacement occurs in short-lived cells such as those in bone marrow, intestines, skin, and hair follicles whose function is tolerant to spatial changes in tissue architecture produced by continuous cell growth and use-dependent removal or exit of cells from the compartment of origin. These rapidly dividing and structurally malleable tissues can turn over several times per year, with nearly 100% of the cells in these compartments turning over in a few days to weeks or months. Other tissues turn over more slowly. Adipose tissue is replaced at a rate of 10% of the cells per year [70], turning over completely in about 10 years. Skeletal muscle turns over at a rate of 6.6% per year [71]. Muscle fiber loss accelerates after age 60, leading to sarcopenia [72]. The kidneys weigh about 150 grams each, shed about 1.7×10^6 epithelial cells/day into the urine [73], and replace this loss and other losses by cell division and remodeling from recruited stem cells [74]. The adult liver weighs about 1400 grams and consists of about 2.4×10^{11} cells and replaces about 1.8% per year as a young adult, but fewer as liver size decreases with age [75]. Heart cells are replaced at a rate of 1% per year at age 25, falling to 0.45% per year at age 75 [76], and pancreatic islet β-cells are long-lived and not replaced after age 30 years [77].

Exposure to physical injury, toxins, or infection adds pathological cell death to the basal level of physiologic cell death described above. When a cell dies physiologically by apoptosis, the inflammatory reactions associated with CDR1 are avoided. Instead, the tissue skips to CDR2 and CDR3 to replace the lost cell and restore normal metabolism and tissue-specific gene expression patterns (Figure 1). The number of cells that die pathologically during a typical viral or bacterial infection will be dependent on the type of microbial pathogen and the severity of the infection, but the exact cell numbers lost in the course of a typical infection are not known. An average child has 5–6 recognized viral or bacterial infections each year for the first 5 years of life, then 2–3 per year throughout adult life [78]. Pathological cell death caused by infection, toxins, or injury will trigger inflammation and entry into the healing cycle by activating CDR1. As a thought experiment, one can imagine a systemic viral infection that might kill 1×10^{10} cells (about 10 grams, or 2 teaspoons), a number equal to the basal loss per day by apoptosis. Somatic DNA mutation rates are about 2.7×10^{-5} per typical 10,000 bp, protein-coding locus per cell division [79]. In this example, one turn of the healing cycle would result in 270,000 cells (10^{10} cells replaced $\times 2.7 \times 10^{-5}$ mutations/gene/division = 270,000 cells) sustaining a mutation that marks that cell as different from neighboring cells in the tissue. Mitotic recombination errors and chromosomal microaneuploidy [80], mobilization of retroelements [81] and endogenous retroviruses [82] in recruited stem cells, and reactivation and suppression cycles of latent DNA virus infections [83] will contribute to genetic variation produced by repetitive activation of the CDR over a lifetime. In addition to DNA mutations, there will also be an even larger number of induced and stochastic changes in transcription, post-transcriptional, and metabolic features that lead to changes in cell development.

With each turn of the healing cycle, cells such as neutrophils move out of capillaries and into tissues, releasing ROS via activated NADPH oxidases such as NOX2. Over time, this cyclic process creates tides of reactive oxygen, metabolic, innate immune, and inflammatory defenses that rise with injury and fall during wellness, ebbing and flowing with each turn of the healing cycle. Sialic acid binding immunoglobulin-type lectins (SIGLECs) include the CD33-related molecules expressed on neutrophils, natural killer (NK), macrophages, and T-cells that modulate ROS production, innate immune cell, and adaptive T-cell activity upon binding to sia-SAMPs (sialoglycan self-associated molecular patterns) [84]. Across species, the number of CD33r-SIGLEC genes is associated with healthy aging in the "wellderly", and decreased numbers accelerate age-related symptoms in mouse models [85]. Many cancers emerge from hypersialated cellular fields that have the effect of dampening the immune response by engaging inhibitory SIGLECs such as Siglec-9 [86]. The ceaseless tides or waves of innate immune cell migration into and out of tissue compartments continues throughout life with each turn of the healing cycle (Figures 1 and 2), and contributes to the progressively dysfunctional mosaics illustrated in Figure 2.

In the brain, a compensatory system of glial–lymphatic (glymphatic) tidal flows occurs by regulated changes in cell volume that are controlled by circadian changes in metabolism [87]. This daily cycle of glympahtic flow helps to remove aggregates of tau and beta amyloid that would otherwise accumulate as the byproducts of CDR activation both from normal learning and from microglial and synaptosomal innate immune activation that increase progressively with aging. Interestingly, a number of molecules that regulate mTOR and other metabolic aspects of the CDR, have recently been found to slow the aging process and extend longevity, while simultaneously protecting against age-related markers of Alzheimer dementia in animal models [88]. These "geroneuroprotecting" drugs were plant-based natural products similar to curcumin and polyphenols such as fisetin. The healing cycle-promoting properties of these drugs prevented the accumulation of tau and beta amyloid markers of Alzheimer dementia even though the drugs were not directed at the protein markers specifically [88].

The importance of innate immune activation and healing in aging is underscored by a recent discovery in Werner syndrome. Werner syndrome is a recessively inherited adult progeria syndrome caused by mutations in the Werner helicase (WRN). In addition to genomic instability caused by WRN mutations, the protein was recently found to play a key role in innate immunity as a transcriptional coactivator of the NFkB-dependent expression of the chemokine IL8 [89]. Werner mutations inhibit NFkB- and IL8-dependent ROS production. ROS production is not only an essential component of an effective CDR1 (Figure 1), but is important for NRF2 induction of long-term anti-oxidant and detoxification defenses [90]. In addition, the WRN protein is important for HIF1 stabilization [91] needed for tissue regeneration [26] during CDR2 (Figure 1).

With each turn of the healing cycle, the induced combination of transcriptional and metabolic changes and the DNA damage response will result in some injured and newly replaced cells being unable to complete the normal stages of the cycle. Perhaps 0.1–1 million cells of the 10^{10} cells that must be replaced after pathological cell death may be left behind, stuck or delayed in one of the three stages, CDR1, CDR2, and CDR3. An even larger number of cells whose function was transiently changed by injury, but were not killed, will contribute to the cells that are left behind in stages of the healing cycle. This process is illustrated as progressively dysfunctional mosaics on the right of the spiral in Figure 2. The ecogenetic interaction of inherited genotype of a given individual with the particular environmental insult (nutritional stress, pathogen, toxin, or physical injury) will determine how many cells are delayed or lost in the three different stages of the CDR. Examples of aging cellular mosaics are illustrated at age 20, 40, 60, and 90 years (Figure 2). Delayed and arrested cells that are incompletely differentiated create physical and metabolic separation—buffer zones—that act like control rods in a nuclear reactor that absorb signals and inhibit communication between previously adjoining, synergistic, and metabolically interdependent cells. Loss of cellular connectivity through functional gap junctions decreases the physiologic reserve capacity and increases vulnerability to stress-related cell death [92].

The propagation of calcium waves from cell to cell in a tissue via the inositol trisphosphate (IP3)-gated release of intracellular calcium stores is a common example of the cooperative response to neuroendocrine signaling, and is highly dependent on fully differentiated mitochondrial function [93]. CD38 increases with age [94] and is used by the cell to synthesize cyclic adenosine diphosphate ribose (cADPR) from NAD+ and nicotinic acid adenine dinucleotide phosphate (NAADP) from NADP+ as IP3-responsive calcium signaling declines [95]. cADPR and NAADP are intracellular ligands that stimulate IP3-independent release of calcium from the endoplasmic reticulum (ER) and lysosomes, respectively [29]. NAD+ and NADP+ are depleted by CD38 in the course of aging [94].

Cellular depletion of NAD+ and NADH leads to a decrease in mitonuclear communication [96], and to progressive declines in mitochondrial electron transport that contribute to aging [97]. Supplementation with the NAD+ precursor, nicotinamide riboside (NR), in animal models improves cell differentiation, restores laminin scaffolding and a more normal cellular mosaic in tissues, attenuates senescent cell formation, and increases longevity by about 10% [98]. In a recent study of the effect of transplanted senescent cells, the authors found that ≥ 1 senescent cell among 10,000 total cells in the body (0.01%), or ≥ 1 in 350 normal cells in a specific tissue (≥0.28%) was enough to produce a dysfunctional cellular mosaic and measurable functional defects such as decreased grip strength and exercise capacity [99].

8. De-Emergence as a Cause of Dysfunction

Complex living systems are comprised of at least seven discrete subsystems: (1) molecules; monomers, metabolites, and other building blocks, (2) polymers requiring energy for synthesis, such as proteins, polysaccharides, nucleic acids, and lipids made of amino acid, monosaccharide, nucleotide, acetyl and isoprene monomers, respectively, (3) polymers that assemble spontaneously such as charged and neutral lipids and hydrophobic proteins that form membranes, lipid droplets, and other lowered free energy structures in aqueous matrices, (4) organelles, (5) cells, (6) tissues, (7) organs. Many of the phenotypes associated with normal development, health, disease, and aging are emergent properties that depend on the function, arrangement, and interaction of the subsystems, but are qualitatively, quantitatively, and chronologically distinct from any single subsystem. For example, long-term memory is an emergent property that requires new protein synthesis that alters several aspects of all the subsystems that make up the brain, but knowledge of any one of the subsystems is insufficient to explain long-term memory. Other emergent properties of health include the ability to walk, talk, think, eat, excrete, mount an immune response, reproduce, detect and respond to a toxin or injury, and to heal. Aging degrades each of these abilities. If health is thought of as a collection of emergent phenotypes, then illness and aging can be thought of the gradual fading away, or de-emergence of the emergent properties that results from the degradation of metabolic and cellular order and the interactions of the subsystems that define health and resilience to environmental change. The images of progressively dysfunctional cellular mosaics shown in Figure 2 illustrate just one level—the cellular level—of the problem. Dysfunctional mosaics occur in every one of the seven subsystems, and mis-repairs accumulate at each level with each turn of the healing cycle.

9. The Hallmarks of Aging Emerge as a Result of Incomplete Healing

The molecular, cellular [2,3,48] and gene expression hallmarks [39] of aging can be organized according to their usage during the healing cycle and the stages of the multisystem CDR [10,11,29] (Figure 3). These hallmarks have been identified in studies of aging that have led naturally to an improved molecular understanding of mitochondrial stress [13,17,100], the integrated cell stress response (ICSR) [12,101,102], and the CDR [10,11,29,103]. Each trait associated with these hallmarks arises naturally from the activation of a molecular feature needed in the healing cycle, followed by the failure to extinguish this normal function or gene expression state of the CDR once it is no longer needed. Persistence of an aging hallmark after a turn of the healing cycle (Figure 2) can also occur in five other ways: (1) when a cell sustains too much genetic damage and becomes senescent, (2) when a cell that has been removed is not replaced and leads to tissue atrophy, (3) when a cell is replaced but fails to re-specialize according to the differentiated needs of the tissue, (4) when a mis-repair such as fibrosis or scarring [104] cannot be removed by remodeling, and (5) when genetic and metabolic changes bypass senescence and lead to cancer (Figure 1). The legacy of each of these events is to add to the dysfunctional mosaic illustrated in Figure 2.

Brain, muscle, nerve, and endocrine tissues are particularly susceptible to the incomplete replacement of lost cells once they have died and have been removed because of their limited regeneration capacity and the high degree of spatial organization required for optimal organ function. The function of these tissues is highly dependent on the spatial organization and metabolic complementarity of the cellular mosaic. This complementarity can only be achieved through cell specialization and chemical cooperativity between neighboring cells, which in turn, depends on the relative position and architecture of layers and columns of cells that must remain fixed over long periods of time. The cost of removing a cell from a chain of connected cells in the cerebral cortex or any other part of the brain has a high price that cannot easily be repaid by recruiting stem cells from another location. However, the cost of converting a cell to the hypersecretory phenotype of a senescent cell [105] is even higher, since just one senescent cell in 350 normal cells is enough to decrease the function of a tissue [99]. The smaller collective volume of cells that occurs in most tissues with age is illustrated in Figure 2. The decreasing volume of tissues in the columns labeled "Merge" and "CDR3" represents the effect of cell loss (atrophy) that occurs with incomplete progress through the healing cycle (Figures 1 and 2). In the example illustrated in Figure 2, more cells have accumulated in CDR2 (illustrated in green) by the ages of 60 and 90 years. Accumulation of cells in CDR2 will increase the risk of diseases such as diabetes, heart disease, and cancer because cells arrested or delayed in CDR2 maintain their proliferative capacity [11] (Figure 1). Proliferative capacity is maintained in part because cells in CDR2 contain more multipotential M0 mitochondria adapted for Warburg metabolism (Table 1). Inherited mutations in the RECQL4 helicase, which interacts with p53 and POLG in mitochondria, increase aerobic glycolysis associated with CDR2, lead to an increased risk of cancer, and to a form of progeria known as Rothmund–Thomson syndrome [106–108].

10. Conclusions

A new model is presented that reframes aging as the result of repeated cycles of incomplete healing. In this model, cycles of incomplete healing stack over time, leading to cellular mosaics that become progressively dysfunctional with age (Figure 2). Cells that accumulate in one of the three stages of the cell danger response (CDR) lead to specific risks of age-related disease. The accumulation of cells in CDR1 leads to chronic inflammatory and pain syndromes, and to susceptibility to chronic viral, bacterial, and fungal infections that require adaptive T-cell immunity for eradication [11]. The accumulation of cells in CDR2 leads to diabetes, heart disease, congestive heart failure, peripheral vascular disease, fibrotic disorders, and cancer risk. The accumulation of cells in CDR3 leads to autoimmune disorders, immune suppression or deficiency, neuropathic pain syndromes, behavioral and mental health disorders, or neurodevelopmental and neurodegenerative disease [11]. It is currently unknown what drugs, devices, or procedures can unblock arrested cells in the mosaic and permit the completion of the healing cycle. However, specific nutritional interventions such as nicotinamide riboside or essential amino acids have been found to facilitate healing [66], longevity [98] or both [64,66] in animal models. A prediction of the incomplete healing model is that differentiation checkpoints exist at each stage of the CDR (Figure 1) and that the molecular signals that are used naturally by cells to facilitate completion of healing may provide fresh clues for both preventing and treating age-related symptoms and disease. One procedure that is well-known to improve physical and psychological well-being, decrease mortality, and decrease the risk of age-related disease, is exercise [109,110]. The effects of exercise are multifaceted, but the increase in autophagy and remodeling of the mitochondrial network, leading to adaptive improvements in quality control and increased reserve capacity, may play key roles [111]. The search for geroneuroprotectants [88] and chemical exercise mimetics [112] has already begun. However, a deeper understanding of the role of natural metabolites as signaling molecules—metabokines [11]—and molecules regulated by exercise—exerkines and exosomes [113–115]—that regulate the completion of the healing cycle holds promise for preventing, slowing, and perhaps reversing [47] some of the effects of aging. Interventions that exploit purinergic [20] and sphingolipid [116] signaling pathways may be particularly powerful since over half of all mitochondrial proteins are regulated by ATP, NAD+, and related nucleotides (Table S1), and many of the molecular hallmarks of aging trace to or regulate the interplay between purines, mitochondria, sphingolipids [117,118], and the nucleus [119].

Supplementary Materials: The following are available online at http://www.mdpi.com/2079-7737/8/2/27/s1, Table S1: The mitochondrial proteome and nucleotide regulation.

Funding: This work was funded in part by philanthropic gifts from the UCSD Christini Fund, the Lennox Foundation, Malone Family Foundation, the Aloe family, the Harb family, Marc Spilo and all the others who contributed to the Aloe family autism research fund, the N of One Autism Research Foundation, the UCSD Mitochondrial Disease Research Fund, the JMS Fund, gifts in memory of Wayne Riggs, and from Linda Clark, Jeanne Conrad, Jeff Ansell, Josh Spears, David Cannistraro, the Kirby and Katie Mano Family, Simon and Evelyn Foo, Wing-kun Tam, Gita and Anurag Gupta, the Brent Kaufman Family, and the Daniel and Kelly White Family. RKN wishes to thank over 2000 individuals who have each provided gifts in the past year to support Naviaux Lab research.

Acknowledgments: RKN thanks the many families with primary mitochondrial disease, autism spectrum disorder (ASD), and myalgic encephalomyelitis/chronic fatigue syndrome (ME/CFS) who have helped make this research possible. RKN thanks Scott McAvoy from the UCSD Digital Media Lab for preparing the art work for the healing spiral in Figure 2, and Vamsi Mootha, Sarah Calvo, and Jonathan Monk for the bioinformatic analysis of mitochondrial proteins and enzymes.

Conflicts of Interest: RKN is an unpaid scientific advisory board member for the Autism Research Institute (ARI) and the Open Medicine Foundation (OMF). Financial supporters for this study had no role in data analysis, interpretation, writing, or publication of this work.

References

1. McCay, C.M.; Crowell, M.F. Prolonging the life span. *Sci. Mon.* **1934**, *39*, 405–414.
2. Finkel, T. The metabolic regulation of aging. *Nat. Med.* **2015**, *21*, 1416–1423. [CrossRef] [PubMed]
3. Lopez-Otin, C.; Galluzzi, L.; Freije, J.M.P.; Madeo, F.; Kroemer, G. Metabolic Control of Longevity. *Cell* **2016**, *166*, 802–821. [CrossRef]
4. Pagliarini, D.J.; Calvo, S.E.; Chang, B.; Sheth, S.A.; Vafai, S.B.; Ong, S.E.; Walford, G.A.; Sugiana, C.; Boneh, A.; Chen, W.K.; et al. A mitochondrial protein compendium elucidates complex I disease biology. *Cell* **2008**, *134*, 112–123. [CrossRef]
5. Calvo, S.E.; Clauser, K.R.; Mootha, V.K. MitoCarta2.0: An updated inventory of mammalian mitochondrial proteins. *Nucleic Acids Res.* **2016**, *44*, D1251–D1257. [CrossRef] [PubMed]
6. Naviaux, R.K.; Le, T.P.; Bedelbaeva, K.; Leferovich, J.; Gourevitch, D.; Sachadyn, P.; Zhang, X.M.; Clark, L.; Heber-Katz, E. Retained features of embryonic metabolism in the adult MRL mouse. *Mol. Genet. Metab.* **2009**, *96*, 133–144. [CrossRef] [PubMed]
7. Ma, S.; Huang, Q.; Tominaga, T.; Liu, C.; Suzuki, K. An 8-Week Ketogenic Diet Alternated Interleukin-6, Ketolytic and Lipolytic Gene Expression, and Enhanced Exercise Capacity in Mice. *Nutrients* **2018**, *10*, 1696. [CrossRef]
8. Ogborn, D.I.; McKay, B.R.; Crane, J.D.; Safdar, A.; Akhtar, M.; Parise, G.; Tarnopolsky, M.A. Effects of age and unaccustomed resistance exercise on mitochondrial transcript and protein abundance in skeletal muscle of men. *Am. J. Physiol. Regul. Integr. Comp. Physiol.* **2015**, *308*, R734–R741. [CrossRef] [PubMed]
9. Winckelmans, E.; Nawrot, T.S.; Tsamou, M.; Den Hond, E.; Baeyens, W.; Kleinjans, J.; Lefebvre, W.; Van Larebeke, N.; Peusens, M.; Plusquin, M.; et al. Transcriptome-wide analyses indicate mitochondrial responses to particulate air pollution exposure. *Environ. Health A Glob. Access Sci. Source* **2017**, *16*, 87. [CrossRef] [PubMed]
10. Naviaux, R.K. Metabolic features of the cell danger response. *Mitochondrion* **2014**, *16*, 7–17. [CrossRef]
11. Naviaux, R.K. Metabolic features and regulation of the healing cycle-A new model for chronic disease pathogenesis and treatment. *Mitochondrion* **2018**. [CrossRef] [PubMed]
12. Quiros, P.M.; Prado, M.A.; Zamboni, N.; D'Amico, D.; Williams, R.W.; Finley, D.; Gygi, S.P.; Auwerx, J. Multi-omics analysis identifies ATF4 as a key regulator of the mitochondrial stress response in mammals. *J. Cell Biol.* **2017**, *216*, 2027–2045. [CrossRef] [PubMed]
13. Nikkanen, J.; Forsstrom, S.; Euro, L.; Paetau, I.; Kohnz, R.A.; Wang, L.; Chilov, D.; Viinamaki, J.; Roivainen, A.; Marjamaki, P.; et al. Mitochondrial DNA Replication Defects Disturb Cellular dNTP Pools and Remodel One-Carbon Metabolism. *Cell Metab.* **2016**, *23*, 635–648. [CrossRef]
14. Silva, J.M.; Wong, A.; Carelli, V.; Cortopassi, G.A. Inhibition of mitochondrial function induces an integrated stress response in oligodendroglia. *Neurobiol. Dis.* **2009**, *34*, 357–365. [CrossRef]

15. Cameron, J.L.; Eagleson, K.L.; Fox, N.A.; Hensch, T.K.; Levitt, P. Social Origins of Developmental Risk for Mental and Physical Illness. *J. Neurosci. Off. J. Soc. Neurosci.* **2017**, *37*, 10783–10791. [CrossRef] [PubMed]
16. Fredrickson, B.L.; Grewen, K.M.; Algoe, S.B.; Firestine, A.M.; Arevalo, J.M.; Ma, J.; Cole, S.W. Psychological well-being and the human conserved transcriptional response to adversity. *PLoS ONE* **2015**, *10*, e0121839. [CrossRef] [PubMed]
17. Picard, M.; McManus, M.J.; Gray, J.D.; Nasca, C.; Moffat, C.; Kopinski, P.K.; Seifert, E.L.; McEwen, B.S.; Wallace, D.C. Mitochondrial functions modulate neuroendocrine, metabolic, inflammatory, and transcriptional responses to acute psychological stress. *Proc. Natl. Acad. Sci. USA* **2015**, *112*, E6614–E6623. [CrossRef] [PubMed]
18. Leday, G.G.R.; Vertes, P.E.; Richardson, S.; Greene, J.R.; Regan, T.; Khan, S.; Henderson, R.; Freeman, T.C.; Pariante, C.M.; Harrison, N.A.; et al. Replicable and Coupled Changes in Innate and Adaptive Immune Gene Expression in Two Case-Control Studies of Blood Microarrays in Major Depressive Disorder. *Biol. Psychiatry* **2018**, *83*, 70–80. [CrossRef] [PubMed]
19. Burnstock, G.; Knight, G.E. Cell culture: Complications due to mechanical release of ATP and activation of purinoceptors. *Cell Tissue Res.* **2017**. [CrossRef]
20. Burnstock, G. The therapeutic potential of purinergic signalling. *Biochem. Pharmacol.* **2018**, *151*, 157–165. [CrossRef]
21. Burnstock, G. Short- and long-term (trophic) purinergic signalling. *Philos. Trans. R. Soc. Lond.* **2016**, *371*, 20150422. [CrossRef]
22. De la Fuente, I.M. Elements of the cellular metabolic structure. *Front. Mol. Biosci.* **2015**, *2*, 16. [CrossRef] [PubMed]
23. Jindrichova, M.; Kuzyk, P.; Li, S.; Stojilkovic, S.S.; Zemkova, H. Conserved ectodomain cysteines are essential for rat P2X7 receptor trafficking. *Purinergic Signal.* **2012**, *8*, 317–325. [CrossRef] [PubMed]
24. Sussman, I.; Erecinska, M.; Wilson, D.F. Regulation of cellular energy metabolism: The Crabtree effect. *Biochim. Biophys. Acta* **1980**, *591*, 209–223. [CrossRef]
25. Dror, E.; Dalmas, E.; Meier, D.T.; Wueest, S.; Thevenet, J.; Thienel, C.; Timper, K.; Nordmann, T.M.; Traub, S.; Schulze, F.; et al. Postprandial macrophage-derived IL-1beta stimulates insulin, and both synergistically promote glucose disposal and inflammation. *Nat. Immunol.* **2017**, *18*, 283–292. [CrossRef]
26. Heber-Katz, E. Oxygen, Metabolism, and Regeneration: Lessons from Mice. *Trends Mol. Med.* **2017**, *23*, 1024–1036. [CrossRef] [PubMed]
27. Barros-Barbosa, A.R.; Oliveira, A.; Lobo, M.G.; Cordeiro, J.M.; Correia-de-Sa, P. Under stressful conditions activation of the ionotropic P2X7 receptor differentially regulates GABA and glutamate release from nerve terminals of the rat cerebral cortex. *Neurochem. Int.* **2018**, *112*, 81–95. [CrossRef]
28. Xiong, Y.; Teng, S.; Zheng, L.; Sun, S.; Li, J.; Guo, N.; Li, M.; Wang, L.; Zhu, F.; Wang, C.; et al. Stretch-induced Ca(2+) independent ATP release in hippocampal astrocytes. *J. Physiol.* **2018**, *596*, 1931–1947. [CrossRef]
29. Naviaux, R.K. Antipurinergic therapy for autism-An in-depth review. *Mitochondrion* **2017**. [CrossRef]
30. Sakaki, H.; Tsukimoto, M.; Harada, H.; Moriyama, Y.; Kojima, S. Autocrine regulation of macrophage activation via exocytosis of ATP and activation of P2Y11 receptor. *PLoS ONE* **2013**, *8*, e59778. [CrossRef] [PubMed]
31. Huang, Z.L.; Zhang, Z.; Qu, W.M. Roles of adenosine and its receptors in sleep-wake regulation. *Int. Rev. Neurobiol.* **2014**, *119*, 349–371. [PubMed]
32. He, S.; Sharpless, N.E. Senescence in Health and Disease. *Cell* **2017**, *169*, 1000–1011. [CrossRef] [PubMed]
33. Wang-Michelitsch, J.; Michelitsch, T.M. Misrepair accumulation theory: A theory for understanding aging, cancer development, longevity, and adaptation. *arXiv* **2015**, arXiv:1505.07016.
34. Wang-Michelitsch, J.; Michelitsch, T.M. Tissue fibrosis: A principal proof for the central role of Misrepair in aging. *arXiv* **2015**, arXiv:1505.01376.
35. Wang, J.; Michelitsch, T.; Wunderlin, A.; Mahadeva, R. Aging as a consequence of misrepair–a novel theory of aging. *arXiv* **2009**, arXiv:0904.0575.
36. Zhao, R.; Liang, D.; Sun, D. Blockade of Extracellular ATP Effect by Oxidized ATP Effectively Mitigated Induced Mouse Experimental Autoimmune Uveitis (EAU). *PLoS ONE* **2016**, *11*, e0155953. [CrossRef]
37. Hultqvist, M.; Olsson, L.M.; Gelderman, K.A.; Holmdahl, R. The protective role of ROS in autoimmune disease. *Trends Immunol.* **2009**, *30*, 201–208. [CrossRef]

38. Alvarez, K.; Vasquez, G. Damage-associated molecular patterns and their role as initiators of inflammatory and auto-immune signals in systemic lupus erythematosus. *Int. Rev. Immunol.* **2017**, *36*, 259–270. [CrossRef] [PubMed]
39. Frenk, S.; Houseley, J. Gene expression hallmarks of cellular ageing. *Biogerontology* **2018**, *19*, 547–566. [CrossRef]
40. Chen, W.; Sandoval, H.; Kubiak, J.Z.; Li, X.C.; Ghobrial, R.M.; Kloc, M. The phenotype of peritoneal mouse macrophages depends on the mitochondria and ATP/ADP homeostasis. *Cell. Immunol.* **2018**, *324*, 1–7. [CrossRef] [PubMed]
41. Liu, P.S.; Ho, P.C. Determining Macrophage Polarization upon Metabolic Perturbation. *Methods Mol. Biol.* **2019**, *1862*, 173–186. [PubMed]
42. Van den Bossche, J.; O'Neill, L.A.; Menon, D. Macrophage Immunometabolism: Where Are We (Going)? *Trends Immunol.* **2017**, *38*, 395–406. [CrossRef] [PubMed]
43. Motori, E.; Puyal, J.; Toni, N.; Ghanem, A.; Angeloni, C.; Malaguti, M.; Cantelli-Forti, G.; Berninger, B.; Conzelmann, K.K.; Gotz, M.; et al. Inflammation-induced alteration of astrocyte mitochondrial dynamics requires autophagy for mitochondrial network maintenance. *Cell Metab.* **2013**, *18*, 844–859. [CrossRef]
44. Roger, A.J.; Munoz-Gomez, S.A.; Kamikawa, R. The Origin and Diversification of Mitochondria. *Curr. Biol.* **2017**, *27*, R1177–R1192. [CrossRef] [PubMed]
45. Chen, H.; Chan, D.C. Mitochondrial Dynamics in Regulating the Unique Phenotypes of Cancer and Stem Cells. *Cell Metab.* **2017**, *26*, 39–48. [CrossRef] [PubMed]
46. Giedt, R.J.; Fumene Feruglio, P.; Pathania, D.; Yang, K.S.; Kilcoyne, A.; Vinegoni, C.; Mitchison, T.J.; Weissleder, R. Computational imaging reveals mitochondrial morphology as a biomarker of cancer phenotype and drug response. *Sci. Rep.* **2016**, *6*, 32985. [CrossRef]
47. Singh, B.; Schoeb, T.R.; Bajpai, P.; Slominski, A.; Singh, K.K. Reversing wrinkled skin and hair loss in mice by restoring mitochondrial function. *Cell Death Dis.* **2018**, *9*, 735. [CrossRef] [PubMed]
48. Atamna, H.; Tenore, A.; Lui, F.; Dhahbi, J.M. Organ reserve, excess metabolic capacity, and aging. *Biogerontology* **2018**, *19*, 171–184. [CrossRef] [PubMed]
49. Verkhratsky, A.; Burnstock, G. Biology of purinergic signalling: Its ancient evolutionary roots, its omnipresence and its multiple functional significance. *BioEssays News Rev. Mol. Cell. Dev. Biol.* **2014**, *36*, 697–705. [CrossRef]
50. Burnstock, G. Intracellular expression of purinoceptors. *Purinergic Signal.* **2015**, *11*, 275–276. [CrossRef] [PubMed]
51. Diaz-Hernandez, J.I.; Sebastian-Serrano, A.; Gomez-Villafuertes, R.; Diaz-Hernandez, M.; Miras-Portugal, M.T. Age-related nuclear translocation of P2X6 subunit modifies splicing activity interacting with splicing factor 3A1. *PLoS ONE* **2015**, *10*, e0123121. [CrossRef]
52. Dietz, W.H. Critical periods in childhood for the development of obesity. *Am. J. Clin. Nutr.* **1994**, *59*, 955–959. [CrossRef]
53. Das, J.K.; Salam, R.A.; Thornburg, K.L.; Prentice, A.M.; Campisi, S.; Lassi, Z.S.; Koletzko, B.; Bhutta, Z.A. Nutrition in adolescents: Physiology, metabolism, and nutritional needs. *Ann. N. Y. Acad. Sci.* **2017**, *1393*, 21–33. [CrossRef]
54. Desbrow, B.; Burd, N.A.; Tarnopolsky, M.; Moore, D.R.; Elliott-Sale, K.J. Nutrition for Special Populations: Young, Female, and Masters Athletes. *Int. J. Sport Nutr. Exerc. Metab.* **2019**, 1–8. [CrossRef]
55. Lopez-Lluch, G.; Irusta, P.M.; Navas, P.; de Cabo, R. Mitochondrial biogenesis and healthy aging. *Exp. Gerontol.* **2008**, *43*, 813–819. [CrossRef] [PubMed]
56. Weir, H.J.; Yao, P.; Huynh, F.K.; Escoubas, C.C.; Goncalves, R.L.; Burkewitz, K.; Laboy, R.; Hirschey, M.D.; Mair, W.B. Dietary Restriction and AMPK Increase Lifespan via Mitochondrial Network and Peroxisome Remodeling. *Cell Metab.* **2017**, *26*, 884–896 e885. [CrossRef]
57. Rabol, R.; Svendsen, P.F.; Skovbro, M.; Boushel, R.; Haugaard, S.B.; Schjerling, P.; Schrauwen, P.; Hesselink, M.K.; Nilas, L.; Madsbad, S.; et al. Reduced skeletal muscle mitochondrial respiration and improved glucose metabolism in nondiabetic obese women during a very low calorie dietary intervention leading to rapid weight loss. *Metabolism* **2009**, *58*, 1145–1152. [CrossRef]

58. Muller, M.J.; Enderle, J.; Pourhassan, M.; Braun, W.; Eggeling, B.; Lagerpusch, M.; Gluer, C.C.; Kehayias, J.J.; Kiosz, D.; Bosy-Westphal, A. Metabolic adaptation to caloric restriction and subsequent refeeding: The Minnesota Starvation Experiment revisited. *Am. J. Clin. Nutr.* **2015**, *102*, 807–819. [CrossRef] [PubMed]
59. Ewald, C.Y.; Castillo-Quan, J.I.; Blackwell, T.K. Untangling Longevity, Dauer, and Healthspan in Caenorhabditis elegans Insulin/IGF-1-Signalling. *Gerontology* **2018**, *64*, 96–104. [CrossRef] [PubMed]
60. Albert, P.S.; Riddle, D.L. Developmental alterations in sensory neuroanatomy of the Caenorhabditis elegans dauer larva. *J. Comp. Neurol.* **1983**, *219*, 461–481. [CrossRef] [PubMed]
61. Lee, D.; Lee, H.; Kim, N.; Lim, D.S.; Lee, J. Regulation of a hitchhiking behavior by neuronal insulin and TGF-beta signaling in the nematode Caenorhabditis elegans. *Biochem. Biophys. Res. Commun.* **2017**, *484*, 323–330. [CrossRef] [PubMed]
62. Lant, B.; Storey, K.B. An overview of stress response and hypometabolic strategies in Caenorhabditis elegans: Conserved and contrasting signals with the mammalian system. *Int. J. Biol. Sci.* **2010**, *6*, 9–50. [CrossRef] [PubMed]
63. Androwski, R.J.; Flatt, K.M.; Schroeder, N.E. Phenotypic plasticity and remodeling in the stress-induced Caenorhabditis elegans dauer. *Wiley Interdiscip. Rev. Dev. Biol.* **2017**, *6*, e278. [CrossRef]
64. D'Antona, G.; Ragni, M.; Cardile, A.; Tedesco, L.; Dossena, M.; Bruttini, F.; Caliaro, F.; Corsetti, G.; Bottinelli, R.; Carruba, M.O.; et al. Branched-chain amino acid supplementation promotes survival and supports cardiac and skeletal muscle mitochondrial biogenesis in middle-aged mice. *Cell Metab.* **2010**, *12*, 362–372. [CrossRef]
65. Bonfili, L.; Cecarini, V.; Cuccioloni, M.; Angeletti, M.; Flati, V.; Corsetti, G.; Pasini, E.; Dioguardi, F.S.; Eleuteri, A.M. Essential amino acid mixtures drive cancer cells to apoptosis through proteasome inhibition and autophagy activation. *FEBS J.* **2017**, *284*, 1726–1737. [CrossRef] [PubMed]
66. Corsetti, G.; Romano, C.; Pasini, E.; Marzetti, E.; Calvani, R.; Picca, A.; Flati, V.; Dioguardi, F.S. Diet enrichment with a specific essential free amino acid mixture improves healing of undressed wounds in aged rats. *Exp. Gerontol.* **2017**, *96*, 138–145. [CrossRef]
67. Fielenbach, N.; Antebi, A. C. elegans dauer formation and the molecular basis of plasticity. *Genes Dev.* **2008**, *22*, 2149–2165. [CrossRef]
68. Sender, R.; Fuchs, S.; Milo, R. Revised Estimates for the Number of Human and Bacteria Cells in the Body. *PLoS Biol.* **2016**, *14*, e1002533. [CrossRef]
69. Renehan, A.G.; Booth, C.; Potten, C.S. What is apoptosis, and why is it important? *BMJ* **2001**, *322*, 1536–1538. [CrossRef]
70. Spalding, K.L.; Arner, E.; Westermark, P.O.; Bernard, S.; Buchholz, B.A.; Bergmann, O.; Blomqvist, L.; Hoffstedt, J.; Naslund, E.; Britton, T.; et al. Dynamics of fat cell turnover in humans. *Nature* **2008**, *453*, 783–787. [CrossRef]
71. Spalding, K.L.; Bhardwaj, R.D.; Buchholz, B.A.; Druid, H.; Frisen, J. Retrospective birth dating of cells in humans. *Cell* **2005**, *122*, 133–143. [CrossRef] [PubMed]
72. Doherty, T.J. Invited review: Aging and sarcopenia. *J. Appl. Physiol.* **2003**, *95*, 1717–1727. [CrossRef]
73. Prescott, L.F. The normal urinary excretion rates of renal tubular cells, leucocytes and red blood cells. *Clin. Sci.* **1966**, *31*, 425–435. [PubMed]
74. Rinkevich, Y.; Montoro, D.T.; Contreras-Trujillo, H.; Harari-Steinberg, O.; Newman, A.M.; Tsai, J.M.; Lim, X.; Van-Amerongen, R.; Bowman, A.; Januszyk, M.; et al. In vivo clonal analysis reveals lineage-restricted progenitor characteristics in mammalian kidney development, maintenance, and regeneration. *Cell Rep.* **2014**, *7*, 1270–1283. [CrossRef] [PubMed]
75. Kim, I.H.; Kisseleva, T.; Brenner, D.A. Aging and liver disease. *Curr. Opin. Gastroenterol.* **2015**, *31*, 184–191. [CrossRef] [PubMed]
76. Bergmann, O.; Bhardwaj, R.D.; Bernard, S.; Zdunek, S.; Barnabe-Heider, F.; Walsh, S.; Zupicich, J.; Alkass, K.; Buchholz, B.A.; Druid, H.; et al. Evidence for cardiomyocyte renewal in humans. *Science* **2009**, *324*, 98–102. [CrossRef]
77. Perl, S.; Kushner, J.A.; Buchholz, B.A.; Meeker, A.K.; Stein, G.M.; Hsieh, M.; Kirby, M.; Pechhold, S.; Liu, E.H.; Harlan, D.M.; et al. Significant human beta-cell turnover is limited to the first three decades of life as determined by in vivo thymidine analog incorporation and radiocarbon dating. *J. Clin. Endocrinol. Metab.* **2010**, *95*, E234–E239. [CrossRef] [PubMed]

78. Fox, J.P.; Hall, C.E.; Cooney, M.K.; Luce, R.E.; Kronmal, R.A. The Seattle virus watch. II. Objectives, study population and its observation, data processing and summary of illnesses. *Am. J. Epidemiol.* **1972**, *96*, 270–285. [CrossRef] [PubMed]
79. Milholland, B.; Dong, X.; Zhang, L.; Hao, X.; Suh, Y.; Vijg, J. Differences between germline and somatic mutation rates in humans and mice. *Nat. Commun.* **2017**, *8*, 15183. [CrossRef]
80. Peterson, S.E.; Westra, J.W.; Paczkowski, C.M.; Chun, J. Chromosomal mosaicism in neural stem cells. *Methods Mol. Biol.* **2008**, *438*, 197–204.
81. Muotri, A.R.; Chu, V.T.; Marchetto, M.C.; Deng, W.; Moran, J.V.; Gage, F.H. Somatic mosaicism in neuronal precursor cells mediated by L1 retrotransposition. *Nature* **2005**, *435*, 903–910. [CrossRef]
82. Romer, C.; Singh, M.; Hurst, L.D.; Izsvak, Z. How to tame an endogenous retrovirus: HERVH and the evolution of human pluripotency. *Curr. Opin. Virol.* **2017**, *25*, 49–58. [CrossRef]
83. Harris, S.A.; Harris, E.A. Herpes Simplex Virus Type 1 and Other Pathogens are Key Causative Factors in Sporadic Alzheimer's Disease. *J. Alzheimer's Dis. JAD* **2015**, *48*, 319–353. [CrossRef]
84. Bornhofft, K.F.; Goldammer, T.; Rebl, A.; Galuska, S.P. Siglecs: A journey through the evolution of sialic acid-binding immunoglobulin-type lectins. *Dev. Comp. Immunol.* **2018**, *86*, 219–231. [CrossRef]
85. Schwarz, F.; Pearce, O.M.; Wang, X.; Samraj, A.N.; Laubli, H.; Garcia, J.O.; Lin, H.; Fu, X.; Garcia-Bingman, A.; Secrest, P.; et al. Siglec receptors impact mammalian lifespan by modulating oxidative stress. *Elife* **2015**, *4*. [CrossRef]
86. Stanczak, M.A.; Siddiqui, S.S.; Trefny, M.P.; Thommen, D.S.; Boligan, K.F.; von Gunten, S.; Tzankov, A.; Tietze, L.; Lardinois, D.; Heinzelmann-Schwarz, V.; et al. Self-associated molecular patterns mediate cancer immune evasion by engaging Siglecs on T cells. *J. Clin. Investig.* **2018**, *128*, 4912–4923. [CrossRef]
87. Bower, N.I.; Hogan, B.M. Brain drains: New insights into brain clearance pathways from lymphatic biology. *J. Mol. Med.* **2018**, *96*, 383–390. [CrossRef]
88. Schubert, D.; Currais, A.; Goldberg, J.; Finley, K.; Petrascheck, M.; Maher, P. Geroneuroprotectors: Effective Geroprotectors for the Brain. *Trends Pharmacol. Sci.* **2018**, *39*, 1004–1007. [CrossRef]
89. Mizutani, T.; Ishizaka, A.; Furuichi, Y. The Werner Protein Acts as a Coactivator of Nuclear Factor kappaB (NF-kappaB) on HIV-1 and Interleukin-8 (IL-8) Promoters. *J. Biol. Chem.* **2015**, *290*, 18391–18399. [CrossRef]
90. Huang, Y.; Li, W.; Su, Z.Y.; Kong, A.N. The complexity of the Nrf2 pathway: Beyond the antioxidant response. *J. Nutr. Biochem.* **2015**, *26*, 1401–1413. [CrossRef]
91. Li, B.; Iglesias-Pedraz, J.M.; Chen, L.Y.; Yin, F.; Cadenas, E.; Reddy, S.; Comai, L. Downregulation of the Werner syndrome protein induces a metabolic shift that compromises redox homeostasis and limits proliferation of cancer cells. *Aging Cell* **2014**, *13*, 367–378. [CrossRef]
92. Blanc, E.M.; Bruce-Keller, A.J.; Mattson, M.P. Astrocytic gap junctional communication decreases neuronal vulnerability to oxidative stress-induced disruption of Ca2+ homeostasis and cell death. *J. Neurochem.* **1998**, *70*, 958–970. [CrossRef] [PubMed]
93. Duchen, M.R. Mitochondria and calcium: From cell signalling to cell death. *J. Physiol.* **2000**, *529 Pt 1*, 57–68. [CrossRef]
94. Camacho-Pereira, J.; Tarrago, M.G.; Chini, C.C.S.; Nin, V.; Escande, C.; Warner, G.M.; Puranik, A.S.; Schoon, R.A.; Reid, J.M.; Galina, A.; et al. CD38 Dictates Age-Related NAD Decline and Mitochondrial Dysfunction through an SIRT3-Dependent Mechanism. *Cell Metab.* **2016**, *23*, 1127–1139. [CrossRef]
95. Igwe, O.J.; Filla, M.B. Aging-related regulation of myo-inositol 1,4,5-trisphosphate signal transduction pathway in the rat striatum. *Brain Res. Mol. Brain Res.* **1997**, *46*, 39–53. [CrossRef]
96. Gomes, A.P.; Price, N.L.; Ling, A.J.; Moslehi, J.J.; Montgomery, M.K.; Rajman, L.; White, J.P.; Teodoro, J.S.; Wrann, C.D.; Hubbard, B.P.; et al. Declining NAD(+) induces a pseudohypoxic state disrupting nuclear-mitochondrial communication during aging. *Cell* **2013**, *155*, 1624–1638. [CrossRef]
97. Chini, C.C.S.; Tarrago, M.G.; Chini, E.N. NAD and the aging process: Role in life, death and everything in between. *Mol. Cell. Endocrinol.* **2017**, *455*, 62–74. [CrossRef] [PubMed]
98. Zhang, H.; Ryu, D.; Wu, Y.; Gariani, K.; Wang, X.; Luan, P.; D'Amico, D.; Ropelle, E.R.; Lutolf, M.P.; Aebersold, R.; et al. NAD(+) repletion improves mitochondrial and stem cell function and enhances life span in mice. *Science* **2016**, *352*, 1436–1443. [CrossRef] [PubMed]
99. Xu, M.; Pirtskhalava, T.; Farr, J.N.; Weigand, B.M.; Palmer, A.K.; Weivoda, M.M.; Inman, C.L.; Ogrodnik, M.B.; Hachfeld, C.M.; Fraser, D.G.; et al. Senolytics improve physical function and increase lifespan in old age. *Nat. Med.* **2018**, *24*, 1246–1256. [CrossRef] [PubMed]

100. Go, Y.M.; Fernandes, J.; Hu, X.; Uppal, K.; Jones, D.P. Mitochondrial network responses in oxidative physiology and disease. *Free Radic. Biol. Med.* **2018**, *116*, 31–40. [CrossRef]
101. Salminen, A.; Kaarniranta, K.; Kauppinen, A. Integrated stress response stimulates FGF21 expression: Systemic enhancer of longevity. *Cell Signal.* **2017**, *40*, 10–21. [CrossRef] [PubMed]
102. Pakos-Zebrucka, K.; Koryga, I.; Mnich, K.; Ljujic, M.; Samali, A.; Gorman, A.M. The integrated stress response. *EMBO Rep.* **2016**, *17*, 1374–1395. [CrossRef] [PubMed]
103. Naviaux, R.K. Oxidative shielding or oxidative stress? *J. Pharmacol. Exp. Ther.* **2012**, *342*, 608–618. [CrossRef] [PubMed]
104. Wang-Michelitsch, J.; Michelitsch, T. Aging as a process of accumulation of Misrepairs. *arXiv* **2015**, arXiv:1503.07163.
105. Hernandez-Segura, A.; de Jong, T.V.; Melov, S.; Guryev, V.; Campisi, J.; Demaria, M. Unmasking Transcriptional Heterogeneity in Senescent Cells. *Curr. Biol.* **2017**, *27*, 2652–2660 e2654. [CrossRef] [PubMed]
106. Kumari, J.; Hussain, M.; De, S.; Chandra, S.; Modi, P.; Tikoo, S.; Singh, A.; Sagar, C.; Sepuri, N.B.; Sengupta, S. Mitochondrial functions of RECQL4 are required for the prevention of aerobic glycolysis-dependent cell invasion. *J. Cell Sci.* **2016**, *129*, 1312–1318. [CrossRef] [PubMed]
107. Gupta, S.; De, S.; Srivastava, V.; Hussain, M.; Kumari, J.; Muniyappa, K.; Sengupta, S. RECQL4 and p53 potentiate the activity of polymerase gamma and maintain the integrity of the human mitochondrial genome. *Carcinogenesis* **2014**, *35*, 34–45. [CrossRef]
108. Croteau, D.L.; Singh, D.K.; Hoh Ferrarelli, L.; Lu, H.; Bohr, V.A. RECQL4 in genomic instability and aging. *Trends Genet.* **2012**, *28*, 624–631. [CrossRef]
109. Hupin, D.; Roche, F.; Gremeaux, V.; Chatard, J.C.; Oriol, M.; Gaspoz, J.M.; Barthelemy, J.C.; Edouard, P. Even a low-dose of moderate-to-vigorous physical activity reduces mortality by 22% in adults aged >/=60 years: A systematic review and meta-analysis. *Br. J. Sports Med.* **2015**, *49*, 1262–1267. [CrossRef]
110. Gries, K.J.; Raue, U.; Perkins, R.K.; Lavin, K.M.; Overstreet, B.S.; D'Acquisto, L.J.; Graham, B.; Finch, W.H.; Kaminsky, L.A.; Trappe, T.A.; et al. Cardiovascular and Skeletal Muscle Health with Lifelong Exercise. *J. Appl. Physiol.* **2018**. [CrossRef]
111. Radak, Z.; Torma, F.; Berkes, I.; Goto, S.; Mimura, T.; Posa, A.; Balogh, L.; Boldogh, I.; Suzuki, K.; Higuchi, M.; et al. Exercise effects on physiological function during aging. *Free Radic. Biol. Med.* **2019**, *132*, 33–41. [CrossRef]
112. Fan, W.; Evans, R.M. Exercise Mimetics: Impact on Health and Performance. *Cell Metab.* **2017**, *25*, 242–247. [CrossRef]
113. Yu, M.; Tsai, S.F.; Kuo, Y.M. The Therapeutic Potential of Anti-Inflammatory Exerkines in the Treatment of Atherosclerosis. *Int. J. Mol. Sci.* **2017**, *18*, 1260. [CrossRef]
114. Whitham, M.; Parker, B.L.; Friedrichsen, M.; Hingst, J.R.; Hjorth, M.; Hughes, W.E.; Egan, C.L.; Cron, L.; Watt, K.I.; Kuchel, R.P.; et al. Extracellular Vesicles Provide a Means for Tissue Crosstalk during Exercise. *Cell Metab.* **2018**, *27*, 237–251 e234. [CrossRef]
115. Safdar, A.; Tarnopolsky, M.A. Exosomes as Mediators of the Systemic Adaptations to Endurance Exercise. *Cold Spring Harb. Perspect. Med.* **2018**, *8*, a029827. [CrossRef]
116. Jesko, H.; Stepien, A.; Lukiw, W.J.; Strosznajder, R.P. The Cross-Talk Between Sphingolipids and Insulin-Like Growth Factor Signaling: Significance for Aging and Neurodegeneration. *Mol. Neurobiol.* **2018**. [CrossRef]
117. Trayssac, M.; Hannun, Y.A.; Obeid, L.M. Role of sphingolipids in senescence: Implication in aging and age-related diseases. *J. Clin. Investig.* **2018**, *128*, 2702–2712. [CrossRef]
118. Jazwinski, S.M. Mitochondria to nucleus signaling and the role of ceramide in its integration into the suite of cell quality control processes during aging. *Ageing Res. Rev.* **2015**, *23*, 67–74. [CrossRef]
119. Fakouri, N.B.; Hansen, T.L.; Desler, C.; Anugula, S.; Rasmussen, L.J. From powerhouse to perpetrator—mitochondria in health and disease. *Biology* **2019**, (in press).

© 2019 by the author. Licensee MDPI, Basel, Switzerland. This article is an open access article distributed under the terms and conditions of the Creative Commons Attribution (CC BY) license (http://creativecommons.org/licenses/by/4.0/).

Review

CoQ10 and Aging

Isabella Peixoto de Barcelos [1] and Richard H. Haas [1,2,*]

[1] Department of Neurosciences, University of California San Diego, San Diego, CA 92093-0935, USA; ibarcelos@ucsd.edu
[2] Department of Pediatrics, University of California San Diego, San Diego, CA 92093-0935, USA
* Correspondence: rhaas@ucsd.edu; Tel.: +1-858-822-6700

Received: 25 February 2019; Accepted: 13 April 2019; Published: 11 May 2019

Abstract: The aging process includes impairment in mitochondrial function, a reduction in anti-oxidant activity, and an increase in oxidative stress, marked by an increase in reactive oxygen species (ROS) production. Oxidative damage to macromolecules including DNA and electron transport proteins likely increases ROS production resulting in further damage. This oxidative theory of cell aging is supported by the fact that diseases associated with the aging process are marked by increased oxidative stress. Coenzyme Q10 (CoQ_{10}) levels fall with aging in the human but this is not seen in all species or all tissues. It is unknown whether lower CoQ_{10} levels have a part to play in aging and disease or whether it is an inconsequential cellular response to aging. Despite the current lay public interest in supplementing with CoQ_{10}, there is currently not enough evidence to recommend CoQ_{10} supplementation as an anti-aging anti-oxidant therapy.

Keywords: coenzyme Q10; aging; age-related diseases; mitochondrial dysfunction

1. Introduction

CoQ_{10} was first described in 1955, named ubiquitous quinone, a small lipophilic molecule located widely in cell membranes [1], and in 1957 its function as an electron carrier in the mitochondrial electron transport chain was reported [2]. The role in human disease was unknown for 20 years until in 1986 a benefit of CoQ_{10} treatment was reported in Kearns–Sayre syndrome [3]. Initially, therapeutic use of CoQ_{10} was focused on the oxidative phosphorylation (OXPHOS) defects in which there is documented CoQ_{10} deficiency [4] and in the group of CoQ_{10} synthesis disorders [5]. These conditions provided evidence for efficacy and safety of treatment with CoQ_{10} [6]. Subsequent larger-scale trials in Parkinson disease [7] and other neurodegenerative diseases have shown safety but no convincing benefit.

In the last decade, CoQ_{10} functions in membranes throughout the cell where antioxidant and signaling roles predominate have been of increasing interest [8]. There is growing evidence that oxidative stress is a major component of cellular senescence [9]. This multifactorial process involves DNA injury [10], protein and lipid damage, and activation of signaling pathways associated with aging [11]. Recently, the CoQ_{10} antioxidant effect has been shown to reduce markers for cardiovascular disease (CVD) and inflammation, the main components of atherosclerotic vascular disease [12].

It is suggested that CoQ_{10} supplementation can improve the symptoms of mitochondrial diseases and of aging because of an improvement in bioenergetics [12,13].

Our objective is to review, from a translational perspective, data regarding the association of CoQ_{10} and aging. Are the aging process and mitochondrial progressive failure related and can CoQ_{10} supplementation decelerate aging?

2. CoQ$_{10}$

2.1. What Is It?

Coenzyme Q10, CoQ$_{10}$ or ubiquinone (2,3 dimethoxy-5-methyl-6-decaprenyl-1,4-benzoquinone) is a small lipophilic structure, composed of a benzoquinone ring and an isoprenoid side-chain and it is found universally in cell membranes. In humans, synthesis occurs utilizing a collection of enzymes (complex Q) located in the mitochondrial matrix membrane [14]. The benzoquinone ring is derived from 4-hydroxybenzoic acid, while 10 isoprenes are derived from mevalonic acid (from the cholesterol synthesis pathway). The quinone ring is the functional group in the molecule, responsible for carrying electrons to complex III. CoQ$_{10}$ (ubiquinone) is reversibly reduced to ubiquinol. The polyisoprenoid tail is very lipophilic and localizes to hydrophobic membranes. The length of the isoprenyl chain is variable between species with 10 isoprenes forming human CoQ$_{10}$ whilst rodents predominantly have CoQ$_9$ [15].

2.2. Function

CoQ$_{10}$ is widely distributed in all cell membranes and forms a critical component of the electron transport chain (ETC) transporting electrons between complexes I/II and III [13]. In rat liver the largest ubiquinone (CoQ$_9$) concentration is found in the Golgi vesicles (2.62 µg/mg) followed by mitochondrial matrix membrane and lysosomes (with levels of 1.86 in each structure) [16].

The major function of ubiquinone is in the mitochondrial ETC. CoQ$_{10}$ accepts electrons from different donors, including complex I (reduced nicotinamide adenine dinucleotide [NADH]-coenzyme Q oxidoreductase), complex II (succinate dehydrogenase), the oxidation of fatty acids and branched-chain amino acids via flavin-linked dehydrogenases and electron transfer factor Q oxidoreductase (ETF-QO) to complex III (ubiquinone-cytochrome c oxidoreductase) [17,18]. CoQ$_{10}$ cycles between its three chemical forms: completely oxidized (ubiquinone), a semi-oxidized intermediate free radical (semiquinone) and a completely reduced form (ubiquinol) as shown in Figure 1 [19,20]. By moving within the mitochondrial membrane, the proton-motive Q cycle allows proton pumping at complex III helping to generate the proton motive force for adenosine triphosphate (ATP) production.

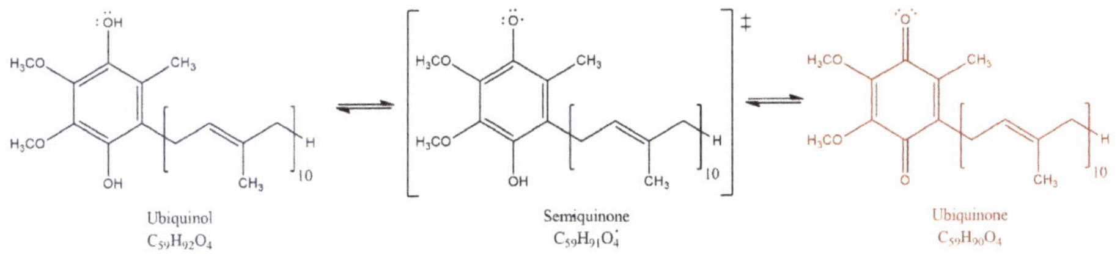

Figure 1. Redox forms of CoQ$_{10}$. ubiquinone (oxidized form), ubiquinol (reduced form), and semiquinone (semi-oxidized). The Q cycle within the matrix membrane allows proton transfer from the mitochondrial matrix to the intermembrane space helping to generate the electrochemical gradient for ATP production.

The interaction of the CoQ pool with the ETC has in recent years been shown to be more complex with the recognition that mitochondrial supercomplexes composed mostly of complexes I/III, I/III/IV, and III/IV interact physically forming respirasomes. It seems likely that both single complexes within the matrix membrane and supercomplexes coexist in a dynamic state. The factors controlling assembly and disassembly of supercomplexes are not known although cardiolipin appears to play a role. There is evidence that in the I/III supercomplex intercomplex binding of CoQ shuttles electrons from complex I to III. This bound CoQ may be in equilibrium with the free CoQ pool. [21].

CoQ10 supplementation has been shown to have epigenetic effects in genes involved with signaling, intermediary metabolism, transport, transcription control, disease mutation, phosphorylation, and embryonal development indicating a role in modulation of gene expression [22,23].

In addition to its major function in the ETC, CoQ_{10} has an important anti-oxidant role stabilizing the plasma membrane and other intracellular membranes protecting membrane phospholipids from peroxidation [13]. Ubiquinone and semiquinone are also involved with recycling of other anti-oxidant molecules, reducing α-tocopherol and ascorbate contributing to redox balance in the cell. Diminished CoQ_{10} levels in aging likely contribute to membrane peroxidation injury. There is evidence that part of its anti-oxidant effect occurs by enhancing the enzymatic activity of the antioxidant proteins superoxide dismutase and glutathione peroxidase [24]. Recent publications associate ubiquinol with protection of plasma low density lipoproteins (LDL) from oxidation, an important anti-atherogenic effect [25]. The pro-oxidant role of CoQ_{10} is a signaling function involved in gene expression but the mechanism of this function is not fully understood [26,27]. Other functions include modulation of the permeability transition pore, thus playing a role in apoptosis [28]. CoQ_{10}'s main functions are summarized in Figure 2.

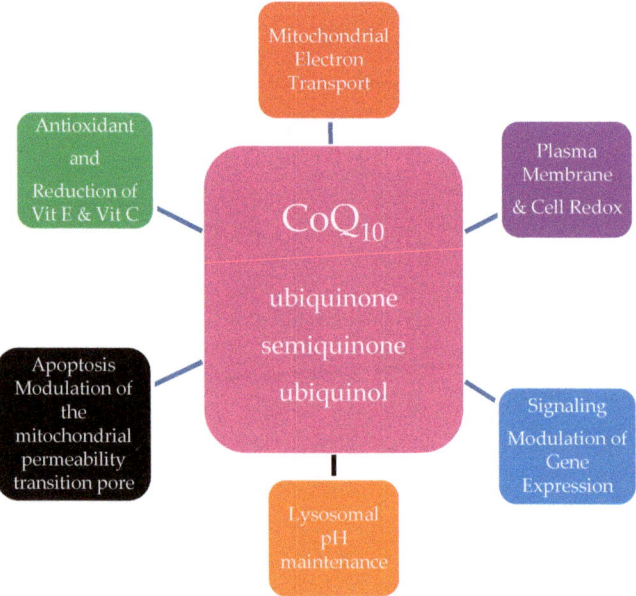

Figure 2. The Multiple Roles of Ubiquinone in the Cell.

Chronic inflammation is a frequent aging-related problem. CoQ_{10}, by reduction of free radicals, reduces the activation of NF-κB (nuclear factor kappa-light-chain-enhancer of activated B cells) cells and consequently reduces the release of pro-inflammatory cytokines mainly tumor necrosis factor alpha (TNF-α) and interleukin 6 (IL-6) [29]. Aging-related reduced CoQ_{10} levels may contribute to inflammation and there is accumulating evidence of secondary anti-inflammatory effects of CoQ_{10} supplementation. A recent meta-analysis provided evidence that CoQ_{10} supplementation significantly reduced the inflammatory markers CRP (C-reactive protein), IL-6 and TNF-α [30]. Another recent publication reported that patients with metabolic diseases (obesity, type 2 diabetes, metabolic syndrome, cardiovascular disease, and nonalcoholic fatty liver disease) had a significant decrease in TNF-α plasma levels with CoQ_{10} supplementation but not CRP or IL-6 [31]. CoQ_{10} was found to have an anti-inflammatory function via epigenetic effects on expression of genes related to NFkappaB1 (NFk-B1) [32]. CoQ_{10} has a hepatoprotective and neuroprotective effect in a rat model of non-alcoholic steatohepatitis [33] apparently through an adenosine 5′ monophosphate-activated protein kinase (AMPK) activation mechanism and in humans a randomized trial showed that supplementation of CoQ_{10} improved biomarkers for inflammation in nonalcoholic fatty liver disease (NAFLD) [34]. CoQ_{10} supplementation in patients with antiphospholipid syndrome has been found to attenuate levels of pro-inflammatory and thrombotic markers, with evidence of endothelial and mitochondrial function improvement [35]. In Down syndrome patients, chronic neuro-inflammatory changes have been proposed as a possible accelerator of Alzheimer disease [36]. These include high levels of interleukin 6 and tumor necrosis factor α along with decreased levels of CoQ_{10}. A positive correlation between CoQ_{10} and intelligence quotient levels was also reported [37]. CoQ_{10} treatment in Down syndrome cells is associated with improved DNA repair mechanisms and DNA protection [38].

Cardiovascular disease is a common aging-related problem. There is a considerable body of evidence supporting a role for CoQ_{10} in cardiovascular function including a correlation of low endomyocardial levels with severity of heart failure and an improvement in cardiac contractility with CoQ_{10} treatment [39]. Improvement in lipid profiles (a major contributor to cardiovascular disease) has been reported with CoQ_{10} treatment [40].

CoQ_{10} has other important functions, participating in metabolic pathways as an electron receptor: (1) CoQ_{10} is a co-factor for dihydro-orotate dehydrogenase, an enzyme involved in the de novo pyrimidine biosynthesis [41]. (2) During the process of sulfide oxidation CoQ_{10} accepts electrons from the enzyme sulfide-quinone reductase to convert sulfide into thiosulfate [42]. (3) The oxidation from choline to glycine is catalyzed by choline dehydrogenase in the inner mitochondrial membrane, and CoQ_{10} is proposed to be the electron acceptor for this reaction [43]. (4) Proline dehydrogenase donates electrons from FAD and NAD+ to CoQ_{10} during process of synthesis of proline and arginine [21,44].

2.3. Sources of CoQ_{10}

2.3.1. Internal Biosynthesis

CoQ_{10} is the only lipid-soluble antioxidant synthetized by the human body [13]. The majority of CoQ_{10} comes from the internal synthesis, from tyrosine or phenylalanine (benzoquinone ring) and mevalonic acid (isoprenoid side-chain). Synthesis occurs in all tissues studied. In humans synthesis occurs utilizing a collection of enzymes (complex Q) located in the mitochondrial matrix membrane and in the endoplasmic reticulum The mevalonic acid pathway is responsible for cholesterol synthesis with 3-hydroxy-3-methyl-glutaryl-coenzyme A (HMG-CoA) reductase (the site of statin inhibition) as the regulatory step. CoQ_{10} derived from dietary intake becomes more important with aging as endogenous production decreases [13].

At least 14 genes are involved in CoQ_{10}'s biosynthesis in yeast with 18 genes so far reported in humans. Many are homologues of genes identified in c. cerevisiae. These synthetic proteins are nuclear encoded and require mitochondrial targeting sequences for entry into the matrix or the inner mitochondrial membrane [14]. There is assembly of many of the components into a 'supercomplex' termed Complex Q in the mammal. Mitochondrial synthesis is thought to occur in all cells containing mitochondria but also in other organelles including the endoplasmic reticulum and peroxisomes [45]. Figure 3 summarizes the mammalian CoQ_{10} biosynthesis pathway.

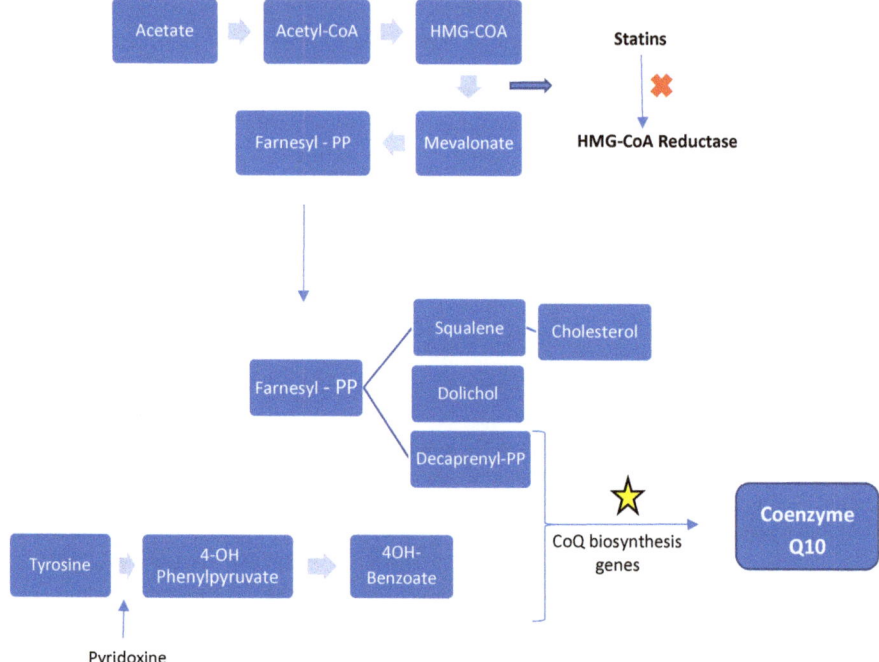

Figure 3. Schematic representation of CoQ biosynthesis. The isoprenoid side-chain derives from the cholesterol and dolichol synthetic pathway from mevalonate, the benzoquinone ring is derived from tyrosine metabolism. The star symbolizes the genes involved in the final synthesis of CoQ$_{10}$.

2.3.2. External Sources

CoQ$_{10}$ is widely found in many animal protein sources (pork, lamb, beef, chicken, fish), vegetables (spinach, pea, broccoli, cauliflower), fruits (orange, strawberry, apple) and cereals (rye, wheat) [13]. Heart, chicken leg, herring, and trout contain particularly high amounts of CoQ$_{10}$. Daily intake between 3 and 5 mg is considered adequate and whilst external supplementation increases plasma levels, supplementation was not thought to increase tissue levels of CoQ$_{10}$ in tissues with normal synthetic capacity [46], although evidence of treatment efficacy in a variety of human diseases does suggest that tissue uptake can occur [13,47].

2.3.3. Absorption and Transport

CoQ$_{10}$ absorption is slow and occurs in the small intestine; in its reduced form, ubiquinol is 3 to 4 times better absorbed than the oxidized form, ubiquinone [48]. The absorption of CoQ$_{10}$ can be increased if administered with food intake, mainly with lipids because of its lipophilic structure [13,49]. After absorption by the enterocytes, CoQ$_{10}$ passes through lymphatic vessels and reaches the plasma, where it circulates bound to lipoproteins (LDL). Because of this, plasma measurements of CoQ$_{10}$ should be corrected for lipoproteins levels. Between 80 and 95% of plasma circulating CoQ$_{10}$ is in the reduced ubiquinol form. [48,50].

2.4. Tissue Levels and Distribution of CoQ$_{10}$

Although the major CoQ$_{10}$ plasma form is ubiquinol, laboratory measurements, in general, report the CoQ$_{10}$ total level [50]. Lymphocyte and platelet levels may give some insight into levels in less accessible tissues such as heart, muscle, and brain [51]. The CoQ$_{10}$ level varies in different tissues.

Tissues with a higher metabolic rate and mitochondrial content (heart, kidney, liver, and muscle) have high levels of CoQ_{10}, for example the level is 8 µg/g in lung and 114 µg/g in heart [13]. Tissue levels largely reflect the results of synthesis and degradation of CoQ_{10}. Following intestinal absorption, liver and lipoprotein concentrations increase, but without a change in the level in heart and kidney noted in early studies [52]. However, there is some evidence that chronic administration can increase tissue levels [53]. In a study where rats were chronically fed a large dose of 150 mg/kg/day of CoQ_{10} for 13 weeks, small but significant increases in both CoQ_9 and CoQ_{10} were found in all tissues measured [54]. No such data exist for young or aged human tissues; however, the accumulated evidence of benefit of CoQ_{10} therapy in human disease states does suggest that tissue levels can be increased by oral administration. The number, size, and structure of the mitochondria in each cell determines the tissue level of CoQ_{10} with highest levels found in tissues with high energy demands and a high mitochondrial content [13,48].

The gold standard for the diagnosis of CoQ_{10} deficiency is based on the measurement of muscle level by high-performance liquid chromatography (HLPC) [55,56]. Plasma levels of CoQ_{10} range between 0.40 and 1.91 µmol/L (0.34–1.65 µg/mL) in controls [57] but do not match tissue levels, reflecting consumption much more than endogenous synthesis. The diagnosis of CoQ_{10} deficiency requires tissue measurement [58]. There is evidence suggesting that the CoQ_{10} level in mononuclear cells can be correlated to muscle measurements [45,56]. Plasma CoQ_{10} levels are increased in some physiological conditions (cold adaptation and exercise) and in some diseases (paraneoplastic nodules, Alzheimer's disease, and prion disease). Levels are decreased in aging. [45].

2.5. Causes of Reduction

Conditions associated with CoQ_{10} deficiency can be divided into three main groups: (1) CoQ_{10} nutritional deficiency, including intake of CoQ_{10} itself and nutrients and vitamins necessary for its synthesis (vitamin B6 is a cofactor in the pathway of CoQ_{10} biosynthesis); (2) CoQ_{10} synthesis genes (*COQ* family genes: *COQ1* and the Complex Q genes including the mammalian homolog of the yeast Coq11 gene; the Complex I subunit *NDUFA9*), and acquired disorders impairing CoQ_{10} synthesis (statin use) [59–61]; and (3) medical conditions associated with decreased levels of CoQ_{10} [45,46]. In this last category, a variety of conditions have been reported with low CoQ_{10} including neurodegenerative disorders Friedreich's ataxia, Nieman–Pick type C disease, and Parkinson Disease. In other disorders such as Alzheimer disease, diabetes, cancer, fibromyalgia, and cardiovascular diseases, elevations in plasma CoQ_{10} levels may be a stress response. [45,46]. Defects in genes which may be associated with reduced CoQ_{10} levels can be included in this group (*APTX, ETFDH, BRAF*).

Special attention should be given to the important observation that CoQ_{10} deficiency is potentially reversible if the supplementation starts before the appearance of the symptoms, when brain and kidney have not sustained permanent damage [46,62].

2.6. Supplementation

2.6.1. Dosing

A recent trend over the last few years has been to supplement adult patients with mitochondrial diseases with high doses of oral CoQ_{10} or ubiquinol, up to 1200 mg/day or higher. For the pediatric population, doses between 5 and 10 mg/kg/day of ubiquinol are recommended [63]. For a dose of 10 mg/kg/day, plasma levels range between 5 and 10 µg/mL 3–4 weeks after the beginning of the supplementation [53,63,64]. As explained above, there are important differences in bioavailability of the different formulations of CoQ_{10} used; ubiquinol (the reduced form) is 3 to 4 times better absorbed than ubiquinone (oxidized form) [48]. Primary CoQ_{10} diseases tend to respond to supplementation but may require very high doses. Also, some patients with secondary CoQ_{10} dysfunction were reported to improve some symptoms with CoQ_{10} supplementation. Examples are cardiomyopathy in organic acidurias (25 mg/kg/day with improvement in the cardiomyopathy) [65], glutaric acidemia type II

(500 mg per day of CoQ_{10} along with riboflavin with improvement in strength, lactate, and creatine kinase levels) [59], ataxia oculomotor apraxia type 1 (200–600 mg with reported improvement in strength, ataxia, and cessation of seizures in one patient) [60], GLUT-1 deficiency (30 mg/kg/day with improvement in the ataxia and nystagmus) [66].

Exogenous administration of CoQ_{10} reportedly does not raise tissue levels above normal in healthy young individuals, except for two tissues (liver and spleen) [13]; however, this traditional view may be wrong given the improvement in multi-organ symptoms in a variety of disorders with mitochondrial dysfunction when treated with CoQ [67].

2.6.2. Safety and Adverse Events

Although the majority of studies have not shown convincing enough scientific evidence to support treatment with CoQ_{10} in specific diseases, they do provide evidence that oral supplementation is safe and well tolerated. One of the largest trials was a phase III randomized, placebo-controlled, double-blind clinical trial at 67 North American sites by the Parkinson Study Group using doses of CoQ_{10} up to 2400 mg/day demonstrated the safety of this dose [7]. Evidence suggests that supplementation does not inhibit endogenous production [13]. Previous studies had reported only mild side effects such as gastrointestinal symptoms, mainly nausea, with CoQ_{10} supplementation [13,68].

3. Aging

3.1. Physiology of Mitochondrial Involvement in the Process of Aging

Human aging is a normal multifactorial process resulting from the interaction of genetic and environmental factors. It is characterized by multi-organ system functional decline in association with the risk of age-related diseases (dementia, neurodegenerative disorders, osteoporosis, arthritis, diabetes, cardiovascular disease, age-related hearing loss, and cancer) [13,68–71].

A common hypothesis to explain some of the pathophysiology of age and degenerative diseases is an oxidative imbalance between the production of reactive oxygen species (hydrogen peroxide: $H2O2$, the oxygen-derived free radicals superoxide: $O2\bullet-$, and hydroxyl radical: $HO\bullet$), and antioxidant mechanisms such as superoxide dismutase, catalase, glutathione peroxidase, ascorbic acid, tocopherol, glutathione, and CoQ_{10}, leading to a state of oxidative stress [72–80]. A number of animal models support this theory with shortened survival in mice lacking superoxide dismutase 1 (SOD1) and a lethal phenotype in mice lacking superoxide dismutase 2 (SOD2) [81,82].

As mitochondria are the main source of reactive oxygen species (ROS) production though OXPHOS supercomplex activity in the cristae of the inner mitochondrial membrane (mainly at complex I and III), this organelle is the major target of ROS damage. Mitochondrial DNA (mtDNA) is particularly vulnerable with a high mutation rate and limited mtDNA repair mechanisms [69,74]. The continuous production and accumulation of mitochondrial ROS is the basis for "the free radical theory of aging" [70,71]. Although the accumulation of ROS has a major effect on DNA (strand breaks, oxidation of bases, damage in sites coding for ETC proteins), other structures of the cell are also damaged: lipids, membranes, proteins (leading to dysfunction of the ETC, inadequate ATP production, and further ROS production) [72–75]. There is evidence that impaired mitochondrial machinery produces more oxidative stress and more ROS production, resulting in a vicious cycle [76,83]. Electrons leaking from impaired OXPHOS react with oxygen molecules to form the free radical superoxide [80]. There is also an effect on mitochondrial dynamics with impairment of fission, contributing to mitochondrial enlargement, which reduces recycling through mitophagy leading to a reduction in ATP generation [73]. Impairment of the mitochondrial–lysosomal axis occurs with aging, with accumulation of lipofuscin inside lysosomes with senescence. Lipofuscin accumulation is postulated to limit the ability of lysosomes to participate in mitophagy [73].

Considerable evidence supports the relationship between ROS accumulation and mitochondrial dysfunction leading to aging (the mitochondrial free radical theory of aging): ROS production increases in aged humans and animals [70,84], imbalance in the levels of pro and anti-oxidant substances occurs [85] with high levels of oxidized and damaged macromolecules (proteins, lipids, and DNA) [86]. There seems to be a convincing relationship between high ROS levels and longevity in humans and animals [87]; however, the exact role of ROS remains unclear as some animal models report a failure to increase longevity with ROS reduction and others link high levels of ROS and longevity. It remains unclear what the contribution of ROS generation is versus epigenetic factors modulating genes related to the protection from effects of aging [77,88,89]. Evidence against the oxidative theory of aging comes from some animal models where longevity is unaffected by increased ROS production. Some of these studies are detailed below.

Growing evidence supports the idea that increased levels of ROS are associated with the specific biochemical pathway that improves longevity, at least in some species [90]. C. elegans, a nematode mutant model clk-1 (*COQ7* equivalent gene in humans), has higher longevity associated with higher ROS production. C. elegans with a point mutation in the gene isp-1, responsible for an iron-sulfur mitochondrial complex, also demonstrate an increase in life span [91]. These observations point out the multiple roles of ROS particularly as signaling molecules triggering protective pathways. Some mouse models of defective CoQ synthesis are difficult to square with the oxidative theory of aging. Mice with only one copy of the gene *Mclk1* (equivalent of mammals *COQ7* and *C. elegans* clk-1 model) have higher production of ROS in the mitochondria (although a normal level of total ROS in the body), a higher level of protection of the immune system (from some infections and also tumorigenesis) and increased longevity [92]. Homozygous knockout of the *Mclk1* gene is embryonic lethal but utilizing a Tamoxifen dependent transgene mouse KO activated at 2 months of age a multisystemic disorder (heart, kidneys, and skeletal muscles) with a decline in ubiquinol levels is produced. At 8 months this "ubiquinol deficit" animal presented normal levels of some factors associated with oxidative stress (catalase, F_2-isoprostanes, DNA oxidative damage, SOD1, and SOD2). Diet supplementation at 9 months with an analogue of the ubiquinol precursor 4-hydroxybenzoic acid rescued the clinical phenotype [93]. Another mouse model heterozygous for the Sod2 gene (with decreased Mn-superoxide dismutase activity) does show oxidative injury (increased tumor incidence and DNA damage) but does not decrease longevity [94]. The concept that mitochondrial dysfunction may be a consequence of aging factors rather than a cause is also supported by observations of sarcopenia in rat and human muscles with other factors such as denervation playing a role [95].

Further studies are needed to understand the balance of ROS as an agent of oxidative injury and its signaling epigenetic role modulating genes related to the protection of effects of aging. It may be that after the saturation of the mechanisms of protection, the ROS-stress protective cascade can no longer prevent oxidative damage [88].

It is not surprising that given the contradictory evidence for a central role for ROS in aging, evidence supporting the utility of anti-oxidant therapies in aging (such as CoQ_{10}) remains unclear.

3.2. CoQ_{10} and Aging, CoQ_{10} Deficiency in Advanced Age, Evidence for Beneficial Supplementation

Published results from various research groups about CoQ levels and lifespan are often at variance, model dependent and do not support a similar pattern in all species [96–98].

3.2.1. C. elegans

Studies in the nematode C. elegans have produced unexpected results compared to mammals. As discussed above, a nematode mutant model clk-1 (mammals *COQ7* equivalent gene), with low production of CoQ_8, has an extension in longevity compared with the wild strain but requires dietary Q supplementation [99]. This diet does trigger a Dauer long-lived larval anaerobic state. Other studies with CoQ gene knockouts confirm that deficiency in CoQ (less than 50%) also leads to an increase in lifespan, and a possible explanation is that less ROS production occurs in the case of moderate CoQ deficiency [100,101], but with more severe depletion of CoQ, longevity would be affected [100,102]. This finding was also confirmed in human cells, with two different studies from the same group reporting that fibroblasts with mutations on *PDSS2* (homologue of yeast coq1) with less than 12 and 20% of CoQ_{10} of control cells had decreased synthesis of ATP without increase in the levels of ROS. However, when the defect was of 30%, with partial defect in the synthesis of ATP the levels of ROS were higher. The explanation proposed centered on the severity of the deficiency of coenzyme Q as an oxphos modulator [103,104].

3.2.2. Rodent Models

A diet supplemented with Ubiquinol-10 in the senescence-accelerated mouse prone 1 (SAMP1) reduced markers of oxidative stress (ratio of reduced and oxidized glutathione—GSH/GSSG), decelerated the normal decline in expression of genes (*Sirt1*, *Sirt3*, and *Pgc-1a*, and *Ppara*), and their respective proteins related to mitochondrial function during aging [105]. This treatment also increased auditory brainstem response hearing loss. The proposed mechanism of CoQ_{10} benefit as an anti-oxidant agent in this aging mouse is through cyclic adenosine monophosphate (cAMP). There is an enhancement in sirtuin genes, and PGC-1α (peroxisome proliferator-activated receptor gamma coactivator 1-alpha) with increased complex I and IV activity and reduce oxidative stress. Despite all these beneficial effects, there was no significant change in the overall lifespans compared to the control animals [105]. These findings of increased cAMP, SIRT1 (sirtuin 1) expression, and PGC-1α in reducing parameters of oxidative stress and increasing mitochondrial function were also confirmed in other experiments with ubiquinol supplementation in senescence-accelerated mice, with added benefits in obesity, insulin resistance, and metabolic syndrome, (hypothesized mechanism in Figure 4 [106]). PPARα (peroxisome proliferator-activated receptors) signaling and lipid metabolism gene expression changes in liver of C57BL6J mice was reported after one-week supplementation with ubiquinol (250 mg/kg BW/day) [23].

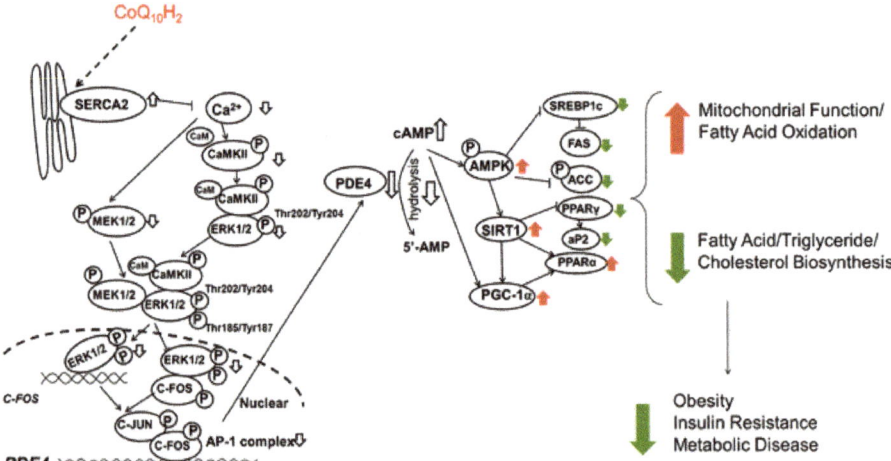

Figure 4. Proposed mechanism by which ubiquinol improves metabolic function and inhibits insulin resistance in KKAy mice (a mouse model of obesity and diabetes). Ubiquinol inhibited phosphorylation of CaMKII (Ca2+/calmodulin-dependent protein kinase II) in the liver resulting in inhibition of C-FOS transcriptional activity and inhibition of PDE4 gene expression. Increased cAMP increases AMPK (AMP-activated protein kinase) activity resulting in SIRT1 and PGC-1α increased mitochondrial function and inhibition of lipid synthesis. (Adapted from Xu H. et al. 2017 [106] with permission).

Interestingly mice with a single copy of the Coq7 synthesis gene demonstrated increased longevity [107]. However the mechanism is unclear and several studies report that dietary supplementation with CoQ_{10} or ubiquinol in various rat models improves mechanisms involved with mitochondrial biogenesis, including parameters of oxidative stress [105,108–111].

3.2.3. Mammals and Tissue CoQ levels

In mammals there is a tendency to ubiquinone to be reduced with age, but this finding depends on the tissue investigated and also the species [97]. Early studies suggested that tissue levels of ubiquinol were endogenously produced with little change with dietary supplementation except in liver and spleen; however, subsequent studies in rodents confirm that oral supplementation with CoQ_{10} does increase tissue levels of both CoQ_{10} and CoQ_9 in skeletal muscle, heart, and kidney [26], and when high doses (200 mg/kg) are used for 2 months in rats, brain levels are increased and are neuroprotective [112].

In humans, there is also a lack of consistent data. One study did not find a relationship between aging and CoQ_{10} plasma levels in elderly women [113]. Others report that plasma and tissue levels change over time, with a peak in pancreas and adrenal by 1 year of age and in the brain, heart, and lung by 20 years. After this peak, levels decrease over time [114]. A decrease in brain CoQ_{10} was confirmed in other studies [115,116]. Only 50% of the myocardial CoQ_{10} endogenous production remains by the age of 80 [114]. Serum total CoQ_{10} and ascorbic acid levels were decreased in centenarians compared with 76-year-old controls. An elevation of the CoQ_{10} binding protein prosaposin was also noted presumably in an attempt to compensate for low CoQ_{10} levels [117]. The authors conclude that CoQ_{10} supplementation could be beneficial for centenarians. However, it is not known if low tissue and plasma CoQ_{10} levels contribute to or are a side effect of aging.

Reduction of CoQ_{10} with age is postulated to result from reduction in biosynthesis coupled with an increase in degradation attributed to age-related modification in lipid membranes, which alters quinone behavior [118]. This reduction of CoQ_{10} levels has a tissue/organ specificity with reported high levels of CoQ_{10} in the brain mitochondria from old rats [26], and reduced level in the muscle [98]. A recent paper showed age-related reduction in mitochondrial respiration parameters and ATP production in epithelial cells, rescued with CoQ_{10} administration. Both cited that parameters decrease as age increases, as seen in Figure 5, with the estimate of 10% reduction in mitochondrial respiration every ten years [70].

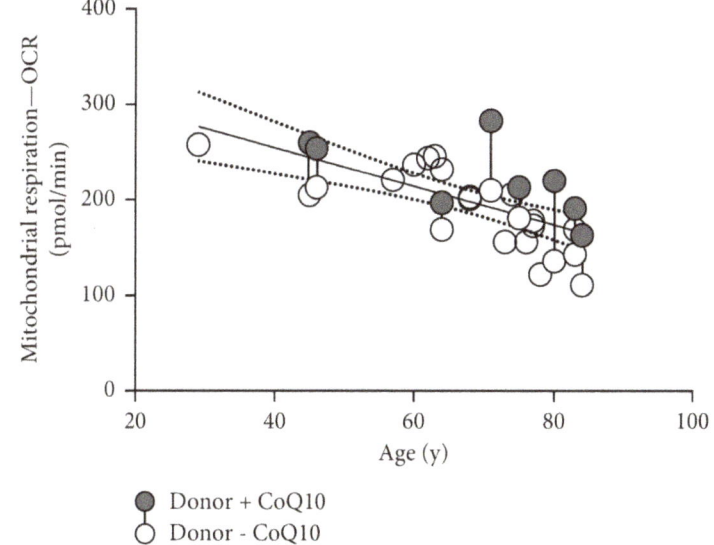

Figure 5. Age-related decline in the oxygen consumption rate in epithelial tissue measured 16 h after collection in a Seahorse XF analyzer. The gray circles show treated samples, the white circles untreated samples, and the connected circles represent samples from the same donor. Significant improvement with 100 µM CoQ_{10} incubation. (Adapted from Schniertshauer et al. 2018 [70] with permission).

High CoQ_{10} concentration was reported to be associated with higher physical activity, lipid peroxidation, and lower oxidized LDL levels in elderly people. The same publication describes higher levels of plasma CoQ_{10} in elderly (more than 50 years old) people compared with young (less than 30 years) [119]. The same group confirmed lower lipid peroxidation and lower oxidized LDL in young adults and also showed that CoQ_{10} is lower in obese elderly patients [120].

In another study, prolonged CoQ_{10} supplementation for 4 years in community-dwelling elderly was associated not only with an improvement in health-related quality of life but more importantly with a lower "more days out of hospital" rate. In this study, subjects were co-supplemented with selenium [121].

Although the association between CoQ_{10} and muscle power is not well established, one study showed a positive relation between CoQ_{10}/cholesterol levels and hand grip, and lower ubiquinol levels in patients with less muscle strength. This study was undertaken to evaluate a possible relationship between low levels of CoQ_{10} and ubiquinol in sarcopenia [122].

Despite the multiple reports on the effects of CoQ_{10} supplementation on aging-related oxidative markers, reduction in biomarkers related to inflammation and in DNA repair mechanisms reports there is a great need for more controlled studies in an older population to determine effectiveness of CoQ_{10} as an anti-aging therapy [97] and also to determine the tissue CoQ_{10} levels in the human species during senescence.

3.3. CoQ_{10} and Specific Conditions Associated with Age

The case for beneficial effects of CoQ_{10} or ubiquinol supplementation is stronger for a number of aging-related diseases many of which have documented mitochondrial dysfunction or CoQ_{10} deficiency. Despite this, the majority of studies proving therapeutic effects of CoQ_{10} supplementation were carried out in animal models [25].

Early studies were carried out with CoQ_{10} although often information on the formulation used is lacking (it is known that ubiquinol can be 3 to 4 times better absorbed than ubiquinone). Also, the doses used in most of the studies were lower than the doses studied for primary CoQ_{10} deficiency (up to 1200 mg/day) and in secondary CoQ_{10} deficiency studies. The formulation and absorption will affect studies of treatment efficacy. A colloidal-Q10 formulation has been shown to have a better enteral absorption and bioavailability in human tissues, and also the vehicle of the active ingredient can impact the results of trials [123].

3.3.1. Neurodegenerative Disorders

Neurodegenerative diseases involve neuro-inflammation and oxidative stress, with ROS accumulation and mitochondrial dysfunction [83,124,125].

The data supporting mitochondrial cofactor supplementation in aging and the aging-related diseases, including Alzheimer disease, Parkinson disease, amyotrophic lateral sclerosis, and multiple sclerosis, are addressed in focused review articles in this supplement. A systematic review published in 2014 reviewed 16 articles and concluded that there is no available data from randomized controlled trials (RCTs) to support the use of mitochondrial supplements for Parkinson disease, atypical Parkinsonism, Huntington disease, and Friedreich ataxia [126].

Previous studies have shown that patients with Lewy's body disease and amyotrophic lateral sclerosis have lower CoQ_{10} plasma levels and Alzheimer disease patients have reduced levels of CoQ_{10} in the cerebral spinal fluid [127,128].

In Huntington disease, increased levels of cortical lactate were observed, which were normalized after 2 months of Coenzyme Q_{10} oral supplementation at a dose of 360 mg/day. In another publication, after discontinuation of the supplement, the lactate levels rose again [67,124]. This study does provide evidence for CNS penetration of administered CoQ_{10} in humans. These and other data lead to a multicenter RCT with 609 patients, which could not identify a benefit in Functional Capacity Score and time to death after 60 months of 2400 mg per day of CoQ_{10} supplementation compared to placebo [129]. Thus, currently, there is insufficient evidence to support CoQ_{10} supplementation as a treatment to delay neurodegeneration in Huntington disease even in the early stages of the disease.

A similar situation occurred in Parkinson disease. An early study found that patients had decreased levels of α-tocopherol and CoQ_{10} in plasma and cerebrospinal fluid with increased levels of lipoprotein oxidation compared to controls in a 2004 study of 161 subjects. [130]. Mitochondrial complexes I and II/III were reduced in platelet mitochondria from early Parkinson disease patients [131]. A more recent study confirmed the association of Parkinson's disease and CoQ_{10} deficiency using Functional Intracellular Assay, when compared to matched controls for age and gender [132]. A phase 2 study in 80 subjects at early or mid-stage Parkinson disease showed a beneficial effect from supplementation with CoQ_{10} in progressive doses of 300, 600, 1200 mg/day for 16 months. There was an improvement in the ADL UPDRS (Activity Of Daily Living Unified Parkinson Disease Rating Scale) and Schwab and England scales compared to placebo (+11.99), with highest dose (+6.69) having the best effect [133]. This lead to a phase III randomized, placebo-controlled, double-blind clinical trial at 67 North American sites by the Parkinson Study Group which used doses up to 2400 mg/day of CoQ_{10} vitamin E in the dose of 1200 IU/d (to enhance the absorption of the lipophilic coenzyme) however this large study did not find convincing evidence to support CoQ_{10} treatment for this disease [7]. A systematic review and meta-analysis from 2016 could not find sufficient evidence in five selected RCTs to support the use of ubiquinone (300 mg/day to 2400 mg/day) to decelerate the progression of Parkinson's disease or improve symptoms [134].

In aged, cognitively impaired mice, CoQ_{10} supplementation does improve special learning and attenuates oxidative damage [135]. In Alzheimer Disease, the Alzheimer's Disease Cooperative Study could not identify benefits in the levels of oxidative stress and neurodegeneration biomarkers in CSF (cerebrospinal fluid) with a dose of 400 mg of CoQ_{10} 3 times/day for 16 weeks for patients [136]. To date, chronic large-scale trials of CoQ_{10} have not been carried out in Alzheimer disease.

3.3.2. Cardiovascular Disease

A meta-analysis reviewed randomized controlled trials in healthy adults; two trials reported reduction in systolic blood pressure, and no evidence for reduction in diastolic blood pressure was reported. Another trial included in this meta-analysis failed to show an effect of CoQ_{10} supplementation on the lipid profile (LDL: low-density lipoprotein cholesterol, HDL: high-density lipoprotein cholesterol, and triglycerides) [137]. Another meta-analysis from 2016 concluded that there is not enough evidence to support CoQ_{10} use to treat hypertension [138]. For dyslipidemia, no relationship was found between CoQ levels in hyperlipidemic and normolipidemic, older women and no association with body mass index was found [139]. Statins are well known to reduce CoQ_{10} levels because of inhibition of 3-hydroxy-3-methylglutaryl-coenzyme A (HMG-CoA) reductase. This enzyme is the first step in the mevalonic acid synthesis pathway, which is necessary for the synthesis of cholesterol but also the isoprenoid side-chain of CoQ_{10} [140].

Patients with mitochondrial myopathy do not tolerate statin treatment, often developing exercise intolerance and myalgia when treated with statins. These symptoms may improve with CoQ_{10} oral supplementation [141–143]. Statins were also found related to a reduction in complex III activity. However, the exact mechanism for the mitochondrial myotoxicity of statins is not well understood and other factors than CoQ_{10} can be involved, such as genetic polymorphisms [144].

There is evidence that cardiac function in the elderly may be improved by CoQ_{10} treatment. This effect may be related to the documented fall in cardiac CoQ_{10} levels in the aged heart [114]. A meta-analysis reviewed eight clinical trials and concluded that patients submitted to cardiopulmonary bypass have a decreased requirement for ionotropic drugs and a reduced chance of ventricular arrhythmia if receiving CoQ_{10} supplementation [145].

The best case for CoQ_{10} supplementation is in cardiovascular disease. It was recently reported that 12 years after a 4-year double-blind treatment protocol in the elderly co-supplemented with selenium (200 µg) and CoQ_{10} (200 mg/day) (ubiquinone in a softgel vehicle of vegetable oils) the treated subjects still had a reduced cardiovascular mortality (38.7% in placebo-treated group versus 28.1% in CoQ-treated group) [146]. The Q-SYMBIO trial enrolled 420 patients and showed that CoQ$_{10}$ 100 mg 3 times daily for 2 years plus standard therapy reduced the risk of cardiovascular events, in this study measured as cardiovascular mortality (16% in placebo and 9% in the treated group) and hospitalization for heart failure compared to placebo, and also a positive change in the New York Heart Association (NYHA) functional classification (26% in placebo and 15% in the treated group) [147]. A recent review of the use of CoQ$_{10}$ in heart failure is recommended [39].

3.3.3. Endothelial Dysfunction

Endothelial function may be a risk factor for coronary artery disease and atherosclerosis. CoQ$_{10}$ together with a Mediterranean diet was shown to improve markers of endothelial function in elderly patients [148,149].

3.3.4. Renal Disease

Although ubiquinol can improve the endothelial dysfunction associated with the diabetic kidney disease (systolic blood pressure and urinary albumin) [150], and a trial with CoQ$_{10}$ supplementation (1200 mg/day in dialysis patients) identified a reduction in a plasma indicator of oxidative stress (F2-isoprostane) [151], a meta-analysis failed to prove CoQ$_{10}$ efficacy in avoiding the progression of diabetic kidney disease [152].

3.3.5. Inflammation

Many aging-related diseases share a common physiologic pathway of chronic inflammation leading to oxidative stress, such as cardiovascular diseases, diabetes, cancer, and chronic kidney disease.

A recent meta-analysis reports a reduction in the plasma inflammatory biomarkers, C-reactive protein, IL-6, and TNF-α, after CoQ$_{10}$ supplementation (60 to 500 mg/day, formulations described as CoQ10 or ubiquinol), for 1 week to 4 months in different inflammatory disorders (cardio and cerebral vascular disease, multiple sclerosis, obesity, renal failure, rheumatoid arthritis, diabetes, and fatty liver disease). The same review reports that CoQ$_{10}$ decreases other biomarkers for inflammation and inflammatory cytokines [30]. Although CoQ$_{10}$ treatment has been shown to improve markers of inflammation, a benefit of chronic treatment for the diseases associated with inflammation has not been demonstrated.

CoQ$_{10}$ (100 mg/day) supplementation for 2 months decreased the levels of TFN-α in rheumatoid arthritis patients compared to placebo, [153]. A more recent systematic review and meta-analysis reported CoQ$_{10}$ supplementation between 60 to 300 mg/day (no extra information about formulations) was associated with a slightly drop in C-reactive protein levels and a significant decrease in in IL-6 levels [154].

Down syndrome patients have an abnormal pro-inflammatory profile (increased IL-6 and TNF-α) [36,37] and reduced CoQ$_{10}$ levels, and supplementation reduced markers of oxidative stress and mitochondrial dysfunction [38,155].

3.3.6. Osteoporosis

Animal and human studies have demonstrated that benefits of CoQ$_{10}$ supplementation have a beneficial profile for osteoporosis [156–158].

3.3.7. Cancer

There is evidence of the relationship between some cancers and reduced CoQ$_{10}$ levels in blood, particularly breast cancer [159], myeloma [160], melanoma [161], and follicular and papillary thyroid carcinomas [162]. There is also evidence that CoQ$_{10}$ supplementation modulates phospholipid hydroperoxide glutathione peroxidase gene expression, free radical production, and can decelerate the growth of tumor cells in a prostate cancer line (PC3 line) [163].

4. Conclusions

There is still much to be learned about the pathophysiology of the process of aging. There are well documented reductions of tissue CoQ_{10} in senescence. It is not known if low CoQ_{10} is an effect of aging, perhaps matching the fall in mitochondrial electron transport function or a contributing cause to the aging process. There is accumulating evidence that some diseases of aging may benefit from supplemental ubiquinol or CoQ_{10} treatment. Studies to date have supported the safety and the potential of CoQ_{10} in reducing oxidative stress biomarkers. There remains a lack of adequate large-scale clinical trials preferably utilizing ubiquinol as the better absorbed form of CoQ_{10}. Despite the lack of evidence, large numbers of people in the population are taking CoQ_{10} and other vitamins and cofactors in the hope that these agents will slow senescence and expand longevity.

Funding: I.P.d.B. is recipient of a postdoctoral fellowship from the North American Mitochondrial Disease Consortium (NAMDC) supported by National Institute of Health (NIH) U54 grant NS078059 and RHH is also supported by this grant.

Acknowledgments: Thank you to the many patients and families who have taught the authors so much including their experiences with CoQ_{10} treatment.

Conflicts of Interest: The authors declare no conflict of interest.

References

1. Festenstein, G.N.; Heaton, F.W.; Lowe, J.S.; Morton, R.A. A constituent of the unsaponifiable portion of animal tissue lipids (λmax. Mμ.). *Biochem. J.* **1955**, *59*, 558–566. [CrossRef] [PubMed]
2. Crane, F.L. Electron transport and cytochromes of sub-cellular particles from cauliflower buds. *Plant Physiol.* **1957**, *32*, 619–625. [CrossRef]
3. Ogasahara, S.; Nishikawa, Y.; Yorifuji, S.; Soga, F.; Nakamura, Y.; Takahashi, M.; Hashimoto, S.; Kono, N.; Tarui, S. Treatment of Kearns-Sayre syndrome with coenzyme Q10. *Neurology* **1986**, *36*, 45–53. [CrossRef]
4. Miles, M.V.; Miles, L.; Tang, P.H.; Horn, P.S.; Steele, P.E.; DeGrauw, A.J.; Wong, B.L.; Bove, K.E. Systematic evaluation of muscle coenzyme Q10 content in children with mitochondrial respiratory chain enzyme deficiencies. *Mitochondrion* **2008**, *8*, 170–180. [CrossRef] [PubMed]
5. Hirano, M.; Garone, C.; Quinzii, C.M. CoQ(10) deficiencies and MNGIE: Two treatable mitochondrial disorders. *Biochim. Biophys. Acta* **2012**, *1820*, 625–631. [CrossRef] [PubMed]
6. Marin, S.E.; Haas, R.H. Coenzyme Q10 and the Treatment of Mitochondrial Disease. In *Coenzyme Q10: From Fact to Fiction*; Hargreaves, I.P., Hargreaves, A.K., Eds.; Nova Science Publishers: Hauppauge, NY, USA, 2015; pp. 85–108.
7. Parkinson Study Group QE3 Investigators; Beal, M.F.; Oakes, D.; Shoulson, I.; Henchcliffe, C.; Galpern, W.R.; Haas, R.; Juncos, J.L.; Nutt, J.G.; Voss, T.S.; et al. A randomized clinical trial of high-dosage coenzyme Q10 in early Parkinson disease: No evidence of benefit. *JAMA Neurol* **2014**, *71*, 543–552. [PubMed]
8. Navas, P.; Villalba, J.M.; de Cabo, R. The importance of plasma membrane coenzyme Q in aging and stress responses. *Mitochondrion* **2007**, *7*, S34–S40. [CrossRef] [PubMed]
9. Huo, J.; Xu, Z.; Hosoe, K.; Kubo, H.; Miyahara, H.; Dai, J.; Mori, M.; Sawashita, J.; Higuchi, K. Coenzyme Q10 prevents senescence and dysfunction caused by oxidative stress in vascular endothelial cells. *Oxid. Med. Cell. Longev.* **2018**, *2018*, 1–15. [CrossRef] [PubMed]
10. Zhang, W.; Hui, R.; Yang, S. Telomeres, cardiovascular aging, and potential intervention for cellular senescence. *Sci. China Life Sci.* **2014**, *57*, 858–862. [CrossRef] [PubMed]
11. Davalli, P.; Mitic, T.; Caporali, A.; Lauriola, A.; D'Arca, D. Ros, cell senescence, and novel molecular mechanisms in aging and age-related diseases. *Oxid. Med. Cell. Longev.* **2016**, *2016*. [CrossRef] [PubMed]
12. Hernández-Camacho, J.D.; Bernier, M.; López-Lluch, G.; Navas, P. Coenzyme Q10 supplementation in aging and disease. *Front. Physiol.* **2018**. [CrossRef] [PubMed]
13. Gutierrez-Mariscal, F.M.; Yubero-Serrano, E.M.; Villalba, J.M.; Lopez-Miranda, J. Coenzyme Q10: From bench to clinic in aging diseases, a translational review. *Crit. Rev. Food Sci. Nutr.* **2018**. [CrossRef]
14. Awad, A.M.; Bradley, M.C.; Fernandez-Del-Rio, L.; Nag, A.; Tsui, H.S.; Clarke, C.F. Coenzyme Q10 deficiencies: Pathways in yeast and humans. *Essays Biochem.* **2018**, *62*, 361–376. [CrossRef]

15. Gale, P.H.; Erickson, R.E.; Page, A.C., Jr.; Folkers, K. Coenzyme Q. Li. New data on the distribution of coenzyme Q in nature. *Arch. Biochem. Biophys.* **1964**, *104*, 169–172. [CrossRef]
16. Kalen, A.; Norling, B.; Appelkvist, E.L.; Dallner, G. Ubiquinone biosynthesis by the microsomal fraction from rat liver. *Biochim. Biophys. Acta* **1987**, *926*, 70–78. [CrossRef]
17. Lenaz, G.; Fato, R.; Di Bernardo, S.; Jarreta, D.; Costa, A.; Genova, M.L.; Parenti Castelli, G. Localization and mobility of coenzyme Q in lipid bilayers and membranes. *BioFactors* **1999**, *9*, 87–93. [CrossRef]
18. Olson, R.E.; Rudney, H. Biosynthesis of ubiquinone. *Vitam Horm* **1983**, *40*, 1–43. [PubMed]
19. Alcazar-Fabra, M.; Navas, P.; Brea-Calvo, G. Coenzyme Q biosynthesis and its role in the respiratory chain structure. *Biochim. Biophys. Acta* **2016**, *1857*, 1073–1078. [CrossRef]
20. Lenaz, G.; Fato, R.; Castelluccio, C.; Genova, M.L.; Bovina, C.; Estornell, E.; Valls, V.; Pallotti, F.; Parenti Castelli, G. The function of coenzyme Q in mitochondria. *Clin. Investig.* **1993**, *71*, S66–S70. [CrossRef]
21. Enriquez, J.A.; Lenaz, G. Coenzyme Q and the respiratory chain: Coenzyme Q pool and mitochondrial supercomplexes. *Mol. Syndromol.* **2014**, *5*, 119–140. [CrossRef] [PubMed]
22. Groneberg, D.A.; Kindermann, B.; Althammer, M.; Klapper, M.; Vormann, J.; Littarru, G.P.; Doring, F. Coenzyme Q10 affects expression of genes involved in cell signalling, metabolism and transport in human caco-2 cells. *Int. J. Biochem. Cell Biol.* **2005**, *37*, 1208–1218. [CrossRef] [PubMed]
23. Schmelzer, C.; Kitano, M.; Hosoe, K.; Döring, F. Ubiquinol affects the expression of genes involved in pparα signalling and lipid metabolism without changes in methylation of CpG promoter islands in the liver of mice. *J. Clin. Biochem. Nutr.* **2012**, *50*, 119–126. [CrossRef]
24. Blatt, T.; Littarru, G.P. Biochemical rationale and experimental data on the antiaging properties of CoQ(10) at skin level. *BioFactors* **2011**, *37*, 381–385. [CrossRef]
25. Lopez-Lluch, G.; Rodriguez-Aguilera, J.C.; Santos-Ocana, C.; Navas, P. Is coenzyme Q a key factor in aging? *Mech. Ageing Dev.* **2010**, *131*, 225–235. [CrossRef] [PubMed]
26. Sohal, R.S.; Kamzalov, S.; Sumien, N.; Ferguson, M.; Rebrin, I.; Heinrich, K.R.; Forster, M.J. Effect of coenzyme Q10 intake on endogenous coenzyme Q content, mitochondrial electron transport chain, antioxidative defenses, and life span of mice. *Free Radic. Biol. Med.* **2006**, *40*, 480–487. [CrossRef]
27. Crane, F.L. Discovery of ubiquinone (coenzyme Q) and an overview of function. *Mitochondrion* **2007**, *7*, S2–S7. [CrossRef]
28. Belliere, J.; Devun, F.; Cottet-Rousselle, C.; Batandier, C.; Leverve, X.; Fontaine, E. Prerequisites for ubiquinone analogs to prevent mitochondrial permeability transition-induced cell death. *J. Bioenerg. Biomembr.* **2012**, *44*, 207–212. [CrossRef] [PubMed]
29. Olivieri, F.; Lazzarini, R.; Babini, L.; Prattichizzo, F.; Rippo, M.R.; Tiano, L.; Di Nuzzo, S.; Graciotti, L.; Festa, R.; Bruge, F.; et al. Anti-inflammatory effect of ubiquinol-10 on young and senescent endothelial cells via mir-146a modulation. *Free Radic. Biol. Med.* **2013**, *63*, 410–420. [CrossRef] [PubMed]
30. Fan, L.; Feng, Y.; Chen, G.C.; Qin, L.Q.; Fu, C.L.; Chen, L.H. Effects of coenzyme Q10 supplementation on inflammatory markers: A systematic review and meta-analysis of randomized controlled trials. *Pharmacol. Res.* **2017**, *119*, 128–136. [CrossRef]
31. Zhai, J.; Bo, Y.; Lu, Y.; Liu, C.; Zhang, L. Effects of coenzyme Q10 on markers of inflammation: A systematic review and meta-analysis. *PLoS ONE* **2017**, *12*, e0170172. [CrossRef]
32. Schmelzer, C.; Lindner, I.; Rimbach, G.; Niklowitz, P.; Menke, T.; Doring, F. Functions of coenzyme Q10 in inflammation and gene expression. *BioFactors* **2008**, *32*, 179–183. [CrossRef]
33. Saleh, D.O.; Ahmed, R.F.; Amin, M.M. Modulatory role of co-enzyme Q10 on methionine and choline deficient diet-induced non-alcoholic steatohepatitis (nash) in albino rats. *Appl. Physiol. Nutr. Metabol.* **2017**, *42*, 243–249. [CrossRef]
34. Farhangi, M.A.; Alipour, B.; Jafarvand, E.; Khoshbaten, M. Oral coenzyme Q10 supplementation in patients with nonalcoholic fatty liver disease: Effects on serum vaspin, chemerin, pentraxin 3, insulin resistance and oxidative stress. *Arch. Med. Res.* **2014**, *45*, 589–595. [CrossRef]
35. Perez-Sanchez, C.; Aguirre, M.A.; Ruiz-Limon, P.; Abalos-Aguilera, M.C.; Jimenez-Gomez, Y.; Arias-de la Rosa, I.; Rodriguez-Ariza, A.; Fernandez-Del Rio, L.; Gonzalez-Reyes, J.A.; Segui, P.; et al. Ubiquinol effects on antiphospholipid syndrome prothrombotic profile: A randomized, placebo-controlled trial. *Arterioscler. Thromb. Vasc. Biol.* **2017**, *37*, 1923–1932. [CrossRef]
36. Wilcock, D.M.; Griffin, W.S. Down's syndrome, neuroinflammation, and Alzheimer neuropathogenesis. *J. Neuroinflamm.* **2013**, *10*, 84. [CrossRef] [PubMed]

37. Zaki, M.E.; El-Bassyouni, H.T.; Tosson, A.M.; Youness, E.; Hussein, J. Coenzyme Q10 and pro-inflammatory markers in children with Down syndrome: Clinical and biochemical aspects. *J. Pediatr.* **2017**, *93*, 100–104. [CrossRef]
38. Tiano, L.; Carnevali, P.; Padella, L.; Santoro, L.; Principi, F.; Bruge, F.; Carle, F.; Gesuita, R.; Gabrielli, O.; Littarru, G.P. Effect of coenzyme Q10 in mitigating oxidative DNA damage in Down syndrome patients, a double blind randomized controlled trial. *Neurobiol. Aging* **2011**, *32*, 2103–2105. [CrossRef]
39. Sharma, A.; Fonarow, G.C.; Butler, J.; Ezekowitz, J.A.; Felker, G.M. Coenzyme Q10 and heart failure: A state-of-the-art review. *Circ. Heart Fail.* **2016**, *9*, e002639. [CrossRef]
40. Jorat, M.V.; Tabrizi, R.; Mirhosseini, N.; Lankarani, K.B.; Akbari, M.; Heydari, S.T.; Mottaghi, R.; Asemi, Z. The effects of coenzyme Q10 supplementation on lipid profiles among patients with coronary artery disease: A systematic review and meta-analysis of randomized controlled trials. *Lipids Health Dis.* **2018**, *17*, 230. [CrossRef]
41. Lucas-Hourani, M.; Munier-Lehmann, H.; El Mazouni, F.; Malmquist, N.A.; Harpon, J.; Coutant, E.P.; Guillou, S.; Helynck, O.; Noel, A.; Scherf, A.; et al. Original 2-(3-alkoxy-1h-pyrazol-1-yl)azines inhibitors of human dihydroorotate dehydrogenase (dhodh). *J. Med. Chem.* **2015**, *58*, 5579–5598. [CrossRef]
42. Ziosi, M.; Di Meo, I.; Kleiner, G.; Gao, X.H.; Barca, E.; Sanchez-Quintero, M.J.; Tadesse, S.; Jiang, H.; Qiao, C.; Rodenburg, R.J.; et al. Coenzyme Q deficiency causes impairment of the sulfide oxidation pathway. *EMBO Mol. Med.* **2017**, *9*, 96–111. [CrossRef]
43. Salvi, F.; Gadda, G. Human choline dehydrogenase: Medical promises and biochemical challenges. *Arch. Biochem. Biophys.* **2013**, *537*, 243–252. [CrossRef] [PubMed]
44. Szabados, L.; Savoure, A. Proline: A multifunctional amino acid. *Trends Plant Sci.* **2010**, *15*, 89–97. [CrossRef]
45. Bentinger, M.; Tekle, M.; Dallner, G. Coenzyme Q–biosynthesis and functions. *Biochem. Biophys. Res. Commun.* **2010**, *396*, 74–79. [CrossRef] [PubMed]
46. Garrido-Maraver, J.; Cordero, M.D.; Oropesa-Avila, M.; Vega, A.F.; de la Mata, M.; Pavon, A.D.; Alcocer-Gomez, E.; Calero, C.P.; Paz, M.V.; Alanis, M.; et al. Clinical applications of coenzyme Q10. *Front. Biosci. (Landmark Ed)* **2014**, *19*, 619–633. [CrossRef]
47. Zozina, V.I.; Covantev, S.; Goroshko, O.A.; Krasnykh, L.M.; Kukes, V.G. Coenzyme Q10 in cardiovascular and metabolic diseases: Current state of the problem. *Curr. Cardiol. Rev.* **2018**, *14*, 164–174. [CrossRef] [PubMed]
48. Bhagavan, H.N.; Chopra, R.K. Plasma coenzyme Q10 response to oral ingestion of coenzyme Q10 formulations. *Mitochondrion* **2007**, *7*, S78–S88. [CrossRef] [PubMed]
49. Ochiai, A.; Itagaki, S.; Kurokawa, T.; Kobayashi, M.; Hirano, T.; Iseki, K. Improvement in intestinal coenzyme Q10 absorption by food intake. *Yakugaku zasshi: J. Pharm. Soc. Jpn.* **2007**, *127*, 1251–1254. [CrossRef]
50. Mohr, D.; Bowry, V.W.; Stocker, R. Dietary supplementation with coenzyme Q10 results in increased levels of ubiquinol-10 within circulating lipoproteins and increased resistance of human low-density lipoprotein to the initiation of lipid peroxidation. *Biochim. Biophys. Acta* **1992**, *1126*, 247–254. [CrossRef]
51. Hargreaves, I.P. Ubiquinone: Cholesterol's reclusive cousin. *Ann. Clin. Biochem.* **2003**, *40*, 207–218. [CrossRef] [PubMed]
52. Ogasahara, S.; Engel, A.G.; Frens, D.; Mack, D. Muscle coenzyme Q deficiency in familial mitochondrial encephalomyopathy. *Proc. Natl. Acad. Sci. USA* **1989**, *86*, 2379–2382. [CrossRef] [PubMed]
53. Miles, M.V. The uptake and distribution of coenzyme Q10. *Mitochondrion* **2007**, *7*, S72–S77. [CrossRef]
54. Kwong, L.K.; Kamzalov, S.; Rebrin, I.; Bayne, A.C.; Jana, C.K.; Morris, P.; Forster, M.J.; Sohal, R.S. Effects of coenzyme Q(10) administration on its tissue concentrations, mitochondrial oxidant generation, and oxidative stress in the rat. *Free Radic. Biol. Med.* **2002**, *33*, 627–638. [CrossRef]
55. Niklowitz, P.; Menke, T.; Wiesel, T.; Mayatepek, E.; Zschocke, J.; Okun, J.G.; Andler, W. Coenzyme Q10 in plasma and erythrocytes: Comparison of antioxidant levels in healthy probands after oral supplementation and in patients suffering from sickle cell anemia. *Clin. Chim. Acta* **2002**, *326*, 155–161. [CrossRef]
56. Duncan, A.J.; Heales, S.J.; Mills, K.; Eaton, S.; Land, J.M.; Hargreaves, I.P. Determination of coenzyme Q10 status in blood mononuclear cells, skeletal muscle, and plasma by HPLC with di-propoxy-coenzyme Q10 as an internal standard. *Clin. Chem.* **2005**, *51*, 2380–2382. [CrossRef]
57. Bhagavan, H.N.; Chopra, R.K. Coenzyme Q10: Absorption, tissue uptake, metabolism and pharmacokinetics. *Free Radic. Res.* **2006**, *40*, 445–453. [CrossRef] [PubMed]

58. Musumeci, O.; Naini, A.; Slonim, A.E.; Skavin, N.; Hadjigeorgiou, G.L.; Krawiecki, N.; Weissman, B.M.; Tsao, C.Y.; Mendell, J.R.; Shanske, S.; et al. Familial cerebellar ataxia with muscle coenzyme Q10 deficiency. *Neurology* **2001**, *56*, 849–855. [CrossRef]
59. Gempel, K.; Topaloglu, H.; Talim, B.; Schneiderat, P.; Schoser, B.G.; Hans, V.H.; Palmafy, B.; Kale, G.; Tokatli, A.; Quinzii, C.; et al. The myopathic form of coenzyme Q10 deficiency is caused by mutations in the electron-transferring-flavoprotein dehydrogenase (ETFDH) gene. *Brain* **2007**, *130*, 2037–2044. [CrossRef] [PubMed]
60. Quinzii, C.M.; Kattah, A.G.; Naini, A.; Akman, H.O.; Mootha, V.K.; DiMauro, S.; Hirano, M. Coenzyme Q deficiency and cerebellar ataxia associated with an aprataxin mutation. *Neurology* **2005**, *64*, 539–541. [CrossRef]
61. Ihara, Y.; Namba, R.; Kuroda, S.; Sato, T.; Shirabe, T. Mitochondrial encephalomyopathy (MELAS): Pathological study and successful therapy with coenzyme Q10 and idebenone. *J. Neurol. Sci.* **1989**, *90*, 263–271. [CrossRef]
62. Quinzii, C.M.; Hirano, M. Primary and secondary CoQ(10) deficiencies in humans. *BioFactors* **2011**, *37*, 361–365. [CrossRef]
63. Hathcock, J.N.; Shao, A. Risk assessment for coenzyme Q10 (ubiquinone). *Regul. Toxicol. Pharmacol.* **2006**, *45*, 282–288. [CrossRef] [PubMed]
64. Balreira, A.; Boczonadi, V.; Barca, E.; Pyle, A.; Bansagi, B.; Appleton, M.; Graham, C.; Hargreaves, I.P.; Rasic, V.M.; Lochmuller, H.; et al. Ano10 mutations cause ataxia and coenzyme Q(1)(0) deficiency. *J. Neurol.* **2014**, *261*, 2192–2198. [CrossRef]
65. Baruteau, J.; Hargreaves, I.; Krywawych, S.; Chalasani, A.; Land, J.M.; Davison, J.E.; Kwok, M.K.; Christov, G.; Karimova, A.; Ashworth, M.; et al. Successful reversal of propionic acidaemia associated cardiomyopathy: Evidence for low myocardial coenzyme Q10 status and secondary mitochondrial dysfunction as an underlying pathophysiological mechanism. *Mitochondrion* **2014**, *17*, 150–156. [CrossRef]
66. Yubero, D.; O'Callaghan, M.; Montero, R.; Ormazabal, A.; Armstrong, J.; Espinos, C.; Rodriguez, M.A.; Jou, C.; Castejon, E.; Aracil, M.A.; et al. Association between coenzyme Q10 and glucose transporter (GLUT1) deficiency. *BMC Pediatr.* **2014**, *14*, 284. [CrossRef] [PubMed]
67. Beal, M.F. Coenzyme Q10 administration and its potential for treatment of neurodegenerative diseases. *BioFactors* **1999**, *9*, 261–266. [CrossRef]
68. Baggio, E.; Gandini, R.; Plancher, A.C.; Passeri, M.; Carmosino, G. Italian multicenter study on the safety and efficacy of coenzyme Q10 as adjunctive therapy in heart failure. CoQ10 drug surveillance investigators. *Mol. Asp. Med.* **1994**, *15*, s287–s294. [CrossRef]
69. Wallace, D.C. A mitochondrial paradigm of metabolic and degenerative diseases, aging, and cancer: A dawn for evolutionary medicine. *Annu. Rev. Genet.* **2005**, *39*, 359–407. [CrossRef] [PubMed]
70. Schniertshauer, D.; Gebhard, D.; Bergemann, J. Age-dependent loss of mitochondrial function in epithelial tissue can be reversed by coenzyme Q10. *J. Aging Res.* **2018**, *2018*, 6354680. [CrossRef] [PubMed]
71. Harman, D. The biologic clock: The mitochondria? *J. Am. Geriatr. Soc.* **1972**, *20*, 145–147. [CrossRef]
72. Krutmann, J.; Schroeder, P. Role of mitochondria in photoaging of human skin: The defective powerhouse model. *J. Investig. Dermatol. Symp. Proc.* **2009**, *14*, 44–49. [CrossRef]
73. Brunk, U.T.; Terman, A. The mitochondrial-lysosomal axis theory of aging: Accumulation of damaged mitochondria as a result of imperfect autophagocytosis. *Eur. J. Biochem.* **2002**, *269*, 1996–2002. [CrossRef]
74. Clayton, D.A.; Doda, J.N.; Friedberg, E.C. The absence of a pyrimidine dimer repair mechanism in mammalian mitochondria. *Proc. Natl. Acad. Sci. USA* **1974**, *71*, 2777–2781. [CrossRef]
75. Demple, B.; Harrison, L. Repair of oxidative damage to DNA: Enzymology and biology. *Annu. Rev. Biochem.* **1994**, *63*, 915–948. [CrossRef]
76. Barja, G. Mitochondrial oxygen consumption and reactive oxygen species production are independently modulated: Implications for aging studies. *Rejuv. Res.* **2007**, *10*, 215–224. [CrossRef]
77. Miquel, J. An update on the oxygen stress-mitochondrial mutation theory of aging: Genetic and evolutionary implications. *Exp. Gerontol.* **1998**, *33*, 113–126. [CrossRef]
78. Sohal, R.S.; Mockett, R.J.; Orr, W.C. Mechanisms of aging: An appraisal of the oxidative stress hypothesis. *Free Radic. Biol. Med.* **2002**, *33*, 575–586. [CrossRef]

79. Maurya, P.K.; Noto, C.; Rizzo, L.B.; Rios, A.C.; Nunes, S.O.; Barbosa, D.S.; Sethi, S.; Zeni, M.; Mansur, R.B.; Maes, M.; et al. The role of oxidative and nitrosative stress in accelerated aging and major depressive disorder. *Prog. Neuro-Psychopharmacol. Biol. Psychiatry* **2016**, *65*, 134–144. [CrossRef] [PubMed]
80. Wallace, D.C.; Fan, W.; Procaccio, V. Mitochondrial energetics and therapeutics. *Annu. Rev. Pathol.* **2010**, *5*, 297–348. [CrossRef]
81. Elchuri, S.; Oberley, T.D.; Qi, W.; Eisenstein, R.S.; Jackson Roberts, L.; Van Remmen, H.; Epstein, C.J.; Huang, T.T. Cuznsod deficiency leads to persistent and widespread oxidative damage and hepatocarcinogenesis later in life. *Oncogene* **2005**, *24*, 367–380. [CrossRef]
82. Li, Y.; Huang, T.T.; Carlson, E.J.; Melov, S.; Ursell, P.C.; Olson, J.L.; Noble, L.J.; Yoshimura, M.P.; Berger, C.; Chan, P.H.; et al. Dilated cardiomyopathy and neonatal lethality in mutant mice lacking manganese superoxide dismutase. *Nat. Genet.* **1995**, *11*, 376–381. [CrossRef]
83. Ma, D.; Stokes, K.; Mahngar, K.; Domazet-Damjanov, D.; Sikorska, M.; Pandey, S. Inhibition of stress induced premature senescence in presenilin-1 mutated cells with water soluble coenzyme Q10. *Mitochondrion* **2014**, *17*, 106–115. [CrossRef]
84. Sohal, R.S.; Sohal, B.H. Hydrogen peroxide release by mitochondria increases during aging. *Mech. Ageing Dev.* **1991**, *57*, 187–202. [CrossRef]
85. Asensi, M.; Sastre, J.; Pallardo, F.V.; Lloret, A.; Lehner, M.; Garcia-de la Asuncion, J.; Vina, J. Ratio of reduced to oxidized glutathione as indicator of oxidative stress status and DNA damage. *Methods Enzymol.* **1999**, *299*, 267–276.
86. Halliwell, B. The wanderings of a free radical. *Free Radic. Biol. Med.* **2009**, *46*, 531–542. [CrossRef]
87. Lambert, A.J.; Boysen, H.M.; Buckingham, J.A.; Yang, T.; Podlutsky, A.; Austad, S.N.; Kunz, T.H.; Buffenstein, R.; Brand, M.D. Low rates of hydrogen peroxide production by isolated heart mitochondria associate with long maximum lifespan in vertebrate homeotherms. *Aging Cell* **2007**, *6*, 607–618. [CrossRef]
88. Barja, G. Updating the mitochondrial free radical theory of aging: An integrated view, key aspects, and confounding concepts. *Antioxid. Redox Signal.* **2013**, *19*, 1420–1445. [CrossRef]
89. Hekimi, S.; Lapointe, J.; Wen, Y. Taking a "good" look at free radicals in the aging process. *Trends Cell Biol.* **2011**, *21*, 569–576. [CrossRef]
90. Wang, Y.; Hekimi, S. Mitochondrial dysfunction and longevity in animals: Untangling the knot. *Science* **2015**, *350*, 1204–1207. [CrossRef]
91. Feng, J.; Bussiere, F.; Hekimi, S. Mitochondrial electron transport is a key determinant of life span in caenorhabditis elegans. *Dev. Cell* **2001**, *1*, 633–644. [CrossRef]
92. Hekimi, S. Enhanced immunity in slowly aging mutant mice with high mitochondrial oxidative stress. *Oncoimmunology* **2013**, *2*, e23793. [CrossRef] [PubMed]
93. Wang, Y.; Oxer, D.; Hekimi, S. Mitochondrial function and lifespan of mice with controlled ubiquinone biosynthesis. *Nat. Commun.* **2015**, *6*, 6393. [CrossRef]
94. Van Remmen, H.; Ikeno, Y.; Hamilton, M.; Pahlavani, M.; Wolf, N.; Thorpe, S.R.; Alderson, N.L.; Baynes, J.W.; Epstein, C.J.; Huang, T.T.; et al. Life-long reduction in mnsod activity results in increased DNA damage and higher incidence of cancer but does not accelerate aging. *Physiol. Genom.* **2003**, *16*, 29–37. [CrossRef] [PubMed]
95. Hepple, R.T. Mitochondrial involvement and impact in aging skeletal muscle. *Front. Aging Neurosci.* **2014**, *6*, 211. [CrossRef]
96. Buffenstein, R.; Edrey, Y.H.; Yang, T.; Mele, J. The oxidative stress theory of aging: Embattled or invincible? Insights from non-traditional model organisms. *Age* **2008**, *30*, 99–109. [CrossRef] [PubMed]
97. Varela-Lopez, A.; Giampieri, F.; Battino, M.; Quiles, J.L. Coenzyme Q and its role in the dietary therapy against aging. *Molecules* **2016**, *21*, 373. [CrossRef]
98. Sohal, R.S.; Forster, M.J. Coenzyme Q, oxidative stress and aging. *Mitochondrion* **2007**, *7*, S103–S111. [CrossRef]
99. Jonassen, T.; Davis, D.E.; Larsen, P.L.; Clarke, C.F. Reproductive fitness and quinone content of caenorhabditis elegans clk-1 mutants fed coenzyme Q isoforms of varying length. *J. Biol. Chem.* **2003**, *278*, 51735–51742. [CrossRef] [PubMed]
100. Asencio, C.; Navas, P.; Cabello, J.; Schnabel, R.; Cypser, J.R.; Johnson, T.E.; Rodriguez-Aguilera, J.C. Coenzyme Q supports distinct developmental processes in caenorhabditis elegans. *Mech. Ageing Dev.* **2009**, *130*, 145–153. [CrossRef]

101. Gavilan, A.; Asencio, C.; Cabello, J.; Rodriguez-Aguilera, J.C.; Schnabel, R.; Navas, P.C. Elegans knockouts in ubiquinone biosynthesis genes result in different phenotypes during larval development. *BioFactors* **2005**, *25*, 21–29. [CrossRef]
102. Hihi, A.K.; Gao, Y.; Hekimi, S. Ubiquinone is necessary for caenorhabditis elegans development at mitochondrial and non-mitochondrial sites. *J. Biol. Chem.* **2002**, *277*, 2202–2206. [CrossRef] [PubMed]
103. Quinzii, C.M.; Lopez, L.C.; Von-Moltke, J.; Naini, A.; Krishna, S.; Schuelke, M.; Salviati, L.; Navas, P.; DiMauro, S.; Hirano, M. Respiratory chain dysfunction and oxidative stress correlate with severity of primary CoQ10 deficiency. *Faseb J.* **2008**, *22*, 1874–1885. [CrossRef] [PubMed]
104. Quinzii, C.M.; López, L.C.; Gilkerson, R.W.; Dorado, B.; Coku, J.; Naini, A.B.; Lagier-Tourenne, C.; Schuelke, M.; Salviati, L.; Carrozzo, R.; et al. Reactive oxygen species, oxidative stress, and cell death correlate with level of CoQ10 deficiency. *Faseb J.* **2010**, *24*, 3733–3743. [CrossRef] [PubMed]
105. Tian, G.; Sawashita, J.; Kubo, H.; Nishio, S.Y.; Hashimoto, S.; Suzuki, N.; Yoshimura, H.; Tsuruoka, M.; Wang, Y.; Liu, Y.; et al. Ubiquinol-10 supplementation activates mitochondria functions to decelerate senescence in senescence-accelerated mice. *Antioxid. Redox Signal.* **2014**, *20*, 2606–2620. [CrossRef] [PubMed]
106. Xu, Z.; Huo, J.; Ding, X.; Yang, M.; Li, L.; Dai, J.; Hosoe, K.; Kubo, H.; Mori, M.; Higuchi, K.; et al. Coenzyme Q10 improves lipid metabolism and ameliorates obesity by regulating camkii-mediated pde4 inhibition. *Sci. Rep.* **2017**, *7*, 8253. [CrossRef] [PubMed]
107. Liu, X.; Jiang, N.; Hughes, B.; Bigras, E.; Shoubridge, E.; Hekimi, S. Evolutionary conservation of the clk-1-dependent mechanism of longevity: Loss of mclk1 increases cellular fitness and lifespan in mice. *Genes Dev.* **2005**, *19*, 2424–2434. [CrossRef]
108. Ochoa, J.J.; Quiles, J.L.; Lopez-Frias, M.; Huertas, J.R.; Mataix, J. Effect of lifelong coenzyme Q10 supplementation on age-related oxidative stress and mitochondrial function in liver and skeletal muscle of rats fed on a polyunsaturated fatty acid (PUFA)-rich diet. *J. Gerontol. Ser. A Biol. Sci. Med. Sci.* **2007**, *62*, 1211–1218. [CrossRef]
109. Quiles, J.L.; Ochoa, J.J.; Battino, M.; Gutierrez-Rios, P.; Nepomuceno, E.A.; Frias, M.L.; Huertas, J.R.; Mataix, J. Life-long supplementation with a low dosage of coenzyme Q10 in the rat: Effects on antioxidant status and DNA damage. *BioFactors* **2005**, *25*, 73–86. [CrossRef]
110. Quiles, J.L.; Pamplona, R.; Ramirez-Tortosa, M.C.; Naudi, A.; Portero-Otin, M.; Araujo-Nepomuceno, E.; Lopez-Frias, M.; Battino, M.; Ochoa, J.J. Coenzyme Q addition to an n-6 PUFA-rich diet resembles benefits on age-related mitochondrial DNA deletion and oxidative stress of a MUFA-rich diet in rat heart. *Mech. Ageing Dev.* **2010**, *131*, 38–47. [CrossRef]
111. Schmelzer, C.; Kubo, H.; Mori, M.; Sawashita, J.; Kitano, M.; Hosoe, K.; Boomgaarden, I.; Doring, F.; Higuchi, K. Supplementation with the reduced form of Coenzyme Q10 decelerates phenotypic characteristics of senescence and induces a peroxisome proliferator-activated receptor-alpha gene expression signature in SAMP1 mice. *Mol. Nutr. Food Res.* **2010**, *54*, 805–815. [CrossRef]
112. Matthews, R.T.; Yang, L.; Browne, S.; Baik, M.; Beal, M.F. Coenzyme Q10 administration increases brain mitochondrial concentrations and exerts neuroprotective effects. *Proc. Natl. Acad. Sci. USA* **1998**, *95*, 8892–8897. [CrossRef]
113. Wolters, M.; Hahn, A. Plasma ubiquinone status and response to six-month supplementation combined with multivitamins in healthy elderly women–results of a randomized, double-blind, placebo-controlled study. *Int. J. Vitam. Nutr. Res.* **2003**, *73*, 207–214. [CrossRef] [PubMed]
114. Kalen, A.; Appelkvist, E.L.; Dallner, G. Age-related changes in the lipid compositions of rat and human tissues. *Lipids* **1989**, *24*, 579–584. [CrossRef]
115. Soderberg, M.; Edlund, C.; Kristensson, K.; Dallner, G. Lipid compositions of different regions of the human brain during aging. *J. Neurochem.* **1990**, *54*, 415–423. [CrossRef] [PubMed]
116. Edlund, C.; Soderberg, M.; Kristensson, K.; Dallner, G. Ubiquinone, dolichol, and cholesterol metabolism in aging and Alzheimer's disease. *Biochem. Cell Biol.* **1992**, *70*, 422–428. [CrossRef]
117. Nagase, M.; Yamamoto, Y.; Matsumoto, N.; Arai, Y.; Hirose, N. Increased oxidative stress and coenzyme Q10 deficiency in centenarians. *J. Clin. Biochem. Nutr.* **2018**, *63*, 129–136. [CrossRef]
118. Bentinger, M.; Brismar, K.; Dallner, G. The antioxidant role of coenzyme Q. *Mitochondrion* **2007**, *7*, S41–S50. [CrossRef] [PubMed]

119. Del Pozo-Cruz, J.; Rodriguez-Bies, E.; Ballesteros-Simarro, M.; Navas-Enamorado, I.; Tung, B.T.; Navas, P.; Lopez-Lluch, G. Physical activity affects plasma coenzyme Q10 levels differently in young and old humans. *Biogerontology* **2014**, *15*, 199–211. [CrossRef]
120. Del Pozo-Cruz, J.; Rodriguez-Bies, E.; Navas-Enamorado, I.; Del Pozo-Cruz, B.; Navas, P.; Lopez-Lluch, G. Relationship between functional capacity and body mass index with plasma coenzyme Q10 and oxidative damage in community-dwelling elderly-people. *Exp. Gerontol.* **2014**, *52*, 46–54. [CrossRef] [PubMed]
121. Johansson, P.; Dahlstrom, O.; Dahlstrom, U.; Alehagen, U. Improved health-related quality of life, and more days out of hospital with supplementation with selenium and coenzyme Q10 combined. Results from a double blind, placebo-controlled prospective study. *J. Nutr. Health Aging* **2015**, *19*, 870–877. [CrossRef]
122. Fischer, A.; Onur, S.; Niklowitz, P.; Menke, T.; Laudes, M.; Rimbach, G.; Doring, F. Coenzyme Q10 status as a determinant of muscular strength in two independent cohorts. *PLoS ONE* **2016**, *11*, e0167124. [CrossRef] [PubMed]
123. Liu, Z.X.; Artmann, C. Relative bioavailability comparison of different coenzyme Q10 formulations with a novel delivery system. *Altern. Ther. Health Med.* **2009**, *15*, 42–46.
124. Koroshetz, W.J.; Jenkins, B.G.; Rosen, B.R.; Beal, M.F. Energy metabolism defects in Huntington's disease and effects of coenzyme Q10. *Ann. Neurol.* **1997**, *41*, 160–165. [CrossRef] [PubMed]
125. Balaban, R.S.; Nemoto, S.; Finkel, T. Mitochondria, oxidants, and aging. *Cell* **2005**, *120*, 483–495. [CrossRef] [PubMed]
126. Liu, J.; Wang, L.N. Mitochondrial enhancement for neurodegenerative movement disorders: A systematic review of trials involving creatine, coenzyme Q10, idebenone and mitoquinone. *CNS Drugs* **2014**, *28*, 63–68. [CrossRef] [PubMed]
127. Molina, J.A.; de Bustos, F.; Ortiz, S.; Del Ser, T.; Seijo, M.; Benito-Leon, J.; Oliva, J.M.; Perez, S.; Manzanares, J. Serum levels of coenzyme Q in patients with Lewy body disease. *J. Neural Transm. (Vienna, Austria : 1996)* **2002**, *109*, 1195–1201. [CrossRef] [PubMed]
128. Isobe, C.; Abe, T.; Terayama, Y. Levels of reduced and oxidized coenzyme Q-10 and 8-hydroxy-2'-deoxyguanosine in the cerebrospinal fluid of patients with living Parkinson's disease demonstrate that mitochondrial oxidative damage and/or oxidative DNA damage contributes to the neurodegenerative process. *Neurosci. Lett.* **2010**, *469*, 159–163.
129. McGarry, A.; McDermott, M.; Kieburtz, K.; de Blieck, E.A.; Beal, F.; Marder, K.; Ross, C.; Shoulson, I.; Gilbert, P.; Mallonee, W.M.; et al. A randomized, double-blind, placebo-controlled trial of coenzyme Q10 in Huntington disease. *Neurology* **2017**, *88*, 152–159. [CrossRef]
130. Buhmann, C.; Arlt, S.; Kontush, A.; Moller-Bertram, T.; Sperber, S.; Oechsner, M.; Stuerenburg, H.J.; Beisiegel, U. Plasma and CSF markers of oxidative stress are increased in Parkinson's disease and influenced by antiparkinsonian medication. *Neurobiol. Dis.* **2004**, *15*, 160–170. [CrossRef] [PubMed]
131. Haas, R.H.; Nasirian, F.; Nakano, K.; Ward, D.; Pay, M.; Hill, R.; Shults, C.W. Low platelet mitochondrial complex I and complex II/III activity in early untreated Parkinson's disease. *Ann. Neurol.* **1995**, *37*, 714–722. [CrossRef]
132. Mischley, L.K.; Allen, J.; Bradley, R. Coenzyme Q10 deficiency in patients with Parkinson's disease. *J. Neurol. Sci.* **2012**, *318*, 72–75. [CrossRef] [PubMed]
133. Shults, C.W.; Oakes, D.; Kieburtz, K.; Beal, M.F.; Haas, R.; Plumb, S.; Juncos, J.L.; Nutt, J.; Shoulson, I.; Carter, J.; et al. Effects of coenzyme Q10 in early Parkinson disease: Evidence of slowing of the functional decline. *Arch. Neurol.* **2002**, *59*, 1541–1550. [CrossRef] [PubMed]
134. Negida, A.; Menshawy, A.; El Ashal, G.; Elfouly, Y.; Hani, Y.; Hegazy, Y.; El Ghonimy, S.; Fouda, S.; Rashad, Y. Coenzyme Q10 for patients with Parkinson's disease: A systematic review and meta-analysis. *CNS Neurol. Disord. Drug Targets* **2016**, *15*, 45–53. [CrossRef]
135. Shetty, R.A.; Forster, M.J.; Sumien, N. Coenzyme Q(10) supplementation reverses age-related impairments in spatial learning and lowers protein oxidation. *Age* **2013**, *35*, 1821–1834. [CrossRef]
136. Galasko, D.R.; Peskind, E.; Clark, C.M.; Quinn, J.F.; Ringman, J.M.; Jicha, G.A.; Cotman, C.; Cottrell, B.; Montine, T.J.; Thomas, R.G.; et al. Antioxidants for Alzheimer disease: A randomized clinical trial with cerebrospinal fluid biomarker measures. *Arch. Neurol.* **2012**, *69*, 836–841. [CrossRef]
137. Flowers, N.; Hartley, L.; Todkill, D.; Stranges, S.; Rees, K. Co-enzyme Q10 supplementation for the primary prevention of cardiovascular disease. *Cochrane Database Syst. Rev.* **2014**, *12*, Cd010405. [CrossRef]

138. Ho, M.J.; Li, E.C.; Wright, J.M. Blood pressure lowering efficacy of coenzyme Q10 for primary hypertension. *Cochrane Database Syst. Rev.* **2016**, *3*, Cd007435. [CrossRef] [PubMed]
139. Hosoe, K.; Kitano, M.; Kishida, H.; Kubo, H.; Fujii, K.; Kitahara, M. Study on safety and bioavailability of ubiquinol (Kaneka QH) after single and 4-week multiple oral administration to healthy volunteers. *Regul. Toxicol. Pharmacol.* **2007**, *47*, 19–28. [CrossRef] [PubMed]
140. Rundek, T.; Naini, A.; Sacco, R.; Coates, K.; DiMauro, S. Atorvastatin decreases the coenzyme Q10 level in the blood of patients at risk for cardiovascular disease and stroke. *Arch. Neurol.* **2004**, *61*, 889–892. [CrossRef] [PubMed]
141. Lamperti, C.; Naini, A.B.; Lucchini, V.; Prelle, A.; Bresolin, N.; Moggio, M.; Sciacco, M.; Kaufmann, P.; DiMauro, S. Muscle coenzyme Q10 levels in statin-related myopathy. *Arch. Neurol.* **2005**, *62*, 1709–1712. [CrossRef]
142. Caso, G.; Kelly, P.; McNurlan, M.A.; Lawson, W.E. Effect of coenzyme Q10 on myopathic symptoms in patients treated with statins. *Am. J. Cardiol.* **2007**, *99*, 1409–1412. [CrossRef]
143. Fedacko, J.; Pella, D.; Fedackova, P.; Hanninen, O.; Tuomainen, P.; Jarcuska, P.; Lopuchovsky, T.; Jedlickova, L.; Merkovska, L.; Littarru, G.P. Coenzyme Q(10) and selenium in statin-associated myopathy treatment. *Can. J. Physiol. Pharmacol.* **2013**, *91*, 165–170. [CrossRef]
144. Schirris, T.J.; Renkema, G.H.; Ritschel, T.; Voermans, N.C.; Bilos, A.; van Engelen, B.G.; Brandt, U.; Koopman, W.J.; Beyrath, J.D.; Rodenburg, R.J.; et al. Statin-induced myopathy is associated with mitochondrial complex III inhibition. *Cell Metab.* **2015**, *22*, 399–407. [CrossRef] [PubMed]
145. de Frutos, F.; Gea, A.; Hernandez-Estefania, R.; Rabago, G. Prophylactic treatment with coenzyme Q10 in patients undergoing cardiac surgery: Could an antioxidant reduce complications? A systematic review and meta-analysis. *Interact. Cardiovasc. Thorac. Surg.* **2015**, *20*, 254–259. [CrossRef] [PubMed]
146. Alehagen, U.; Aaseth, J.; Alexander, J.; Johansson, P. Still reduced cardiovascular mortality 12 years after supplementation with selenium and coenzyme Q10 for four years: A validation of previous 10-year follow-up results of a prospective randomized double-blind placebo-controlled trial in elderly. *PLoS ONE* **2018**, *13*, e0193120. [CrossRef] [PubMed]
147. Mortensen, S.A.; Rosenfeldt, F.; Kumar, A.; Dolliner, P.; Filipiak, K.J.; Pella, D.; Alehagen, U.; Steurer, G.; Littarru, G.P. The effect of coenzyme Q10 on morbidity and mortality in chronic heart failure: Results from Q-SYMBIO: A randomized double-blind trial. *JACC Heart Fail.* **2014**, *2*, 641–649. [CrossRef]
148. Yubero-Serrano, E.M.; Gonzalez-Guardia, L.; Rangel-Zuniga, O.; Delgado-Lista, J.; Gutierrez-Mariscal, F.M.; Perez-Martinez, P.; Delgado-Casado, N.; Cruz-Teno, C.; Tinahones, F.J.; Villalba, J.M.; et al. Mediterranean diet supplemented with coenzyme Q10 modifies the expression of proinflammatory and endoplasmic reticulum stress-related genes in elderly men and women. *J. Gerontol. Ser. A Biol. Sci. Med. Sci.* **2012**, *67*, 3–10. [CrossRef]
149. Tsai, H.Y.; Lin, C.P.; Huang, P.H.; Li, S.Y.; Chen, J.S.; Lin, F.Y.; Chen, J.W.; Lin, S.J. Coenzyme Q10 attenuates high glucose-induced endothelial progenitor cell dysfunction through AMP-activated protein kinase pathways. *J. Diabetes Res.* **2016**, *2016*, 6384759. [CrossRef] [PubMed]
150. Ishikawa, A.; Kawarazaki, H.; Ando, K.; Fujita, M.; Fujita, T.; Homma, Y. Renal preservation effect of ubiquinol, the reduced form of coenzyme Q10. *Clin. Exp. Nephrol.* **2011**, *15*, 30–33. [CrossRef] [PubMed]
151. Rivara, M.B.; Yeung, C.K.; Robinson-Cohen, C.; Phillips, B.R.; Ruzinski, J.; Rock, D.; Linke, L.; Shen, D.D.; Ikizler, T.A.; Himmelfarb, J. Effect of coenzyme Q10 on biomarkers of oxidative stress and cardiac function in hemodialysis patients: The CoQ10 biomarker trial. *Am. J. Kidney Dis.* **2017**, *69*, 389–399. [CrossRef] [PubMed]
152. Bolignano, D.; Cernaro, V.; Gembillo, G.; Baggetta, R.; Buemi, M.; D'Arrigo, G. Antioxidant agents for delaying diabetic kidney disease progression: A systematic review and meta-analysis. *PLoS ONE* **2017**, *12*, e0178699. [CrossRef]
153. Abdollahzad, H.; Aghdashi, M.A.; Asghari Jafarabadi, M.; Alipour, B. Effects of coenzyme Q10 supplementation on inflammatory cytokines (tnf-alpha, il-6) and oxidative stress in rheumatoid arthritis patients: A randomized controlled trial. *Arch. Med. Res.* **2015**, *46*, 527–533. [CrossRef] [PubMed]
154. Mazidi, M.; Kengne, A.P.; Banach, M. Effects of coenzyme Q10 supplementation on plasma c-reactive protein concentrations: A systematic review and meta-analysis of randomized controlled trials. *Pharmacol. Res.* **2018**, *128*, 130–136. [CrossRef] [PubMed]

155. Tiano, L.; Busciglio, J. Mitochondrial dysfunction and Down's syndrome: Is there a role for coenzyme Q(10)? *BioFactors* **2011**, *37*, 386–392. [CrossRef]
156. Zhang, X.X.; Qian, K.J.; Zhang, Y.; Wang, Z.J.; Yu, Y.B.; Liu, X.J.; Cao, X.T.; Liao, Y.H.; Zhang, D.Y. Efficacy of coenzyme Q10 in mitigating spinal cord injury-induced osteoporosis. *Mol. Med. Rep.* **2015**, *12*, 3909–3915. [CrossRef] [PubMed]
157. Moon, H.J.; Ko, W.K.; Jung, M.S.; Kim, J.H.; Lee, W.J.; Park, K.S.; Heo, J.K.; Bang, J.B.; Kwon, I.K. Coenzyme Q10 regulates osteoclast and osteoblast differentiation. *J. Food Sci.* **2013**, *78*, H785–H891. [CrossRef]
158. Zheng, D.; Cui, C.; Yu, M.; Li, X.; Wang, L.; Chen, X.; Lin, Y. Coenzyme Q10 promotes osteoblast proliferation and differentiation and protects against ovariectomy-induced osteoporosis. *Mol. Med. Rep.* **2018**, *17*, 400–407. [CrossRef]
159. Jolliet, P.; Simon, N.; Barre, J.; Pons, J.Y.; Boukef, M.; Paniel, B.J.; Tillement, J.P. Plasma coenzyme Q10 concentrations in breast cancer: Prognosis and therapeutic consequences. *Int. J. Clin. Pharmacol. Ther.* **1998**, *36*, 506–509.
160. Folkers, K.; Osterborg, A.; Nylander, M.; Morita, M.; Mellstedt, H. Activities of vitamin Q10 in animal models and a serious deficiency in patients with cancer. *Biochem. Biophys. Res. Commun.* **1997**, *234*, 296–299. [CrossRef]
161. Rusciani, L.; Proietti, I.; Paradisi, A.; Rusciani, A.; Guerriero, G.; Mammone, A.; De Gaetano, A.; Lippa, S. Recombinant interferon alpha-2b and coenzyme Q10 as a postsurgical adjuvant therapy for melanoma: A 3-year trial with recombinant interferon-alpha and 5-year follow-up. *Melanoma Res.* **2007**, *17*, 177–183. [CrossRef]
162. Mano, T.; Iwase, K.; Hayashi, R.; Hayakawa, N.; Uchimura, K.; Makino, M.; Nagata, M.; Sawai, Y.; Oda, N.; Hamada, M.; et al. Vitamin E and coenzyme Q concentrations in the thyroid tissues of patients with various thyroid disorders. *Am. J. Med. Sci.* **1998**, *315*, 230–232. [PubMed]
163. Quiles, J.L.; Farquharson, A.J.; Ramirez-Tortosa, M.C.; Grant, I.; Milne, L.; Huertas, J.R.; Battino, M.; Mataix, J.; Wahle, K.W. Coenzyme Q differentially modulates phospholipid hydroperoxide glutathione peroxidase gene expression and free radicals production in malignant and non-malignant prostate cells. *BioFactors* **2003**, *18*, 265–270. [CrossRef]

© 2019 by the authors. Licensee MDPI, Basel, Switzerland. This article is an open access article distributed under the terms and conditions of the Creative Commons Attribution (CC BY) license (http://creativecommons.org/licenses/by/4.0/).

*biology*

Review
Mitochondria's Role in Skin Ageing

Roisin Stout and Mark Birch-Machin *

Dermatological Sciences, Institute of Cellular Medicine, Medical School, Newcastle University, Newcastle upon Tyne NE2 4HH, UK; r.stout2@newcastle.ac.uk
* Correspondence: mark.birch-machin@newcastle.ac.uk

Received: 21 December 2018; Accepted: 7 February 2019; Published: 11 May 2019

Abstract: Skin ageing is the result of a loss of cellular function, which can be further accelerated by external factors. Mitochondria have important roles in skin function, and mitochondrial damage has been found to accumulate with age in skin cells, but also in response to solar light and pollution. There is increasing evidence that mitochondrial dysfunction and oxidative stress are key features in all ageing tissues, including skin. This is directly linked to skin ageing phenotypes: wrinkle formation, hair greying and loss, uneven pigmentation and decreased wound healing. The loss of barrier function during skin ageing increases susceptibility to infection and affects wound healing. Therefore, an understanding of the mechanisms involved is important clinically and also for the development of antiageing skin care products.

Keywords: mitochondria; skin; ageing; reactive oxygen species; photoageing

1. Skin Structure

Skin is the largest organ of the human body and made up of three distinct layers: the epidermis, the dermis and subcutaneous fat. It functions as a barrier against the environment, providing protection against microbes as well as fluid and temperature homeostasis. The epidermis is a thin layer of densely packed keratinised epithelial cells (keratinocytes) which contains no nerves or blood vessels and relies on the thick dermal layer underneath for metabolism. The dermis is the main living tissue in the skin, consisting of fibroblast cells in an extracellular matrix interspersed with sweat glands, hair follicles, muscle, capillaries and nerve endings. The epidermal basal layer is the innermost layer of the epidermis dispersed with melanocytes [1]. It separates the outer layers of the epidermis from the nutrient-providing papillary dermis and the thick supporting reticular dermis layer below, all of which rest on a layer of subcutaneous fat. With skin taking a large environmental insult, the keratinocytes undergo constant turnover by epidermal stem cells to replace damaged cells [2]. Melanocytes produce melanin, which is transported to keratinocytes to produce skin pigmentation and provide some solar protection. Structural integrity, repair and strength within the dermal layer of the skin are provided by type I collagen fibres, produced from procollagen bodies. Collagen regulation occurs through synthesis promotion by cytokine TGF-β, inhibition by transcription factor AP-1 and active degradation by the collagenase enzymes matrix metalloproteinases (MMPs) [3].

2. Mitochondria's Role in Skin

Mitochondria play a vital role in the skin. While the energy requirement may not be as great as other organs, such as skeletal muscle, it is still integral for processes like cell signalling, wound healing, pigmentation, vasculature homeostasis and hair growth. They are critical in microbial defence; glycolysis and ATP production have been found to rapidly increase in response to *Staphylococcus aureus* infection on the skin, in response to hypoxia induced metabolic stress [4]. This implements mitochondrial reactive oxygen species (ROS) signalling in the defence against skin infection through hypoxia-inducible factor 1-alpha (HIF1α) activation and immune cell recruitment [5]. Mitochondrial function and ROS production aid the regulation of stem cell differentiation, and have further been specifically linked to epidermal homeostasis and hair follicle development [6]. Reactive oxygen species signalling via mitochondria is therefore specifically involved in skin structure and function.

In patients with genetic mitochondrial disease, skin manifestations are often neglected as the disease predominantly affects the neuromuscular system, due to its high energy requirements. Various abnormalities in mitochondrial function, such as mutations of mitochondrial repair genes and haem synthesis, have been directly linked to a multitude of skin aberrations [7]. Lipomas and pigmentation disorders are the most commonly recorded skin complaints in mitochondrial disease patients [7,8]. This is likely due to mitochondrial defects in brown fat [9] and the direct effect of mitochondrial function in pigment production, respectively. Skin complaints can therefore be directly linked to genetic mitochondrial dysfunction and through complex ROS signalling. However, the correlation between the skin and mitochondria should be approached with caution, many other skin disorders are a result of secondary factors in diseases which do not primarily affect the mitochondrion itself, such as skin blistering in epidermolysis bullosa simplex [10].

3. Hallmarks of Skin Ageing

Wrinkles are one of the first features thought of when considering facial appearance and ageing. Intrinsic ageing is the result of chronological, inevitable senescence of the skin cells which varies depending on ethnicity, hormones and the anatomical region of affected skin. Extrinsic skin ageing is a result of all the external factors that can induce skin ageing, such as lifestyle, smoking, UV exposure and the environment, which have a cumulative effect over time [11]. Fine lines, breakdown of the skin structure, reddening due to increased visible vasculature and a reduction of elasticity are the main clinical features of intrinsic ageing; extrinsic ageing produces much deeper wrinkles, dryer skin, spider veins and uneven pigmentation [12].

As skin is constantly defending against environmental insult, it is important to maintain its integrity: ageing skin has reduced wound healing capacity and increased water loss. This increases susceptibility to cuts and infection, and makes it more prone to irritation and dermatoses [13]. It is essential to maintain an adequate skin barrier and understand the mechanisms involved in its loss to protect against age related dysregulation.

4. Mitochondria and Ageing

The "Free Radical Theory of Ageing" was first proposed by Harman in the 1950s [14]. It states that mutations acquired in mitochondrial DNA (mtDNA) during life, both spontaneously and through stress, can disrupt cellular metabolism like oxidative phosphorylation in the mitochondria and ultimately increase ROS. This, in turn, results in the oxidation of cellular components including proteins, lipids, DNA and RNA, which creates a cycle of altered metabolism and further damage. Ultimately, this results in the subsequent decline of cellular function seen in ageing and degenerative diseases.

Since it was first proposed, there have been a multitude of studies attempting to corroborate this theory by analysing mtDNA damage in ageing. As of yet there is no consensus, but there is a correlation between mtDNA damage, increasing oxidative stress and ageing. In skin samples, an accumulation of mtDNA deletions has been found to not only increase with age, but also in sun exposed areas compared to protected areas [15]. In this study by Ray et al., epidermal skin had a significantly greater increase in mtDNA deletions with chronic sun exposure compared to dermal skin, which had no significant change in mtDNA deletion quantity to protected skin. This was higher than those found in normal ageing. Additionally, an increase in point mutations has been observed in aged human fibroblasts which suggest that there are different types of mtDNA damage that could contribute to skin ageing [16]. Mitochondrial mutations and deletions have also been found to increase with ageing in other tissues. An accumulation of somatic mtDNA point mutations has been observed in the noncoding region of muscle tissue [17,18], and an increase in the frequency of a common 4977 bp deletion has been observed with increasing age in both breast and brain tissue [19,20]. This specific deletion has been found to increase in an age dependent manner in whole human skin [21], however more recent studies suggests that this increase only correlates with sun exposure [22] and particularly in the dermis [23]. The T414G transversion mutation in the mtDNA promoter region has a strong correlation with dermal fibroblast ageing, which is significantly increased in UV exposed regions [24]. This mutation has been found in tissues affected by age-related conditions such as in the brain of Alzheimer's patients [25], precancerous cells in the colon [26] and an age-dependent increase in this mutation has also been found in muscle tissue [27]. This demonstrates that mtDNA damage increases with age in a variety of tissues and UV stress can accelerate this damage in the skin. However, it has been difficult to determine whether these mutations are generated by or induce ROS production, which is the centre of many debates surrounding this theory [28,29].

UV-induced oxidative stress and its role in signalling during extrinsic ageing have been well documented in the skin [30], but oxidative stress can also have ageing effects elsewhere. Reducing calorie intake by 10 to 50% is believed to decrease metabolic stress and scavenge ROS. It has been shown to increase lifespan in many organisms, including yeast [31] and mice [32], and decrease ageing biomarkers in nonhuman primates [33], making it a good candidate as a model to test Harman's theory. Altogether, tissue-specific increase in mitochondrial efficiency and the oxidative stress response [34] and consequential reduction in ROS and oxidative stress have been shown following CR; and while not all CR study models increased in lifespan, there is evidence to suggest improved ageing health [35]. The focus of these studies is primarily lifespan and effects on organs other than the skin. However, a report on CR in Rhesus monkeys presented with subjective increase in hair loss in the ad libitum control compared to the CR monkey [36], therefore CR and the resulting metabolic changes could also have an impact on the skin.

Skin changes with CR were further investigated by Forni et al. [37] in mice. Caloric restriction resulted in increased mitochondrial function in the dermis, but not the epidermis, and fur remodelling and impaired vasoconstriction relating to thermoregulation. This implicates mitochondria as energy providers for cold adaptation in the skin, which agrees with previous observations of mitochondrial uncoupling as a preventative mechanism to cold stress [38]. In terms of skin ageing, CR promotes epidermal thickening and increases hair follicle stem cell pool which could indicate a rejuvenation of the skin by preventing skin thinning and hair loss indicative of skin ageing. This is coupled by evidence that mice without mitochondrial matrix antioxidant superoxide dismutase SOD2, which converts superoxide anon to hydrogen peroxide for further metabolism, exhibit normal ageing phenotypes at a younger age, particularly cellular senescence in the skin [39]. Therefore, the reduced capacity to break down superoxide anion results in higher oxidative stress and accelerated skin ageing in this model.

Another important discovery in mice looked at the induction of mtDNA depletion by an amino acid substitution in POLG1. Mice with this mutation showed a significant reduction in all OXPHOS complex activities in the skin and phenotypical ageing symptoms such as hair greying and loss, curvature of the spine, reduced movement and wrinkling of the skin [40]. The skin and hair changes are attributed to deformities of the hair follicle, increased epidermal cell proliferation causing deep wrinkles and epidermal thickening. Skin is not often commented on when addressing symptoms of POLG mutations in humans, so a satisfactory comparison cannot yet be made.

So far, research has demonstrated an increase of mtDNA mutations in ageing and a link between lower metabolic stress and increased ageing health, but does not show cause and effect. The problem with these models is two-fold. Firstly, is the amount of mutant mtDNA present in the cell enough to cause a pathogenic effect and secondly, do the observed age-related mutations dysregulate the respiratory chain? This can be partially answered with an observed decrease in mitochondrial function and metabolism in ageing muscle tissue in parallel with an increasing mutation load with ageing [41,42]. This does not rule out other mechanisms of mitochondrial dysfunction unrelated to the observed mutations [43], nor does it prove that ROS is a confounding factor in the pathogenesis.

5. Pigmentation

Melanin pigment is formed in response to oxidation reactions in melanocytes. Pheomelanin is the yellow-red pigment associated with red head and freckles and eumelanin which is more prevalent in dark haired individuals, the ratio of these is regulated by the melanocortin 1 receptor (MCR1) gene which has been previously reviewed [44]. Pheomelanin has been shown to have higher prooxidant effects on the cell as it sequesters cysteine and glutathione antioxidant during synthesis, and the loss-of-function in MCR1 results in the inability to produce eumelanin in response to α-melanocyte-stimulating-hormone (α-MSH), the process of which produces some antioxidant properties [45]. In terms of skin ageing, loss of MCR1 function has been linked to increased perceived age by an average of two years and heterozygous variants without complete loss-of-function had an increased perceived age of one year on average [46]. It has been suggested that this could be the effect of oxidative effects of pheomelanin production or changes to fibroblast function in the absence of MCR1 activity. In relation to mitochondrial function, melanocytes pretreated with α-MSH have been shown to have a protective effect on mtDNA copy number in response to UVB light. This could infer that stimulation of eumelanin rather than pheomelanin could have photoprotective properties; however, no quantification pheomelanin/eumelanin was performed [47]. Links between mitochondrial function, mtDNA and pheomelanin/eumelanin ratio have not yet been formally researched, which could provide an insight into the mechanisms involved in skin ageing though oxidative stress.

Uneven pigmentation is one of the hallmarks of skin ageing. Mitochondria have even been implicated in the biosynthesis of melanin in melanocytes, required to create pigmentation in response to UV light. Prohibitin proteins found localised in the inner mitochondrial membrane were found to bind directly to melanogenin, a synthetic pigmentation promotor. Prohibitin silencing directly interfered with melanogenin activity and is thought to be involved in the regulation of rate limiting melanin enzyme tyrosinase [48]. In addition to this, mitochondria in melanocytes have been shown to interact with melanosomes, suggesting a functional role in melanosome biogenesis and therefore melanin production [49].

Melatonin is a hormone predominantly produced in the brain to regulate sleep, and its metabolism relies heavily on mitochondria and ROS signalling [50]. It exhibits antioxidant effects and can inhibit melanogenesis in animal models at high concentrations, which can help modulate coat colour [51]. Studies observing changes in human skin pigmentation with intake of oral melatonin found no effects over a 30-day period in patients with hyperpigmentation of varying causes [52], melanoma patients or a control group [53]. However, melatonin and its metabolites have been detected in human skin in vivo, and all were found to inhibit melanocyte proliferation and tyrosinase activity in vitro [54]. All of the above suggests that mitochondrial function is a modulator of skin pigmentation, and also that dysfunction could interfere with melanin production both directly and indirectly through excessive ROS signalling and melatonin production.

These findings correspond with the observation from individuals with vitiligo—a condition which results in areas of the skin with inactive melanocytes—of reduced energy production in cultured melanocytes with depleted skin pigmentation, compared to healthy melanocytes [55], and the increased presence of vitiligo in patients with genetic mitochondrial dysfunction [56].

6. Photoageing

Sun damage is a well-known cause of skin cancer and ageing, and photoageing is the process of chronic sun exposure leading to extrinsic skin ageing. Solar light is composed of ultraviolet radiation (UVR) (10–380 nm), visible light (380–780 nm) and infrared radiation (IR) (above 780 nm). UVR is the most notorious for causing skin damage, so protection against both UVA and UVB is found in most sunscreens. UVB radiation was once believed to be the only contributor to photoageing due to higher energy levels than UVA, but it is now proven that UVA is the key player, though each can affect skin in different ways. UVA makes up the majority of solar UVR but only affects DNA indirectly, whereas UVB radiation represents a small portion but causes direct DNA damage [57].

The epidermis is the first line of defence against UVB damage and absorbs the majority of the radiation. The level of which depends on ethnicity, site of skin, hydration and many other factors [58]. Pigmentation from melanin plays a large part in the initial protection against UVR [59], and higher melanin has an inverse correlation with DNA lesions in humans [60] and this is also seen in other species, such as whales [61]. UVA, visible and infrared light can penetrate deeper into the skin than UVB (Figure 1), and it has been shown that dermal fibroblasts are in fact more susceptible to DNA damage from longer wavelengths of light (>300 nm) [62]. There is also evidence of visible and infrared light induced skin damage, and even a synergistic effect of all wavelengths [63].

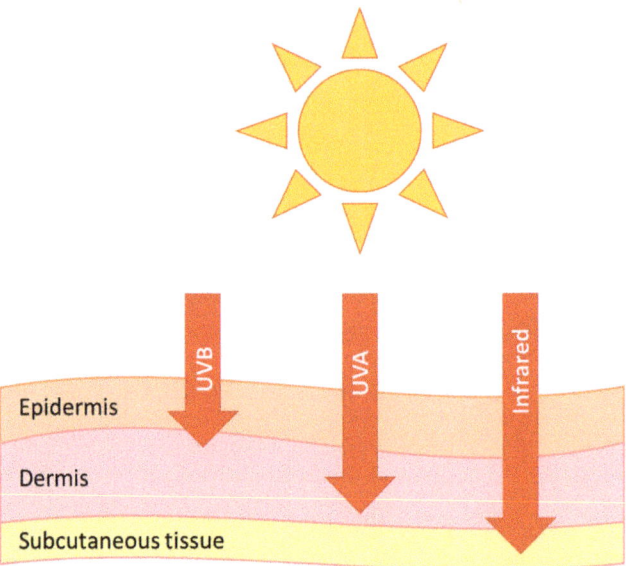

Figure 1. Penetration of UVB, UVA and infrared radiation into the skin.

Ageing is a natural process, and even without UV insult would occur over time due to the gradual shortening of telomere caps in nuclear DNA, loss-of-functionality in ageing and consequent cellular senescence [64]. This ageing damage profile has been found in a number of age related diseases including dementia and atherosclerosis [65,66]. It can also be induced in vitro in skin fibroblasts using UVA radiation [67] and in keratinocytes using UVB exposure [68], which further demonstrates a role for UVR in ageing. In addition to this, mtDNA damage is repeatedly seen in higher levels in photoexposed skin leading to an accumulation of cellular damage and a decrease in mitochondrial activity in the skin [24,69–71].

Although the precise mechanisms of photoageing are still being researched, collagen, mtDNA damage and increased ROS production are all key features. Inflammation induces the breakdown of matrix proteins to help the recruitment and migration of immune cells, therefore chronic inflammation can result in detrimental breakdown of tissues [72]. Inflammation due to UVA and UVB exposure has been shown to induce the enzymatic activity of MMPs in the dermis and epidermis, respectively, through ROS signalling and the activation of transcription factor AP-1 and NFκB [73–76]. AP-1 production reduces synthesis of procollagen type I and III, thereby disrupting the formation of new collagen at the same time as degrading it [77]. This is supported by the presence of ultrastructural changes and degradation of collagen fibre bundles from the dermis in photoexposed areas of human skin and UV irradiated mouse models [78]. Infrared radiation can also induce MMPs in dermal fibroblasts [79]. Therefore, sun exposure is directly linked to the active breakdown of collagen and the reduction of collagen synthesis through ROS signalling, leading to the formation of deep wrinkles and extrinsic skin ageing. Hypothetically, the mtDNA damage observed in photoexposed areas could cause dysregulation of OXPHOS and an increase in ROS production, as theorised in the 'Free Radical Theory of Ageing', and this could then lead to accelerated skin ageing.

The process of photodegradation of melanin results in singlet oxygen production, which has been shown in synthetic models [80], and UVA irradiation of pheomelanin exhibited conformational changes in catalase antioxidant which was attenuated with a singlet oxygen quencher [81]. In humans, a six-fold increase in photoageing was observed in those with homozygous loss of MC1R gene variants compared to those with wild type gene [82]. This all suggests that photodegradation of melanin could also be an important mechanism in the oxidative stress and mtDNA damage exhibited in photoexposed skin.

7. Pollution

With climate change a major danger to human health, research into the effects of air pollution has received growing attention. Outdoor pollutants predominantly originate from vehicle emissions, combustion of fossil fuels and industrial processes. Small particulate matter ($PM_{2.5}$) and ozone has been shown to increase DNA and protein damage through ROS and oxidative stress in vitro and in mouse models [83,84], and have a positive association with skin ageing though pigmentation spots and skin wrinkling in Chinese [85] and German populations [86]. Activation of the aryl hydrocarbon receptor (AhR) is key in modulating the effects of pollution. Its activation by ligands, including ozone and $PM_{2.5}$ [87,88], increases the expression of cytochrome P450, which can metabolise polycyclic aromatic hydrocarbons into carcinogenic substances capable of inducing DNA damage. These mechanisms have been further reviewed elsewhere [12,89,90].

Lentigines are hyperpigmented lesions in the skin as a consequence of excess melanocytes, which often become more abundant during ageing. Solar lentigines are associated with the level of sun exposure of the skin [91], and are the product of excess melanin production. Traffic-associated particles and soot have also been associated with the formation of these pigment spots on the forehead and cheeks [86,92], suggesting a common mechanism between UVR and pollution stressors. It has been suggested that one factor in this could be due to the age-dependent decline of mitochondrial complex II: succinate dehydrogenase [93]. Complex II is entirely transcribed by nuclear DNA and has been shown to decrease it's activity in comparison to complex I and IV in an age dependent manner in senescent cells, predominantly in the epidermis [69]. Inducing hyperpigmentation in melanoma cells results in the inhibition of complex II activity and an initial increase in superoxide ROS formation, and increasing melanin concentration was capable of attenuating further formation [94]. Therefore, lentigine formation could be a protective mechanism against the increase ROS production from senescent skin cells, due to a decline in complex II function.

8. Hair Follicles

Hair loss and greying is another key observable ageing phenotype. Unfortunately, this is one ageing phenotype which is poorly understood and as yet cannot be prevented. Hair follicles are situated within the dermis, and melanocytes and keratinocytes are responsible for the colour and growth of hair; it is therefore important to mention hair within the context of skin and ageing.

One theory of hair greying involves the acquisition of mitochondrial DNA damage with increased oxidative stress; the 'common' 4977 bp deletion has been discovered in higher levels in greying and unpigmented hair follicles (40% and 20%, respectively) than in pigmented (5%) [95]. Greying hair can be characterised by the gradual loss of melanocytes in the pigmentary units by apoptosis, until the hair follicle is unpigmented [96]. This pigmentary unit in greying hair was found to have the highest level of oxidative stress in melanocytes and a loss of ROS scavenging Bcl-2 activity. Bcl-2 is an inhibitor of the mitochondrial pathway of apoptosis, so the diminished expression can explain the increased apoptosis of melanocytes in response to cell stressors. These findings are in line with the grey hair phenotype of Bcl2−/− mice [97]. This details the mechanisms associated with mitochondrial function;

but all of three theorised mechanisms, as reviewed by SK Jo et al. (2018) [98], point to a role of oxidative or genotoxic stress induced damage, which supports mitochondrial involvement.

Hair loss can occur though ageing, nutrient deficiency and temporary or permanent alopecia, all of which present differently. Like hair greying, hair loss is little understood. Mice with PolgA mutations leading to a faulty mtDNA repair mechanism, have an increased mtDNA mutation abundance and an accelerated ageing phenotype which exhibits a reduction in hair density [99]. Age-related hair loss in humans is predominantly androgenic with a higher prevalence in men [100]. Dermal papilla cells from balding regions have been found to express higher levels of senescence markers and have antioxidant superoxide dismutase expression localised in the nuclear region. This is in contrast to the cytoplasmic or mitochondrial localisation found in nonbalding counterparts [101]. This translocation of superoxide dismutase in response to oxidative stress in the cell is a protective mechanism as it is suggested to promote oxidative resistance and repair and function as a transcription factor [22].

Mice with mitochondrial transcription factor A (TFAM) knockdown, which causes mtDNA depletion and loss of electron transport chain (ETC) complexes, in the epidermis and hair follicle epithelium show a reduction in proliferation, increase in apoptosis and exhibit abnormalities in melanin production and function in the hair follicles [102]. ROS signalling from mitochondrial function is required for hair growth, as shown by high mitochondrial membrane potential and a burst of ROS production at the matrix epidermal cell and hair shaft interface in growing cultured bovine and human hair follicles [103]. These epithelial cells subsequently underwent mitochondrial de-polarisation as they converted to hair shaft matrix. Both these studies demonstrate critical mitochondrial function and ROS signalling in the morphogenesis of hair follicles and hair shaft elongation, and the importance of functional ETC in these processes.

Although there is a limited understanding of hair ageing, there is evidence to support mitochondrial involvement in both, predominantly through the increase of oxidative stress and a reduced capacity to cope.

9. Summary

Mitochondria are important for skin function and mtDNA mutations; functional decline is linked to skin ageing through stress-induced wrinkle formation, pigmentation and hair greying and loss. Many of the mechanisms contributing to skin ageing are not fully understood, and it is likely that mitochondrial function is part of a complex set of processes that lead to tissue functional decline and ageing. It is important scientifically to understand the mechanisms behind skin ageing to clinically develop prevention strategies, but it is also useful in industry to improve antiageing formulas and those targeted to mature skin.

Funding: This research was funded by the NIHR Newcastle Biomedical Research Centre awarded to the Newcastle upon Tyne Hospitals NHS Foundation Trust and Newcastle University.

Conflicts of Interest: The authors declare no conflict of interest.

Abbreviations

UVA	Ultraviolet A
UVB	Ultraviolet B
IR	Infrared
MMPs	Matrix metalloproteinases
ROS	Reactive oxygen species
AP-1	activator protein-1
NFκB	Nuclear factor kappa-light-chain-enhancer of activated B cells
TGF-β	Transforming growth factor beta
$PM_{2.5}$	Small particulate matter
AhR	Aryl hydrocarbon receptor
Bcl-2	B-cell lymphoma 2
ETC	Electron transport chain

References

1. Thingnes, J.; Lavelle, T.J.; Hovig, E.; Omholt, S.W. Understanding the Melanocyte Distribution in Human Epidermis: An Agent-Based Computational Model Approach. *PLoS ONE* **2012**, *7*, e40377. [CrossRef]
2. Koster, M.I. Making an epidermis. *Ann. N. Y. Acad. Sci.* **2009**, *1170*, 7–10. [CrossRef] [PubMed]
3. Poon, F.; Kang, S.; Chien, A.L. Mechanisms and treatments of photoaging. *Photodermatol. Photoimmunol. Photomed.* **2015**, *31*, 65–74. [CrossRef]
4. Wickersham, M.; Wachtel, S.; Wong Fok Lung, T.; Soong, G.; Jacquet, R.; Richardson, A.; Parker, D.; Prince, A. Metabolic Stress Drives Keratinocyte Defenses against Staphylococcus aureus Infection. *Cell Rep.* **2017**, *18*, 2742–2751. [CrossRef] [PubMed]
5. Hamanaka, R.B.; Chandel, N.S. Mitochondrial reactive oxygen species regulate hypoxic signaling. *Curr. Opin. Cell Biol.* **2009**, *21*, 894–899. [CrossRef]
6. Hamanaka, R.B.; Chandel, N.S. Mitochondrial metabolism as a regulator of keratinocyte differentiation. *Cell. Logist.* **2013**, *3*, e25456. [CrossRef]
7. Feichtinger, R.G.; Sperl, W.; Bauer, J.W.; Kofler, B. Mitochondrial dysfunction: A neglected component of skin diseases. *Exp. Dermatol.* **2014**, *23*, 607–614. [CrossRef] [PubMed]
8. Birch-Machin, M.A. Mitochondria and skin disease. *Clin. Exp. Dermatol.* **2000**, *25*, 141–146. [CrossRef]
9. Plummer, C.; Spring, P.J.; Marotta, R.; Chin, J.; Taylor, G.; Sharpe, D.; Athanasou, N.A.; Thyagarajan, D.; Berkovic, S.F. Multiple Symmetrical Lipomatosis—A mitochondrial disorder of brown fat. *Mitochondrion* **2013**, *13*, 269–276. [CrossRef]
10. Boulton, S.J.; Bowman, A.; Koohgoli, R.; Birch-Machin, M.A. Skin manifestations of mitochondrial dysfunction: More important than previously thought. *Exp. Dermatol.* **2015**, *24*, 12–13. [CrossRef]
11. Farage, M.A.; Miller, K.W.; Elsner, P.; Maibach, H.I. Intrinsic and extrinsic factors in skin ageing: A review. *Int. J. Cosmet. Sci.* **2008**, *30*, 87–95. [CrossRef]
12. Naidoo, K.; Birch-Machin, M. Oxidative Stress and Ageing: The Influence of Environmental Pollution, Sunlight and Diet on Skin. *Cosmetics* **2017**, *4*, 4. [CrossRef]
13. Wey, S.-J.; Chen, D.-Y. Common cutaneous disorders in the elderly. *J. Clin. Gerontol. Geriatr.* **2010**, *1*, 36–41. [CrossRef]
14. Harman, D. Aging: A Theory Based on Free Radical and Radiation Chemistry. *J. Gerontol.* **1956**, *11*, 298–300. [CrossRef] [PubMed]
15. Ray, A.J.; Turner, R.; Nikaido, O.; Rees, J.L.; Birch-Machin, M.A. The Spectrum of Mitochondrial DNA Deletions is a Ubiquitous Marker of Ultraviolet Radiation Exposure in Human Skin. *J. Investig. Dermatol.* **2000**, *115*, 674–679. [CrossRef] [PubMed]
16. Michikawa, Y.; Mazzucchelli, F.; Bresolin, N.; Scarlato, G.; Attardi, G. Aging-Dependent Large Accumulation of Point Mutations in the Human mtDNA Control Region for Replication. *Science (New York N. Y.)* **1999**, *286*, 774–779. [CrossRef]
17. Wang, Y.; Michikawa, Y.; Mallidis, C.; Bai, Y.; Woodhouse, L.; Yarasheski, K.E.; Miller, C.A.; Askanas, V.; Engel, W.K.; Bhasin, S.; et al. Muscle-specific mutations accumulate with aging in critical human mtDNA control sites for replication. *Proc. Natl. Acad. Sci. USA* **2001**, *98*, 4022–4027. [CrossRef] [PubMed]
18. Del Bo, R.; Crimi, M.; Sciacco, M.; Malferrari, G.; Bordoni, A.; Napoli, L.; Prelle, A.; Biunno, I.; Moggio, M.; Bresolin, N.; et al. High mutational burden in the mtDNA control region from aged muscles: A single-fiber study. *Neurobiol. Aging* **2003**, *24*, 829–838. [CrossRef]
19. Pavicic, W.H.; Richard, S.M. Correlation analysis between mtDNA 4977-bp deletion and ageing. *Mutat. Res./Fundam. Mol. Mech. Mutagen.* **2009**, *670*, 99–102. [CrossRef] [PubMed]
20. Corral-Debrinski, M.; Horton, T.; Lott, M.T.; Shoffner, J.M.; Flint Beal, M.; Wallace, D.C. Mitochondrial DNA deletions in human brain: Regional variability and increase with advanced age. *Nat. Genet.* **1992**, *2*, 324. [CrossRef]
21. Yang, J.H.; Lee, H.C.; Lin, K.J.; Wei, Y.H. A specific 4977-bp deletion of mitochondrial DNA in human ageing skin. *Arch. Dermatol. Res.* **1994**, *286*, 386–390. [CrossRef] [PubMed]
22. Powers, J.M.; Murphy, G.; Ralph, N.; O'Gorman, S.M.; Murphy, J.E.J. Mitochondrial DNA deletion percentage in sun exposed and non sun exposed skin. *J. Photochem. Photobiol. B Biol.* **2016**, *165*, 277–282. [CrossRef] [PubMed]

23. Eshaghian, A.; Vleugels, R.A.; Canter, J.A.; McDonald, M.A.; Stasko, T.; Sligh, J.E. Mitochondrial DNA Deletions Serve as Biomarkers of Aging in the Skin, but Are Typically Absent in Nonmelanoma Skin Cancers. *J. Investig. Dermatol.* **2006**, *126*, 336–344. [CrossRef] [PubMed]
24. Birket, M.J.; Birch-Machin, M.A. Ultraviolet radiation exposure accelerates the accumulation of the aging-dependent T414G mitochondrial DNA mutation in human skin. *Aging Cell* **2007**, *6*, 557–564. [CrossRef]
25. Coskun, P.E.; Beal, M.F.; Wallace, D.C. Alzheimer's brains harbor somatic mtDNA control-region mutations that suppress mitochondrial transcription and replication. *Proc. Natl. Acad. Sci. USA* **2004**, *101*, 10726–10731. [CrossRef] [PubMed]
26. Kassem, A.M.; El-Guendy, N.; Tantawy, M.; Abdelhady, H.; El-Ghor, A.; Wahab, A.H.A. Mutational Hotspots in the Mitochondrial D-Loop Region of Cancerous and Precancerous Colorectal Lesions in Egyptian Patients. *DNA Cell Biol.* **2011**, *30*, 899–906. [CrossRef]
27. Murdock, D.G.; Christacos, N.C.; Wallace, D.C. The age-related accumulation of a mitochondrial DNA control region mutation in muscle, but not brain, detected by a sensitive PNA-directed PCR clamping based method. *Nucleic Acids Res.* **2000**, *28*, 4350–4355. [CrossRef]
28. Zhang, R.; Wang, Y.; Ye, K.; Picard, M.; Gu, Z. Independent impacts of aging on mitochondrial DNA quantity and quality in humans. *BMC Genom.* **2017**, *18*, 890. [CrossRef]
29. Payne, B.A.I.; Chinnery, P.F. Mitochondrial dysfunction in aging: Much progress but many unresolved questions. *Biochim. Biophys. Acta* **2015**, *1847*, 1347–1353. [CrossRef]
30. Rinnerthaler, M.; Bischof, J.; Streubel, M.K.; Trost, A.; Richter, K. Oxidative stress in aging human skin. *Biomolecules* **2015**, *5*, 545–589. [CrossRef]
31. Lin, S.-J.; Kaeberlein, M.; Andalis, A.A.; Sturtz, L.A.; Defossez, P.-A.; Culotta, V.C.; Fink, G.R.; Guarente, L. Calorie restriction extends Saccharomyces cerevisiae lifespan by increasing respiration. *Nature* **2002**, *418*, 344. [CrossRef]
32. Bartke, A.; Wright, J.C.; Mattison, J.A.; Ingram, D.K.; Miller, R.A.; Roth, G.S. Extending the lifespan of long-lived mice. *Nature* **2001**, *414*, 412. [CrossRef]
33. Lane, M.A.; Ingram, D.K.; Ball, S.S.; Roth, G.S. Dehydroepiandrosterone Sulfate: A Biomarker of Primate Aging Slowed by Calorie Restriction. *J. Clin. Endocrinol. Metab.* **1997**, *82*, 2093–2096. [CrossRef]
34. Lanza, I.R.; Zabielski, P.; Klaus, K.A.; Morse, D.M.; Heppelmann, C.J.; Bergen, H.R., 3rd; Dasari, S.; Walrand, S.; Short, K.R.; Johnson, M.L.; et al. Chronic caloric restriction preserves mitochondrial function in senescence without increasing mitochondrial biogenesis. *Cell Metab.* **2012**, *16*, 777–788. [CrossRef]
35. Mattison, J.A.; Roth, G.S.; Beasley, T.M.; Tilmont, E.M.; Handy, A.M.; Herbert, R.L.; Longo, D.L.; Allison, D.B.; Young, J.E.; Bryant, M.; et al. Impact of caloric restriction on health and survival in rhesus monkeys from the NIA study. *Nature* **2012**, *489*, 318. [CrossRef]
36. Colman, R.J.; Anderson, R.M.; Johnson, S.C.; Kastman, E.K.; Kosmatka, K.J.; Beasley, T.M.; Allison, D.B.; Cruzen, C.; Simmons, H.A.; Kemnitz, J.W.; et al. Caloric Restriction Delays Disease Onset and Mortality in Rhesus Monkeys. *Science (New York N. Y.)* **2009**, *325*, 201–204. [CrossRef]
37. Forni, M.F.; Peloggia, J.; Braga, T.T.; Chinchilla, J.E.O.; Shinohara, J.; Navas, C.A.; Camara, N.O.S.; Kowaltowski, A.J. Caloric Restriction Promotes Structural and Metabolic Changes in the Skin. *Cell Rep.* **2017**, *20*, 2678–2692. [CrossRef]
38. Stier, A.; Bize, P.; Habold, C.; Bouillaud, F.; Massemin, S.; Criscuolo, F. Mitochondrial uncoupling prevents cold-induced oxidative stress: A case study using UCP1 knockout mice. *J. Exp. Biol.* **2014**, *217*, 624–630. [CrossRef]
39. Weyemi, U.; Parekh, P.R.; Redon, C.E.; Bonner, W.M. SOD2 deficiency promotes aging phenotypes in mouse skin. *Aging* **2012**, *4*, 116–118. [CrossRef]
40. Singh, B.; Schoeb, T.R.; Bajpai, P.; Slominski, A.; Singh, K.K. Reversing wrinkled skin and hair loss in mice by restoring mitochondrial function. *Cell Death Dis.* **2018**, *9*, 735. [CrossRef]
41. McCully, K.K.; Fielding, R.A.; Evans, W.J.; Leigh, J.S., Jr.; Posner, J.D. Relationships between in vivo and in vitro measurements of metabolism in young and old human calf muscles. *J. Appl. Physiol. (Bethesda Md. 1985)* **1993**, *75*, 813–819. [CrossRef]
42. Short, K.R.; Bigelow, M.L.; Kahl, J.; Singh, R.; Coenen-Schimke, J.; Raghavakaimal, S.; Nair, K.S. Decline in skeletal muscle mitochondrial function with aging in humans. *Proc. Natl. Acad. Sci. USA* **2005**, *102*, 5618–5623. [CrossRef]

43. Niyazov, D.M.; Kahler, S.G.; Frye, R.E. Primary Mitochondrial Disease and Secondary Mitochondrial Dysfunction: Importance of Distinction for Diagnosis and Treatment. *Mol. Syndromol.* **2016**, *7*, 122–137. [CrossRef]
44. Denat, L.; Kadekaro, A.L.; Marrot, L.; Leachman, S.A.; Abdel-Malek, Z.A. Melanocytes as instigators and victims of oxidative stress. *J. Investig. Dermatol.* **2014**, *134*, 1512–1518. [CrossRef]
45. Song, X.; Mosby, N.; Yang, J.; Xu, A.; Abdel-Malek, Z.; Kadekaro, A.L. alpha-MSH activates immediate defense responses to UV-induced oxidative stress in human melanocytes. *Pigment Cell Melanoma Res.* **2009**, *22*, 809–818. [CrossRef]
46. Liu, F.; Hamer, M.A.; Deelen, J.; Lall, J.S.; Jacobs, L.; van Heemst, D.; Murray, P.G.; Wollstein, A.; de Craen, A.J.; Uh, H.W.; et al. The MC1R Gene and Youthful Looks. *Curr. Biol. CB* **2016**, *26*, 1213–1220. [CrossRef]
47. Böhm, M.; Hill, H.Z. Ultraviolet B, melanin and mitochondrial DNA: Photo-damage in human epidermal keratinocytes and melanocytes modulated by alpha-melanocyte-stimulating hormone. *F1000 Res.* **2016**, *5*, 881. [CrossRef]
48. Snyder, J.R.; Hall, A.; Ni-Komatsu, L.; Khersonsky, S.M.; Chang, Y.-T.; Orlow, S.J. Dissection of Melanogenesis with Small Molecules Identifies Prohibitin as a Regulator. *Chem. Biol.* **2005**, *12*, 477–484. [CrossRef]
49. Daniele, T.; Hurbain, I.; Vago, R.; Casari, G.; Raposo, G.; Tacchetti, C.; Schiaffino, M.V. Mitochondria and melanosomes establish physical contacts modulated by Mfn2 and involved in organelle biogenesis. *Curr. Biol. CB* **2014**, *24*, 393–403. [CrossRef]
50. Slominski, A.T.; Semak, I.; Fischer, T.W.; Kim, T.-K.; Kleszczyński, K.; Hardeland, R.; Reiter, R.J. Metabolism of melatonin in the skin: Why is it important? *Exp. Dermatol.* **2017**, *26*, 563–568. [CrossRef]
51. Slominski, A.; Pruski, D. Melatonin inhibits proliferation and melanogenesis in rodent melanoma cells. *Exp. Cell Res.* **1993**, *206*, 189–194. [CrossRef]
52. Lerner, A.B.; Nordlund, J.J. The Effects of Oral Melatonin on Skin Color and on the Release of Pituitary Hormones. *J. Clin. Endocrinol. Metab.* **1977**, *45*, 768–774. [CrossRef]
53. McElhinney, D.B.; Hoffman, S.J.; Robinson, W.A.; Ferguson, J.A.N. Effect of Melatonin on Human Skin Color. *J. Investig. Dermatol.* **1994**, *102*, 258–259. [CrossRef]
54. Kim, T.-K.; Lin, Z.; Tidwell, W.J.; Li, W.; Slominski, A.T. Melatonin and its metabolites accumulate in the human epidermis in vivo and inhibit proliferation and tyrosinase activity in epidermal melanocytes in vitro. *Mol. Cell. Endocrinol.* **2015**, *404*, 1–8. [CrossRef]
55. Dell'Anna, M.L.; Ottaviani, M.; Kovacs, D.; Mirabilii, S.; Brown, D.A.; Cota, C.; Migliano, E.; Bastonini, E.; Bellei, B.; Cardinali, G.; et al. Energetic mitochondrial failing in vitiligo and possible rescue by cardiolipin. *Sci. Rep.* **2017**, *7*, 13663. [CrossRef]
56. Karvonen, S.L.; Haapasaari, K.M.; Kallioinen, M.; Oikarinen, A.; Hassinen, I.E.; Majamaa, K. Increased prevalence of vitiligo, but no evidence of premature ageing, in the skin of patients with bp 3243 mutation in mitochondrial DNA in the mitochondrial encephalomyopathy, lactic acidosis and stroke-like episodes syndrome (MELAS). *Br. J. Dermatol.* **1999**, *140*, 634–639. [CrossRef]
57. Rünger, T.M.; Epe, B.; Möller, K. Processing of Directly and Indirectly Ultraviolet-Induced DNA Damage in Human Cells. *Recent Results Cancer Res.* **1995**, *139*, 31–42.
58. Sklar, L.R.; Almutawa, F.; Lim, H.W.; Hamzavi, I. Effects of ultraviolet radiation, visible light, and infrared radiation on erythema and pigmentation: A review. *Photochem. Photobiol. Sci.* **2013**, *12*, 54–64. [CrossRef]
59. Swalwell, H.; Latimer, J.; Haywood, R.M.; Birch-Machin, M.A. Investigating the role of melanin in UVA/UVB- and hydrogen peroxide-induced cellular and mitochondrial ROS production and mitochondrial DNA damage in human melanoma cells. *Free Radic. Biol. Med.* **2012**, *52*, 626–634. [CrossRef]
60. Wenczl, E.; van der Schans, G.P.; Roza, L.; Kolb, R.; Smit, N.; Schothorst, A.A. UVA-induced oxidative DNA damage in human melanocytes is related to their (pheo)melanin content. *J. Dermatol. Sci.* **1998**, *16*, S221. [CrossRef]
61. Martinez-Levasseur, L.M.; Birch-Machin, M.A.; Bowman, A.; Gendron, D.; Weatherhead, E.; Knell, R.J.; Acevedo-Whitehouse, K. Whales use distinct strategies to counteract solar ultraviolet radiation. *Sci. Rep.* **2013**, *3*, 2386. [CrossRef]
62. Latimer, J.A.; Lloyd, J.J.; Diffey, B.L.; Matts, P.J.; Birch-Machin, M.A. Determination of the Action Spectrum of UVR-Induced Mitochondrial DNA Damage in Human Skin Cells. *J. Investig. Dermatol.* **2015**, *135*, 2512–2518. [CrossRef]

63. Rashdan, E.; Wilkinson, S.; Birch-Machin, M. British Society for Investigative Dermatology Annual Meeting 2018 26–28 March 2018 Blizard Institute, Barts and The London School of Medicine and Dentistry. *Br. J. Dermatol.* **2018**. [CrossRef]
64. Harley, C.B.; Futcher, A.B.; Greider, C.W. Telomeres shorten during ageing of human fibroblasts. *Nature* **1990**, *345*, 458. [CrossRef]
65. Saretzki, G.; Zglinicki, T. Replicative Aging, Telomeres, and Oxidative Stress. *Ann. N. Y. Acad. Sci.* **2006**, *959*, 24–29. [CrossRef]
66. Samani, N.J.; Boultby, R.; Butler, R.; Thompson, J.R.; Goodall, A.H. Telomere shortening in atherosclerosis. *Lancet* **2001**, *358*, 472–473. [CrossRef]
67. Ma, H.M.; Liu, W.; Zhang, P.; Yuan, X.Y. Human skin fibroblast telomeres are shortened after ultraviolet irradiation. *J. Int. Med. Res.* **2012**, *40*, 1871–1877. [CrossRef]
68. Nistico, S.; Ehrlich, J.; Gliozzi, M.; Maiuolo, J.; Del Duca, E.; Muscoli, C.; Mollace, V. Telomere and telomerase modulation by bergamot polyphenolic fraction in experimental photoageing in human keratinocytes. *J. Biol. Regul. Homeost. Agents* **2015**, *29*, 723–728.
69. Bowman, A.; Birch-Machin, M.A. Age-Dependent Decrease of Mitochondrial Complex II Activity in Human Skin Fibroblasts. *J. Investig. Dermatol.* **2016**, *136*, 912–919. [CrossRef]
70. Krishnan, K.J.; Harbottle, A.; Birch-Machin, M.A. The Use of a 3895 bp Mitochondrial DNA Deletion as a Marker for Sunlight Exposure in Human Skin. *J. Investig. Dermatol.* **2004**, *123*, 1020–1024. [CrossRef]
71. Berneburg, M.; Gattermann, N.; Stege, H.; Grewe, M.; Vogelsang, K.; Ruzicka, T.; Krutmann, J. Chronically Ultraviolet-exposed Human Skin Shows a Higher Mutation Frequency of Mitochondrial DNA as Compared to Unexposed Skin and the Hematopoietic System. *Photochem. Photobiol.* **1997**, *66*, 271–275. [CrossRef]
72. Manicone, A.M.; McGuire, J.K. Matrix metalloproteinases as modulators of inflammation. *Semin. Cell Dev. Biol.* **2008**, *19*, 34–41. [CrossRef]
73. Vicentini, F.T.; He, T.; Shao, Y.; Fonseca, M.J.; Verri, W.A., Jr.; Fisher, G.J.; Xu, Y. Quercetin inhibits UV irradiation-induced inflammatory cytokine production in primary human keratinocytes by suppressing NF-kappaB pathway. *J. Dermatol. Sci.* **2011**, *61*, 162–168. [CrossRef]
74. Kang, S.; Chung, J.H.; Lee, J.H.; Fisher, G.J.; Wan, Y.S.; Duell, E.A.; Voorhees, J.J. Topical N-Acetyl Cysteine and Genistein Prevent Ultraviolet-Light-Induced Signaling That Leads to Photoaging in Human Skin in vivo. *J. Investig. Dermatol.* **2003**, *120*, 835–841. [CrossRef]
75. Tewari, A.; Grys, K.; Kollet, J.; Sarkany, R.; Young, A.R. Upregulation of MMP12 and its activity by UVA1 in human skin: Potential implications for photoaging. *J. Investig. Dermatol.* **2014**, *134*, 2598–2609. [CrossRef]
76. Chiang, H.M.; Chen, H.C.; Lin, T.J.; Shih, I.C.; Wen, K.C. Michelia alba extract attenuates UVB-induced expression of matrix metalloproteinases via MAP kinase pathway in human dermal fibroblasts. *Food Chem. Toxicol.* **2012**, *50*, 4260–4269. [CrossRef]
77. Fisher, G.J.; Datta, S.; Wang, Z.; Li, X.Y.; Quan, T.; Chung, J.H.; Kang, S.; Voorhees, J.J. c-Jun-dependent inhibition of cutaneous procollagen transcription following ultraviolet irradiation is reversed by all-trans retinoic acid. *J. Clin. Investig.* **2000**, *106*, 663–670. [CrossRef]
78. Edwards, C.; Pearse, A.; Marks, R.; Nishimori, Y.; Matsumoto, K.; Kawai, M. Degenerative Alterations of Dermal Collagen Fiber Bundles in Photodamaged Human Skin and UV-Irradiated Hairless Mouse Skin: Possible Effect on Decreasing Skin Mechanical Properties and Appearance of Wrinkles. *J. Investig. Dermatol.* **2001**, *117*, 1458–1463. [CrossRef]
79. Schieke, S.M.; Stege, H.; Kürten, V.; Grether-Beck, S.; Sies, H.; Krutmann, J. Infrared-A Radiation-Induced Matrix Metalloproteinase 1 Expression is Mediated Through Extracellular Signal-regulated Kinase 1/2 Activation in Human Dermal Fibroblasts. *J. Investig. Dermatol.* **2002**, *119*, 1323–1329. [CrossRef]
80. Szewczyk, G.; Zadlo, A.; Sarna, M.; Ito, S.; Wakamatsu, K.; Sarna, T. Aerobic photoreactivity of synthetic eumelanins and pheomelanins: Generation of singlet oxygen and superoxide anion. *Pigment Cell Melanoma Res.* **2016**, *29*, 669–678. [CrossRef]
81. Maresca, V.; Flori, E.; Briganti, S.; Camera, E.; Cario-André, M.; Taïeb, A.; Picardo, M. UVA-Induced Modification of Catalase Charge Properties in the Epidermis Is Correlated with the Skin Phototype. *J. Investig. Dermatol.* **2006**, *126*, 182–190. [CrossRef]
82. Elfakir, A.; Ezzedine, K.; Latreille, J.; Ambroisine, L.; Jdid, R.; Galan, P.; Hercberg, S.; Gruber, F.; Malvy, D.; Tschachler, E.; et al. Functional MC1R-Gene Variants Are Associated with Increased Risk for Severe Photoaging of Facial Skin. *J. Investig. Dermatol.* **2010**, *130*, 1107–1115. [CrossRef]

83. Piao, M.J.; Ahn, M.J.; Kang, K.A.; Ryu, Y.S.; Hyun, Y.J.; Shilnikova, K.; Zhen, A.X.; Jeong, J.W.; Choi, Y.H.; Kang, H.K.; et al. Particulate matter 2.5 damages skin cells by inducing oxidative stress, subcellular organelle dysfunction, and apoptosis. *Arch. Toxicol.* **2018**, *92*, 2077–2091. [CrossRef]
84. Valacchi, G.; van der Vliet, A.; Schock, B.C.; Okamoto, T.; Obermuller-Jevic, U.; Cross, C.E.; Packer, L. Ozone exposure activates oxidative stress responses in murine skin. *Toxicology* **2002**, *179*, 163–170. [CrossRef]
85. Ding, A.; Yang, Y.; Zhao, Z.; Hüls, A.; Vierkötter, A.; Yuan, Z.; Cai, J.; Zhang, J.; Gao, W.; Li, J.; et al. Indoor PM2.5 exposure affects skin aging manifestation in a Chinese population. *Sci. Rep.* **2017**, *7*, 15329. [CrossRef]
86. Vierkötter, A.; Schikowski, T.; Ranft, U.; Sugiri, D.; Matsui, M.; Krämer, U.; Krutmann, J. Airborne Particle Exposure and Extrinsic Skin Aging. *J. Investig. Dermatol.* **2010**, *130*, 2719–2726. [CrossRef]
87. Gualtieri, M.; Øvrevik, J.; Mollerup, S.; Asare, N.; Longhin, E.; Dahlman, H.-J.; Camatini, M.; Holme, J.A. Airborne urban particles (Milan winter-PM2.5) cause mitotic arrest and cell death: Effects on DNA, mitochondria, AhR binding and spindle organization. *Mutat. Res./Fundam. Mol. Mech. Mutagen.* **2011**, *713*, 18–31. [CrossRef]
88. Afaq, F.; Zaid, M.A.; Pelle, E.; Khan, N.; Syed, D.N.; Matsui, M.S.; Maes, D.; Mukhtar, H. Aryl Hydrocarbon Receptor Is an Ozone Sensor in Human Skin. *J. Investig. Dermatol.* **2009**, *129*, 2396–2403. [CrossRef]
89. Koohgoli, R.; Hudson, L.; Naidoo, K.; Wilkinson, S.; Chavan, B.; Birch-Machin, M.A. Bad air gets under your skin. *Exp. Dermatol.* **2017**, *26*, 384–387. [CrossRef]
90. Shimada, T.; Fujii-Kuriyama, Y. Metabolic activation of polycyclic aromatic hydrocarbons to carcinogens by cytochromes P450 1A1 and 1B1. *Cancer Sci.* **2004**, *95*, 1–6. [CrossRef]
91. Bastiaens, M.; Hoefnagel, J.; Westendorp, R.; Vermeer, B.-J.; Bouwes Bavinck, J.N. Solar Lentigines are Strongly Related to Sun Exposure in Contrast to Ephelides. *Pigment Cell Res.* **2004**, *17*, 225–229. [CrossRef]
92. Hüls, A.; Vierkötter, A.; Gao, W.; Krämer, U.; Yang, Y.; Ding, A.; Stolz, S.; Matsui, M.; Kan, H.; Wang, S.; et al. Traffic-Related Air Pollution Contributes to Development of Facial Lentigines: Further Epidemiological Evidence from Caucasians and Asians. *J. Investig. Dermatol.* **2016**, *136*, 1053–1056. [CrossRef]
93. Chang, A.L.S. Expanding Our Understanding of Human Skin Aging. *J. Investig. Dermatol.* **2016**, *136*, 897–899. [CrossRef] [PubMed]
94. Boulton, S.J.; Birch-Machin, M.A. Impact of hyperpigmentation on superoxide flux and melanoma cell metabolism at mitochondrial complex II. *FASEB J.* **2015**, *29*, 346–353. [CrossRef] [PubMed]
95. Arck, P.C.; Overall, R.; Spatz, K.; Liezman, C.; Handjiski, B.; Klapp, B.F.; Birch-Machin, M.A.; Peters, E.M.J. Towards a "free radical theory of graying": Melanocyte apoptosis in the aging human hair follicle is an indicator of oxidative stress induced tissue damage. *FASEB J.* **2006**, *20*, 1567–1569. [CrossRef]
96. Commo, S.; Gallard, O.; Bernard, B.A. Human hair greying is linked to a specific depletion of hair follicle melanocytes affecting both the bulb and the outer root sheath. *Br. J. Dermatol.* **2004**, *150*, 435–443.
97. Veis, D.J.; Sorenson, C.M.; Shutter, J.R.; Korsmeyer, S.J. Bcl-2-deficient mice demonstrate fulminant lymphoid apoptosis, polycystic kidneys, and hypopigmented hair. *Cell* **1993**, *75*, 229–240. [CrossRef]
98. Jo, S.K.; Lee, J.Y.; Lee, Y.; Kim, C.D.; Lee, J.-H.; Lee, Y.H. Three Streams for the Mechanism of Hair Graying. *Ann. Dermatol.* **2018**, *30*, 397–401. [CrossRef]
99. Trifunovic, A.; Wredenberg, A.; Falkenberg, M.; Spelbrink, J.N.; Rovio, A.T.; Bruder, C.E.; Bohlooly, -Y.M.; Gidlöf, S.; Oldfors, A.; Wibom, R.; et al. Premature ageing in mice expressing defective mitochondrial DNA polymerase. *Nature* **2004**, *429*, 417. [CrossRef]
100. Trüeb, R.M. Oxidative stress in ageing of hair. *Int. J. Trichol.* **2009**, *1*, 6–14. [CrossRef]
101. Bahta, A.W.; Farjo, N.; Farjo, B.; Philpott, M.P. Premature Senescence of Balding Dermal Papilla Cells In Vitro Is Associated with p16INK4a Expression. *J. Investig. Dermatol.* **2008**, *128*, 1088–1094. [CrossRef]
102. Kloepper, J.E.; Baris, O.R.; Reuter, K.; Kobayashi, K.; Weiland, D.; Vidali, S.; Tobin, D.J.; Niemann, C.; Wiesner, R.J.; Paus, R. Mitochondrial Function in Murine Skin Epithelium Is Crucial for Hair Follicle Morphogenesis and Epithelial–Mesenchymal Interactions. *J. Investig. Dermatol.* **2015**, *135*, 679–689. [CrossRef]
103. Lemasters, J.J.; Ramshesh, V.K.; Lovelace, G.L.; Lim, J.; Wright, G.D.; Harland, D.; Dawson, T.L. Compartmentation of Mitochondrial and Oxidative Metabolism in Growing Hair Follicles: A Ring of Fire. *J. Investig. Dermatol.* **2017**, *137*, 1434–1444. [CrossRef]

© 2019 by the authors. Licensee MDPI, Basel, Switzerland. This article is an open access article distributed under the terms and conditions of the Creative Commons Attribution (CC BY) license (http://creativecommons.org/licenses/by/4.0/).

Review

Vitamin D Deficiency: Effects on Oxidative Stress, Epigenetics, Gene Regulation, and Aging

Sunil J. Wimalawansa

Professor of Medicine, Endocrinology & Nutrition, Cardio Metabolic and Endocrine Institute, NJ 08873, USA; suniljw@hotmail.com

Received: 1 January 2019; Accepted: 18 March 2019; Published: 11 May 2019

Abstract: Recent advances in vitamin D research indicate that this vitamin, a secosteroid hormone, has beneficial effects on several body systems other than the musculoskeletal system. Both 25 dihydroxy vitamin D [$25(OH)_2D$] and its active hormonal form, 1,25-dihydroxyvitamin D [$1,25(OH)_2D$] are essential for human physiological functions, including damping down inflammation and the excessive intracellular oxidative stresses. Vitamin D is one of the key controllers of systemic inflammation, oxidative stress and mitochondrial respiratory function, and thus, the aging process in humans. In turn, molecular and cellular actions form $1,25(OH)_2D$ slow down oxidative stress, cell and tissue damage, and the aging process. On the other hand, hypovitaminosis D impairs mitochondrial functions, and enhances oxidative stress and systemic inflammation. The interaction of $1,25(OH)_2D$ with its intracellular receptors modulates vitamin D–dependent gene transcription and activation of vitamin D-responsive elements, which triggers multiple second messenger systems. Thus, it is not surprising that hypovitaminosis D increases the incidence and severity of several age-related common diseases, such as metabolic disorders that are linked to oxidative stress. These include obesity, insulin resistance, type 2 diabetes, hypertension, pregnancy complications, memory disorders, osteoporosis, autoimmune diseases, certain cancers, and systemic inflammatory diseases. Vitamin D adequacy leads to less oxidative stress and improves mitochondrial and endocrine functions, reducing the risks of disorders, such as autoimmunity, infections, metabolic derangements, and impairment of DNA repair; all of this aids a healthy, graceful aging process. Vitamin D is also a potent anti-oxidant that facilitates balanced mitochondrial activities, preventing oxidative stress-related protein oxidation, lipid peroxidation, and DNA damage. New understandings of vitamin D-related advances in metabolomics, transcriptomics, epigenetics, in relation to its ability to control oxidative stress in conjunction with micronutrients, vitamins, and antioxidants, following normalization of serum $25(OH)D$ and tissue $1,25(OH)_2D$ concentrations, likely to promise cost-effective better clinical outcomes in humans.

Keywords: $25(OH)D$; $1,25(OH)_2D$; aging; cytokines; inflammation; morbidity and mortality; prevention; reactive oxygen species; ultraviolet

1. Introduction

Vitamin D is a micronutrient that is metabolized into a multifunctional secosteroid hormone that is essential for human health. Globally, its deficiency is a major public health problem affecting all ages and ethnic groups; it has surpassed iron deficiency as the most common nutritional deficiency in the world. The increasing prevalence of vitamin D deficiency and its associated complications are prominent in countries furthest from the equator. However, incidence is also high among those who live within 1,000 km of the equator (e.g., in Sri Lanka, India, and Far Eastern, Middle Eastern, Central American, and Persian Gulf countries) because of a combination of climatic conditions, ethnic and cultural habits, and having darker skin color [1–4].

Most of the vitamin D the human body requires can be generated by an individual's exposure to summer-like sunlight, with dietary sources playing a supporting role when sunlight exposure is limited or ineffective for vitamin D production. Despite the above, more than 50% of the population in the group of countries mentioned has vitamin D deficiency [5,6]. If effective public health guidelines are implemented, vitamin D deficiency can be easily and cost-effectively treated and prevented, saving millions of dollars and lives. Although excessive sun exposure does not cause hypervitaminosis D, it can cause other harm because of dermal cell DNA damage [7–9]. Thus, guidelines for safe sun exposure are needed for each country.

During the past decade, many advances in the understanding of the physiology and biology of vitamin D and its receptor ecology have emerged [10]. Vitamin D metabolism and functions are modulated by many factors. Accumulating evidence supports biological associations of vitamin D with disease risk reduction and improved physical and mental functions. The field is rapidly advancing, including the knowledge of the physiology of vitamin D-vitamin D receptor (VDR) interactions and the biology and metabolism of vitamin D and their effects on vitamin D axis and gene polymorphisms [11]. Together these data have facilitated our understanding of new pathways to intervene to prevent and treat human diseases.

However, what is lacking is the adequately powered, conducted for sufficient duration, well-designed randomized controlled clinical studies (RCTs) conducted in an unbiased manner with the nutrient vitamin D as the key intervention and having predefined hard endpoints/primary outcomes [12]. Moreover, such studies must recruit persons with vitamin D deficiency [i.e., serum 25(OH)D concentrations less than 20 ng/mL (50 nmol/L)] and achieve a pre-determined target serum 25(OH)D concentration through daily oral administration and/or safe exposure to ultraviolet rays; not merely by relying on oral administered doses. The goal of this review is to explore the effects from vitamin D-modulated gene interactions and the effects of hypovitaminosis-induced, mitochondria-based oxidative stress on aging.

1.1. Extrarenal Generation of $1,25(OH)_2D$

The active form of vitamin D, 1,25-dihydroxyvitamin D [$1,25(OH)_2D$], is generated not only in renal tubular cells (endocrine functions as a hormone) but also in extrarenal target tissue cells, providing autocrine and paracrine functions. However, because it remains within the target tissue cells, the intra-cellular concentrations achieved are unclear. In addition, the catabolic activity of 24-hydroxylase in target tissues plays an important part in regulating both 25-hydroxy vitamin D [25(OH)D; calcidiol], and $1,25(OH)_2D$ (calcitriol) concentrations and their availability.

The amounts of $1,25(OH)_2D$ generated in renal tubules and target cells can vary from person to person and day to day and are hard to quantify. Although the calcitriol in the circulation is modulated by parathyroid hormone (PTH) and serum ionized calcium concentrations [13], the intracellular content is regulated largely through serum 25(OH)D availability and calcidiol and calcitriol catabolism through hydroxylation at the molecular positions C-24 and C-23 by a specific 24-hydroxylase (CYP24A1) [14].

1.2. Excess Sun Exposure Does Not Cause Hypervitaminosis D

After exposure to ultraviolet B (UVB) rays, dermal cells actively synthesize vitamin D. As a feedback mechanism, excess precursors produced are catabolized within the dermal cells by the same UVB rays. In addition, skin also contains the inactivating enzyme 24-hydroxylase, which also prevents over-production of vitamin D by 24-hydroxylation of vitamin D [15]. This process of homeostasis is regulated by UVB, PTH, and serum ionized calcium concentrations [16]. When an individual is overexposed to ultraviolet rays, the mentioned built-in protective mechanism prevents excessive retention of vitamin D in the skin. Therefore, sun exposure does not raise serum 25(OH)D to pathological levels, cause hypervitaminosis D or its complications, such as hypercalcemia.

In addition, vitamin D synthesized in the skin from UVB exposure in excess of need is catabolized in part through 20-hydroxylation by the cholesterol side chain cleavage enzyme CYP11A1 [17]. Thus, there are multiple intrinsic mechanisms present to prevent excess vitamin D from reaching the circulation. The efficiency of vitamin D synthesis in the skin is affected by many factors, including the density of melanin pigment, condition of the skin and age, and the use of sunscreen and UV-blocking makeup, creams, and ointments and clothing. In addition, older age or scarred skin, as well as time of day or year and the duration of sun exposure affect vitamin D synthesis in the skin [18–21].

Although solar radiation is the source of vitamin D generation in the skin, excessive exposure, particularly in those who are genetically vulnerable, may cause skin cancers [22–24]. On the other hand, optimal vitamin D status protects against several types of internal cancers, melanoma, and several other diseases. Therefore, one needs to balance sun exposure in favor of benefits while avoiding potential harmful effects [25–27].

2. Vitamin D and Gene Regulation

$1,25(OH)_2D$ regulates many genes within the human genome [28]. Tissue vitamin D concentrations and its receptor gene polymorphisms, not only influence mechanism of actions and modulate second messenger systems, but also modulate the ability of the ligand-bound VDR to bind to vitamin D response elements (VDREs) on promoter regions in genes and initiate second messenger systems [29]. The National Human Genome Research Institute (NHGRI) launched a public research consortium, ENCODE [Encyclopedia of DNA Elements; http://www.genome.gov/10005107] to answer pertinent questions [30].

ENCODE has demonstrated the genome-wide actions of $1,25(OH)_2D_3$ on the formation rates of proteins, such as the insulator protein CTCF (transcriptional repressor CTCF; 11-zinc finger protein, CCCTC-binding factor, etc.) and the VDR [31,32]. These findings suggest the presence of numerous functional VDRE regions across the human genome [29,33]. In addition, it has been suggested that the expression of these genes could be used as biomarkers for different actions of vitamin D in varied tissues and cells and assessment of vulnerabilities [34].

Hormone, $1,25(OH)_2D$ modulates cell proliferation through direct and indirect pathways. For example, vitamin D inhibits the pathways related to transcription factor NF-κB [35]. People with chronic non-communicable diseases, such as cardiovascular disease, type 2 diabetes, autoimmune diseases, arthritis, and osteoporosis are reported to have chronically elevated NF-κB [36]. NF-κB enhances the oxidative stress and cellular responses to inflammation and injury; including following head injury [37]. Whereas, $1,25(OH)_2D$ (calcitriol) suppresses NF-κB and thereby reduces chronic diffuse somatic inflammation [38,39]. In addition, calcitriol also reduce cell proliferation and enhance cell differentiation—key anti-cancer effects of vitamin D [40].

2.1. Epigenetic Mechanisms Influence Cancer Genesis

Epigenetic mechanisms influence cancer genesis, growth, dissemination, and aging phenomena [41–44]. For example, the epigenetic modifications of VDR–$1,25(OH)_2D$ effects can be mediated through complex processes involving CYP27A1 and CYP27B1 and via the vitamin D-catabolizing enzyme CYP24 [14,44]. These actions can be favorably influenced by modifications of VDREs across the genome modulated by both histone acetylases and deacetylases [28,29,45].

Epigenetic regulation of vitamin D metabolism influences several physiological mechanisms and modulate outcomes of some human diseases. Example of diseases include adenocarcinoma of the lung [44], specific gene mutations in Asians with advanced non-small cell lung cancer [41], and genetic alterations in the effectiveness of systemic therapy for lung cancer induced by cigarette smoking [42]. In severely obese children, low 25(OH)D concentrations are associated with increased markers of oxidative and nitrosative stress, inflammation, and endothelial over-activation [46].

CYP27B1-mediated target tissue production of $1,25(OH)D$ is critically important for the paracrine and autocrine functions of calcitriol to obtain the full biological potential of vitamin D. Taken together, the benefits of having adequate serum 25(OH)D concentrations and maintaining vitamin D repletion in the long run and considering the overall health benefits of vitamin D, there is an urgent need to create national policies to combat hypovitaminosis D. The savings derived from reducing the risks and severity of infectious and parasitic diseases alone would pay-off the cost of this public health approach. This can be achieved through targeting to raise the population serum 25(OH)D concentration, leading to a tangible positive impact on humans and on the economy.

2.2. Epigenetics and Molecular Genetics of Vitamin D

Molecular and genetic studies confirm that vitamin D also modulates risks of several other human diseases, including autoimmune disorders such as multiple sclerosis [47]. Although the predominant cause of cancer is modulation of the underlying metabolic abnormalities through genes, such as p53 and c-myc modifying the metastatic risks, the responsiveness to therapy is in part determined by epigenetic modifications of genes.

Tumor-related key metabolic abnormalities include imbalance between glucose fermentation and oxidative phosphorylation (under aerobic and anaerobic conditions—the Warburg effect); dysregulation of metabolic enzymes, such as pyruvate kinase, fumarate hydratase, and succinate dehydrogenase; isocitrate dehydrogenase mutations; and alterations of gene expression levels linked to tumorigenesis that are influenced by the vitamin D status [48].

Examples related to activity of the vitamin D axis include epigenetic changes that affect the expression of the CYP24A1 gene and VDR polymorphisms. Although epigenetic enhancement can occur through methylation and repression by histone-modifications of DNA, vitamin D markedly influences the regulation of cell replication [42,44]. This substantiates targeting of CYP24A1 to optimize the antiproliferative effects of $1,25(OH)_2D$ in a target-specific manner [49].

In addition, gene activation following the interaction of $1,25(OH)_2D$ with VDR is important for mitochondrial integrity and respiration, and many other physiological activities. Moreover, the vitamin D signaling pathway plays a central role in protecting cells from elevated mitochondrial respiration and associated damage and overproduction of reactive oxygen species (ROS), which can lead to cellular and DNA damage [50].

3. Vitamin D–Oxidative Stress

$1,25(OH)_2D$ is involved in many intracellular genomic activities and biochemical and enzymatic reactions, whereas 25(OH)D concentrations are important in overcoming inflammation, the destruction of invading microbes and parasites, the minimization of oxidative stress following the day-to-day exposure to toxic agents, and controlling the aging process [51,52].

For example, the presence of a physiologic 25(OH)D concentration enhances the expression of the nuclear factor, erythroid-2(Nf-E2)-related factor 2(Nrf2) [53–55] and enhances Klotho, a phosphate regulating hormone and also an antiaging protein [56,57]. It also facilitates protein stabilization [11]. Klotho also regulates cellular signaling systems, including the formation of antioxidants [58]. Consequently, in mice, functional abnormalities of the Klotho gene or removal of it through gene knock-out procedures induce premature aging syndrome [59]. In animal studies, inefficient FGF23 and/or Klotho expression have shown to cause premature aging. Figure 1 is a schematic representation of various key factors and their interactions that influence aging and death.

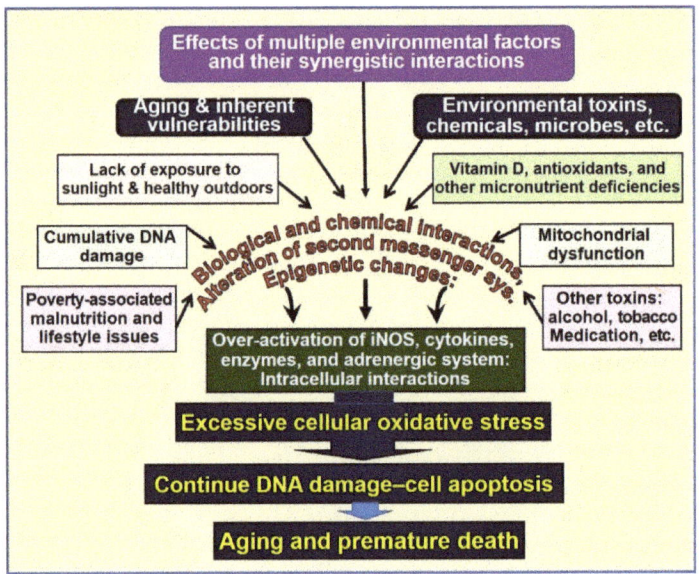

Figure 1. Environmental, microbial, biological and chemical interactions that modify the DNA and mitochondrial functions and epigenetics, which modifies the aging process. Vitamin D deficiency is one of the factors that enhances this oxidative-stress cycle and accelerating premature cell death [abbreviations used: DNA = deoxyribonucleic acid; iNOS = inducible nitric oxide enzyme].

Influences of Vitamin D on Oxidative Stress

When vitamin D status is adequate, many of the intracellular oxidative stress-related activities are downregulated. Having suboptimal concentrations of serum 25(OH)D fails to subdue oxidative stress conditions, augment intracellular oxidative damage and the rate of apoptosis. The intracellular Nrf2 level is inversely correlated with the accumulation of mitochondrial ROS [51,60] and the consequent escalation of oxidative stress. Thus, Nrf2 plays a key role in protecting cells against oxidative stress; this is modulated by vitamin D [61,62].

In addition, vitamin D supports cellular oxidation and reduction (redox) control by maintaining normal mitochondrial functions [63–65]. Loss in the redox control of the cell cycle may lead to aberrant cell proliferation, cell death, the development of neurodegenerative diseases, and accelerated aging [65–68]. Peroxisome proliferator-activated receptor-coactivator 1α (PGC-1α) is bound to mitochondrial deacetylase (SIRT3). PGC-1α directly couples to the oxidative stress cycle [69] and interacts with Nrf2. This complex regulates the expression of SIRT3; this process is influenced by vitamin D metabolites [70]. In addition, the activation of the mitochondrial Nrf2/PGC-1α-SIRT3 path is dependent on intracellular calcitriol concentrations.

Calcitriol has overarching beneficial effects in upregulating the expression of certain antioxidants and anti-inflammatory cytokines [71], thereby protecting the tissues from toxins, micronutrient deficiency-related abnormalities, and parasitic and intracellular microbe-induced harm [72]. It regulates ROS levels through its anti-inflammatory effects and mitochondrial-based expression of antioxidants through cell-signaling pathways [67,73].

4. Role of Vitamin D in Neutralization of Toxins and Aging-Related Compounds

Following 1,25(OH)$_2$D—VDR interaction, the transcription factor Nrf2 translocates from the cytoplasm to the nucleus. Nrf2 activates the expression of several genes that have antioxidant activity [52,54,67]. When Nrf2 activity is insufficient, risks from oxidative stress-related tissue damage increases [61,74]. The resultant excessive ROS formation by dysregulated mitochondria leads to a pathologic oxidative stress cycle, a key cause of toxin-induced and age-related cell death [68,75,76].

Meanwhile, the Nrf2 activity in part is controlled by the cytosolic protein Keap1 [77], another transcription factor and a negative regulator of Nrf2 [55,61]. Keap1 also controls the subcellular distribution of Nrf2 that correlates with its antioxidant activity [53,54]. When confronted with intracellular oxidative and/or electrophilic stresses, as a protective mechanism, the Nrf2-antioxidant response path is activated. This response enhances gene transcription and translation of protein products that are necessary to eliminate and/or neutralize toxins, ROS, and cumulating aging-related products through conjugation [78,79].

4.1. The Concept and the Process of Aging

Aging generally refers to the biological process of growing older, also known as cellular senescence, and is a complex process. Advancing age is, especially after adulthood, is associated with a gradual decline of physiological functions and capacities [80]. Aging has also been quantified from mortality curves using mathematical modeling; for example, by using Gompertz equation $m(t) = Ae^{Gt}$, for which, $m(t)$ = the mortality rate as a function of time or age (t); A = extrapolated constant to birth or maturity; G = the exponential (Gompertz) mortality rate coefficient [79].

Moreover, efficiency and the functions of the body decline after sexual maturity, suggesting a connection between the aging process after fulfilling the procreation needs. Most age-related functions are irreversible, in part due to accumulation of oxidative stress-related toxic products, methylation of DNA, and mitochondrial damage, leading to reduced viability of cells and consequent accelerated cell death [81]. There is also a parallel decline in the immune system functions (i.e., immune-senescence) and an increase in inflammation, demonstrable with increased circulating pro-inflammatory cytokines [82,83]. These are likely to contribute to several age-related disorders, such as Alzheimer's disease, cardiovascular and pulmonary diseases, and susceptibility to autoimmunity and infections [82,83].

Many bodily functions slow with aging, including response and reaction time; access to and the capacity of memory; pulmonary, gastrointestinal, and cardiovascular capacities; and even the ability to generate vitamin D in the skin. While age is perhaps the strongest risk factors for death, age-related disorders are the number one cause of death among the adults. This scenario is aggravated in the presence of vitamin D deficiency.

Chronic hypovitaminosis D is associated with cardiovascular and metabolic dysfunctions and premature deaths [84], even among children [85]. Overall data suggest that vitamin D deficiency could be considered an important comorbidity or a risk factor for premature death [84–87]. In fact, inverse relationships have been reported with vitamin D adequacy, with reduced all-cause mortality [88–90], and cancer [90–93].

4.2. Effects of Vitamin D on Apoptosis and Aging

The generalized inflammatory process is known to cause cellular damage and increase apoptosis [94], as in the case of interstitial tubular cell damage in chronic kidney disease, and thus is a part of the aging process [52,55,95]. In addition, hypovitaminosis D and dysfunctional mitochondrial activity increase inflammation [73,96,97]. Thus, the anti-inflammatory effects from having adequate, physiological vitamin D concentrations are important [95,98]. Hypovitaminosis D increases the expression of inflammatory cytokines [71,99] such as tumor necrosis factor-α (TNF-α), increasing the expression of the InsP3Rs and resulting in increased intracellular Ca^{2+} and accelerating cellular damage, apoptosis, and aging [66,75].

Many of the genes in the Klotho–Nrf2 regulatory system have multiple functions that are regulated by calcitriol [57,62,65]. These include, increasing intracellular antioxidant concentration, maintaining the redox homeostasis and, normal intracellular-reduced environment by removing excess ROS, and thereby down-regulating the oxidative stress [100]. In addition, the vitamin D-dependent expression of γ-glutamyl transpeptidase, glutamate cysteine ligase, and glutathione reductase contribute to the synthesis of the key redox agent glutathione (an essential antioxidant of low–molecular-weight thiol) [99,101].

Vitamin D also upregulates the expression of glutathione peroxidase that converts the ROS molecule H_2O_2 to water [101]. Vitamin D also effect the formation of glutathione through activation of the enzyme glucose-6-phosphate dehydrogenase [101]—which downregulates nitrogen oxide (NOx), a potent precursor for generating ROS that converts O_2^- to H_2O_2 and upregulating superoxide dismutase (SOD). These vitamin D-related actions collectively reduce the burden of intracellular ROS.

Telomeres are repetitive DNA sequences that caps end of linear chromosomes protecting DNA molecules [102]. Aging is associated with shortening of telomeres, including in stem cells. The amount of telomerase present is gradually become too short to maintain its protective effects on DNA during cell division, and thus cell apoptosis. While vitamin D deficiency increases inflammation and the intracellular oxidative stress, the latter enhances the rate of telomere shortening during cell proliferation, resulting in genomic instability [36].

4.3. Hypovitaminosis D Leads to Deranged Mitochondrial Respiration

Activated vitamin D is an essential component for maintaining physiological respiratory chain activity in mitochondria, facilitating the generation of energy [103,104]. In addition, 25(OH)D regulates the expression of the uncoupling protein that is attached to the inner membrane of mitochondria that regulates thermogenesis [105–107]. Chronic vitamin D deficiency reduces the capacity of mitochondrial respiration through modulating nuclear mRNA [108–110]. The latter also downregulates the expression of complex I of the electron transport chain and thus reduces the formation of adenosine triphosphate (ATP) [67,75], another mechanism that increases cancer risks. Consequently, a low level of electron transport chain increases the formation of ROS and oxidative stress, a common phenomenon following acute and chronic exposure to toxins and many chronic diseases and seen in aging [66,111,112].

The accumulation of intracellular toxins and/or age-related products disrupts signaling pathways, including the G protein–coupled systems, caspases, mitochondria, and the death receptor-linked mechanisms, triggering cell apoptosis and causing premature cell death [113,114]. The process is aggravated by stimulating G proteins, leading to activation of downstream pathways, including protein kinase A and C (PKA and PKC), phosphatidylinositol 3 kinase (PI3-kinase), Ca^{2+} and MAP kinase-dependent systems, tyrosine phosphorylation [75,114], and work additively, aiding cancer genesis [93] and accelerating the aging process.

4.4. Calcitriol Protects Mitochondrial Functions

Toxins, chronic metabolic abnormalities, and the aging process are known to cause mitochondrial dysfunction [66,106–108,115]. Abnormal mitochondria produce suboptimal amounts of ATP while generating excess ROS, creating a vicious cycle of enhanced and persisting the effects from excessive oxidative stress [106,107,116]. These events cause DNA damage (and impair DNA repair systems), premature cell death, and accelerated aging [62,66]. Data are accumulating that suggest that mitochondrial dysfunction is likely fueled by sustained intracellular inflammation, as in the case with vitamin D deficiency [79,95,97,117].

Based on animal studies, researchers have reported that mitochondrial decay, a part of the aging process, can be slowed by micronutrient supplementation (e.g., by lipoic acid, acetyl carnitine, vitamin K, and vitamin D) and by boosting coenzyme levels through high doses of vitamin B, such as pantothenic acid [118]. The NAD^+-dependent protein deacetylases sirtuins function as antiaging proteins that also neutralize excess ROS [11]. For example, sirtuin 1 (SIRT) is essential to maintaining normal mitochondrial functions; meanwhile, calcitriol and SIRT1 work synergistically to regenerate mitochondria [119,120]. Their actions facilitate the removal and neutralization of toxins and thereby reduce the rate and the effects of aging [121,122].

Dysfunctional mitochondria also have reduced intracellular Ca^{2+} buffering capacity, resulting in increased (and fluctuating) intracellular Ca^{2+} levels, which are cytotoxic and contribute to sustenance of several chronic diseases [108,115]. Sub physiological concentrations of calcitriol, at least in part, enhance and maintain oxidative stress, autophagy, inflammation, mitochondrial dysfunction, epigenetic changes, DNA damage, intracellular Ca^{2+}, and generation and signaling of ROS. Therefore, sustained, adequate serum 25(OH)D concentrations should allow target tissues to keep many of these harmful processes under control [68,71,73]. Multiple benefits of controlling excessive oxidative stress are illustrated in Figure 2.

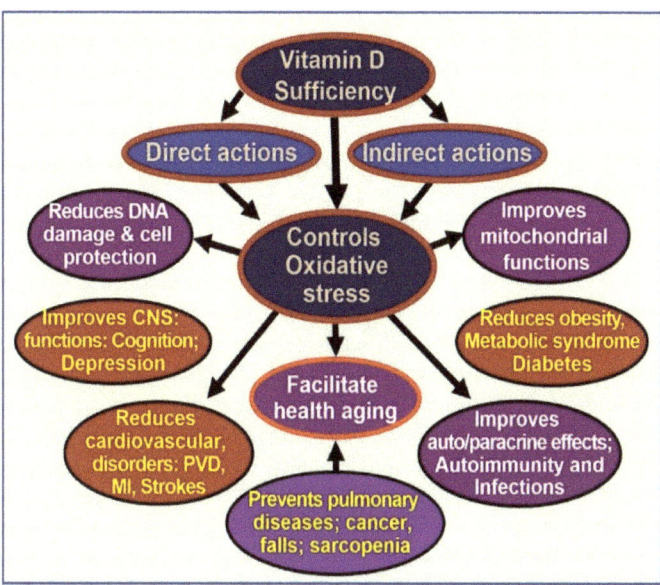

Figure 2. Oxidative stress is harmful to cells. Controlling oxidative stresses through vitamin D adequacy leads to cellular and organ protection and reduces the effects of aging [abbreviations used: CNS = central nervous system; DNA = deoxyribonucleic acid; MI = myocardial infarction; PVD = peripheral vascular diseases].

5. Discussion

The proper functioning of the vitamin D endocrine, paracrine, and autocrine systems is essential for many human physiological functions. Vitamin D deficiency, as determined by serum 25(OH)D concentrations of less than 30 ng/mL, is associated with increased risks of illnesses and disorders and increased all-cause mortality even among apparently healthy individuals, including those with normal serum 1,25(OH)$_2$D. Some of the key functions of vitamin D include subduing oxidative stress and chronic inflammation and maintaining mitochondrial respiratory functions. Through its targeted mitochondrial activity and subduing of ROS through multiple mechanisms, vitamin D has key beneficial effects on controlling oxidative stress, inflammation, and energy metabolism.

Normal serum concentrations of both 25(OH)D and 1,25(OH)$_2$D are essential for optimal cellular function and protect from the excessive oxidative stress-related DNA damage. However, increased risk for illnesses and reduced longevity can occur despite the presence of physiologic concentrations of calcitriol because this is not the only mechanism protecting cells from oxidative stress. Physiologic serum 25(OH)D and 1,25(OH)$_2$D levels in target tissues allow exertion of the homeostatic modulatory effects on enzymatic reactions, mitochondrial activities, and functioning of optimal second messenger systems. These are essential parts of the actions of vitamin D mediated through the mentioned mechanisms.

Vitamin D metabolism and functions are modulated by many factors, including physical activities and lifestyles, certain medications, environmental pollutants, and epigenetics, all of which also modify the balance between energy intake and expenditure through mitochondrial metabolic control [123]. For reductions in the incidence of diseases, longer-term maintenance of a steady state of the serum 25(OH)D concentration is necessary [124]. The minimal level is considered to be 30 ng/mL (50 nmol/L).

After correction of vitamin D deficiency through loading doses of oral vitamin D (or safe sun exposure), adequate maintenance doses of vitamin D_3 are needed. This can be achieved in approximately 90% of the adult population with vitamin D supplementation between 1000 to 4000 IU/day, 10,000 IU twice a week, or 50,000 IU twice a month [10,125]. On a population basis, such doses would allow approximately 97% of people to maintain their serum 25(OH)D concentrations above 30 ng/mL [19,126]. Others, such as persons with obesity, those with gastrointestinal disorders, and during pregnancy and lactation, are likely to require doses of 6,000 IU/day [127,128].

Funding: This research received no funding.

Acknowledgments: The author greatly appreciates the helpful comments of Eugene L. Heyden.

Conflicts of Interest: The author declares no conflict of interest.

Abbreviations

1,25(OH)$_2$D	1,25-dihydroxyvitamin D
25(OH)D	25-hydroxy vitamin D
Nrf2	Nuclear factor erythroid 2 (Nf-E2)-related factor 2
PTH	Parathyroid hormone
ROS	Reactive oxygen species
UVB	Ultraviolet B
VDR	Vitamin D receptor
VDRE	Vitamin D response element

References

1. Van Schoor, N.M.; Lips, P. Worldwide vitamin D status. *Best Pract. Res. Clin. Endocrinol. Metab.* **2011**, *25*, 671–680. [CrossRef] [PubMed]
2. Eggemoen, A.R.; Knutsen, K.V.; Dalen, I.; Jenum, A.K. Vitamin D status in recently arrived immigrants from Africa and Asia: A cross-sectional study from Norway of children, adolescents and adults. *BMJ Open* **2013**, *3*, e003293. [CrossRef] [PubMed]
3. Garland, C.F.; Gorham, E.D.; Mohr, S.B.; Garland, F.C. Vitamin D for cancer prevention: Global perspective. *Ann. Epidemiol.* **2009**, *19*, 468–483. [CrossRef]
4. Hilger, J.; Friedel, A.; Herr, R.; Rausch, T.; Roos, F.; Wahl, D.A.; Pierroz, D.D.; Weber, P.; Hoffmann, K. A systematic review of vitamin D status in populations worldwide. *Br. J. Nutr.* **2014**, *111*, 23–45. [CrossRef]
5. Haq, A.; Wimalawansa, S.J.; Carlberg, C. Highlights from the 5th International Conference on Vitamin D Deficiency, Nutrition and Human Health, Abu Dhabi, United Arab Emirates, March 24-25, 2016. *J. Steroid Biochem. Mol. Biol.* **2018**, *175*, 1–3. [CrossRef]
6. Pludowski, P.; Holick, M.F.; Grant, W.B.; Konstantynowicz, J.; Mascarenhas, M.R.; Haq, A.; Povoroznyuk, V.; Balatska, N.; Barbosa, A.P.; Karonova, T.; et al. Vitamin D supplementation guidelines. *J. Steroid Biochem. Mol. Biol.* **2018**, *175*, 125–135. [CrossRef]
7. Felton, S.J.; Cooke, M.S.; Kift, R.; Berry, J.L.; Webb, A.R.; Lam, P.M.; de Gruijl, F.R.; Vail, A.; Rhodes, L.E. Concurrent beneficial (vitamin D production) and hazardous (cutaneous DNA damage) impact of repeated low-level summer sunlight exposures. *Br. J. Dermatol.* **2016**, *175*, 1320–1328. [CrossRef] [PubMed]
8. Gordon-Thomson, C.; Tongkao-on, W.; Song, E.J.; Carter, S.E.; Dixon, K.M.; Mason, R.S. Protection from ultraviolet damage and photocarcinogenesis by vitamin D compounds. *Adv. Exp. Med. Biol.* **2014**, *810*, 303–328. [PubMed]
9. Petersen, B.; Wulf, H.C.; Triguero-Mas, M.; Philipsen, P.A.; Thieden, E.; Olsen, P.; Heydenreich, J.; Dadvand, P.; Basagana, X.; Liljendahl, T.S.; et al. Sun and ski holidays improve vitamin D status, but are associated with high levels of DNA damage. *J. Investig. Dermatol.* **2014**, *134*, 2806–2813. [CrossRef] [PubMed]

10. Wimalawansa, S.J. Vitamin D. What clinicians would like to know. *Sri Lanka Journal of Diabetes, Endocrinology and Metabolism* **2012**, *1*, 73–88. [CrossRef]
11. Mark, K.A.; Dumas, K.J.; Bhaumik, D.; Schilling, B.; Davis, S.; Oron, T.R.; Sorensen, D.J.; Lucanic, M.; Brem, R.B.; Melov, S.; et al. Vitamin D Promotes Protein Homeostasis and Longevity via the Stress Response Pathway Genes skn-1, ire-1, and xbp-1. *Cell Rep.* **2016**, *17*, 1227–1237. [CrossRef] [PubMed]
12. Grant, W.B.; Boucher, B.J.; Bhattoa, H.P.; Lahore, H. Why vitamin D clinical trials should be based on 25-hydroxyvitamin D concentrations. *J. Steroid Biochem. Mol. Biol.* **2018**, *177*, 266–269. [CrossRef] [PubMed]
13. Kroll, M.H.; Bi, C.; Garber, C.C.; Kaufman, H.W.; Liu, D.; Caston-Balderrama, A.; Zhang, K.; Clarke, N.; Xie, M.; Reitz, R.E.; et al. Temporal relationship between vitamin D status and parathyroid hormone in the United States. *PLoS ONE* **2015**, *10*, e0118108. [CrossRef] [PubMed]
14. Jones, G.; Prosser, D.E.; Kaufmann, M. 25-Hydroxyvitamin D-24-hydroxylase (CYP24A1): Its important role in the degradation of vitamin D. *Arch. Biochem. Biophys.* **2012**, *523*, 9–18. [CrossRef] [PubMed]
15. Webb, A.R.; DeCosta, B.R.; Holick, M.F. Sunlight regulates the cutaneous production of vitamin D3 by causing its photodegradation. *J. Clin. Endocrinol. Metab.* **1989**, *68*, 882–887. [CrossRef]
16. Armbrecht, H.J.; Hodam, T.L.; Boltz, M.A.; Partridge, N.C.; Brown, A.J.; Kumar, V.B. Induction of the vitamin D 24-hydroxylase (CYP24) by 1,25-dihydroxyvitamin D3 is regulated by parathyroid hormone in UMR106 osteoblastic cells. *Endocrinology* **1998**, *139*, 3375–3381. [CrossRef]
17. Miller, W.L. Genetic disorders of Vitamin D biosynthesis and degradation. *J. Steroid Biochem. Mol. Biol.* **2017**, *165*, 101–108. [CrossRef]
18. Mithal, A.; Wahl, D.A.; Bonjour, J.P.; Burckhardt, P.; Dawson-Hughes, B.; Eisman, J.A.; El-Hajj Fuleihan, G.; Josse, R.G.; Lips, P.; Morales-Torres, J. Global vitamin D status and determinants of hypovitaminosis D. *Osteoporos. Int.* **2009**, *20*, 1807–1820. [CrossRef]
19. Wimalawansa, S.J. Vitamin D in the new millennium. *Curr. Osteoporos. Rep.* **2012**, *10*, 4–15. [CrossRef]
20. Grant, W.B.; Wimalawansa, S.J.; Holick, M.F. Vitamin D supplements and reasonable solar UVB should be recommended to prevent escalating incidence of chronic diseases. *Br. Med. J.* **2015**, *350*, h321.
21. Holick, M.F. High prevalence of vitamin D inadequacy and implications for health. *Mayo Clin. Proc.* **2006**, *81*, 353–373. [CrossRef]
22. Calzavara-Pinton, P.; Ortel, B.; Venturini, M. Non-melanoma skin cancer, sun exposure and sun protection. *Giornale Italiano di Dermatologia e Venereologia* **2015**, *150*, 369–378.
23. Cascinelli, N.; Krutmann, J.; MacKie, R.; Pierotti, M.; Prota, G.; Rosso, S.; Young, A. European School of Oncology advisory report. Sun exposure, UVA lamps and risk of skin cancer. *Eur. J. Cancer* **1994**, *30A*, 548–560. [PubMed]
24. Moan, J.; Porojnicu, A.C.; Dahlback, A.; Setlow, R.B. Addressing the health benefits and risks, involving vitamin D or skin cancer, of increased sun exposure. *Proc. Natl. Acad. Sci. USA* **2008**, *105*, 668–673. [CrossRef]
25. Omura, Y. Clinical Significance of Human Papillomavirus Type 16 for Breast Cancer & Adenocarcinomas of Various Internal Organs and Alzheimer's Brain with Increased beta-amyloid (1-42); Combined Use of Optimal Doses of Vitamin D3 and Taurine 3 times/day Has Significant Beneficial Effects of Anti-Cancer, Anti-Ischemic Heart, and Memory & Other Brain Problems By Significant Urinary Excretion of Viruses, Bacteria, and Toxic Metals & Substances. *Acupunct. Electro-Ther. Res.* **2016**, *41*, 127–134.
26. Gilad, L.A.; Bresler, T.; Gnainsky, J.; Smirnoff, P.; Schwartz, B. Regulation of vitamin D receptor expression via estrogen-induced activation of the ERK 1/2 signaling pathway in colon and breast cancer cells. *J. Endocrinol.* **2005**, *185*, 577–592. [CrossRef] [PubMed]
27. Robsahm, T.E.; Tretli, S.; Dahlback, A.; Moan, J. Vitamin D3 from sunlight may improve the prognosis of breast-, colon- and prostate cancer (Norway). *Cancer Causes Control* **2004**, *15*, 149–158. [CrossRef] [PubMed]
28. Jeon, S.M.; Shin, E.A. Exploring vitamin D metabolism and function in cancer. *Exp. Mol. Med.* **2018**, *50*, 20. [CrossRef] [PubMed]
29. Valdivielso, J.M. The physiology of vitamin D receptor activation. *Contrib. Nephrol.* **2009**, *163*, 206–212. [PubMed]
30. Farnham, P.J. Thematic minireview series on results from the ENCODE Project: Integrative global analyses of regulatory regions in the human genome. *J. Biol. Chem.* **2012**, *287*, 30885–30887. [CrossRef]
31. Washington, S.D.; Edenfield, S.I.; Lieux, C.; Watson, Z.L.; Taasan, S.M.; Dhummakupt, A.; Bloom, D.C.; Neumann, D.M. Depletion of the insulator protein CTCF results in HSV-1 reactivation in vivo. *J. Virol.* **2018**. [CrossRef]

32. MacPherson, M.J.; Sadowski, P.D. The CTCF insulator protein forms an unusual DNA structure. *BMC Mol. Biol.* **2010**, *11*, 101. [CrossRef] [PubMed]
33. Carlberg, C. Genome-wide (over)view on the actions of vitamin D. *Front. Physiol.* **2014**, *5*, 167. [CrossRef]
34. Narvaez, C.J.; Matthews, D.; LaPorta, E.; Simmons, K.M.; Beaudin, S.; Welsh, J. The impact of vitamin D in breast cancer: Genomics, pathways, metabolism. *Front. Physiol.* **2014**, *5*, 213. [CrossRef]
35. Tilstra, J.S.; Robinson, A.R.; Wang, J.; Gregg, S.Q.; Clauson, C.L.; Reay, D.P.; Nasto, L.A.; St Croix, C.M.; Usas, A.; Vo, N.; et al. NF-kappaB inhibition delays DNA damage-induced senescence and aging in mice. *J. Clin. Investig.* **2012**, *122*, 2601–2612. [CrossRef]
36. Pusceddu, I.; Farrell, C.J.; Di Pierro, A.M.; Jani, E.; Herrmann, W.; Herrmann, M. The role of telomeres and vitamin D in cellular aging and age-related diseases. *Clin. Chem. Lab. Med.* **2015**, *53*, 1661–1678. [CrossRef] [PubMed]
37. Tang, H.; Hua, F.; Wang, J.; Sayeed, I.; Wang, X.; Chen, Z.; Yousuf, S.; Atif, F.; Stein, D.G. Progesterone and vitamin D: Improvement after traumatic brain injury in middle-aged rats. *Horm. Behav.* **2013**, *64*, 527–538. [CrossRef] [PubMed]
38. Wang, Q.; He, Y.; Shen, Y.; Zhang, Q.; Chen, D.; Zuo, C.; Qin, J.; Wang, H.; Wang, J.; Yu, Y. Vitamin D inhibits COX-2 expression and inflammatory response by targeting thioesterase superfamily member 4. *J. Biol. Chem.* **2014**, *289*, 11681–11694. [CrossRef] [PubMed]
39. Myszka, M.; Klinger, M. The immunomodulatory role of Vitamin D. *Postepy Hig. Med. Dosw.* **2014**, *68*, 865–878. [CrossRef]
40. Watanabe, R.; Inoue, D. Current Topics on Vitamin D. Anti-cancer effects of vitamin D. *Clin. Calcium* **2015**, *25*, 373–380.
41. Shi, Y.; Au, J.S.; Thongprasert, S.; Srinivasan, S.; Tsai, C.M.; Khoa, M.T.; Heeroma, K.; Itoh, Y.; Cornelio, G.; Yang, P.C. A prospective, molecular epidemiology study of EGFR mutations in Asian patients with advanced non-small-cell lung cancer of adenocarcinoma histology (PIONEER). *J. Thorac. Oncol.* **2014**, *9*, 154–162. [CrossRef]
42. O'Malley, M.; King, A.N.; Conte, M.; Ellingrod, V.L.; Ramnath, N. Effects of cigarette smoking on metabolism and effectiveness of systemic therapy for lung cancer. *J. Thorac. Oncol.* **2014**, *9*, 917–926. [CrossRef] [PubMed]
43. Iswariya, G.T.; Paital, B.; Padma, P.R.; Nirmaladevi, R. MicroRNAs: Epigenetic players in cancer and aging. *Front. Biosci. (Schol Ed)* **2019**, *11*, 29–55.
44. Ramnath, N.; Nadal, E.; Jeon, C.K.; Sandoval, J.; Colacino, J.; Rozek, L.S.; Christensen, P.J.; Esteller, M.; Beer, D.G.; Kim, S.H. Epigenetic regulation of vitamin D metabolism in human lung adenocarcinoma. *J. Thorac. Oncol.* **2014**, *9*, 473–482. [CrossRef]
45. Karlic, H.; Varga, F. Impact of vitamin D metabolism on clinical epigenetics. *Clin. Epigenet.* **2011**, *2*, 55–61. [CrossRef] [PubMed]
46. Codoner-Franch, P.; Tavarez-Alonso, S.; Simo-Jorda, R.; Laporta-Martin, P.; Carratala-Calvo, A.; Alonso-Iglesias, E. Vitamin D status is linked to biomarkers of oxidative stress, inflammation, and endothelial activation in obese children. *J. Pediatr.* **2012**, *161*, 848–854. [CrossRef]
47. Zeitelhofer, M.; Adzemovic, M.Z.; Gomez-Cabrero, D.; Bergman, P.; Hochmeister, S.; N'Diaye, M.; Paulson, A.; Ruhrmann, S.; Almgren, M.; Tegner, J.N.; et al. Functional genomics analysis of vitamin D effects on CD4+ T cells in vivo in experimental autoimmune encephalomyelitis. *Proc. Natl. Acad. Sci. USA* **2017**, *114*, E1678–E1687. [CrossRef] [PubMed]
48. Wu, W.; Zhao, S. Metabolic changes in cancer: Beyond the Warburg effect. *Acta Biochim. Biophys. Sin.* **2013**, *45*, 18–26. [CrossRef]
49. Garcia-Quiroz, J.; Garcia-Becerra, R.; Barrera, D.; Santos, N.; Avila, E.; Ordaz-Rosado, D.; Rivas-Suarez, M.; Halhali, A.; Rodriguez, P.; Gamboa-Dominguez, A.; et al. Astemizole synergizes calcitriol antiproliferative activity by inhibiting CYP24A1 and upregulating VDR: A novel approach for breast cancer therapy. *PLoS ONE* **2012**, *7*, e45063. [CrossRef]
50. Ricca, C.; Aillon, A.; Bergandi, L.; Alotto, D.; Castagnoli, C.; Silvagno, F. Vitamin D Receptor Is Necessary for Mitochondrial Function and Cell Health. *Int. J. Mol. Sci.* **2018**, *19*, 1672. [CrossRef]
51. Holmes, S.; Abbassi, B.; Su, C.; Singh, M.; Cunningham, R.L. Oxidative stress defines the neuroprotective or neurotoxic properties of androgens in immortalized female rat dopaminergic neuronal cells. *Endocrinology* **2013**, *154*, 4281–4292. [CrossRef]

52. Petersen, K.S.; Smith, C. Ageing-Associated Oxidative Stress and Inflammation Are Alleviated by Products from Grapes. *Oxid. Med. Cell. Longev.* **2016**, *2016*, 6236309. [CrossRef]
53. Nakai, K.; Fujii, H.; Kono, K.; Goto, S.; Kitazawa, R.; Kitazawa, S.; Hirata, M.; Shinohara, M.; Fukagawa, M.; Nishi, S. Vitamin D activates the Nrf2-Keap1 antioxidant pathway and ameliorates nephropathy in diabetic rats. *Am. J. Hypertens.* **2014**, *27*, 586–595. [CrossRef] [PubMed]
54. Lewis, K.N.; Mele, J.; Hayes, J.D.; Buffenstein, R. Nrf2, a guardian of healthspan and gatekeeper of species longevity. *Integr. Comp. Biol.* **2010**, *50*, 829–843. [CrossRef] [PubMed]
55. Tullet, J.M.A.; Green, J.W.; Au, C.; Benedetto, A.; Thompson, M.A.; Clark, E.; Gilliat, A.F.; Young, A.; Schmeisser, K.; Gems, D. The SKN-1/Nrf2 transcription factor can protect against oxidative stress and increase lifespan in C. elegans by distinct mechanisms. *Aging Cell* **2017**, *16*, 1191–1194. [CrossRef] [PubMed]
56. Forster, R.E.; Jurutka, P.W.; Hsieh, J.C.; Haussler, C.A.; Lowmiller, C.L.; Kaneko, I.; Haussler, M.R.; Kerr Whitfield, G. Vitamin D receptor controls expression of the anti-aging klotho gene in mouse and human renal cells. *Biochem. Biophys. Res. Commun.* **2011**, *414*, 557–562. [CrossRef]
57. Berridge, M.J. Vitamin D: A custodian of cell signalling stability in health and disease. *Biochem. Soc. Trans.* **2015**, *43*, 349–358. [CrossRef]
58. Razzaque, M.S. FGF23, klotho and vitamin D interactions: What have we learned from in vivo mouse genetics studies? *Adv. Exp. Med. Biol.* **2012**, *728*, 84–91.
59. Kuro-o, M. Klotho and aging. *Biochim. Biophys. Acta* **2009**, *1790*, 1049–1058. [CrossRef]
60. Tseng, A.H.; Shieh, S.S.; Wang, D.L. SIRT3 deacetylates FOXO3 to protect mitochondria against oxidative damage. *Free Radic. Biol. Med.* **2013**, *63*, 222–234. [CrossRef]
61. Wang, L.; Lewis, T.; Zhang, Y.L.; Khodier, C.; Magesh, S.; Chen, L.; Inoyama, D.; Chen, Y.; Zhen, J.; Hu, L.; et al. The identification and characterization of non-reactive inhibitor of Keap1-Nrf2 interaction through HTS using a fluorescence polarization assay. In *Probe Reports from the NIH Molecular Libraries Program*; National Center for Biotechnology Information (US): Bethesda, MD, USA, 2010.
62. Berridge, M.J. Vitamin D deficiency: Infertility and neurodevelopmental diseases (attention deficit hyperactivity disorder, autism, and schizophrenia). *Am. J. Physiol. Cell Physiol.* **2018**, *314*, C135–C151. [CrossRef]
63. Ryan, Z.C.; Craig, T.A.; Folmes, C.D.; Wang, X.; Lanza, I.R.; Schaible, N.S.; Salisbury, J.L.; Nair, K.S.; Terzic, A.; Sieck, G.C.; et al. 1alpha,25-Dihydroxyvitamin D3 Regulates Mitochondrial Oxygen Consumption and Dynamics in Human Skeletal Muscle Cells. *J. Biol. Chem.* **2016**, *291*, 1514–1528. [CrossRef]
64. Bouillon, R.; Verstuyf, A. Vitamin D, mitochondria, and muscle. *J. Clin. Endocrinol. Metab.* **2013**, *98*, 961–963. [CrossRef]
65. Sarsour, E.H.; Kumar, M.G.; Chaudhuri, L.; Kalen, A.L.; Goswami, P.C. Redox control of the cell cycle in health and disease. *Antioxid. Redox Signal.* **2009**, *11*, 2985–3011. [CrossRef]
66. Lin, M.T.; Beal, M.F. Mitochondrial dysfunction and oxidative stress in neurodegenerative diseases. *Nature* **2006**, *443*, 787–795. [CrossRef]
67. Berridge, M.J. Vitamin D cell signalling in health and disease. *Biochem. Biophys. Res. Commun.* **2015**, *460*, 53–71. [CrossRef]
68. Ureshino, R.P.; Rocha, K.K.; Lopes, G.S.; Bincoletto, C.; Smaili, S.S. Calcium signaling alterations, oxidative stress, and autophagy in aging. *Antioxid. Redox Signal.* **2014**, *21*, 123–137. [CrossRef]
69. Chen, Y.; Zhang, J.; Lin, Y.; Lei, Q.; Guan, K.L.; Zhao, S.; Xiong, Y. Tumour suppressor SIRT3 deacetylates and activates manganese superoxide dismutase to scavenge ROS. *EMBO Rep.* **2011**, *12*, 534–541. [CrossRef]
70. Song, C.; Fu, B.; Zhang, J.; Zhao, J.; Yuan, M.; Peng, W.; Zhang, Y.; Wu, H. Sodium fluoride induces nephrotoxicity via oxidative stress-regulated mitochondrial SIRT3 signaling pathway. *Sci. Rep.* **2017**, *7*, 672. [CrossRef]
71. Wei, R.; Christakos, S. Mechanisms Underlying the Regulation of Innate and Adaptive Immunity by Vitamin D. *Nutrients* **2015**, *7*, 8251–8260. [CrossRef]
72. George, N.; Kumar, T.P.; Antony, S.; Jayanarayanan, S.; Paulose, C.S. Effect of vitamin D3 in reducing metabolic and oxidative stress in the liver of streptozotocin-induced diabetic rats. *Br. J. Nutr.* **2012**, *108*, 1410–1418. [CrossRef]
73. Shelton, R.C.; Claiborne, J.; Sidoryk-Wegrzynowicz, M.; Reddy, R.; Aschner, M.; Lewis, D.A.; Mirnics, K. Altered expression of genes involved in inflammation and apoptosis in frontal cortex in major depression. *Mol. Psychiatry* **2011**, *16*, 751–762. [CrossRef]

74. Ramsey, C.P.; Glass, C.A.; Montgomery, M.B.; Lindl, K.A.; Ritson, G.P.; Chia, L.A.; Hamilton, R.L.; Chu, C.T.; Jordan-Sciutto, K.L. Expression of Nrf2 in neurodegenerative diseases. *J. Neuropathol. Exp. Neurol.* **2007**, *66*, 75–85. [CrossRef] [PubMed]
75. Petersen, O.H.; Verkhratsky, A. Calcium and ATP control multiple vital functions. *Philos. Trans. R. Soc. Lond. B Biol. Sci.* **2016**, *371*, 20150418. [CrossRef]
76. Petrosillo, G.; Di Venosa, N.; Moro, N.; Colantuono, G.; Paradies, V.; Tiravanti, E.; Federici, A.; Ruggiero, F.M.; Paradies, G. In vivo hyperoxic preconditioning protects against rat-heart ischemia/reperfusion injury by inhibiting mitochondrial permeability transition pore opening and cytochrome c release. *Free Radic. Biol. Med.* **2011**, *50*, 477–483. [CrossRef]
77. McMahon, M.; Itoh, K.; Yamamoto, M.; Hayes, J.D. Keap1-dependent proteasomal degradation of transcription factor Nrf2 contributes to the negative regulation of antioxidant response element-driven gene expression. *J. Biol. Chem.* **2003**, *278*, 21592–21600. [CrossRef] [PubMed]
78. Nguyen, T.; Nioi, P.; Pickett, C.B. The Nrf2-antioxidant response element signaling pathway and its activation by oxidative stress. *J. Biol. Chem.* **2009**, *284*, 13291–13295. [CrossRef]
79. Finch, C.E.; Pike, M.C. Maximum life span predictions from the Gompertz mortality model. *J. Gerontol. A Biol. Sci. Med. Sci.* **1996**, *51*, B183–B194. [CrossRef]
80. Macedo, J.C.; Vaz, S.; Logarinho, E. Mitotic Dysfunction Associated with Aging Hallmarks. *Adv. Exp. Med. Biol.* **2017**, *1002*, 153–188.
81. Jallali, N.; Ridha, H.; Thrasivoulou, C.; Underwood, C.; Butler, P.E.; Cowen, T. Vulnerability to ROS-induced cell death in ageing articular cartilage: The role of antioxidant enzyme activity. *Osteoarthr. Cartil.* **2005**, *13*, 614–622. [CrossRef]
82. Fulop, T.; Larbi, A.; Dupuis, G.; Le Page, A.; Frost, E.H.; Cohen, A.A.; Witkowski, J.M.; Franceschi, C. Immunosenescence and Inflamm-Aging As Two Sides of the Same Coin: Friends or Foes? *Front. Immunol.* **2017**, *8*, 1960. [CrossRef]
83. Franceschi, C.; Bonafe, M.; Valensin, S.; Olivieri, F.; De Luca, M.; Ottaviani, E.; De Benedictis, G. Inflamm-aging. An evolutionary perspective on immunosenescence. *Ann. N. Y. Acad. Sci.* **2000**, *908*, 244–254. [CrossRef]
84. Pilz, S.; Marz, W.; Wellnitz, B.; Seelhorst, U.; Fahrleitner-Pammer, A.; Dimai, H.P.; Boehm, B.O.; Dobnig, H. Association of vitamin D deficiency with heart failure and sudden cardiac death in a large cross-sectional study of patients referred for coronary angiography. *J. Clin. Endocrinol. Metab.* **2008**, *93*, 3927–3935. [CrossRef] [PubMed]
85. Cohen, M.C.; Offiah, A.; Sprigg, A.; Al-Adnani, M. Vitamin D deficiency and sudden unexpected death in infancy and childhood: A cohort study. *Pediatr. Dev. Pathol.* **2013**, *16*, 292–300. [CrossRef] [PubMed]
86. Grant, W.B. Vitamin D Deficiency May Explain Comorbidity as an Independent Risk Factor for Death Associated with Cancer in Taiwan. *Asia Pac. J. Public Health* **2015**, *27*, 572–573. [CrossRef] [PubMed]
87. Scorza, F.A.; Albuquerque, M.; Arida, R.M.; Terra, V.C.; Machado, H.R.; Cavalheiro, E.A. Benefits of sunlight: Vitamin D deficiency might increase the risk of sudden unexpected death in epilepsy. *Med. Hypotheses* **2010**, *74*, 158–161. [CrossRef]
88. Autier, P.; Mullie, P.; Macacu, A.; Dragomir, M.; Boniol, M.; Coppens, K.; Pizot, C.; Boniol, M. Effect of vitamin D supplementation on non-skeletal disorders: A systematic review of meta-analyses and randomised trials. *Lancet Diabetes Endocrinol.* **2017**, *5*, 986–1004. [CrossRef]
89. Brenner, H.; Jansen, L.; Saum, K.U.; Holleczek, B.; Schottker, B. Vitamin D Supplementation Trials Aimed at Reducing Mortality Have Much Higher Power When Focusing on People with Low Serum 25-Hydroxyvitamin D Concentrations. *J. Nutr.* **2017**, *147*, 1325–1333. [CrossRef]
90. Autier, P.; Boniol, M.; Pizot, C.; Mullie, P. Vitamin D status and ill health: A systematic review. *Lancet Diabetes Endocrinol.* **2014**, *2*, 76–89. [CrossRef]
91. Bjelakovic, G.; Gluud, L.L.; Nikolova, D.; Whitfield, K.; Krstic, G.; Wetterslev, J.; Gluud, C. Vitamin D supplementation for prevention of cancer in adults. *Cochrane Database Syst. Rev.* **2014**. [CrossRef] [PubMed]
92. Pilz, S.; Tomaschitz, A.; Marz, W.; Drechsler, C.; Ritz, E.; Zittermann, A.; Cavalier, E.; Pieber, T.R.; Lappe, J.M.; Grant, W.B.; et al. Vitamin D, cardiovascular disease and mortality. *Clin. Endocrinol.* **2011**, *75*, 575–584. [CrossRef]
93. Van der Reest, J.; Gottlieb, E. Anti-cancer effects of vitamin C revisited. *Cell Res.* **2016**, *26*, 269–270. [CrossRef]

94. Da Luz Dias, R.; Basso, B.; Donadio, M.V.F.; Pujol, F.V.; Bartrons, R.; Haute, G.V.; Gassen, R.B.; Bregolin, H.D.; Krause, G.; Viau, C.; et al. Leucine reduces the proliferation of MC3T3-E1 cells through DNA damage and cell senescence. *Toxicol. In Vitro* **2018**, *48*, 1–10. [CrossRef]
95. Cevenini, E.; Caruso, C.; Candore, G.; Capri, M.; Nuzzo, D.; Duro, G.; Rizzo, C.; Colonna-Romano, G.; Lio, D.; Di Carlo, D.; et al. Age-related inflammation: The contribution of different organs, tissues and systems. How to face it for therapeutic approaches. *Curr. Pharm. Des.* **2010**, *16*, 609–618. [CrossRef]
96. Talmor, Y.; Bernheim, J.; Klein, O.; Green, J.; Rashid, G. Calcitriol blunts pro-atherosclerotic parameters through NFkappaB and p38 in vitro. *Eur. J. Clin. Investig.* **2008**, *38*, 548–554. [CrossRef]
97. Morris, G.; Maes, M. Mitochondrial dysfunctions in myalgic encephalomyelitis/chronic fatigue syndrome explained by activated immuno-inflammatory, oxidative and nitrosative stress pathways. *Metab. Brain Dis.* **2014**, *29*, 19–36. [CrossRef]
98. Berk, M.; Williams, L.J.; Jacka, F.N.; O'Neil, A.; Pasco, J.A.; Moylan, S.; Allen, N.B.; Stuart, A.L.; Hayley, A.C.; Byrne, M.L.; et al. So depression is an inflammatory disease, but where does the inflammation come from? *BMC Med.* **2013**, *11*, 200. [CrossRef]
99. Beilfuss, J.; Berg, V.; Sneve, M.; Jorde, R.; Kamycheva, E. Effects of a 1-year supplementation with cholecalciferol on interleukin-6, tumor necrosis factor-alpha and insulin resistance in overweight and obese subjects. *Cytokine* **2012**, *60*, 870–874. [CrossRef]
100. Calton, E.K.; Keane, K.N.; Soares, M.J.; Rowlands, J.; Newsholme, P. Prevailing vitamin D status influences mitochondrial and glycolytic bioenergetics in peripheral blood mononuclear cells obtained from adults. *Redox Biol.* **2016**, *10*, 243–250. [CrossRef]
101. Liu, Y.; Hyde, A.S.; Simpson, M.A.; Barycki, J.J. Emerging regulatory paradigms in glutathione metabolism. *Adv. Cancer Res.* **2014**, *122*, 69–101.
102. Weipoltshammer, K.; Schofer, C.; Almeder, M.; Philimonenko, V.V.; Frei, K.; Wachtler, F.; Hozak, P. Intranuclear anchoring of repetitive DNA sequences: Centromeres, telomeres, and ribosomal DNA. *J. Cell Biol.* **1999**, *147*, 1409–1418. [CrossRef] [PubMed]
103. Consiglio, M.; Viano, M.; Casarin, S.; Castagnoli, C.; Pescarmona, G.; Silvagno, F. Mitochondrial and lipogenic effects of vitamin D on differentiating and proliferating human keratinocytes. *Exp. Dermatol.* **2015**, *24*, 748–753. [CrossRef] [PubMed]
104. Wyckelsma, V.L.; Levinger, I.; McKenna, M.J.; Formosa, L.E.; Ryan, M.T.; Petersen, A.C.; Anderson, M.J.; Murphy, R.M. Preservation of skeletal muscle mitochondrial content in older adults: Relationship between mitochondria, fibre type and high-intensity exercise training. *J. Physiol.* **2017**, *595*, 3345–3359. [CrossRef] [PubMed]
105. Constantinescu, A.A.; Abbas, M.; Kassem, M.; Gleizes, C.; Kreutter, G.; Schini-Kerth, V.; Mitrea, I.L.; Toti, F.; Kessler, L. Differential influence of tacrolimus and sirolimus on mitochondrial-dependent signaling for apoptosis in pancreatic cells. *Mol. Cell. Biochem.* **2016**, *418*, 91–102. [CrossRef] [PubMed]
106. Petrosillo, G.; Matera, M.; Casanova, G.; Ruggiero, F.M.; Paradies, G. Mitochondrial dysfunction in rat brain with aging Involvement of complex I, reactive oxygen species and cardiolipin. *Neurochem. Int.* **2008**, *53*, 126–131. [CrossRef] [PubMed]
107. Petrosillo, G.; Matera, M.; Moro, N.; Ruggiero, F.M.; Paradies, G. Mitochondrial complex I dysfunction in rat heart with aging: Critical role of reactive oxygen species and cardiolipin. *Free Radic. Biol. Med.* **2009**, *46*, 88–94. [CrossRef] [PubMed]
108. Yin, F.; Sancheti, H.; Patil, I.; Cadenas, E. Energy metabolism and inflammation in brain aging and Alzheimer's disease. *Free Radic. Biol. Med.* **2016**, *100*, 108–122. [CrossRef] [PubMed]
109. Prior, S.; Kim, A.; Yoshihara, T.; Tobita, S.; Takeuchi, T.; Higuchi, M. Mitochondrial respiratory function induces endogenous hypoxia. *PLoS ONE* **2014**, *9*, e88911. [CrossRef] [PubMed]
110. Scaini, G.; Rezin, G.T.; Carvalho, A.F.; Streck, E.L.; Berk, M.; Quevedo, J. Mitochondrial dysfunction in bipolar disorder: Evidence, pathophysiology and translational implications. *Neurosci. Biobehav. Rev.* **2016**, *68*, 694–713. [CrossRef] [PubMed]
111. Zhu, Q.S.; Xia, L.; Mills, G.B.; Lowell, C.A.; Touw, I.P.; Corey, S.J. G-CSF induced reactive oxygen species involves Lyn-PI3-kinase-Akt and contributes to myeloid cell growth. *Blood* **2006**, *107*, 1847–1856. [CrossRef] [PubMed]

112. Lin, Y.; Berg, A.H.; Iyengar, P.; Lam, T.K.; Giacca, A.; Combs, T.P.; Rajala, M.W.; Du, X.; Rollman, B.; Li, W.; et al. The hyperglycemia-induced inflammatory response in adipocytes: The role of reactive oxygen species. *J. Biol. Chem.* **2005**, *280*, 4617–4626. [CrossRef] [PubMed]
113. Agalakova, A.A.; Gusev, G.P. Molecular mechanisms of cytotoxicity and apoptosis induced by inorganic fluoride. *ISRN Cell Biol.* **2012**, *2012*, 403835. [CrossRef]
114. Agalakova, N.I.; Gusev, G.P. Fluoride induces oxidative stress and ATP depletion in the rat erythrocytes in vitro. *Environ. Toxicol. Pharmacol.* **2012**, *34*, 334–337. [CrossRef]
115. Yin, F.; Sancheti, H.; Liu, Z.; Cadenas, E. Mitochondrial function in ageing: Coordination with signalling and transcriptional pathways. *J. Physiol.* **2016**, *594*, 2025–2042. [CrossRef] [PubMed]
116. Marzetti, E.; Calvani, R.; Cesari, M.; Buford, T.W.; Lorenzi, M.; Behnke, B.J.; Leeuwenburgh, C. Mitochondrial dysfunction and sarcopenia of aging: From signaling pathways to clinical trials. *Int. J. Biochem. Cell Biol.* **2013**, *45*, 2288–2301. [CrossRef]
117. Morris, G.; Berk, M. The many roads to mitochondrial dysfunction in neuroimmune and neuropsychiatric disorders. *BMC Med.* **2015**, *13*, 68. [CrossRef]
118. Ames, B.N. Optimal micronutrients delay mitochondrial decay and age-associated diseases. *Mech. Ageing Dev.* **2010**, *131*, 473–479. [CrossRef]
119. Manna, P.; Achari, A.E.; Jain, S.K. Vitamin D supplementation inhibits oxidative stress and upregulate SIRT1/AMPK/GLUT4 cascade in high glucose-treated 3T3L1 adipocytes and in adipose tissue of high fat diet-fed diabetic mice. *Arch. Biochem. Biophys.* **2017**, *615*, 22–34. [CrossRef]
120. Marampon, F.; Gravina, G.L.; Festuccia, C.; Popov, V.M.; Colapietro, E.A.; Sanita, P.; Musio, D.; De Felice, F.; Lenzi, A.; Jannini, E.A.; et al. Vitamin D protects endothelial cells from irradiation-induced senescence and apoptosis by modulating MAPK/SirT1 axis. *J. Endocrinol. Investig.* **2016**, *39*, 411–422. [CrossRef]
121. Grabowska, W.; Sikora, E.; Bielak-Zmijewska, A. Sirtuins, a promising target in slowing down the ageing process. *Biogerontology* **2017**, *18*, 447–476. [CrossRef]
122. Imai, S.; Guarente, L. Ten years of NAD-dependent SIR2 family deacetylases: Implications for metabolic diseases. *Trends Pharmacol. Sci.* **2010**, *31*, 212–220. [CrossRef] [PubMed]
123. Binkley, N.; Ramamurthy, R.; Krueger, D. Low vitamin D status: Definition, prevalence, consequences, and correction. *Rheum. Dis. Clin. N. Am.* **2012**, *38*, 45–59. [CrossRef] [PubMed]
124. Hollis, B.W.; Wagner, C.L. Clinical review: The role of the parent compound vitamin D with respect to metabolism and function: Why clinical dose intervals can affect clinical outcomes. *J. Clin. Endocrinol. Metab.* **2013**, *98*, 4619–4628. [CrossRef] [PubMed]
125. Holick, M.F.; Binkley, N.C.; Bischoff-Ferrari, H.A.; Gordon, C.M.; Hanley, D.A.; Heaney, R.P.; Murad, M.H.; Weaver, C.M.; Endocrine, S. Evaluation, treatment, and prevention of vitamin D deficiency: An Endocrine Society clinical practice guideline. *J. Clin. Endocrinol. Metab.* **2011**, *96*, 1911–1930. [CrossRef] [PubMed]
126. Wimalawansa, S.J. *Vitamin D: Everything You Need to Know*; Karunaratne & Sons: Homagama, Sri Lanka, 2012; ISBN 978-955-9098-94-2.
127. Heyden, E.L.; Wimalawansa, S.J. Vitamin D: Effects on human reproduction, pregnancy, and fetal well-being. *J. Steroid Biochem. Mol. Biol.* **2018**, *180*, 41–50. [CrossRef] [PubMed]
128. Hewison, M.; Wagner, C.L.; Hollis, B.W. Vitamin D Supplementation in Pregnancy and Lactation and Infant Growth. *N. Engl. J. Med.* **2018**, *379*, 1880–1881.

 © 2019 by the author. Licensee MDPI, Basel, Switzerland. This article is an open access article distributed under the terms and conditions of the Creative Commons Attribution (CC BY) license (http://creativecommons.org/licenses/by/4.0/).

Review

Mitochondrial Dysfunction in the Aging Retina

Janis T. Eells

Department of Biomedical Sciences, University of Wisconsin-Milwaukee, Milwaukee, WI 53211, USA; jeells@uwm.edu; Tel.: +1-608-215-5405

Received: 12 April 2019; Accepted: 9 May 2019; Published: 11 May 2019

Abstract: Mitochondria are central in retinal cell function and survival and they perform functions that are critical to cell function. Retinal neurons have high energy requirements, since large amounts of ATP are needed to generate membrane potentials and power membrane pumps. Mitochondria over the course of aging undergo a number of changes. Aged mitochondria exhibit decreased rates of oxidative phosphorylation, increased reactive oxygen species (ROS) generation and increased numbers of mtDNA mutations. Mitochondria in the neural retina and the retinal pigment epithelium are particularly susceptible to oxidative damage with aging. Many age-related retinal diseases, including glaucoma and age-related macular degeneration, have been associated with mitochondrial dysfunction. Therefore, mitochondria are a promising therapeutic target for the treatment of retinal disease.

Keywords: aging; mitochondria; retina; optic nerve; diabetic retinopathy; age-related macular degeneration; glaucoma

1. Retinal Anatomy and Physiology

The fundamental organization of the retina is conserved across vertebrates. The retina contains five major neuronal cell classes (photoreceptors, bipolar cells, amacrine cells, horizontal cells, and retinal ganglion cells) with Müller glial cells and retinal pigment epithelial cells providing metabolic and homeostatic support [1] (Figure 1). The interaction of light with light-sensitive pigments in the outer segments of rod and cone photoreceptors initiates signaling mechanisms that convert light energy to neural activity. Photoreceptors synapse with bipolar cells. Bipolar cells synapse with the retinal ganglion cells that transmit the visual signal along their axons (optic nerve) and, ultimately, to the brain. The retinal pigment epithelium envelopes the photoreceptors, which facilitates the diffusion of nutrients, key metabolites, and oxygen from the choroidal vessels. Horizontal cells, amacrine cells, and Muller cells mediate and support the synapses between the retinal neurons. Unique to the Muller cell, its processes span the inner retina and most of the outer retina, assisting in the maintenance of multiple synapses throughout the retina. Pericytes surround the vascular endothelial cells in all retinal layers, forming the blood-retinal barrier.

The retina is the highest oxygen-consuming organ in the human body [2–6]. The inner retinal neurons have the highest metabolic rate of all central nervous tissue, and the oxygen consumption rate of the photoreceptors is several times higher again. The inner segments of the photoreceptor cells are rich in mitochondria providing the ATP that are required by the ionic pumps that drive the 'dark current'. In addition, photoreceptors also metabolize glucose through aerobic glycolysis [2–6].

The key role of aerobic glycolysis in the vertebrate retina has been long recognized and well established [2,5,6]. The majority of aerobic glycolysis occurs in the photoreceptors. Although aerobic glycolysis dominates energy production in the photoreceptors, the retina requires both glycolysis and oxidative phosphorylation to initiate vision [4]. This has been demonstrated in studies showing that the responses of neurons downstream of photoreceptors are abolished by inhibiting glycolysis or by removing O_2 [2]. Recent studies have proposed a model in which glucose from the choroidal blood passes through the retinal pigment epithelium to the retina, where photoreceptors convert it to lactate [6]. Photoreceptors then export the lactate as fuel for the retinal pigment epithelium and for neighboring Müller glial cells. Aerobic glycolysis not only enables enhanced anabolic metabolism in photoreceptors, but also maximizes their function and adaptability to nutrient stress conditions.

Photoreceptors consume more oxygen per gram of tissue weight than any cell in the body [2,5]. This intense degree of oxidative phosphorylation in their inner segments, coupled with high concentrations of polyunsaturated fatty acids in their outer segments, renders the retina susceptible to oxidative stress and lipid peroxidation [2–4]. The retina is exposed to visible light and it contains abundant photosensitizers [1]. Normally, endogenous antioxidants and repair systems minimize oxidative damage. However, with aging and retinal disease, there is an increase in mitochondrial dysfunction and oxidative damage and corresponding decrements in antioxidants and repair systems, resulting in retinal dysfunction and retinal cell loss, which leads to visual impairment [7–13].

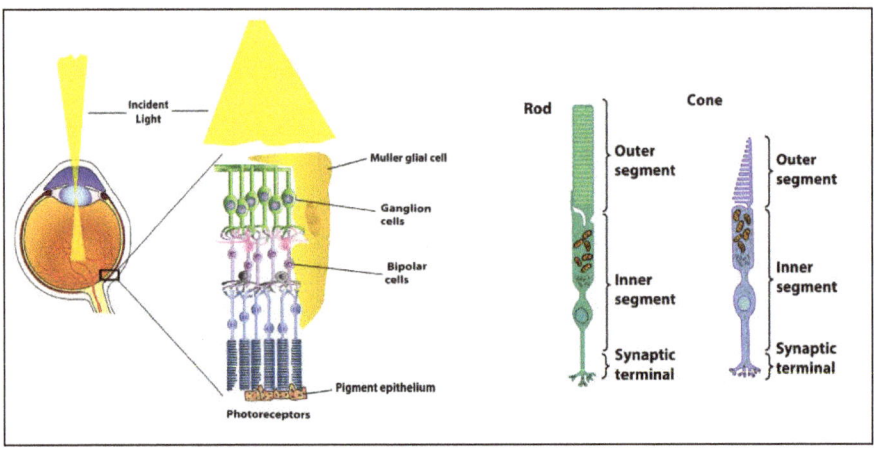

Figure 1. Anatomy of the retina and the structure of rod and cone photoreceptors.

2. Mitochondrial Dysfunction with Aging and Disease

2.1. Age-Related Mitochondrial Changes

Mitochondria are central to retinal cell function and survival and they perform a variety of functions critical to cell metabolism, including oxidative phosphorylation, beta-oxidation of fatty acids, calcium homeostasis, and the regulation of neuronal excitability [7–13]. Retinal neurons have high energy requirements and large amounts of ATP are needed to generate membrane potentials and power membrane pumps. It has been estimated that 90% of mitochondria-generated ATP is used to maintain transmembrane ion gradients [10].

Mitochondria in the aging retina show a decline in the function of the electron transport chain and a significant increase in the generation of reactive oxygen species (ROS) (Figure 2). This increased ROS production leads to increased oxidative damage to mitochondrial DNA, lipids, and proteins [5,6,14–16]. With aging, there is an accumulation of mtDNA mutations and a disruption of mitochondrial structure. Mitochondrial genomic instability has been postulated to play an important role in age-related retinal pathophysiology [16]. However, murine studies have shown that the majority of mtDNA mutations in aged cells appear to be caused by replication errors early in life, rather than by oxidative damage [15]. Mitochondrial DNA repair pathways are limited and they begin to fail with aging, leading to additional mtDNA damage and contributing to the development of retinal degenerative diseases [8,9].

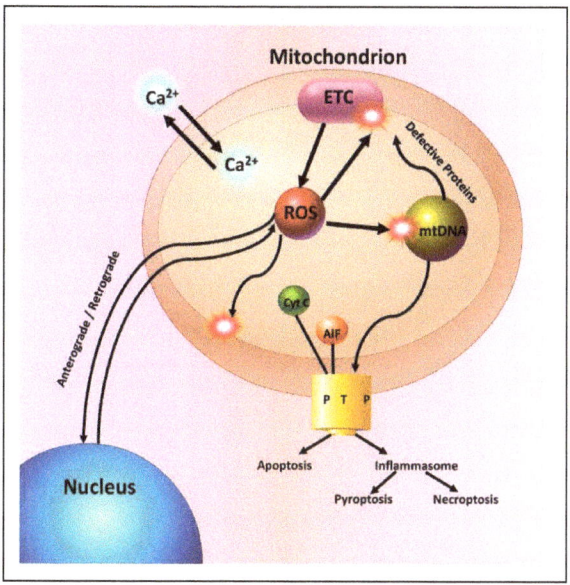

Figure 2. Mitochondria in the aging retina. Mitochondria are essential for many cellular functions including: (1) the synthesis of ATP by oxidative phosphorylation, (2) the regulation of intracellular calcium homeostasis, (3) anterograde and retrograde signaling between the nucleus and mitochondria, (4) the generation of reactive oxygen species (ROS) from the electron transport chain (ETC). ROS act as signaling molecules in low concentrations or as toxic molecules in higher concentrations. ROS oxidize mitochondrial lipids, proteins and DNA and (5) the regulation of apoptosis. Excess ROS or intramitochondrial calcium can lead to the activation of cell death pathways by opening the mitochondrial permeability transition pore (PTP).

2.2. Age-Related Macular Degeneration

Age-related macular degeneration (AMD) is the leading cause of blindness in individuals over 65 in developed countries [17]. As the name implies, AMD primarily affects the central retina or macula. It is characterized by the development of drusen, extracellular lipoprotein deposits under the retinal pigment epithelium (RPE), in the early stages of the disease, followed by the loss of photoreceptors and RPE. Choroidal neovascularization (CNV) develops during later stages of the disease (wet AMD) [17]. Choroidal neovascularization (CNV) is the growth of new blood vessels that originate from the choroid through a break in Bruch's membrane into the subretinal space or sub-retinal pigment epithelium. Dry or atrophic AMD is the most common form of AMD and there are no treatments for this form of AMD.

Aging is a recognized factor that is associated with mitochondrial dysfunction in the outer retina. Other important risk factors for the development of AMD are genetics and environmental stressors, including smoking and exposure to ultraviolet and blue light [18–24]. All of these factors are likely to result in mtDNA damage in the RPE.

Genetic risk factors also contribute to the development of AMD. Genetic polymorphisms of complement factor H (CFH) contribute to AMD pathology. An important relationship between mitochondrial dysfunction and CFH has been recently established [24,25]. These studies have demonstrated mitochondrial abnormalities in Complement factor H null (Cfh$^{-/-}$) mice that are characterized by disrupted mitochondrial morphology and decreased ATP production [24,25]. These findings suggest that CFH plays an important role in retinal oxidative metabolism. Genome wide associated studies have implicated several genes in the cholesterol pathway, including apolipoprotein

E (APOE), in the development of AMD [17]. Recent studies by Lee et al. (2016) have shown that mutations in the transmembrane 135 (Tmem 135) gene that codes for a mitochondrial protein located in RPE and photoreceptors results in accelerated retinal aging in a mouse model of AMD [26].

Although in vitro studies have unequivocally demonstrated that either UV or blue light damage RPE cells [22], epidemiological evidence of the association between exposure to sunlight and AMD is mixed [23]. The ability of the ocular media (cornea, lens, and vitreous humor) to block UV and blue light and the protection of the retina by antioxidant systems, including superoxide dismutase and glutathione peroxidase and macular pigments, including melanin and flavoproteins likely explain these differences [17].

Recent studies suggest that a bioenergetic crisis in the RPE contributes to the pathology of AMD, as illustrated in Figure 3 [20]. Analysis of retinal tissue from human donors with AMD have documented a reduction in the mitochondrial number, a disruption in mitochondrial structure and an increase in mtDNA damage in the RPE that correlates with the severity of disease [15–18]. Recent studies in the primary cultures of RPE derived from human donors with and without AMD have shown decreased rates of ATP synthesis from oxidative phosphorylation and glycolysis in AMD donors consistent with a bioenergetic crisis. Fisher and Ferrington (2018) have suggested that mitochondrial damage is limited to the RPE, because these cells rely nearly exclusively on mitochondrial oxidative phosphorylation, whereas photoreceptors utilize aerobic glycolysis [4–7]. In addition to disruptions in mitochondrial bioenergetics, mitochondrial dysfunction has been shown to disrupt calcium homeostasis and mitochondrial nuclear signaling. Several studies have suggested that therapeutic approaches that target RPE mitochondria may provide an effective early treatment strategy for AMD [18–22].

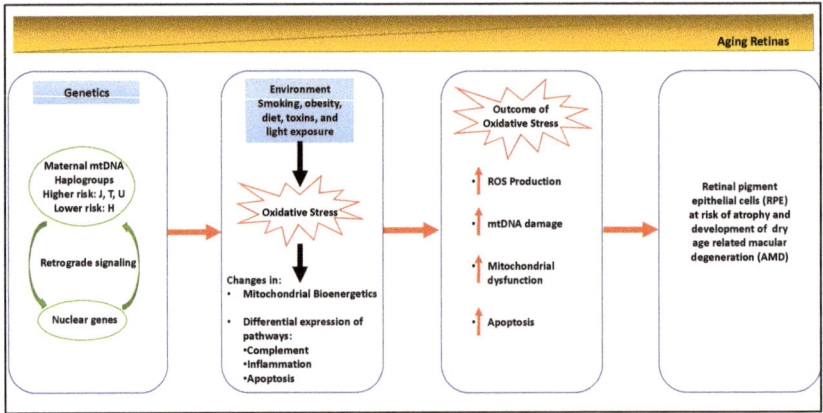

Figure 3. Mitochondrial involvement in dry age-related macular degeneration (AMD).

2.3. Diabetic Retinopathy

Diabetic Retinopathy (DR) is the most common complication of diabetes mellitus. DR is a leading cause of blindness in many developed countries [27]. Chronic hyperglycemia coupled with other risk factors, including hypertension and dyslipidemia, are postulated to trigger a cascade of biochemical and pathophysiological changes that lead to microvascular damage and retinal dysfunction. Mitochondrial dysfunction is a key component of this cascade.

The pathogenesis of diabetic retinopathy is complex, involving the disruption of multiple intracellular pathways [27]. Early pathologic changes that were observed in diabetic retinopathy include mitochondrial dysfunction, oxidative stress, and inflammation [27–30]. Oxidative stress results from the increased production of reactive oxygen species (ROS), including superoxide and hydrogen peroxide. ROS oxidize intracellular proteins, lipids, and nucleic acids, disrupting normal signaling and culminating in disease. Under physiological conditions, small amounts of superoxide leak from the electron transport chain and they are converted to hydrogen peroxide by mitochondrial superoxide dismutase. Hydrogen peroxide diffuses out of the mitochondria into the cytosol and it serves an important signaling molecule. Intracellular ROS concentrations are regulated by a complex array of antioxidant systems to maintain low intracellular ROS concentrations [29]. Under hyperglycemic conditions, the excess production of ROS overwhelms the antioxidant systems, resulting in oxidative damage [11,28,29].

Hyperglycemia is known to disrupt several metabolic pathways that are involved in the pathogenesis of diabetic retinopathy. These pathways include protein kinase C (PKC) activation, accumulation of advanced glycation products (AGEs), the polyol pathway, and activation of the hexosamine pathway [27,28]. Experimental models have shown that, in the etiology of this progressive disease, the activation of NADPH oxidase (Nox) increases cytosolic ROS and this sustained accumulation of ROS damages mitochondria, which further increases oxidative stress [30]. Superoxide radicals (O_2^-) that are produced by the mitochondrial electron transport chain activate these major pathways, and activation, in turn, can damage the mitochondria propagating the vicious cycle of ROS.

A major source of the excess ROS production under hyperglycemic conditions is the mitochondrial electron transport chain (ETC) [28–30]. Under normal physiological conditions, the electrons are donated to complex I or complex II of the ETC and are then passed on to coenzyme Q, complex III, cytochrome C, complex IV, and ultimately to molecular oxygen, the final electron acceptor. The transfer of electrons generates a proton gradient across the mitochondrial intermembrane, which drives the synthesis of ATP by ATP synthase. During this process, small amounts of superoxide are produced, but the cell via antioxidant defense mechanisms easily clears them [20–30].

Complex I of the ETC is generally regarded as the primary source of superoxide production. However, the 2-oxoglutarate dehydrogenase (OGDH), branched-chain 2-oxoacid dehydrogenase (BCKDH), and pyruvate dehydrogenase (PDH) complexes are also capable of considerable superoxide/H_2O_2 production. Moreover, mitochondria can generate ROS, superoxide, or hydrogen peroxide, from at least ten distinct sites on the ETC or associated pathways [31–34].

The precise cause of mitochondrial dysfunction in a diabetic state is unclear. Some studies suggest that the increased glucose concentration overwhelms the electron transport chain by increasing the rate of oxidative phosphorylation [16–19], thus generating ROS. Other studies suggest that mitochondrial DNA is damaged, which results in dysfunctional proteins that are required for ETC complexes, also causing excess ROS and mitochondrial dysfunction [27,28].

Mitochondrial function, structure, and mtDNA are damaged in the retina in diabetic retinopathy [29]. Mitochondrial biogenesis and mtDNA repair mechanisms are compromised [35,36]. The ETC becomes dysfunctional and the import of nuclear DNA encoded proteins is disrupted. These diabetes-induced abnormalities in mitochondria persist after the removal of the hyperglycemic insult and they are implicated in the metabolic memory phenomenon associated with the continued progression of diabetic retinopathy [36,37]. Hyperglycemic insult also results in epigenetic modifications [38]. These epigenetic changes in histones and DNA methylation can be passed to the next generation. Epigenetic modifications also contribute to mitochondrial damage and disrupt mitochondrial homeostasis [38]. Therapeutic strategies directed at mitochondrial homeostasis and epigenetic modifications have the potential to halt or slow the progression of diabetic retinopathy.

2.4. Glaucoma

Glaucoma is a leading cause of blindness, which is characterized by the accelerated death of retinal ganglion cells (RGC), leading to vision loss [39–42]. Age and elevated intraocular pressure (IOP) are two major risk factors in developing glaucoma. Therapies reducing IOP have been shown to slow the progression of glaucoma in many but not all patients. Aging may increase the vulnerability of the optic nerve to various stressors, ultimately resulting in RGC death and optic nerve degeneration [39–41].

A growing body of evidence supports a key role for mitochondrial dysfunction in the pathogenesis of glaucoma [39–51]. Mitochondrial dysfunction has been demonstrated in RGC loss in animal and cultured cell experimental models of glaucoma [47,51]. Mutations in mitochondrial and nuclear genes encoding mitochondrial proteins are known to cause primary optic neuropathies, including Leber's Hereditary Optic Neuropathy (LHON) and Autosomal Dominant Optic Atrophy (AODA) [51]. Recent studies indicate that the bioenergetic consequences of mtDNA and nuclear DNA mutations contribute to the development of Primary Open Angle Glaucoma (POAG) and Normal Tension Glaucoma (NTG) [40–43].

In addition to genetic factors, retinal ganglion cells are profoundly susceptible to mitochondrial dysfunction [42,51]. A combination of the acute energy demand of RGCs and their unique morphology appears to underlie the susceptibility of these neurons to mitochondrial dysfunction. The RGC cell body is located in the ganglion cell layer of the retina. RGCs possess elaborate dendritic arbors that project into the inner plexiform layer to synapse with other retinal neurons. These dendritic arbors are packed with mitochondria that are needed to supply the ATP required to maintain these connections. The long axons of the RGCs form the optic nerve and extend into the brain. The RGC axons make a 90-degree turn at the optic nerve head and pass through the lamina cribrosa, a series of perforated collagen plates, before forming the optic nerve. After passing through the lamina cribrosa, the optic nerve becomes laminated and continues to the visual cortex. The optic nerve has one of the highest oxygen consumption rates and energy demands of any tissue in the body [51]. The mitochondrial density is greater in the unmyelinated region of the optic nerve than in the myelinated region. The unmyelinated prelaminar and laminar regions of the optic nerve require more energy than the myelinated segment of the optic nerve due to the absence of saltatory conduction. This region of the optic nerve has a high density of voltage gated sodium channels and it requires more energy to restore ion gradients.

The combination of genetic susceptibility due to inherited or acquired mutations, energy demand, and unique morphology of retinal ganglion cells in an aging population all contribute to the pathogenesis and incidence of glaucoma (Figure 4).

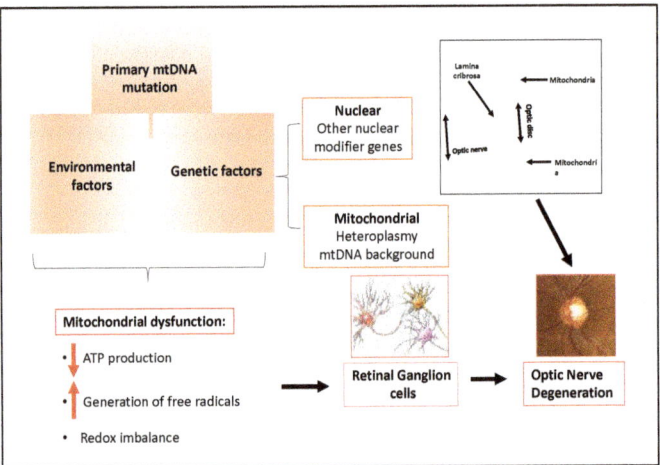

Figure 4. Mitochondrial dysfunction in glaucoma.

3. Conclusions

Experimental evidence has shown a link between mitochondrial dysfunction and the loss of vision. Most mitochondrial diseases exhibit some form of visual impairment. Many retinal and optic nerve diseases, including AMD, diabetic retinopathy, and glaucoma are also characterized by mitochondrial dysfunction. The development of therapeutic approaches that target retinal mitochondrial dysfunction has the potential to profoundly impact the treatment of retinal and optic nerve disease.

Funding: This work was supported in part by the National Institutes of Health [NIH/NEI R43-EY025892 and NIH/NEI P30-EY01931].

Conflicts of Interest: The author declares no conflict of interest.

References

1. Sung, C.H.; Chuang, J.Z. The cell biology of vision. *J. Cell Biol.* **2010**, *190*, 953–963. [CrossRef] [PubMed]
2. Winkler, B.S. Glycolytic and oxidative metabolism in relation to retinal function. *J. Gen. Physiol.* **1981**, *77*, 667–692. [CrossRef]
3. Warburg, O. Uber die klassifizierung tierischer gewebe nach ihrem stoffwechsel. *Biochem. Z.* **1927**, *184*, 484–488.
4. Ames, A., 3rd; Li, Y.Y.; Heher, E.C.; Kimble, C.R. Energy metabolism of rabbit retina as related to function: High cost of sodium transport. *J. Neurosci.* **1992**, *12*, 840–853. [CrossRef]
5. Medrano, C.J.; Fox, D.A. Oxygen consumption in the rat outer and inner retina: Light- and pharmacologically-induced inhibition. *Exp. Eye Res.* **1995**, *61*, 273–284. [CrossRef]
6. Hurley, J.B.; Lindsay, K.J.; Du, J. Glucose, lactate and shuttling of metabolites in vertebrate retinas. *J. Neurosci. Res.* **2015**, *93*, 1079–1092. [CrossRef] [PubMed]
7. Yu, D.Y.; Cringle, S.J. Retinal degeneration and local oxygen metabolism. *Exp. Eye Res.* **2005**, *80*, 745–751. [CrossRef]
8. Barot, M.; Gokulgandhi, M.R.; Mitra, A.K. Mitochondrial dysfunction in retinal diseases. *Curr. Eye Res.* **2011**, *36*, 1069–1077. [CrossRef]
9. Wang, A.L.; Lukas, T.J.; Yuan, M.; Neufeld, A.H. Age-related increase in mitochondrial DNA damage and loss of DNA repair capacity in the neural retina. *Neurobiol. Aging* **2010**, *31*, 2002–2010. [CrossRef]
10. Barron, M.J.; Johnson, M.A.; Andrews, R.M.; Clarke, M.P.; Griffiths, P.G.; Bristow, E.; He, L.-P.; Durham, S.; Turnbull, D.M. Mitochondrial abnormalities in aging macular photoreceptors. *Investig. Ophthalmol. Vis. Sci.* **2001**, *42*, 3016–3022.
11. Country, M.W. Retinal metabolism: A comparative look as energetics in the retina. *Brain Res.* **2017**, *1672*, 50–57. [CrossRef] [PubMed]
12. Lefevere, E.; Toft-Kehler, A.K.; Vohra, R.; Kolko, M.; Moons, L.; van Hovea, I. Mitochondrial dysfunction underlying outer retinal diseases. *Mitochondrion* **2017**, *36*, 66–76. [CrossRef]
13. Jarrett, S.G.; Lin, H.; Godley, B.F.; Boultonc, M.E. Mitochondrial DNA damage and its potential role in retinal degeneration. *Prog. Retin. Eye Res.* **2008**, *27*, 596–607. [CrossRef]
14. Kennedy, S.R.; Salk, J.J.; Schmitt, M.W.; Loeb, L.A. Ultra-sensitive sequencing reveals an age-related increase in somatic mitochondrial mutations that are inconsistent with oxidative damage. *PLoS Genet.* **2013**, *9*, e1003794. [CrossRef]
15. Ameur, A.; Stewart, J.B.; Freyer, C.; Hagström, E.; Ingman, M.; Larsson, N.-G.; Gyllensten, U. Ultra-deep sequencing of mouse mitochondrial DNA: Mutational patterns and their origins. *PLoS Genet.* **2011**, *7*, e1002028. [CrossRef] [PubMed]
16. Lopez-Otin, C.; Blasco, M.A.; Partridge, L.; Serrano, M.; Kroemer, G. The hallmarks of aging. *Cell* **2013**, *153*, 1194–1217. [CrossRef]
17. Lim, L.S.; Mitchell, P.; Seddon, J.M.; Holz, F.G.; Wong, T.Y. Age-related macular degeneration. *Lancet* **2012**, *379*, 1728–1738. [CrossRef]
18. Feher, J.; Kovacs, I.; Artico, M.; Cavallotti, C.; Papale, A.; Gabrieli, C.B. Mitochondrial alterations of retinal pigment epithelium in age-related macular degeneration. *Neurobiol. Aging* **2006**, *27*, 983–993. [CrossRef] [PubMed]
19. Ferrington, D.A.; Ebeling, M.C.; Kapphahn, R.J.; Terluk, M.R.; Fisher, C.R.; Polanco, J.R.; Roehrich, H.; Leary, M.M.; Geng, Z.; Dutton, J.R.; et al. Altered bioenergetics and enhanced resistance to oxidative stress in human retinal pigment epithelial cells from donors with age-related macular degeneration. *Redox Biol.* **2017**, *13*, 255–265. [CrossRef]
20. Fisher, C.R.; Ferrington, D.A. Perspective on AMD pathobiology: A bioenergetic crisis in the RPE. *Investig. Ophthalmol. Vis. Sci.* **2018**, *59*, 41–55. [CrossRef]

21. Turluk, M.R.; Kapphahn, R.J.; Soukup, L.M.; Gong, H.; Gallardo, C.; Montezuma, S.R.; Ferrington, D.A. Investigating mitochondria as a target for treating age-related macular degeneration. *J. Neurosci.* **2015**, *35*, 7304–7311. [CrossRef]
22. King, A.; Gottlieb, E.; Brooks, D.G.; Murphy, M.P.; Dunaief, J.L. Mitochondria-derived reactive oxygen species mediate blue light-induced death of retinal pigment epithelial cells. *Photochem. Photobiol.* **2004**, *79*, 470475. [CrossRef]
23. Zhou, H.; Zhang, H.; Yu, A.; Xie, J. Association between sunlight exposure and risk of age-related macular degeneration: A meta-analysis. *BMC Ophthalmol.* **2018**, *18*, 331–338. [CrossRef]
24. Sivapathasuntharam, C.; Hayes, M.J.; Shinhmar, H.; Kam, J.H.; Sivaprasad, S.; Jeffery, G. Complement factor H regulates retinal development and its absence may establish a footprint for age related macular degeneration. *Sci. Rep.* **2019**, *9*, 1082–1092. [CrossRef]
25. Calaza, K.C.; Kam, J.H.; Hogg, C.; Jeffery, G. Mitochondrial decline precedes phenotype development in the complement factor H mouse model of retinal degeneration but can be corrected by near infrared light. *Neurobiol. Aging* **2015**, *36*, 2869–2876. [CrossRef]
26. Lee, W.-H.; Higuchi, H.; Ikeda, S.; Macke, E.L.; Takimoto, T.; Pattnaik, B.R.; Liu, C.; Chu, L.-F.; Siepka, S.M.; Krentz, K.J.; et al. Mouse Tmem135 mutation reveals a mechanism involving mitochondrial dynamics that leads to age-dependent retinal pathologies. *eLife* **2016**, *5*, e19264. [CrossRef] [PubMed]
27. Antonetti, D.A.; Klein, R.; Gardner, T.W. Mechanisms of disease diabetic retinopathy. *N. Engl. J. Med.* **2012**, *366*, 1227–1239. [CrossRef]
28. Brownlee, M. The pathobiology of diabetic complications; a unifying mechanism. *Diabetes* **2005**, *54*, 1615–1625. [CrossRef]
29. Kowluru, R.A.; Mirsha, M. Oxidative stress, mitochondrial damage and diabetic retinopathy. *Biochim. Biophys. Acta* **2015**, *1852*, 2474–2483. [CrossRef]
30. Roy, S.; Kern, T.S.; Song, B.; Stuebe, C. Mechanistic insights into pathological changes in the diabetic retina: Implications for targeting diabetic retinopathy. *Am. J. Pathol.* **2017**, *187*, 9–19. [CrossRef]
31. St-Pierre, J.; Buckingham, J.A.; Roebuck, S.J.; Brand, M.D. Topology of superoxide production from different sites in the mitochondrial electron transport chain. *J. Biol. Chem.* **2002**, *277*, 44784–44790. [CrossRef]
32. Brand, M.D. The sites and topology of mitochondrial superoxide production. *Exp. Gerontol.* **2010**, *45*, 466–472. [CrossRef]
33. Murphy, M.P. How mitochondria produce reactive oxygen species. *Biochem. J.* **2009**, *417*, 1–13. [CrossRef] [PubMed]
34. Quinlan, C.L.; Goncalves, R.L.S.; Hey-Mogensen, M.; Yadava, N.; Bunik, V.I.; Brand, M.D. The 2-oxoacid dehydrogenase complexes in mitochondria can produce superoxide/hydrogen peroxide at much higher rates than complex I. *J. Biol. Chem.* **2014**, *289*, 8312–8325. [CrossRef]
35. Kowluru, R.A. Diabetic retinopathy, metabolic memory and epigenetic modifications. *Vis. Res.* **2017**, *139*, 30–38. [CrossRef]
36. Kowluru, R.A. Mitochondria damage in the pathogenesis of diabetic retinopathy and in the metabolic memory associated with its continued progression. *Curr. Med. Chem.* **2013**, *20*, 3226–3233. [CrossRef]
37. Kowluru, R.A.; Abbas, S.N. Diabetes-induced mitochondrial dysfunction in the retina. *Investig. Ophthalmol. Vis. Sci.* **2003**, *12*, 5327–5334. [CrossRef]
38. Kowluru, R.A.; Kowluru, A.; Mishra, M.; Kumar, B. Oxidative stress and epigenetic modifications in the pathogenesis of diabetic retinopathy. *Prog. Retin. Eye Res.* **2015**, *48*, 40–61. [CrossRef] [PubMed]
39. Ju, W.K.; Liu, Q.; Kim, K.-Y.; Crowston, J.G.; Lindsey, J.D.; Agarwal, N.; Ellisman, M.H.; Perkins, G.A.; Weinreb, R.N. Elevated hydrostatic pressure triggers mitochondrial fission and decreases cellular ATP in differentiated RGC-5 cells. *Investig. Ophthalmol. Vis. Sci.* **2007**, *48*, 2145–2151. [CrossRef]
40. Abu-Amero, K.K.; Morales, J.; Bosley, T.M. Mitochondrial abnormalities in patients with primary open-angle glaucoma. *Investig. Ophthalmol. Vis. Sci.* **2006**, *47*, 2533–2541. [CrossRef]
41. Moreno, M.C.; Campanelli, J.; Sande, P.; Sáenz, D.A.; Sarmiento, M.I.K.; Rosenstein, R.E. Retinal oxidative stress induced by high intraocular pressure. *Free Radic. Biol. Med.* **2004**, *37*, 803–812. [CrossRef] [PubMed]
42. Kong, G.Y.; Van Bergen, N.J.; Trounce, I.A.; Crowston, J.G. Mitochondrial dysfunction and glaucoma. *J. Glaucoma* **2009**, *18*, 93–100. [CrossRef]

43. Harun-or-Rashid, M.; Pappenhagen, N.; Palmer, P.G.; Smith, M.A.; Gevorgyan, V.; Wilson, G.N.; Crish, S.D.; Inman, D.M. Structural and functional rescue of chronic metabolically stressed optic nerves through respiration. *J. Neurosci.* **2018**, *38*, 5122–5139. [CrossRef] [PubMed]
44. Libby, R.T.; Anderson, M.G.; Pang, I.H.; Robinson, Z.H.; Savinova, O.V.; Cosma, I.M.; Snow, A.; Wilson, L.A.; Smith, R.S.; Clark, A.F.; et al. Inherited glaucoma in DBA/2J mice: Pertinent disease features for studying the neurodegeneration. *Vis. Neurosci.* **2005**, *22*, 637–648. [CrossRef] [PubMed]
45. Baltan, S.; Inman, D.M.; Danilov, C.A.; Morrison, R.S.; Calkins, D.J.; Horner, P.J. Metabolic vulnerability disposes retinal ganglion cell axons to dysfunction in a model of glaucomatous degeneration. *J. Neurosci.* **2010**, *30*, 5644–5652. [CrossRef]
46. Kleesattel, D.; Crish, S.D.; Inman, D.M. Decreased energy capacity and increased autophagic activity in optic nerve axons with defective anterograde transport. *Investig. Ophthalmol. Vis. Sci.* **2015**, *56*, 8215–8227. [CrossRef] [PubMed]
47. Ju, W.-K.; Kim, K.-Y.; Lindsey, J.D.; Angert, M.; Duong-Polk, K.X.; Scott, R.T.; Kim, J.J.; Kukhmazov, I.; Ellisman, M.H.; Perkins, G.A.; et al. Intraocular pressure elevation induces mitochondrial fission and triggers OPA1 release in glaucomatous optic nerve. *Investig. Ophthalmol. Vis. Sci.* **2008**, *49*, 4903–4911. [CrossRef]
48. Coughlin, L.; Morrison, R.S.; Horner, P.J.; Inman, D.M. Mitochondrial morphology differences and mitophagy deficit in murine glaucomatous optic nerve. *Investig. Ophthalmol. Vis. Sci.* **2015**, *56*, 1437–1446. [CrossRef]
49. Williams, P.A.; Harder, J.M.; Foxworth, N.E.; Cochran, K.E.; Philip, V.M.; Porciatti, V.; Smithies, O.; John, S.W.M. Vitamin B3 modulates mitochondrial vulnerability and prevents glaucoma in aged mice. *Science* **2017**, *760*, 756–760. [CrossRef]
50. Ju, W.K.; Kim, K.Y.; Noh, Y.H.; Hoshijima, M.; Lukas, T.J.; Ellisman, M.H.; Weinreb, R.N.; Perkins, G.A. Increased mitochondrial fission and volume density by blocking glutamate excitotoxicity protect glaucomatous optic nerve head astrocytes. *Glia* **2015**, *63*, 736–753. [CrossRef]
51. Lee, S. Mitochondrial dysfunction in glaucoma and emerging bioenergetic therapies. *Exp. Eye Res.* **2011**, *93*, 204–212. [CrossRef]

© 2019 by the author. Licensee MDPI, Basel, Switzerland. This article is an open access article distributed under the terms and conditions of the Creative Commons Attribution (CC BY) license (http://creativecommons.org/licenses/by/4.0/).

Review

Drug-Induced Mitochondrial Toxicity in the Geriatric Population: Challenges and Future Directions

Yvonne Will [1,*], Jefry E. Shields [1] and Kendall B. Wallace [2]

1 SCIENTIA LLC, Ledyard, CT 06339, USA; jefshields@yahoo.ca
2 Department of Biomedical Sciences, University of Minnesota Medical School, Duluth, MN 55455, USA; kwallace@d.umn.edu
* Correspondence: yvonnewillrider@gmail.com

Received: 22 December 2018; Accepted: 12 February 2019; Published: 11 May 2019

Abstract: Mitochondrial function declines with age, leading to a variety of age-related diseases (metabolic, central nervous system-related, cancer, etc.) and medication usage increases with age due to the increase in diseases. Drug-induced mitochondrial toxicity has been described for many different drug classes and can lead to liver, muscle, kidney and central nervous system injury and, in rare cases, to death. Many of the most prescribed medications in the geriatric population carry mitochondrial liabilities. We have demonstrated that, over the past decade, each class of drugs that demonstrated mitochondrial toxicity contained drugs with both more and less adverse effects on mitochondria. As patient treatment is often essential, we suggest using medication(s) with the best safety profile and the avoidance of concurrent usage of multiple medications that carry mitochondrial liabilities. In addition, we also recommend lifestyle changes to further improve one's mitochondrial function, such as weight loss, exercise and nutrition.

Keywords: drug-induced mitochondrial toxicity; polypharmacy; aging

1. Introduction

Known as the powerhouse of the cell, mitochondria are recognized for their ability to produce large amounts of a cell's energy currency, ATP (Adenosine Triphosphate), as well as being the only other organelle that contains DNA apart from the nucleus. Virtually all eukaryotic cells employ mitochondria to produce the majority of ATP needed to perform cellular functions (e.g., replication, protein synthesis, muscle contraction, etc.) and maintain cellular viability. The electron transport chain (ETC) and ATP synthase are interdependent systems in the mitochondria that generate ATP by coupling the electrochemical potential across the inner mitochondrial membrane created by the ETC to the phosphorylation of ADP by ATP synthase. The protein complexes comprising the ETC shuttle electrons down an energy gradient to ultimately reduce molecular oxygen to water, driving oxidative phosphorylation for ATP synthesis. Compromising the function of any of these protein complexes reduces the efficiency of ATP production and endangers the cell's ability to perform its function and can lead to cell death. Additionally, uncoupling the membrane potential from ATP synthesis through the loss of the inner mitochondrial membrane impermeability dissipates the harvested energy as heat and bypasses ATP synthesis and greatly impairs the cell's viability. Mitochondria contain their own DNA (mtDNA) and both inherited and acquired mutations in mtDNA have been linked to well over 100 mitochondrial syndromes such as Leber's hereditary optic neuropathy, mitochondrial myopathy, encephalomyopathy, lactic acidosis, stroke-like symptoms and myoclonic epilepsy with ragged red fibers [1]. MtDNA is susceptible to oxidative stress due to its proximity to the ETC and the fact that it does not contain histones or a repair mechanism.

As we age, mitochondrial function declines and it is described as the "mitochondrial theory of aging", leading to a variety of age-related diseases (Figure 1).

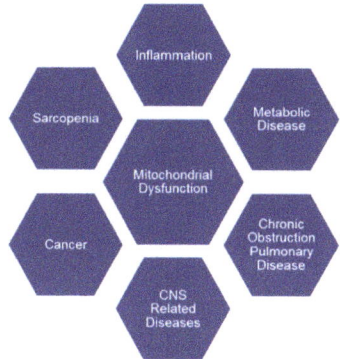

Figure 1. Mitochondrial dysfunction is implicated in many age-related diseases such as metabolic diseases (T2DM, obesity, cardiovascular and cerebrovascular disease, Non-Alcoholic Fatty Liver Disease (NAFDL)) [2–7], CNS-related diseases (Parkinson's, Alzheimer's and Huntington's disease, hearing loss, cataracts) [8–11], inflammation (osteoarthritis) [12], cancer [13,14], sarcopenia [15] and chronic obstructive pulmonary disease (COPD) [16].

These age-related diseases are often treated with multiple medications, some of which are known to cause drug-induced mitochondrial toxicity. In this review, we postulate that drug-induced mitochondrial dysfunction can increase in frequency and severity in the geriatric population due to two independent factors associated with aging, compromised mitochondrial function and polypharmacy.

2. The Mitochondrial Theory of Aging

The "Mitochondrial Theory of Aging" has gained considerable prominence in discussions of longevity, inherent vitality and the biology of mortality. First proposed in his book published in 1928, Raymond Pearl provided a lengthy account of the evidence leading him to posit that longevity is inversely proportional to "the rate of living"; by that, he is referring to the expenditure of energy. His theory is that one is born with a finite quantum of energy, the expenditure of which defines his or her mortality [17]. Although the tenants of this association may hold well for drosophila and perhaps other poikilotherms, in mammals this theory is essentially the antithesis for vitality. There is an abundance of evidence, much of which is current and clinical, demonstrating that energy expenditure is actually vital to delaying mortality and extending one's longevity [18–20]. This is all summarized in the pronouncements that exercise and diet are fundamental to a long and healthy life, perhaps delaying the natural biological regression that is associated with aging.

It is a well-established fact that mitochondrial numbers, function and bioenergetic capacity decline progressively with age [21,22]; Harman summarized this when he coined the term "Mitochondrial Clock" in 1972 [23], which is rooted in his seminal thesis dating from 1956 on the free radical theory of aging [24,25]. This theory is based on the demonstrated accumulation of oxidative damage to lipid, protein, and nucleic acids in aged tissues [26], which Harman attributed to the increased free radical generation observed with aging. The thought is that as biochemical efficiency and physiological function secede with age, the rate of free radical generation increases, leading to increased levels of oxidative damage. This line of thinking continues by prospecting that increased oxidative damage to the ETC within the mitochondrial compartment leads to a progressive loss of coupling efficiency and increased rates of free radical liberation from the electron transport chain. Accordingly, rather than being the consequence of aging, mitochondria are also implicated as the progenitor of the aging process [25]. It becomes quite apparent that the "Mitochondrial Theory of Aging" and the "Free Radical Theory of Aging" are closely intertwined, and it is subject to debate as to which is the cause and which is the consequence of aging (the chicken vs. egg conundrum).

Regardless of whether it is the cause or consequence, hallmarks of mitochondrial aging [19] include a decrease in the number and an increase in the size of mitochondria [27], a net decrease in the cellular mitochondrial membrane potential, a decreased ADP:O ratio (oxidative phosphorylation) and respiratory control index (RCI), and a loss of individual ETC enzyme activities [21,22]; the steady-state [ATP] concentration, however, tends to be unaffected.

Although each of these factors can individually contribute to the progressive loss of mitochondrial function with age, it is far more plausible that they reflect a much broader and concerted decline in homeostatic regulation directed from a molecular level [28]. Molecular hallmarks associated with mitochondrial aging include a decreased mtDNA copy number [29] accompanied by the accumulation of mutations to the mitochondrial genome [30,31], reduced abundance of both nuclear and mitochondrial encoded mRNA [22,32], decreased mitochondrial protein synthesis [33–36], and the dysregulation of the homeostatic balance between mitochondrial biogenesis, fission and mitophagy [19,28].

Ristow and Zarse [20] were first to apply the term "mitohormesis" to conceptualize the loss of the adaptive response of mitochondria by means of regulating mitochondrial homeostasis, contributing to cell and organismal senescence. This has been followed by a number of reviews suggesting that the age-related accumulation of defective mitochondria, accompanied by the loss of ability to clear these from the cell by way of mitophagy and replace them with healthy mitochondria through mitochondrial fission and biogenesis, underpins the mitochondrial theory of aging [19,28,37–39]. This loss of hormesis or mitochondrial homeostasis is equated to a progressive decline in the processes devoted to quality control (QC^{mt}) to sustain mitochondrial fidelity and capacity within the aging cell [19,28,40].

Two major surveillance mechanisms exist within the cell for sensing dysfunctional mitochondria: (1) bioenergetic or nutrient signaling and (2) mitochondrial proteostasis [19,28]. Nutrient sensing is largely by way of activating AMP kinase (AMPK), SIRT1, and the sestrins for a negative nutrient balance, and mTOR for nutrient surplus (Figure 2).

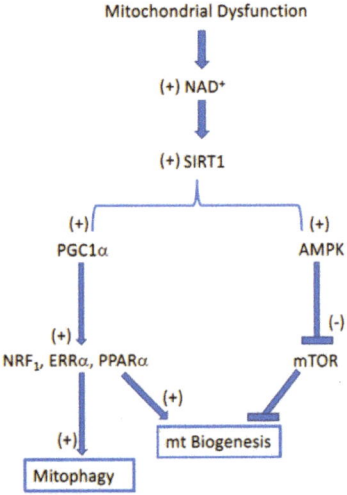

Figure 2. Flow diagram illustrating the interrelationships governing mitochondrial homeostasis in response to the loss of mitochondrial function, such as that which occurs with aging. The Sirt1-dependent regulation of both PGC1α and AMPK provides a well-controlled integration of the disposal of dysfunctional mitochondria (mitophagy) and their replacement with new, supposedly fully functional, mitochondria (biogenesis).

Pyridine nucleotide redox status (NAD$^+$/NADH) is a key factor linking the nutritional status of the cell to both free radical generation and mitochondrial homeostasis. Nutritional deficiency (i.e., mitochondrial dysfunction) is associated with a high NAD$^+$/NADH ratio and low rates of free radical generation. Under these same conditions, the accumulation of high [NAD$^+$] is also associated with high rates of mitochondrial proliferation (biogenesis) and disposal of defective mitochondria (mitophagy), both of which occur by way of the NAD$^+$-dependent activation of SIRT1. The activation of SIRT1 by NAD$^+$ stimulates mitochondrial biogenesis, both by stimulating the PGC1α-dependent activation of mitochondrial biogenesis combined with the AMPK-dependent inhibition of the negative regulator mTOR. The second surveillance pathway, mitochondrial proteostasis, actuates the unfolded protein response (UPR^{mt}) indigenous to mitochondria, where AMPK and mTOR are competing regulators. Collectively, these nutrient sensor pathways regulate both the disposal of defective mitochondria and their replenishment with supposedly fully functional mitochondria.

There is growing evidence that these various mitochondrial signaling pathways are compromised with aging. It may be as simple as the age-associated decrease of [NAD$^+$] [41,42] that is responsible for the lower rates of both mitochondrial biogenesis and mitophagy in aged tissues. Likewise, AMPK and the sirtuins are down-regulated with aging [43–45] as are the concentrations of PGC1α and the rate of mitophagic elimination of dysfunctional mitochondria [46]. These changes suggest loss of molecular mechanisms regulating quality control as being responsible for the ineffective nutrient sensing of mitochondria in aging tissues.

Regardless of the cause, the bioenergetic phenotype of mitochondria from most tissues of aged individuals is very different from that of young adults, which may be a significant factor accounting for the higher rates of drug-induced adverse events in geriatric populations. There is growing appreciation of the possibility that the decline in molecular regulation of mitochondrial hormesis is the primary factor that impedes the ability of aging mitochondria to withstand or adapt to external inputs [26,28,47]. It may be this loss of homeostatic regulation that underlies the sometimes-greater susceptibility of elderly patients to drug-induced adverse events.

3. Drug-Induced Mitochondrial Toxicity

Drug-induced mitochondrial toxicity has been studied in academic settings for well over 50 years. Indeed, many such studies relied on xenobiotic inhibitors, such as rotenone, antimycin and oligomycin, to deduce the function of the ETC, and uncouplers such as 2,4 dinitrophenol, to understand how mitochondria generate ATP. As a result, we should not be surprised that pharmaceutical xenobiotics intended to be therapeutics could also have deleterious 'side effects' on mitochondrial function. Drugs can inhibit mitochondrial function in many different ways such as through the inhibition of ETC protein complexes, inhibition of ATP synthase, inhibition of enzymes of the citric acid cycle, inhibition of various mitochondrial transporters, inhibition of the mitochondrial transcription and translational machinery, as well as through the uncoupling of the ETC from ATP synthase. However, most of these side effects were not detected in preclinical animal studies. This is due to the fact that in vivo toxicity studies are usually done in drug-naïve, young adult animals that have robust mitochondrial reserves; lack of sufficient genetic diversity to allow for idiosyncratic responses; absence of environmental factors; co-medication; insensitivity of histopathology for revealing mitochondrial failure. Much progress has been made in the past decade to develop a variety of high-throughput applicable organelle based and in vitro cell models preclinically [48]. Whereas, traditionally, cells were cultured in high glucose and were unresponsive to mitochondrial toxicity due to shifting to glycolysis for energy production, recently developed cell models grown in galactose can correctly identify potential drug-induced mitochondrial toxicity with greater sensitivity [49,50]. Drugs can also now be tested for potentially causing mitochondrial toxicity in 96- and 384-well formats using solid and soluble oxygen sensors [51,52].

Drug-induced mitochondrial toxicity has been recognized to cause organ toxicity to the liver, skeletal muscle, kidney, heart and the central nervous system. Drug classes identified to cause mitochondrial toxicity are anti-diabetic drugs (thiazolidinediones, fibrates, biguanides), cholesterol lowering drugs (statins), anti-depressants (SARIs), pain medications (NSAIDs), certain antibiotics (fluroquinolones, macrolide), and anti-cancer drugs (kinase inhibitors and anthracyclins) [48]. Most of these observations have been made through studies in isolated mitochondria and cell lines [48].

Over the past decade, our laboratories examined the effects of many different drug classes and found that in each drug class, some members of the class displayed greater potency for adverse effects in vitro than others and the rank order of toxicity observed mimicked the safety profile reported in patients. For example, the anti-diabetic drug Resulin (triglitazone), which was discontinued from the market for liver toxicity, caused greater mitochondrial toxicity when tested in isolated mitochondria than Actos (pioglitazone), which is on the market with a much better safety profile [53]. The same is true for the anti-depressant Zerzone (nefazadone), which belongs to the class of serotonin antagonist reuptake inhibitors. Whereas Zerzone was attried due to liver toxicity caused at least in part by mitochondrial toxicity, Buspar (buspirone) is still on the market and is well tolerated [54]. Another class of anti-diabetic drugs is the biguanides. Phenformin was discontinued because it caused death by lactic acidosis which is considered a hallmark of mitochondrial toxicity. Glucophage (metformin), which causes much less mitochondrial toxicity is widely used in the clinic and only in rare occasions causes lactic acidosis in most likely already predisposed patients [55]. Table 1 provides additional examples of drugs and drug classes studied by our labs.

Table 1. Each drug class contains drugs with more and less observed mitochondrial toxicity.

Drug Class	Rank order of Toxicity Observed (High to Low)	Target Organ
Anti-diabetic (thiazolidinediones)	Trovan * (troglitazone), Avandia (rosiglitazone), Actos (pioglitazone)	Liver [53]
Cholesterol lowering (statins)	Baycol * (cerivastatin), Zocor (simvastatin), Lipitor (atorvastatin), Lescol (fluvastatin)	Muscle [53]
Anti-diabetic (biguanides)	Phenformin * (N-phenethylbiguanide), Buformin *(1-butylbiguanide), Glucophage (metformin)	Lactic acidosis [55]
Anti-depressant/anxiety (SARIs)	Zerzone * (nefazodone), Desyrel (trazodone), Buspar (buspirone)	Liver [54]
Anti-lipidemic (fibrates)	Lopid (gemfibrozil), Lipanor (ciprofibrate), Trilepix (fenofibrate)	Liver [53]
Pain medication (NSAIDs)	Avalanche * (celebrex), Mobic (meloxicam), Voltaran (dichlofenac), Felden (piroxicam), Aspirin (acetylsalicylic acid)	Liver, intestine [56]
Antibiotics (fluoroquinolones)	Trovan * (trovafloxicin), Levaquin (levafloxicin), Cetraxal (ciprofloxicin)	Liver [57]
Anti-cancer (topoisomerase inhibitors)	Adriamycin (doxorubicin)	Heart [58–61]

* withdrawn from the market.

4. Polypharmacy in the Geriatric Population

Geriatric patients are not only predisposed by having lower mitochondrial function due to age but are often also experiencing one or more of the age-related diseases mentioned above (Figure 1). In addition, they are also often taking multiple medications (polypharmacy). In the United States, a 2010 and 2011 survey found that 87% of a representative sampling of 2206 adults aged 62 through 85, used at least one prescription medication and that more than one-third of the group were taking five or more prescription medications [62]. Additionally, it was found that 38% of those surveyed were using over-the-counter medications. A further study by Saraf et al. showed that following acute illness or injury, an average of 14 prescriptions were given to geriatric patients discharged from hospitals to skilled nursing facilities and that over one-third of these prescriptions included side-effects that could aggravate underlying geriatric conditions [63].

Of the most commonly used prescription and over-the-counter drugs in the US for older adults [64], many are known to cause mitochondrial toxicity such as the cholesterol lowering drugs (Zocor, Lipitor, Pravacol, Crestor), pain medication (Aspirin, Tylenol, Aleve) and heartburn medication (Prilosec). Table 2 lists references for mitochondrial dysfunction reported for these prescription and over-the-counter (OTC) medications.

It is important to note that most studies have been conducted using isolated mitochondria and cell systems and often these systems are not of human nature. Also, it is important to understand that the safety margin for a particular drug will depend on exposure (Cmax) [54,97] as well as other contributing mechanistic toxicities [98]. For example, Trovan and Serzone (troglitazone and nefazodone) not only have inhibitory effects on mitochondrial function, but also cause additional toxicities through inhibition of the bile salt efflux pump (BSEP) and the formation of reactive metabolites.

Cholesterol lowering statins are the most commonly prescribed drugs in the geriatric population. Almost 50% of the geriatric population was taking a statin in 2011–2014 versus only 20% in 1999–2002. Statins can cause myopathy in 10–15% of the patients. The adverse events range from mild myalgia and fatigue, to life threatening rhabdomyolysis. Baycol was removed from the market in 2001 because it caused death in 52 patients by rhabdomyolysis, which led to kidney failure [99]. Women, frail individuals and those with low body index as well as patients with increased alcohol consumption are at higher risk. Harper and Jacobson, 2010, noted that Zocor (simvastatin) has greater muscle toxicity and drug interactions and should be avoided if the patient has had adverse events and should be substituted with Lescol (fluvastatin) or Crestor (rosuvastatin) [100].

Mitochondrial toxicity has been postulated as a contributing, if not causal, factor of the observed muscle toxicity. It has been postulated that a deficit in CoQ10 is the cause. CoQ10 is an essential electron carrier in the ETC and the pathway for its synthesis has commonality with that of the cholesterol pathway. A decrease in circulatory CoQ10 levels of 27–50% has been reported [101]. However, since circulating CoQ10 is carried by low-density lipoproteins (LDL) the observed decrease may simply reflect the decrease in circulating levels of LDL [101]. Very few studies have actually examined the level of CoQ10 in muscle, probably due to the invasive nature of human muscle biopsies. While these studies did demonstrate significant decreases (30%) in muscle CoQ10 levels, they failed to demonstrate a decrease in ATP synthesis and no myopathic side effects were reported [102,103]. Since muscle toxicity seems to be patient specific and somewhat rare (the same is true for liver toxicity caused by troglitazone and nefazodone), it is most likely that patients will have either underlying mitochondrial dysfunction (for example, a silent mitochondrial disease) or have different levels of transporter expression (MCT4 in the case of statin accumulation) due to polymorphism in enzymes involved in statin metabolism [104].

Table 2. Most Commonly Used Prescription and Over-the-Counter Medications (OTC) in US Older Adults. Modified from [64].

Medication	Brand Name/Drug Name	% Prescribed	References	Mitochondrial Toxicity Reported
Pain medication (OTC)	Aspirin (acetylsalisylic acid)	40	[65–70]	Inhibition of Respiration, Uncoupling of Oxidative Phosphorylation, Opening of MPT Pore, Inhibition of ATPase, Alteration of Glutathione Status
Cholesterol lowering	Zocor (simvastatin)	22	[53,71–77]	Inhibition of Respiration, Uncoupling of oxidative Phosphorylation, Inhibition of ETC Complexes I, IV, V, Decrease in Membrane Potential, Increase Ca^{++} Release, Decrease ATP Levels
Blood pressure medication	Zestril (lisinopril)	20	No reports	
Diuretic	Microzide (hydrochlorothiazidine)	19	No reports	
Thyroid medication	Synthroid (levothyroxin)	15	No reports	
Heart medication	Lopressor (metoprolol)	15	No reports	
Heartburn (OTC)	Prilosec (omeprazole)	10	[78]	Inhibition of Carnitine/Acylcarnitie Transporter
Cholesterol lowering	Lipitor (atorvastatin)	9	[79–81]	Inhibits Mitochondrial Respiration in Pancreatic, Cardiomyocytes and Endothelial Cells
Pain medication (OTC)	Tylenol (acetaminophen)	9	[82–85]	Opening of MPT Pore, Formation of Reactive Oxygen Species, Depletion of mtDNA
Heart medication	Tenormin (atenolol)	8	No reports	
Diuretic	Lasix (furosemide)	7	[86–88]	Inhibition of Respiration, Uncoupling of Oxidative Phosphorylation
Blood thinner	Plavix (clopidogrel)	7	[89–91]	Inhibition of Respiration, Depletion of Glutathione, Induction of Oxidative Stress, Reduction in Membrane Potential
Blood thinner	Coumadin (warfarin)	6	No reports	
Heart medication	Coreg (carvedilol)	5	[92]	Mixed Reports of Mitochondrial Toxicity vs. Preventive Mode
Cholesterol lowering	Pravachol (pravastatin)	5	[93]	Opening MPT Pore, Inhibition of Respiration at High Concentrations
Cholesterol lowering	Crestor (rosuvastatin)	5	[94]	Mixed Reports of Mitochondrial Toxicity vs. Preventive Mode
Pain medication (OTC)	Aleve (naproxen)	5	[95,96]	Inhibition of Respiration, Inhibition of Ca^{++} Flux
Cholesterol lowering	Zetia (ezetimibe)	5	No reports	

Beyond the effect on CoQ10, statins have been reported to have direct effects on the ETC, especially the lipophilic statins, simvastatin and fluvastatin. In addition, statins have the ability to uncouple oxidative phosphorylation, inhibit fatty acid beta oxidation, induce mitochondrial permeability transition, and cause oxidative stress and decrease mtDNA levels by thus far unidentified mechanisms [105].

It has been reported that elderly patients taking cholesterol lowering drugs (Zocor, Lipitor, etc.) have shown drug–drug interactions (DDI), particularly with the anti-lipidemic drug Lopid (gemfibrozil [53,106]), but also with certain antibiotic classes, such as macrolides [107,108] and fluoroquinolones [109,110] and the heart medication, Nexteron (amiodarone) [64,111]. All of these drugs have been shown to cause mitochondrial toxicity by targeting a variety of, as well as multiple, mitochondrial mechanisms/targets.

While it is well documented that this is mainly due to changes in the drug metabolizing function in elderly patients, we hypothesize that polypharmacy with multiple drugs affecting mitochondrial function could contribute to the toxicities observed. We already spoke of the general mitochondrial decay with age and in age-related diseases.

How do we imagine an individual's lifestyle affects their mitochondrial health and energy production, or their response to medications that carry a potential mitochondrial liability (Table 3)? Differences in body mass index (BMI) have shown that obesity correlates with mitochondrial dysfunction [112]. Additionally, diets of high fat and high sucrose have been shown to be detrimental to mitochondrial function [113,114]. Mitochondria are also weakened by inactive lifestyles and can be revitalized with exercise [115,116]. Studies by DiNicolantonio et al. show deleterious effects on mitochondrial function when consuming large quantities of sugar-containing foods and drinks [117], while other labs have shown that imbibing red wine in moderation [118] and drinking green tea [119] have an advantageous effect on mitochondrial function. Even the choice of area of residency can lead to mitochondrial impairment. Lim et al. have studied the effects of incidental ingestion of pesticides/herbicides and have shown a correlation between the area of residence and increased exposure to exogenous agricultural chemicals which have adverse effects on mitochondrial function [120].

Table 3. Comparison of two geriatric people.

Variable	Patient 1	Patient 2
Age	75	75
Body Mass Index	18	28
Breakfast	Yogurt with granola, fresh fruit, poached egg, green tea	Eggs, bacon and potatoes
Exercise	1 h walk daily, swimming once a week	From the car to the house, around the grocery store
Lunch	Hummus and vegetables, banana, oat meal cookie	French fries and hot dogs, ice cream sundae
Medications	None	Zocor, Lopid, Prilosec, Voltaren (oral), Lasix, Nexteron
Dinner	Wild caught salmon and salad	Pasta with sausage and cheese
Drink Consumption	Red wine	Coke

Now, we imagine how different individuals with ideal vs. less than ideal lifestyles might tolerate mitochondrial insults from medications. "Patient 1" has a lifestyle that promotes mitochondrial health while "Patient 2's" lifestyle predisposes them to mitochondrial decay and toxicity in a number of ways. Taking medications that impair mitochondria is far more concerning for Patient 2 than for Patient 1. Patient 2's healthcare providers could examine their lifestyle and medication regiment and act in unison to come up with a medication plan that minimizes mitochondrial liabilities with the drugs they are prescribing for them, opting for drugs with the best safety profile. As mentioned previously, the combination of Zocor with Lopid should at least be substituted with Crestor. An improvement in Patient 2's diet could eliminate the daily use of Prilosec and Patient 2's pain medication of choice should be Aspirin or Tylenol in moderation.

The question arises if mitochondrial dysfunction/toxicity caused by these medications could be prevented or counteracted by natural compounds or drugs recently developed to treat mitochondrial diseases. A recent review by Rai et al., 2016, provides an overview of a variety of pharmaceuticals which either target mitochondria or maintain its homeostasis [121]. While the majority of these drugs target diseases caused by mutations (MELAS, LHON, MERFF, etc.), one therapy is being developed for improving skeletal muscle function in elderly patients. Elamipretide (MTP-131) has been shown to prevent peroxidation of cardiolipin by binding to the molecule and preventing its interaction with cytochrome c, which prevents the latter from functioning as a peroxidase rather than an electron carrier. The drug is currently in Phase III [121]. Another compound is Idebenone, which is based structurally upon CoQ10. Idebanone is currently being explored for LHON and MELAS [121], but not as a supplemental drug in statin therapy. There are a host of natural products and supplements that have been discussed to improve mitochondrial function such as resveratrol, curcumin, NAD and N-acetyl cysteine (NAC), but there is no clinical evidence of clear success.

5. Conclusions

As patient treatment is often essential, we suggest using medication(s) with the best safety profile and the avoidance of concurrent usage of multiple medications that carry mitochondrial liabilities. Our hope is that future therapies will be devoid of mitochondrial liabilities as screening paradigms can be deployed preclinically. In addition, we also recommend lifestyle changes to further improve one's mitochondrial function, such as weight loss, exercise and nutrition. Another major point is that the cause for mitochondrial dysfunction with age appears to be one of molecular homeostatic control of mitochondrial integrity or functional capacity, not simply an acute depletion of bioenergetic substrates, such as ATP. Consequently, bolus supplementation with mitochondrial cofactors has less potential than long-term lifestyle changes in extending vitality and drug tolerance in aging populations.

Funding: This research received no external funding.

Conflicts of Interest: The authors declare no conflict of interest.

References

1. Dimauro, S.; Davidzon, G. Mitochondrial DNA and disease. *Ann. Med.* **2005**, *37*, 222–232. [CrossRef]
2. Blake, R.; Trounce, I.A. Mitochondrial dysfunction and complications associated with diabetes. *Biochim. Biophys. Acta* **2014**, *1840*, 1404–1412. [CrossRef] [PubMed]
3. Peng, K.Y.; Watt, M.J.; Rensen, S.; Greve, J.W.; Huynh, K.; Jayawardana, K.S.; Meikle, P.J.; Meex, R.C.R. Mitochondrial dysfunction-related lipid changes occur in non-alcoholic fatty liver disease progression. *J. Lipid Res.* **2018**. [CrossRef] [PubMed]
4. Bonora, M.; Wieckowski, M.R.; Sinclair, D.A.; Kroemer, G.; Pinton, P.; Galluzzi, L. Targeting mitochondria for cardiovascular disorders: Therapeutic potential and obstacles. *Nat. Rev. Cardiol.* **2019**, *16*, 33–55. [CrossRef]
5. Hoppel, C.L.; Lesnefsky, E.J.; Chen, Q.; Tandler, B. Mitochondrial Dysfunction in Cardiovascular Aging. *Adv. Exp. Med. Biol.* **2017**, *982*, 451–464. [PubMed]
6. Grimm, A.; Eckert, A. Brain aging and neurodegeneration: From a mitochondrial point of view. *J. Neurochem.* **2017**, *143*, 418–431. [CrossRef] [PubMed]
7. Anupama, N.; Sindhu, G.; Raghu, K.G. Significance of mitochondria on cardiometabolic syndromes. *Fundam. Clin. Pharmacol.* **2018**, *32*, 346–356. [CrossRef] [PubMed]
8. Macdonald, R.; Barnes, K.; Hastings, C.; Mortiboys, H. Mitochondrial abnormalities in Parkinson's disease and Alzheimer's disease: Can mitochondria be targeted therapeutically? *Biochem. Soc. Trans.* **2018**, *46*, 891–909. [CrossRef]
9. Franco-Iborra, S.; Vila, M.; Perier, C. Mitochondrial Quality Control in Neurodegenerative Diseases: Focus on Parkinson's Disease and Huntington's Disease. *Front. Neurosci.* **2018**, *12*, 342. [CrossRef]

10. Falah, M.; Houshmand, M.; Najafi, M.; Balali, M.; Mahmoudian, S.; Asghari, A.; Emamdjomeh, H.; Farhadi, M. The potential role for use of mitochondrial DNA copy number as predictive biomarker in presbycusis. *Ther. Clin. Risk Manag.* **2016**, *12*, 1573–1578. [CrossRef]
11. Leruez, S.; Marill, A.; Bresson, T.; de Saint Martin, G.; Buisset, A.; Muller, J.; Tessier, L.; Gadras, C.; Verny, C.; Gohier, P.; et al. A Metabolomics Profiling of Glaucoma Points to Mitochondrial Dysfunction, Senescence, and Polyamines Deficiency. *Investig. Ophthalmol. Vis. Sci.* **2018**, *59*, 4355–4361. [CrossRef]
12. Bolduc, J.A.; Collins, J.A.; Loeser, R.F. Reactive Oxygen Species, Aging and Articular Cartilage Homeostasis. *Free Radic. Biol. Med.* **2019**, *132*, 73–82. [CrossRef]
13. Ng Kee Kwong, F.; Nicholson, A.G.; Harrison, C.L.; Hansbro, P.M.; Adcock, I.M.; Chung, K.F. Is mitochondrial dysfunction a driving mechanism linking COPD to nonsmall cell lung carcinoma? *Eur. Respir. Rev.* **2017**, *26*. [CrossRef]
14. Zhu, Y.; Dean, A.E.; Horikoshi, N.; Heer, C.; Spitz, D.R.; Gius, D. Emerging evidence for targeting mitochondrial metabolic dysfunction in cancer therapy. *J. Clin. Investig.* **2018**, *128*, 3682–3691. [CrossRef]
15. Always, S.E.; Mohamed, J.S.; Myers, M.J. Mitochondria Initiate and Regulate Sarcopenia. *Exerc. Sport Sci. Rev.* **2017**, *45*, 58–69. [CrossRef]
16. Hara, H.; Kuwano, K.; Araya, J. Mitochondrial Quality Control in COPD and IPF. *Cells* **2018**, *7*, 86. [CrossRef]
17. Raymond, P. *The Rate of Living*; Johns Hopkins University: New York, NY, USA, 1928.
18. Lanza, I.R.; Nair, K.S. Mitochondrial function as a determinant of life span. *Pflugers Arch.* **2010**, *459*, 277–289. [CrossRef]
19. Lopez-Otin, C.; Galluzzi, L.; Freije, J.M.P.; Madeo, F.; Kroemer, G. Metabolic Control of Longevity. *Cell* **2016**, *166*, 802–821. [CrossRef]
20. Ristow, M.; Zarse, K. How increased oxidative stress promotes longevity and metabolic health: The concept of mitochondrial hormesis (mitohormesis). *Exp. Gerontol.* **2010**, *45*, 410–418. [CrossRef]
21. Conley, K.E.; Jubrias, S.A.; Esselman, P.C. Oxidative capacity and ageing in human muscle. *J. Physiol.* **2000**, *526*, 203–210. [CrossRef]
22. Short, K.R.; Bigelow, M.L.; Kahl, J.; Singh, R.; Coenen-Schimke, J.; Raghavakaimal, S.; Nair, K.S. Decline in skeletal muscle mitochondrial function with aging in humans. *Proc. Natl. Acad. Sci. USA* **2005**, *102*, 5618–5623. [CrossRef]
23. Harman, D. The biological clock: The mitochondria? *J. Am. Geriatr. Soc.* **1972**, *20*, 145–147. [CrossRef]
24. Harman, D. Aging: A theory based on free radical and radiation chemistry. *J. Gerontol.* **1956**, *11*, 298–300. [CrossRef]
25. Harman, D. Free radical theory of aging: Consequences of mitochondrial aging. *Age* **1983**, *6*, 86–94. [CrossRef]
26. Chistiakov, D.A.; Sobenin, I.A.; Revin, V.V.; Orekhov, A.N.; Bobryshev, Y.V. Mitochondrial aging and age-related dysfunction of mitochondria. *Biomed. Res. Int.* **2014**, *2014*, 238463. [CrossRef]
27. Tauchi, H.; Sato, T. Age changes in size and number of mitochondria of human hepatic cells. *J. Gerontol.* **1968**, *23*, 454–461. [CrossRef]
28. Lopez-Lluch, G. Mitochondrial activity and dynamics changes regarding metabolism in ageing and obesity. *Mech. Ageing Dev.* **2017**, *162*, 108–121. [CrossRef]
29. Welle, S.; Bhatt, K.; Shah, B.; Needler, N.; Delehanty, J.M.; Thornton, C.A. Reduced amount of mitochondrial DNA in aged human muscle. *J. Appl. Physiol. (1985)* **2003**, *94*, 1479–1484. [CrossRef]
30. Larsson, N.G. Somatic mitochondrial DNA mutations in mammalian aging. *Annu. Rev. Biochem.* **2010**, *79*, 683–706. [CrossRef]
31. Bratic, A.; Larsson, N.G. The role of mitochondria in aging. *J. Clin. Investig.* **2013**, *123*, 951–957. [CrossRef]
32. Lanza, I.R.; Short, D.K.; Short, K.R.; Raghavakaimal, S.; Basu, R.; Joyner, M.J.; McConnell, J.P.; Nair, K.S. Endurance exercise as a countermeasure for aging. *Diabetes* **2008**, *57*, 2933–2942. [CrossRef]
33. Bailey, P.J.; Webster, G.C. Lowered rates of protein synthesis by mitochondria isolated from organisms of increasing age. *Mech. Ageing Dev.* **1984**, *24*, 233–241. [CrossRef]
34. Marcus, D.L.; Ibrahim, N.G.; Freedman, M.L. Age-related decline in the biosynthesis of mitochondrial inner membrane proteins. *Exp. Gerontol.* **1982**, *17*, 333–341. [CrossRef]
35. Rooyackers, O.E.; Adey, D.B.; Ades, P.A.; Nair, K.S. Effect of age on in vivo rates of mitochondrial protein synthesis in human skeletal muscle. *Proc. Natl. Acad. Sci. USA* **1996**, *93*, 15364–15369. [CrossRef]
36. Rooyackers, O.E.; Kersten, A.H.; Wagenmakers, A.J. Mitochondrial protein content and in vivo synthesis rates in skeletal muscle from critically ill rats. *Clin. Sci. (Lond.)* **1996**, *91*, 475–481. [CrossRef]

37. Mammucari, C.; Rizzuto, R. Signaling pathways in mitochondrial dysfunction and aging. *Mech. Ageing Dev.* **2010**, *131*, 536–543. [CrossRef]
38. Vendelbo, M.H.; Nair, K.S. Mitochondrial longevity pathways. *Biochim. Biophys. Acta* **2011**, *1813*, 634–644. [CrossRef]
39. Palikaras, K.; Lionaki, E.; Tavernarakis, N. Coordination of mitophagy and mitochondrial biogenesis during ageing in *C. elegans*. *Nature* **2015**, *521*, 525–528. [CrossRef]
40. Tatsuta, T.; Langer, T. Quality control of mitochondria: Protection against neurodegeneration and ageing. *EMBO J.* **2008**, *27*, 306–314. [CrossRef]
41. Verdin, E. NAD(+) in aging, metabolism, and neurodegeneration. *Science* **2015**, *350*, 1208–1213. [CrossRef]
42. Gomes, A.P.; Price, N.L.; Ling, A.J.; Moslehi, J.J.; Montgomery, M.K.; Rajman, L.; White, J.P.; Teodoro, J.S.; Wrann, C.D.; Hubbard, B.P.; et al. Declining NAD(+) induces a pseudohypoxic state disrupting nuclear-mitochondrial communication during aging. *Cell* **2013**, *155*, 1624–1638. [CrossRef] [PubMed]
43. Lee, J.H.; Budanov, A.V.; Karin, M. Sestrins orchestrate cellular metabolism to attenuate aging. *Cell Metab.* **2013**, *18*, 792–801. [CrossRef] [PubMed]
44. Chang, H.C.; Guarente, L. SIRT1 mediates central circadian control in the SCN by a mechanism that decays with aging. *Cell* **2013**, *153*, 1448–1460. [CrossRef] [PubMed]
45. Guarente, L. Introduction: Sirtuins in aging and diseases. *Methods Mol. Biol.* **2013**, *1077*, 3–10. [PubMed]
46. Chabi, B.; Ljubicic, V.; Menzies, K.J.; Huang, J.H.; Saleem, A.; Hood, D.A. Mitochondrial function and apoptotic susceptibility in aging skeletal muscle. *Aging Cell* **2008**, *7*, 2–12. [CrossRef] [PubMed]
47. Weber, T.A.; Reichert, A.S. Impaired quality control of mitochondria: Aging from a new perspective. *Exp. Gerontol.* **2010**, *45*, 503–511. [CrossRef] [PubMed]
48. Will, Y.; Dykens, J. Mitochondrial toxicity assessment in industry–a decade of technology development and insight. *Expert Opin. Drug Metab. Toxicol.* **2014**, *10*, 1061–1067. [CrossRef]
49. Swiss, R.; Will, Y. Assessment of mitochondrial toxicity in HepG2 cells cultured in high-glucose- or galactose-containing media. *Curr. Protoc. Toxicol.* **2011**, *49*, 2.20.1–2.20.14. [CrossRef]
50. Kamalian, L.; Chadwick, A.E.; Bayliss, M.; French, N.S.; Monshouwer, M.; Snoeys, J.; Park, B.K. The utility of HepG2 cells to identify direct mitochondrial dysfunction in the absence of cell death. *Toxicol. In Vitro* **2015**, *29*, 732–740. [CrossRef]
51. Hynes, J.; Swiss, R.L.; Will, Y. High-Throughput Analysis of Mitochondrial Oxygen Consumption. *Methods Mol. Biol.* **2018**, *1782*, 71–87.
52. Nadanaciva, S.; Rana, P.; Beeson, G.C.; Chen, D.; Ferrick, D.A.; Beeson, C.C.; Will, Y. Assessment of drug-induced mitochondrial dysfunction via altered cellular respiration and acidification measured in a 96-well platform. *J. Bioenerg. Biomembr.* **2012**, *44*, 421–437. [CrossRef]
53. Nadanaciva, S.; Dykens, J.A.; Bernal, A.; Capaldi, R.A.; Will, Y. Mitochondrial impairment by PPAR agonists and statins identified via immunocaptured OXPHOS complex activities and respiration. *Toxicol. Appl. Pharmacol.* **2007**, *223*, 277–287. [CrossRef]
54. Dykens, J.A.; Jamieson, J.D.; Marroquin, L.D.; Nadanaciva, S.; Xu, J.J.; Dunn, M.C.; Smith, A.R.; Will, Y. In vitro assessment of mitochondrial dysfunction and cytotoxicity of nefazodone, trazodone, and buspirone. *Toxicol. Sci.* **2008**, *103*, 335–345. [CrossRef]
55. Dykens, J.A.; Jamieson, J.; Marroquin, L.; Nadanaciva, S.; Billis, P.A.; Will, Y. Biguanide-induced mitochondrial dysfunction yields increased lactate production and cytotoxicity of aerobically-poised HepG2 cells and human hepatocytes in vitro. *Toxicol. Appl. Pharmacol.* **2008**, *233*, 203–210. [CrossRef]
56. Nadanaciva, S.; Aleo, M.D.; Strock, C.J.; Stedman, D.B.; Wang, H.; Will, Y. Toxicity assessments of nonsteroidal anti-inflammatory drugs in isolated mitochondria, rat hepatocytes, and zebrafish show good concordance across chemical classes. *Toxicol. Appl. Pharmacol.* **2013**, *272*, 272–280. [CrossRef]
57. Nadanaciva, S.; Dillman, K.; Gebhard, D.F.; Shrikhande, A.; Will, Y. High-content screening for compounds that affect mtDNA-encoded protein levels in eukaryotic cells. *J. Biomol. Screen.* **2010**, *15*, 937–948. [CrossRef]
58. Solem, L.E.; Henry, T.R.; Wallace, K.B. Disruption of mitochondrial calcium homeostasis following chronic doxorubicin administration. *Toxicol. Appl. Pharmacol.* **1994**, *129*, 214–222. [CrossRef]
59. Palmeira, C.M.; Serrano, J.; Kuehl, D.W.; Wallace, K.B. Preferential oxidation of cardiac mitochondrial DNA following acute intoxication with doxorubicin. *Biochim. Biophys. Acta* **1997**, *1321*, 101–106. [CrossRef]
60. Wallace, K.B. Doxorubicin-induced cardiac mitochondrionopathy. *Pharmacol. Toxicol.* **2003**, *93*, 105–115. [CrossRef]

61. Sardao, V.A.; Oliveira, P.J.; Holy, J.; Oliveira, C.R.; Wallace, K.B. Doxorubicin-induced mitochondrial dysfunction is secondary to nuclear p53 activation in H9c2 cardiomyoblasts. *Cancer Chemother. Pharmacol.* **2009**, *64*, 811–827. [CrossRef]
62. Qato, D.M.; Wilder, J.; Schumm, L.P.; Gillet, V.; Alexander, G.C. Changes in Prescription and Over-the-Counter Medication and Dietary Supplement Use among Older Adults in the United States, 2005 vs 2011. *JAMA Intern. Med.* **2016**, *176*, 473–482. [CrossRef]
63. Saraf, A.A.; Petersen, A.W.; Simmons, S.F.; Schnelle, J.F.; Bell, S.P.; Kripalani, S.; Myers, A.P.; Mixon, A.S.; Long, E.A.; Jacobsen, J.M.; et al. Medications associated with geriatric syndromes and their prevalence in older hospitalized adults discharged to skilled nursing facilities. *J. Hosp. Med.* **2016**, *11*, 694–700. [CrossRef]
64. Merel, S.E.; Paauw, D.S. Common Drug Side Effects and Drug-Drug Interactions in Elderly Adults in Primary Care. *J. Am. Geriatr. Soc.* **2017**, *65*, 1578–1585. [CrossRef]
65. Chatterjee, S.S.; Stefanovich, V. Influence of anti-inflammatory agents on rat liver mitochondrial ATPase. *Arzneimittel-Forschung* **1976**, *26*, 499–502.
66. Nulton-Persson, A.C.; Szweda, L.I.; Sadek, H.A. Inhibition of cardiac mitochondrial respiration by salicylic acid and acetylsalicylate. *J. Cardiovasc. Pharmacol.* **2004**, *44*, 591–595. [CrossRef]
67. Raza, H.; John, A. Implications of altered glutathione metabolism in aspirin-induced oxidative stress and mitochondrial dysfunction in HepG2 cells. *PLoS ONE* **2012**, *7*, e36325. [CrossRef]
68. Deschamps, D.; Fisch, C.; Fromenty, B.; Berson, A.; Degott, C.; Pessayre, D. Inhibition by salicylic acid of the activation and thus oxidation of long chain fatty acids. Possible role in the development of Reye's syndrome. *J. Pharmacol. Exp. Ther.* **1991**, *259*, 894–904.
69. Oh, K.W.; Qian, T.; Brenner, D.A.; Lemasters, J.J. Salicylate enhances necrosis and apoptosis mediated by the mitochondrial permeability transition. *Toxicol. Sci.* **2003**, *73*, 44–52. [CrossRef]
70. Zimmerman, H.J. Effects of aspirin and acetaminophen on the liver. *Arch. Intern. Med.* **1981**, *141*, 333–342. [CrossRef]
71. Bonifacio, A.; Mullen, P.J.; Mityko, I.S.; Navegantes, L.C.; Bouitbir, J.; Krahenbuhl, S. Simvastatin induces mitochondrial dysfunction and increased atrogin-1 expression in H9c2 cardiomyocytes and mice in vivo. *Arch. Toxicol.* **2016**, *90*, 203–215. [CrossRef]
72. Fisar, Z.; Hroudova, J.; Singh, N.; Koprivova, A.; Maceckova, D. Effect of Simvastatin, Coenzyme Q10, Resveratrol, Acetylcysteine and Acetylcarnitine on Mitochondrial Respiration. *Folia Biol.* **2016**, *62*, 53–66.
73. van Diemen, M.P.J.; Berends, C.L.; Akram, N.; Wezel, J.; Teeuwisse, W.M.; Mik, B.G.; Kan, H.E.; Webb, A.; Beenakker, J.W.M.; Groeneveld, G.J. Validation of a pharmacological model for mitochondrial dysfunction in healthy subjects using simvastatin: A randomized placebo-controlled proof-of-pharmacology study. *Eur. J. Pharmacol.* **2017**, *815*, 290–297. [CrossRef]
74. Busanello, E.N.B.; Figueira, T.R.; Marques, A.C.; Navarro, C.D.C.; Oliveira, H.C.F.; Vercesi, A.E. Facilitation of Ca(2+)-induced opening of the mitochondrial permeability transition pore either by nicotinamide nucleotide transhydrogenase deficiency or statins treatment. *Cell Biol. Int.* **2018**, *42*, 742–746. [CrossRef]
75. Sirvent, P.; Bordenave, S.; Vermaelen, M.; Roels, B.; Vassort, G.; Mercier, J.; Raynaud, E.; Lacampagne, A. Simvastatin induces impairment in skeletal muscle while heart is protected. *Biochem. Biophys. Res. Commun.* **2005**, *338*, 1426–1434. [CrossRef]
76. Sirvent, P.; Mercier, J.; Vassort, G.; Lacampagne, A. Simvastatin triggers mitochondria-induced Ca2+ signaling alteration in skeletal muscle. *Biochem. Biophys. Res. Commun.* **2005**, *329*, 1067–1075. [CrossRef]
77. Wagner, B.K.; Kitami, T.; Gilbert, T.J.; Peck, D.; Ramanathan, A.; Schreiber, S.L.; Golub, T.R.; Mootha, V.K. Large-scale chemical dissection of mitochondrial function. *Nat. Biotechnol.* **2008**, *26*, 343–351. [CrossRef]
78. Tonazzi, A.; Giangregorio, N.; Console, L.; Indiveri, C. Mitochondrial carnitine/acylcarnitine translocase: Insights in structure/function relationships. Basis for drug therapy and side effects prediction. *Mini Rev. Med. Chem.* **2015**, *15*, 396–405. [CrossRef]
79. Urbano, F.; Bugliani, M.; Filippello, A.; Scamporrino, A.; Di Mauro, S.; Di Pino, A.; Scicali, R.; Noto, D.; Rabuazzo, A.M.; Averna, M.; et al. Atorvastatin but Not Pravastatin Impairs Mitochondrial Function in Human Pancreatic Islets and Rat beta-Cells. Direct Effect of Oxidative Stress. *Sci. Rep.* **2017**, *7*, 11863. [CrossRef]
80. Godoy, J.C.; Niesman, I.R.; Busija, A.R.; Kassan, A.; Schilling, J.M.; Schwarz, A.; Alvarez, E.A.; Dalton, N.D.; Drummond, J.C.; Roth, D.M.; et al. Atorvastatin, but not pravastatin, inhibits cardiac Akt/mTOR signaling and disturbs mitochondrial ultrastructure in cardiac myocytes. *FASEB J.* **2019**, *33*, 1209–1225. [CrossRef]

81. Broniarek, I.; Jarmuszkiewicz, W. Atorvastatin affects negatively respiratory function of isolated endothelial mitochondria. *Arch. Biochem. Biophys.* **2018**, *637*, 64–72. [CrossRef]
82. Hinson, J.A.; Reid, A.B.; McCullough, S.S.; James, L.P. Acetaminophen-induced hepatotoxicity: Role of metabolic activation, reactive oxygen/nitrogen species, and mitochondrial permeability transition. *Drug Metab. Rev.* **2004**, *36*, 805–822. [CrossRef]
83. Masubuchi, Y.; Suda, C.; Horie, T. Involvement of mitochondrial permeability transition in acetaminophen-induced liver injury in mice. *J. Hepatol.* **2005**, *42*, 110–116. [CrossRef]
84. Prill, S.; Bavli, D.; Levy, G.; Ezra, E.; Schmalzlin, E.; Jaeger, M.S.; Schwarz, M.; Duschl, C.; Cohen, M.; Nahmias, Y. Real-time monitoring of oxygen uptake in hepatic bioreactor shows CYP450-independent mitochondrial toxicity of acetaminophen and amiodarone. *Arch. Toxicol.* **2016**, *90*, 1181–1191. [CrossRef]
85. Moles, A.; Torres, S.; Baulies, A.; Garcia-Ruiz, C.; Fernandez-Checa, J.C. Mitochondrial-Lysosomal Axis in Acetaminophen Hepatotoxicity. *Front. Pharmacol.* **2018**, *9*, 453. [CrossRef]
86. Manuel, M.A.; Weiner, M.W. Effects of ethacrynic acid and furosemide on isolated rat kidney mitochondria: Inhibition of electron transport in the region of phosphorylation site II. *J. Pharmacol. Exp. Ther.* **1976**, *198*, 209.
87. Orita, Y.; Fukuhara, Y.; Yanase, M.; Ando, A.; Okada, N.; Abe, H. Effect of furosemide on mitochondrial electron transport system and oxidative phosphorylation. *Arzneimittel-Forschung* **1983**, *33*, 1446–1450.
88. Wong, S.G.W.; Card, J.W.; Racz, W.J. The role of mitochondrial injury in bromobenzene and furosemide induced hepatotoxicity. *Toxicol. Lett.* **2000**, *116*, 171–181. [CrossRef]
89. Tai, Y.K.; Cheong, Y.M.; Almsherqi, Z.A.; Chia, S.H.; Deng, Y.; McLachlan, C.S. High dose clopidogrel decreases mice liver mitochondrial respiration function in vitro. *Int. J. Cardiol.* **2009**, *133*, 250–252. [CrossRef]
90. Maseneni, S.; Donzelli, M.; Brecht, K.; Krahenbuhl, S. Toxicity of thienopyridines on human neutrophil granulocytes and lymphocytes. *Toxicology* **2013**, *308*, 11–19. [CrossRef]
91. Zahno, A.; Bouitbir, J.; Maseneni, S.; Lindinger, P.W.; Brecht, K.; Krahenbuhl, S. Hepatocellular toxicity of clopidogrel: Mechanisms and risk factors. *Free Radic. Biol. Med.* **2013**, *65*, 208–216. [CrossRef]
92. Felix, L.; Oliveira, M.M.; Videira, R.; Maciel, E.; Alves, N.D.; Nunes, F.M.; Alves, A.; Almeida, J.M.; Domingues, M.R.; Peixoto, F.P. Carvedilol exacerbate gentamicin-induced kidney mitochondrial alterations in adult rat. *Exp. Toxicol. Pathol.* **2017**, *69*, 83–92. [CrossRef]
93. Busanello, E.N.B.; Marques, A.C.; Lander, N.; de Oliveira, D.N.; Catharino, R.R.; Oliveira, H.C.F.; Vercesi, A.E. Pravastatin Chronic Treatment Sensitizes Hypercholesterolemic Mice Muscle to Mitochondrial Permeability Transition: Protection by Creatine or Coenzyme Q10. *Front. Pharmacol.* **2017**, *8*, 185. [CrossRef]
94. Westwood, F.R.; Scott, R.C.; Marsden, A.M.; Bigley, A.; Randall, K. Rosuvastatin: Characterization of induced myopathy in the rat. *Toxicol. Pathol.* **2008**, *36*, 345–352. [CrossRef]
95. van Leeuwen, J.S.; Unlu, B.; Vermeulen, N.P.; Vos, J.C. Differential involvement of mitochondrial dysfunction, cytochrome P450 activity, and active transport in the toxicity of structurally related NSAIDs. *Toxicol. In Vitro* **2012**, *26*, 197–205. [CrossRef]
96. Moreno-Sanchez, R.; Bravo, C.; Vasquez, C.; Ayala, G.; Silveira, L.H.; Martinez-Lavin, M. Inhibition and uncoupling of oxidative phosphorylation by nonsteroidal anti-inflammatory drugs: Study in mitochondria, submitochondrial particles, cells, and whole heart. *Biochem. Pharmacol.* **1999**, *57*, 743–752. [CrossRef]
97. Porceddu, M.; Buron, N.; Roussel, C.; Labbe, G.; Fromenty, B.; Borgne-Sanchez, A. Prediction of liver injury induced by chemicals in human with a multiparametric assay on isolated mouse liver mitochondria. *Toxicol. Sci.* **2012**, *129*, 332–345. [CrossRef]
98. Aleo, M.D.; Luo, Y.; Swiss, R.; Bonin, P.D.; Potter, D.M.; Will, Y. Human drug-induced liver injury severity is highly associated with dual inhibition of liver mitochondrial function and bile salt export pump. *Hepatology* **2014**, *60*, 1015–1022. [CrossRef]
99. Mendes, P.; Robles, P.G.; Mathur, S. Statin-induced rhabdomyolysis: A comprehensive review of case reports. *Physiother. Can.* **2014**, *66*, 124–132. [CrossRef]
100. Harper, C.R.; Jacobson, T.A. Evidence-based management of statin myopathy. *Curr. Atheroscler. Rep.* **2010**, *12*, 322–330. [CrossRef]
101. Hargreaves, I.P.; Duncan, A.J.; Heales, S.J.; Land, J.M. The effect of HMG-CoA reductase inhibitors on coenzyme Q10: Possible biochemical/clinical implications. *Drug Saf.* **2005**, *28*, 659–676. [CrossRef]
102. Paiva, H.; Thelen, K.M.; Van Coster, R.; Smet, J.; De Paepe, B.; Mattila, K.M.; Laakso, J.; Lehtimaki, T.; von Bergmann, K.; Lutjohann, D.; et al. High-dose statins and skeletal muscle metabolism in humans: A randomized, controlled trial. *Clin. Pharmacol. Ther.* **2005**, *78*, 60–68. [CrossRef]

103. Avis, H.J.; Hargreaves, I.P.; Ruiter, J.P.; Land, J.M.; Wanders, R.J.; Wijburg, F.A. Rosuvastatin lowers coenzyme Q10 levels, but not mitochondrial adenosine triphosphate synthesis, in children with familial hypercholesterolemia. *J. Pediatr.* **2011**, *158*, 458–462. [CrossRef]
104. Neergheen, V.; Dyson, A.; Wainwright, L.; Hargreaves, I.P. Statin and Fibrate-induced Dichotomy of Mitochondrial Function. In *Mitochondrial Dysfunction Caused by Drugs and Environmental Toxicants*; Will, Y., Dykens, J., Eds.; Wiley: New York, NY, USA, 2018; pp. 459–473.
105. Hargreaves, I.P.; Al Shahrani, M.; Wainwright, L.; Heales, S.J. Drug-Induced Mitochondrial Toxicity. *Drug Saf.* **2016**, *39*, 661–674. [CrossRef]
106. Johnson, T.E.; Zhang, X.; Shi, S.; Umbenhauer, D.R. Statins and PPARalpha agonists induce myotoxicity in differentiated rat skeletal muscle cultures but do not exhibit synergy with co-treatment. *Toxicol. Appl. Pharmacol.* **2005**, *208*, 210–221. [CrossRef]
107. Salimi, A.; Eybagi, S.; Seydi, E.; Naserzadeh, P.; Kazerouni, N.P.; Pourahmad, J. Toxicity of macrolide antibiotics on isolated heart mitochondria: A justification for their cardiotoxic adverse effect. *Xenobiotica* **2016**, *46*, 82–93. [CrossRef]
108. Mathis, A.; Wild, P.; Boettger, E.C.; Kapel, C.M.; Deplazes, P. Mitochondrial ribosome as the target for the macrolide antibiotic clarithromycin in the helminth Echinococcus multilocularis. *Antimicrob. Agents Chemother.* **2005**, *49*, 3251–3255. [CrossRef]
109. Ding, L.; Zang, L.; Zhang, Y.; Zhang, Y.; Wang, X.; Ai, W.; Ding, N.; Wang, H. Joint toxicity of fluoroquinolone and tetracycline antibiotics to zebrafish (Danio rerio) based on biochemical biomarkers and histopathological observation. *J. Toxicol. Sci.* **2017**, *42*, 267–280. [CrossRef]
110. Kaur, K.; Fayad, R.; Saxena, A.; Frizzell, N.; Chanda, A.; Das, S.; Chatterjee, S.; Hegde, S.; Baliga, M.S.; Ponemone, V.; et al. Fluoroquinolone-related neuropsychiatric and mitochondrial toxicity: A collaborative investigation by scientists and members of a social network. *J. Community Support. Oncol.* **2016**, *14*, 54–65. [CrossRef]
111. Finsterer, J.; Zarrouk-Mahjoub, S. Mitochondrial toxicity of cardiac drugs and its relevance to mitochondrial disorders. *Expert Opin. Drug Metab. Toxicol.* **2015**, *11*, 15–24. [CrossRef]
112. De Mello, A.H.; Costa, A.B.; Engel, J.D.G.; Rezin, G.T. Mitochondrial dysfunction in obesity. *Life Sci.* **2018**, *192*, 26–32. [CrossRef]
113. Miotto, P.M.; LeBlanc, P.J.; Holloway, G.P. High-Fat Diet Causes Mitochondrial Dysfunction as a Result of Impaired ADP Sensitivity. *Diabetes* **2018**, *67*, 2199–2205. [CrossRef]
114. Jorgensen, W.; Rud, K.A.; Mortensen, O.H.; Frandsen, L.; Grunnet, N.; Quistorff, B. Your mitochondria are what you eat: A high-fat or a high-sucrose diet eliminates metabolic flexibility in isolated mitochondria from rat skeletal muscle. *Physiol. Rep.* **2017**, *5*. [CrossRef] [PubMed]
115. Gram, M.; Dahl, R.; Dela, F. Physical inactivity and muscle oxidative capacity in humans. *Eur. J. Sport Sci.* **2014**, *14*, 376–383. [CrossRef]
116. Kim, Y.; Triolo, M.; Hood, D.A. Impact of Aging and Exercise on Mitochondrial Quality Control in Skeletal Muscle. *Oxid. Med. Cell. Longev.* **2017**, *2017*, 3165396. [CrossRef]
117. DiNicolantonio, J.J.; Berger, A. Added sugars drive nutrient and energy deficit in obesity: A new paradigm. *Open Heart* **2016**, *3*, e000469. [CrossRef]
118. Chen, W.J.; Du, J.K.; Hu, X.; Yu, Q.; Li, D.X.; Wang, C.N.; Zhu, X.Y.; Liu, Y.J. Protective effects of resveratrol on mitochondrial function in the hippocampus improves inflammation-induced depressive-like behavior. *Physiol. Behav.* **2017**, *182*, 54–61. [CrossRef] [PubMed]
119. Assuncao, M.; Andrade, J.P. Protective action of green tea catechins in neuronal mitochondria during aging. *Front. Biosci.* **2015**, *20*, 247–262.
120. Lim, S.; Ahn, S.Y.; Song, I.C.; Chung, M.H.; Jang, H.C.; Park, K.S.; Lee, K.U.; Pak, Y.K.; Lee, H.K. Chronic exposure to the herbicide, atrazine, causes mitochondrial dysfunction and insulin resistance. *PLoS ONE* **2009**, *4*, e5186. [CrossRef]
121. Rai, P.K.; Russell, O.M.; Lightowlers, R.N.; Turnbull, D.M. Potential compounds for the treatment of mitochondrial disease. *Br. Med. Bull.* **2015**, *116*, 5–18. [CrossRef]

© 2019 by the authors. Licensee MDPI, Basel, Switzerland. This article is an open access article distributed under the terms and conditions of the Creative Commons Attribution (CC BY) license (http://creativecommons.org/licenses/by/4.0/).

Review

Mitochondrial Dysfunction and Diabetes: Is Mitochondrial Transfer a Friend or Foe?

Magdalene K Montgomery

Department of Physiology, School of Biomedical Sciences, University of Melbourne, Melbourne 3010, Australia; magdalene.montgomery@unimelb.edu.au; Tel.: +61-422-059-907

Received: 9 November 2018; Accepted: 20 December 2018; Published: 11 May 2019

Abstract: Obesity, insulin resistance and type 2 diabetes are accompanied by a variety of systemic and tissue-specific metabolic defects, including inflammation, oxidative and endoplasmic reticulum stress, lipotoxicity, and mitochondrial dysfunction. Over the past 30 years, association studies and genetic manipulations, as well as lifestyle and pharmacological invention studies, have reported contrasting findings on the presence or physiological importance of mitochondrial dysfunction in the context of obesity and insulin resistance. It is still unclear if targeting mitochondrial function is a feasible therapeutic approach for the treatment of insulin resistance and glucose homeostasis. Interestingly, recent studies suggest that intact mitochondria, mitochondrial DNA, or other mitochondrial factors (proteins, lipids, miRNA) are found in the circulation, and that metabolic tissues secrete exosomes containing mitochondrial cargo. While this phenomenon has been investigated primarily in the context of cancer and a variety of inflammatory states, little is known about the importance of exosomal mitochondrial transfer in obesity and diabetes. We will discuss recent evidence suggesting that (1) tissues with mitochondrial dysfunction shed their mitochondria within exosomes, and that these exosomes impair the recipient's cell metabolic status, and that on the other hand, (2) physiologically healthy tissues can shed mitochondria to improve the metabolic status of recipient cells. In this context the determination of whether mitochondrial transfer in obesity and diabetes is a friend or foe requires further studies.

Keywords: mitochondrial dysfunction; insulin resistance; type 2 diabetes; mitochondrial transfer; exosomes

1. Introduction

The dysregulation of carbohydrate and lipid metabolism due to unbalanced diets (i.e., 'fast food') in western societies has led to a dramatic rise in the worldwide prevalence of obesity, with more than 1.9 billion adults worldwide considered overweight or obese in 2016 [1]. Undoubtedly, this number has increased even further within the last three years. Obesity is one of the most common causes of the development of insulin resistance, type 2 diabetes (T2D), and other related cardio-metabolic risk factors. As outlined by the National Institute for Diabetes and Digestive and Kidney Diseases (NIDDK), T2D is characterized by hyperglycaemia and is predominantly prevalent in middle-aged and older individuals (>45 years of age) [2]. Insulin resistance (i.e., pre-diabetes) and T2D are accompanied by a multitude of systemic and tissue-specific metabolic defects, including inflammation, oxidative and endoplasmic reticulum stress, and lipotoxicity (i.e ectopic lipid accumulation in peripheral tissues), as well as mitochondrial dysfunction [3]. Whether these metabolic aspects are a cause or consequence of insulin resistance remains a matter of debate, with studies pointing towards both scenarios, as reviewed in detail in [4–6]. This review will focus on mitochondrial dysfunction in T2D, with a special focus on the systemic implications of mitochondrial dysfunction occurring in peripheral metabolic tissues. Specifically, we will review evidence suggesting that dysfunctional mitochondria and/or mitochondrial

DNA can be packaged into extracellular vesicles, secreted from the respective tissue and can impact other tissues in a paracrine or endocrine manner.

2. Mitochondrial Function and Dysfunction

Mitochondria have a multitude of roles, with the most prominent one being the generation of ATP for the maintenance of cellular processes. In addition, mitochondria play a role in reactive oxygen species (ROS)-mediated signalling [7–10], apoptosis [11,12], calcium signalling [8,13], and haem [14,15] and steroid synthesis [16], just to name a few. Due to the importance of mitochondria in cellular energy metabolism, defects in the processes mentioned above have significant outcomes on tissue and systemic level. In this respect, the term 'mitochondrial dysfunction' was firstly mentioned in 1975 in the context of glucose intolerance [17]. While the term 'mitochondrial dysfunction' is nowadays commonly used in the scientific literature, the definitions and methods of assessment vary between studies. Mitochondrial dysfunction has been assessed as changes in gene expression of mitochondrial markers [18,19], protein content, or enzymatic activities of mitochondrial proteins [18,20], changes in mitochondrial size and shape [20,21], as well as functional assessment of mitochondrial oxidative capacity [22] and ROS generation [23]. For details on these studies and for mechanistic insights into possible causes of mitochondrial dysfunction, please refer to our recent review on 'mitochondrial dysfunction and insulin resistance' [24].

3. Mitochondrial Dysfunction in Obesity and Type 2 Diabetes

Since the first mention of mitochondrial dysfunction in the context of glucose intolerance in 1975 [17], the importance of mitochondrial function/dysfunction in insulin resistance has been disputed for many years. While some studies in both humans and rodents describe a decrease in mitochondrial function (decreased mitochondrial enzyme activities, decreased lipid metabolism, reduced mitochondrial size and number) in obese and insulin resistant individuals [25,26], others show no correlation [27,28] or even a compensatory increase in mitochondrial capacity with lipid oversupply [29,30], with most studies focusing on skeletal muscle (reviewed in detail in [24]). The discrepancies between studies in both rodents and humans suggest that mitochondrial dysfunction is not a requisite feature of insulin resistance and T2D, and that study outcomes (i.e., if mitochondrial dysfunction is present) are likely dependent on methodological approaches, the model system examined (human or rodent) and the definition of the term mitochondrial dysfunction in the context of each study. In addition, differences in mitochondrial defects between metabolic tissues need to be considered. While skeletal muscle is the most common tissue examined in the context of T2D and mitochondrial function [31], the importance of mitochondrial function in other tissues, such as heart, liver and pancreas, and the impact on systemic homeostasis, should not be underestimated.

4. Tissue-Specific Effects of Mitochondrial Function and T2D

Skeletal muscle is packed with mitochondria, which are required for ATP generation during muscle contraction. With the number of mitochondria in a cell being proportional to the energy demand of the cell, it is not surprising that skeletal muscle and heart are the two tissues with the highest mitochondrial abundance [32]. We have reviewed changes in mitochondrial function in skeletal muscle in the presence of insulin resistance and T2D in detail previously [24], with the major 'conclusion' being that skeletal muscle-related investigations into changes in mitochondrial function in the presence of insulin resistance or T2D have been 'inconclusive' (i.e., decreased, unchanged, or compensatory increased in mitochondrial function in insulin resistance have been suggested). Here we briefly describe changes in mitochondrial function and capacity in the heart, due to the hearts high rate of ATP production and turnover, and its reliance on mitochondrial metabolism to produce energy.

A large number of studies points towards mitochondrial defects being present in the heart in individuals with insulin resistance and diabetes [33–37] as well as in rodent models [38–43]. Markers examined in these studies included mitochondrial oxygen consumption, ROS and ATP generation, changes in mitochondrial size, and structure and calcium handling, as well as mRNA, protein, or enzymatic activity levels of markers of mitochondrial biogenesis, content, and oxidative capacity. Of note, whereas studies in humans are of an associative nature and limit investigations of cause-or-consequence scenarios, a majority of rodent models investigating cardiac function in the presence of diabetes use streptozotocin, a compound that shows high toxicity towards the pancreas and destruction of insulin-producing beta cells, and that is used for the generation of type 1 diabetes phenotypes [44]. The use of dietary interventions that more closely mimic human obesity and insulin resistance, as well as rapid improvements in generation and accessibility to transgenic rodent models are crucial for delineating cause-and-consequence relationships. For example, Turkieh and colleagues showed recently that apolipoprotein O (APOO) is overexpressed in diabetic hearts, and, using transgenic APOO mice, show that APOO links impaired mitochondrial function and the onset of cardiomyopathy [45]. Furthermore, using genetic approaches a recent study showed that myocardial function depends on balanced mitochondrial fusion and fission, and defined a critical regulator of cardiomyocyte survival, the mitochondrial metalloendopeptidase OMA1 [46]. In addition, expression of extracellular signal-regulated protein kinase 5 (Erk5) was shown to be lost in diabetic hearts, and cardiac-specific deletion of Erk5 in mice substantiated mitochondrial abnormalities and decreased fuel oxidation [47]. Lastly, epigenetic regulation of metabolic functions was shown to be important in the preservation of cardiac function. Specifically, an N-terminal proteolytically derived fragment of histone deacetylase 4 (HDAC4) was shown to be decreased in failing mouse hearts and that overexpression of this HDAC4 fragment improved calcium handling and contractile function [48].

While diving into the topic of cardiac dysfunction, I noticed that alterations in mitochondrial energy metabolism are a common feature of various forms of heart disease [49,50]. Most predominantly, diabetic cardiomyopathies (ventricular dysfunction in patients with diabetes mellitus) are associated with dysregulated oxidative substrate selection. In humans and rodents, obesity and insulin resistance are associated with increased myocardial fatty acid uptake and fatty acid oxidation [51,52], with a simultaneous decrease in glucose oxidation [53,54]. As fatty acid oxidation produces less ATP (per mol oxygen consumed), this scenario makes the heart less energy efficient [52]. In addition, increased fatty acid supply to the heart is also associated with oxidative stress [55] and accumulation of bioactive lipid intermediates that directly interfere with insulin signalling [56], and lead to a greater decline in heart function with age [57]. Of interest, therapeutic approaches to increase glucose utilization of the diabetic heart, either actively through activation of pyruvate dehydrogenase [58–60] or passively by inhibiting fatty acid oxidation [61,62], were shown to improve heart function.

5. Can Mitochondrial Dysfunction be Reversed?—Insight into Genetic and Pharmacological Interventions

Targeting mitochondrial oxidative capacity as a means to improve insulin sensitivity is questionable with contradicting findings in various genetic mouse models. For example, inhibition of mitochondrial fatty acid uptake (through inhibition of CPT1) improves insulin sensitivity in muscle of diet-induced obese mice, healthy male subjects and in primary human myotubes [63,64]. In addition, activating heat shock protein 72 (HSP72) in skeletal muscle was found to improve mitochondrial oxidative capacity and insulin sensitivity in the setting of lipotoxicity [65], whereas cardiac-specific overexpression showed no metabolic improvements [66]. Furthermore, increasing the capacity of fatty acid transport protein (FATP) 1 in skeletal muscle resulted in increased channelling of fatty acids into oxidative pathways, however without changes in muscle or systemic insulin sensitivity [67].

Similarly, contradicting findings come from lifestyle or pharmacological intervention studies. The major lifestyle interventions are diet and exercise, with both dietary restriction and physical activity shown to reduce the risk of developing insulin resistance and T2D [68,69], and to improve/restore muscle mitochondrial function [70,71]. While the impact of exercise on mitochondrial biogenesis and oxidative capacity is well accepted (as recently reviewed in [72]), the picture is not that clear in respect to mitochondrial improvements after dietary interventions. In addition to the studies above, other reports suggest that caloric restriction improves insulin sensitivity without changes (or even with decreases) in skeletal muscle mitochondrial oxidative capacity [73,74].

Although exercise and dietary interventions represent a good and some might say 'easy' way to improveinsulin sensitivity, long-term adherence to these lifestyle interventions is problematic and remains a public health challenge. In this respect, finding an ideal pharmacological agent that improves metabolic health, by improving mitochondrial function, has shown some promise in rodent models and in the clinic. One of the most well described modulators of oxidative metabolism are peroxisome proliferator-activated receptors (PPARs). For example, PPAR agonists were shown to increase mitochondrial content and oxidative capacity, in the livers of pre-diabetic mice [75], in adipose tissue [76] and skeletal muscle of insulin resistant and T2D patients [76,77] and mice [78], and these mitochondrial adaptations were accompanied by improvements in systemic insulin sensitivity. According to the NIH Clinial Trials register, PPAR agonists are now being trialled for cardiovascular disease, hypertension, pre-diabetes, and diabetes, and various PPAR agonists (e.g. pioglitazone, rosiglitazone, fibrates) have been FDA-approved for such therapeutic applications. In addition to PPAR ligands, a variety of other common anti-diabetic drugs have pronounced effects on mitochondrial function, with improvements in mitochondrial metabolism likely contributing to the systemic improvements in insulin sensitivity and glycaemic control [79]. While insulin therapy has been shown to improve mitochondrial oxidative phosphorylation [80] and ATP synthesis [81], SGLT2 inhibitors, which primarily act on inhibiting glucose reuptake in the kidney, also improve mitochondrial function in the heart [82–84] and in the brain [85]. However, the positive mitochondrial effects have not been supported by other studies, pointing towards potential inhibitory effects of SGLT2 inhibitors on the mitochondrial respiratory chain [86,87]. Furthermore, conflicting findings have also been reported for one of the most commonly prescribed anti-diabetic agents, the biguanide metformin. While some studies point towards metformin having beneficial effects on mitochondrial oxidative capacity [88] and mitochondrial fission [89], others show the opposite [90–93], with metformin being named 'an energy disruptor' [90]. Lastly, thiazolidinediones (TZDs) such as rosiglitazone have been shown to increase mitochondrial biogenesis [94,95]. For a detailed summary of the effects of anti-diabetic drugs on mitochondrial function, please refer to [79,96].

6. Intercellular Mitochondrial Transfer

In addition to 'boosting' mitochondrial oxidative capacity in the presence of mitochondrial dysfunction through genetic, lifestyle of pharmacological interventions, would it be possible to improve mitochondrial function in a target cell or tissue through mitochondrial transfer? Recent evidence suggests that intact mitochondria, mitochondrial DNA or other mitochondrial components (proteins, lipids or metabolites) can be found in the circulation [97–100] and can be transferred between cells or tissues [101–103]. Mitochondrial transfer between cells was shown in the context of leukaemia [101], acute lung injury [104], asthma [105], and acute respiratory distress syndrome (ARDS) [103], through the use of tunnelling nanotube-like structures between donor and recipient cells. Most importantly, recipient cells showed metabolic impairments which were improved upon transfer of functional mitochondria; e.g., acute myeloid leukaemia (AML) cells were less prone to mitochondrial depolarization after chemotherapy [101]. Other researchers turned to intravenous injections of intact mitochondria as a therapeutic approach for mitochondrial replenishment after cardiac injury [106]. While these studies highlight the protective effects of mitochondrial transfer in a variety of disease conditions, other studies point towards exogenous mitochondria having pro-inflammatory effects [107,108]. In the context of cancer, mitochondrial transfer from benign donor to recipient carcinoma cells elicited a chemo-resistant phenotype [109] and contributed to tumour proliferation [110]. Little is known about mitochondrial transfer in conditions of mitochondrial dysfunction and insulin resistance. Could transfer of intact mitochondria or mitochondrial DNA be a potential therapeutic means to improve mitochondrial dysfunction and subsequently insulin resistance? Could mitochondrial transfer on the other hand further worsen already substantiated metabolic defects? While intercellular transfer using nanotubes-like structures (as described above) is a potential transfer mechanism for cells in close proximity, this pathway is unlikely to 'remodel' whole organs and tissues, as would be required for example for skeletal muscle mitochondrial dysfunction. However, of interest, mitochondrial DNA and mitochondrial proteins have also been shown to be present in extracellular vesicles, particularly exosomes, and the significance of these findings will be discussed below.

7. Exosomes

It has been known for more than 40 years that cells secrete vesicles during apoptosis [111]. More recent studies suggest that also healthy cells have the capacity to release vesicles into the extracellular environment, with exosomes having received substantial attention over the past couple of years, primarily due to early findings such as that dendritic cells secrete antigen-presenting exosomes [112] and that tumour cells can transfer exosomes to dendritic cells to subsequently induce potent antitumor effects [113]. The term exosome is used to distinguish extracellular vesicles with a diameter of 50 to 150 nm, that originate from the late endosomal pathway and are released in the extracellular space upon fusion of multivesicular bodies (MVBs) with the plasma membrane [114] (Figure 1). For detailed info on biogenesis of exosomes, mechanisms of exosome secretion and detailed information on cargo, please refer to a recent review on this topic [115]. Exosomes are lipid-bound vesicles that carry lipids, proteins and nucleic acids. Interestingly, up to 10 percent of exosomal proteins have been described as mitochondrial proteins [116]. However, it should be noted that the relative high number of mitochondrial proteins identified within exosomal preparations could be also due to "contamination" during the preparation procedure. Exosomes are released from a multitude of tissues and cell types, including metabolic tissues, such as skeletal muscle [117], liver [118] and adipose tissue [119,120]. The amount and cargo of released microvesicles differs in obesity and type 2 diabetes [121–124], with adipose-derived exosomes from obese individuals and mouse models being associated with insulin resistance [125,126] and plasma exosomes in obese, T2D individuals showing enrichment in molecules involved in inflammation and immune efficiency [127]. Also, plasma exosomes from obese rats, but not lean rats, induced significant oxidative stress and vascular cell adhesion protein 1 (VCAM-1) expression in endothelial cells, indicative of a pro-inflammatory vesicle phenotype [123].

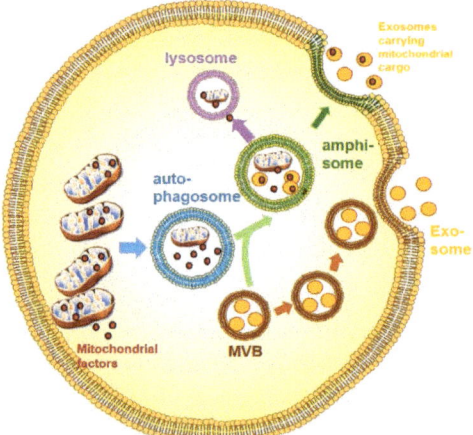

Figure 1. Exosomal secretion of intact mitochondria or mitochondrial factors. Mitochondria or mitochondrial components are packaged into autophagosomes for intracellular degradation. Autophagosomes can fuse with multi-vesicular bodies (MVBs) carrying exosomes, forming amphisomes. Amphisome contents can be either shuttled towards degradation in lysosomes or towards the cell surface where exosomes are released.

Importantly, obesity-associated exosomes released from adipose tissue, macrophages or erythrocytes can transform healthy cells into metabolically defective cells, with robust effects on tissue-specific and systemic insulin sensitivity and overall glucose homeostasis [124,126,128]. Overall, it seems that extracellular vesicles reflect the diverse functional and dysfunctional states of the releasing cells, and it is therefore not surprising that specific vesicle/exosomal signatures are present in distinctive disease states, making exosomes useful biomarkers in the future [129].

8. Exosomes with Mitochondrial Cargo

As mentioned above, up to 10% of exosomal proteins are mitochondria-derived [116]. Initially, vesicles containing mitochondrial cargo were described as being transported within the cytosol towards peroxisomes (i.e., inter-organellar communication) [130], however other vesicle sub-types were also shown fuse with the late endosome or with multi-vesicular bodies for degradation [131], with selective enrichment of oxidized cargo [132]. Of interest, the stress-induced mitochondria-associated proteins Parkin and Pink1 are involved in the generation of these particular multi-vesicular bodies [133], highlighting increased vesicle generation upon autophagy-related stress signals, potentially as a means of mitochondrial quality control. While these studies suggest that these vesicle sub-types and mitochondrial components are destined for degradation, they can also be routed towards the cell surface and fuse with the plasma membrane [134] (Figure 1). It is unclear if increased mitophagy is associated with increased exosomal secretion of mitochondrial components, potentially as a means to get rid of oxidized proteins, in addition to lysosomal degradation.

9. Exosomes with Mitochondrial Cargo—Friend of Foe?

Exosomes containing increased content of mitochondrial lipids, proteins and nucleic acids were found to be released from adipose tissue of obese diabetic and obese non-diabetic rats [135], from pulmonary cells after cigarette smoke injury [136], and are found in plasma upon infection with the human T-lymphotropic retrovirus type 1 (HTLV-1) [99] and in circulating exosomes of breast cancer patients [137], highlighting the broad range of pathological states that are characterized by increased circulating exosomes carrying mitochondrial cargo. Some studies suggest that the cells and tissues releasing these particular exosomes show mitochondrial impairments. For example, it has been shown that mesenchymal stem cells target depolarized mitochondria to the plasma membrane, and that they similarly shed microRNA-containing exosomes, which have the capacity to inhibit macrophage activation, thereby de-sensitizing macrophages to the ingested mitochondria [138]. Cells might use exosomal secretion as a way of quality control, to restore cellular homeostasis and preserve cell viability, to some extent in the presence of mitochondrial dysfunction [139]. This would suggest that mitochondrial cargo within circulating exosomes is largely metabolically defective, and when fusing with recipient cells could either transfer these mitochondrial defects to the recipient cells, or could induce a physiologically adverse response. In support of this hypothesis, it has been shown that leukaemia cells, when challenged with chemotherapeutic drugs that induce oxidative stress, transfer their dysfunctional mitochondria to neighbouring bone marrow mesenchymal stem cells, leading to chemo-resistance [140]. In addition, exosomes can carry mitochondrial electron transport chain (ETC) complexes. Some of these ETC subunits are encoded by mitochondrial DNA. Translation within mitochondria occurs using a formylated initiating methionine (a mechanism still conserved from bacterial origins), and presence of extracellular formylated proteins can lead to an immune response and cytokine release [107].

While mitochondrial proteins are most commonly investigated in the context of exosomal transport, exosomes also carry lipids (a recent paper identified 1,961 lipid species in an exosomal and microvesicle screen [141]), with exosomes from hepatocellular carcinoma cells (Huh7) and human bone marrow-derived mesenchymal stem cells (MSC) specifically enriched in the mitochondrial inner membrane lipid cardiolipin [141]. This enrichment of cardiolipin, but exclusion of other lipid species, such as sphingolipids, within these particular exosomes suggests that cardiolipin might be actively sorted into exosomes of Huh7 cells and MSCs [141]. While the reason for this phenomenon is unknown, the authors suggest that the high concentration of cardiolipin might help in the stability and curvature of the vesicle membrane bilayer. On the other hand, increased secretion of cardiolipin (or other mitochondrial lipids) could impair mitochondrial function of the donor cells, as cardiolipin plays an important role in maintaining optimal mitochondrial function. In this respect, a loss of cardiolipin has been described in the diabetic heart [142–144] and during heart failure [145], while exercise leads to increased cardiolipin content in skeletal muscle [146]. If this is due to differences in synthesis, degradation or exosomal secretion requires further investigation. As mentioned above, exosomal secretion has been considered a way of quality control [139], with the donor cells 'getting rid' of damaged or oxidized cellular components. During conditions of increased mitochondrial ROS production and oxidative stress, as is the case in the presence of insulin resistance and T2D [147], polyunsaturated fatty acids in mitochondrial membranes (such as cardiolipins) are the primary targets of oxidative damage, which may lead to mitochondrial dysfunction [148]. While oxidized cardiolipin can be repaired enzymatically [149], exosomal export is also a likely pathway for clearance of oxidized cellular factors.

Looking at the other side of the coin, a recent study published in the journal *Blood* investigated the metabolic cross-talk between cancer cells and their microenvironment, and found that 'healthy' bone marrow stromal cells (BMSC) were made to transfer their mitochondria to neighbouring acute myeloid leukaemia (AML) cells, supporting the cancer cells' growth [102]. An associated press release in Science Daily termed this phenomenon quite adequately as "Stealing from the body: How cancer recharges its batteries" [150]. A different study identified the complete mitochondrial genome within circulating extracellular vesicles from metastatic breast cancer patients, and showed that these extracellular vesicles can in turn transfer their mtDNA to cells with impaired metabolism, leading to restoration of metabolic activity [151]. The authors suggested that the transfer of mtDNA plays a role in mediating resistance to hormone therapy in these patients.

It seems that, depending on the tissue/cell type and the pathological state examined, mitochondrial cargo can be either transferred from a cell with mitochondrial dysfunction to a cell with 'healthy' metabolic state, leading to metabolic deterioration of the recipient cells; or, on the other hand mitochondrial cargo can be transferred from a 'healthy' cell to a recipient cell with mitochondrial dysfunction, leading to the recipient's metabolic improvement (Figure 2). The impact or physiological importance of exosomal transfer of mitochondrial cargo in the context of mitochondrial dysfunction in insulin resistant and T2D individuals is not known. Does skeletal muscle, heart or liver (or major metabolic tissues in general) have the capacity to shed mitochondria in the presence of mitochondrial dysfunction to rescue the donor's cell energetic state? Could on the other hand mitochondrial cargo be transferred to metabolically deficient cells, to improve mitochondrial dysfunction in the recipient tissue (Figure 2)?

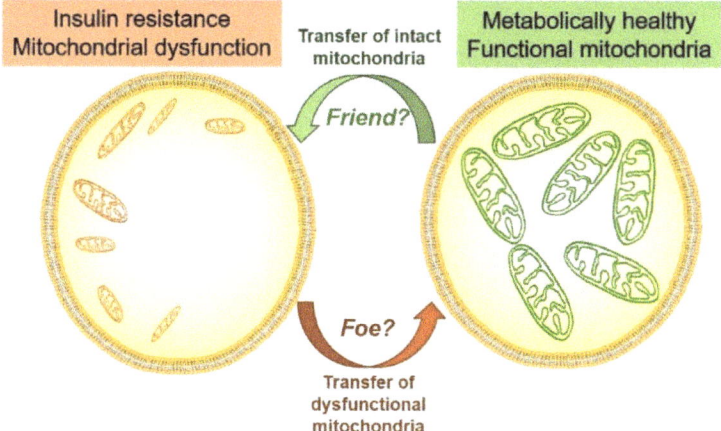

Figure 2. Is mitochondrial transfer in states of insulin resistance and mitochondrial dysfunction a friend or foe? In other pathological conditions, it has been shown that cells with mitochondrial impairments have the capacity to secrete mitochondrial cargo within exosomes that then impairs metabolic state of the recipient cells (i.e., foe). Other studies suggest that 'healthy' cells secrete mitochondrial cargo to improve the recipients cell metabolism (i.e., friend). Future studies will have to show if these phenomena are present states of mitochondrial dysfunction and insulin resistance.

10. Mitochondrial Dysfunction, T2D and Exosomal Transfer of Mitochondrial Cargo

Very little is known about exosomal transfer of mitochondrial cargo in the presence of mitochondrial dysfunction during the development of insulin resistance and T2D. It has been shown that in obese diabetic rats adipose-derived exosomes carry more mitochondrial lipids, proteins and nucleic acids [135]. Furthermore, lower circulating mtDNA content is associated with T2D [152] and severe proliferative diabetic retinopathy [153], with reduced peripheral blood mtDNA content potentially increasing the risk of impaired glucose-stimulated β cell function [152]. In addition, HbA1c, fasting plasma glucose and age of T2D onset are the major factors affecting mtDNA content [154]. While these studies assessed changes in mtDNA content in the circulation, this was not investigated in the context of exosomal transport. In a related matter, point mutations in the mitochondrial genome and decreases in mtDNA copy number have been linked to the pathogenesis of type 2 diabetes [155,156]. Compared with nuclear DNA repair, mtDNA repair mechanisms are significantly less efficient [157] and mtDNA is more susceptible to oxidative stress and mutations [158]. While the secretion of mtDNA within microvesicles has been described previously [159], little is known if mutated mtDNA can be secreted within exosomes and taken up by other cells.

Future studies will have to determine whether metabolic tissues, such as skeletal muscle or liver, have the capacity to shed defective mitochondrial components within exosomes, and if this process is affected in obese T2D individuals. Could on the other hand mitochondrial transfer be used as a means to improve mitochondrial defects (Figure 2)? It would be of interest to assess if targeted transfer of mitochondrial proteins or metabolites, or even fully functioning intact mitochondria, could improve mitochondrial dysfunction and have effects on insulin sensitivity and glucose homeostasis.

11. Conclusions and Future Directions

Mitochondrial impairments have been described in the context of obesity, insulin resistance and T2D, in a variety of metabolic tissues. However, as studies reported opposing findings ranging from decreased mitochondrial function, to unaffected mitochondrial capacity, and even to an increase in mitochondrial oxidative metabolism in individuals and rodent models with insulin resistance or T2D, the requisite for the presence of mitochondrial dysfunction in these pathological states is unclear.

In addition, genetic manipulations of mitochondrial proteins, as well as lifestyle and pharmacological interventions resulted in contradicting findings, and it is still unclear if targeting mitochondrial capacity is a useful therapeutic approach for the treatment of insulin resistance and glycaemic control. In the past years, it has become evident that metabolic tissues secrete microvesicles, such as exosomes, containing mitochondrial cargo, and that these exosomes have the capacity to fuse in a selective targeted manner with recipient cells and tissues, and influence the recipients' cell metabolic status. Future studies will have to show if tissues with mitochondrial impairments in states of obesity and T2D have the capacity to 'shed' their dysfunctional mitochondria to improve their metabolic status, or if dysfunctional tissues have the capacity to 'ingest' exosomes with mitochondrial cargo to improve their oxidative metabolism.

Author Contributions: M.K.M. researched and wrote the manuscript.

Funding: M.K.M. is supported by a Career Development Fellowship from the National Health and Medical Research Council of Australia (ID: APP1143224).

Conflicts of Interest: The author declares no conflict of interest.

References

1. Obesity and Overweight. Available online: http://www.who.int/news-room/fact-sheets/detail/obesity-and-overweight (accessed on 9 September 2018).
2. Type 2 Diabetes. Available online: https://www.niddk.nih.gov/health-information/diabetes/overview/what-is-diabetes/type-2-diabetes (accessed on 8 September 2018).
3. Ye, J. Mechanisms of insulin resistance in obesity. *Front. Med.* **2013**, *7*, 14–24. [CrossRef] [PubMed]
4. Blüher, M. Adipose tissue inflammation: a cause or consequence of obesity-related insulin resistance? *Clin. Sci.* **2016**, *130*, 1603–1614. [CrossRef] [PubMed]
5. Takamura, T.; Misu, H.; Ota, T.; Kaneko, S. Fatty liver as a consequence and cause of insulin resistance: Lessons from type 2 diabetic liver. *Endocr. J.* **2012**, *59*, 745–763. [CrossRef] [PubMed]
6. Keane, K.N.; Cruzat, V.F.; Carlessi, R.; de Bittencourt, P.I.H.; Newsholme, P. Molecular events linking oxidative stress and inflammation to insulin resistance and β-cell dysfunction. *Oxid. Med. Cell Longev.* **2015**, *2015*, 181643. [CrossRef] [PubMed]
7. Daiber, A.; Di Lisa, F.; Oelze, M.; Kröller-Schön, S.; Steven, S.; Schulz, E.; Münzel, T. Crosstalk of mitochondria with NADPH oxidase via reactive oxygen and nitrogen species signalling and its role for vascular function. *Br. J. Pharmacol.* **2017**, *174*, 1670–1689. [CrossRef] [PubMed]
8. Angelova, P.R.; Abramov, A.Y. Functional role of mitochondrial reactive oxygen species in physiology. *Free Radic. Biol. Med.* **2016**, *100*, 81–85. [CrossRef] [PubMed]
9. Owusu-Ansah, E.; Banerjee, U. Reactive oxygen species prime Drosophila haematopoietic progenitors for differentiation. *Nature* **2009**, *461*, 537–541. [CrossRef]
10. Wang, W.; Fang, H.; Groom, L.; Cheng, A.; Zhang, W.; Liu, J.; Wang, X.; Li, K.; Han, P.; Zheng, M.; et al. Superoxide flashes in single mitochondria. *Cell* **2008**, *134*, 279–290. [CrossRef]
11. Redza-Dutordoir, M.; Averill-Bates, D.A. Activation of apoptosis signalling pathways by reactive oxygen species. *Biochim. Biophys. Acta* **2016**, *1863*, 2977–2992. [CrossRef]
12. Estaquier, J.; Vallette, F.; Vayssiere, J.-L.; Mignotte, B. The Mitochondrial Pathways of Apoptosis. In *Advances in Mitochondrial Medicine*; Scatena, R., Bottoni, P., Giardina, B., Eds.; Springer: Dordrecht, The Netherlands; pp. 157–183.
13. Rizzuto, R.; De Stefani, D.; Raffaello, A.; Mammucari, C. Mitochondria as sensors and regulators of calcium signalling. *Nat. Rev. Mol. Cell Biol.* **2012**, *13*, 566. [CrossRef]
14. Chiabrando, D.; Mercurio, S.; Tolosano, E. Heme and erythropoieis: more than a structural role. *Haematologica* **2014**, *99*, 973–983. [CrossRef] [PubMed]
15. Ryu, M.-S.; Zhang, D.; Protchenko, O.; Shakoury-Elizeh, M.; Philpott, C.C. PCBP1 and NCOA4 regulate erythroid iron storage and heme biosynthesis. *J. Clin. Invest.* **2017**, *127*, 1786–1797. [CrossRef] [PubMed]
16. Martin, L.A.; Kennedy, B.E.; Karten, B. Mitochondrial cholesterol: mechanisms of import and effects on mitochondrial function. *J. Bioenerg. Biomembr.* **2016**, *48*, 137–151. [CrossRef] [PubMed]

17. Yamada, T.; Ida, T.; Yamaoka, Y.; Ozawa, K.; Takasan, H.; Honjo, I. Two distinct patterns of glucose intolerance in icteric rats and rabbits. Relationship to impaired liver mitochondria function. *J. Lab. Clin. Med.* **1975**, *86*, 38–45. [PubMed]
18. Heilbronn, L.K.; Gan, S.K.; Turner, N.; Campbell, L.V.; Chisholm, D.J. Markers of mitochondrial biogenesis and metabolism are lower in overweight and obese insulin-resistant subjects. *J. Clin. Endocrinol. Metab.* **2007**, *92*, 1467–1473. [CrossRef] [PubMed]
19. Mootha, V.K.; Lindgren, C.M.; Eriksson, K.F.; Subramanian, A.; Sihag, S.; Lehar, J.; Puigserver, P.; Carlsson, E.; Ridderstrale, M.; Laurila, E.; et al. PGC-1 alpha-responsive genes involved in oxidative phosphorylation are coordinately downregulated in human diabetes. *Nat. Rev. Genet.* **2003**, *34*, 267–273. [CrossRef] [PubMed]
20. Ritov, V.B.; Menshikova, E.V.; He, J.; Ferrell, R.E.; Goodpaster, B.H.; Kelley, D.E. Deficiency of subsarcolemmal mitochondria in obesity and type 2 diabetes. *Diabetes* **2005**, *54*, 8–14. [CrossRef]
21. Kelley, D.E.; He, J.; Menshikova, E.V.; Ritov, V.B. Dysfunction of mitochondria in human skeletal muscle in type 2 diabetes. *Diabetes* **2002**, *51*, 2944–2950. [CrossRef]
22. Kelley, D.E.; Goodpaster, B.; Wing, R.R.; Simoneau, J.A. Skeletal muscle fatty acid metabolism in association with insulin resistance, obesity, and weight loss. *Am. J. Physiol.* **1999**, *277*, E1130–E1141. [CrossRef]
23. Anderson, E.J.; Lustig, M.E.; Boyle, K.E.; Woodlief, T.L.; Kane, D.A.; Lin, C.-T.; Price, J.W., III; Kang, L.; Rabinovitch, P.S.; Szeto, H.H.; et al. Mitochondrial H_2O_2 emission and cellular redox state link excess fat intake to insulin resistance in both rodents and humans. *J. Clin. Invest.* **2009**, *119*, 573–581. [CrossRef]
24. Montgomery, M.K.; Turner, N. Mitochondrial dysfunction and insulin resistance: an update. *Endocr. Connect.* **2015**, *4*, R1–R15. [CrossRef] [PubMed]
25. Boyle, K.E.; Canham, J.P.; Consitt, L.A.; Zheng, D.; Koves, T.R.; Gavin, T.P.; Holbert, D.; Neufer, P.D.; Ilkayeva, O.; Muoio, D.M.; et al. A high-fat diet elicits differential responses in genes coordinating oxidative metabolism in skeletal muscle of lean and obese individuals. *J. Clin. Endocrinol. Metab.* **2011**, *96*, 775–781. [CrossRef] [PubMed]
26. Hwang, H.; Bowen, B.P.; Lefort, N.; Flynn, C.R.; De Filippis, E.A.; Roberts, C.; Smoke, C.C.; Meyer, C.; Højlund, K.; Yi, Z.; et al. Proteomics analysis of human skeletal muscle reveals novel abnormalities in obesity and type 2 diabetes. *Diabetes* **2010**, *59*, 33–42. [CrossRef] [PubMed]
27. Bandyopadhyay, G.K.; Yu, J.G.; Ofrecio, J.; Olefsky, J.M. Increased malonyl-coa levels in muscle from obese and type 2 diabetic subjects lead to decreased fatty acid oxidation and increased lipogenesis; thiazolidinedione treatment reverses these defects. *Diabetes* **2006**, *55*, 2277–2285. [CrossRef] [PubMed]
28. Trenell, M.I.; Hollingsworth, K.G.; Lim, E.L.; Taylor, R. Increased daily walking improves lipid oxidation without changes in mitochondrial function in type 2 diabetes. *Diabetes Care* **2008**, *31*, 1644–1649. [CrossRef] [PubMed]
29. Garcia-Roves, P.; Huss, J.M.; Han, D.-H.; Hancock, C.R.; Iglesias-Gutierrez, E.; Chen, M.; Holloszy, J.O. Raising plasma fatty acid concentration induces increased biogenesis of mitochondria in skeletal muscle. *Proc. Natl. Acad. Sci. USA* **2007**, *104*, 10709–10713. [CrossRef] [PubMed]
30. Hancock, C.R.; Han, D.H.; Chen, M.; Terada, S.; Yasuda, T.; Wright, D.C.; Holloszy, J.O. High-fat diets cause insulin resistance despite an increase in muscle mitochondria. *Proc. Natl. Acad. Sci. USA* **2008**. [CrossRef]
31. Ruegsegger, G.N.; Creo, A.L.; Cortes, T.M.; Dasari, S.; Nair, K.S. Altered mitochondrial function in insulin-deficient and insulin-resistant states. *J. Clin. Invest.* **2018**, *128*, 3671–3681. [CrossRef]
32. Fernández-Vizarra, E.; Enríquez, J.A.; Pérez-Martos, A.; Montoya, J.; Fernández-Silva, P. Tissue-specific differences in mitochondrial activity and biogenesis. *Mitochondrion* **2011**, *11*, 207–213. [CrossRef]
33. Montaigne, D.; Marechal, X.; Coisne, A.; Debry, N.; Modine, T.; Fayad, G.; Potelle, C.; El Arid, J.; Mouton, S.; Sebti, Y.; et al. Myocardial contractile dysfunction is associated with impaired mitochondrial function and dynamics in type 2 diabetic but not in obese patients. *Circulation* **2014**, *130*, 554–564. [CrossRef]
34. Dhalla, N.S.; Takeda, N.; Rodriguez-Leyva, D.; Elimban, V. Mechanisms of subcellular remodeling in heart failure due to diabetes. *Heart Fail. Rev.* **2014**, *19*, 87–99. [CrossRef] [PubMed]
35. Mackenzie, R.M.; Salt, I.P.; Miller, W.H.; Logan, A.; Ibrahim, H.A.; Degasperi, A.; Dymott, J.A.; Hamilton, C.A.; Murphy, M.P.; Delles, C.; et al. Mitochondrial reactive oxygen species enhance AMP-activated protein kinase activation in the endothelium of patients with coronary artery disease and diabetes. *Clin. Sci.* **2013**, *124*, 403–411. [CrossRef] [PubMed]

36. Croston, T.L.; Thapa, D.; Holden, A.A.; Tveter, K.J.; Lewis, S.E.; Shepherd, D.L.; Nichols, C.E.; Long, D.M.; Olfert, I.M.; Jagannathan, R.; et al. Functional deficiencies of subsarcolemmal mitochondria in the type 2 diabetic human heart. *Am. J. Physiol. Heart Circ. Physiol.* **2014**, *307*, H54–H65. [CrossRef] [PubMed]
37. Anderson, E.J.; Rodriguez, E.; Anderson, C.A.; Thayne, K.; Chitwood, W.R.; Kypson, A.P. Increased propensity for cell death in diabetic human heart is mediated by mitochondrial-dependent pathways. *Am. J. Physiol. Heart Circ. Physiol.* **2011**, *300*, H118–H124. [CrossRef] [PubMed]
38. Marciniak, C.; Marechal, X.; Montaigne, D.; Neviere, R.; Lancel, S. Cardiac contractile function and mitochondrial respiration in diabetes-related mouse models. *Cardiovasc. Diabetol.* **2014**, *13*, 118. [CrossRef] [PubMed]
39. Pham, T.; Loiselle, D.; Power, A.; Hickey, A.J.R. Mitochondrial inefficiencies and anoxic ATP hydrolysis capacities in diabetic rat heart. *Am. J. Physiol. Cell Physiol.* **2014**, *307*, C499–C507. [CrossRef]
40. Hicks, S.; Labinskyy, N.; Piteo, B.; Laurent, D.; Mathew, J.E.; Gupte, S.A.; Edwards, J.G. Type II diabetes increases mitochondrial DNA mutations in the left ventricle of the Goto-Kakizaki diabetic rat. *Am. J. Physiol. Heart Circ. Physiol.* **2013**, *304*, H903–H915. [CrossRef] [PubMed]
41. Yan, W.; Zhang, H.; Liu, P.; Wang, H.; Liu, J.; Gao, C.; Liu, Y.; Lian, K.; Yang, L.; Sun, L.; et al. Impaired mitochondrial biogenesis due to dysfunctional adiponectin-AMPK-PGC-1α signaling contributing to increased vulnerability in diabetic heart. *Basic Res. Cardiol.* **2013**, *108*, 329. [CrossRef]
42. Luo, M.; Guan, X.; Luczak, E.D.; Lang, D.; Kutschke, W.; Gao, Z.; Yang, J.; Glynn, P.; Sossalla, S.; Swaminathan, P.D.; et al. Diabetes increases mortality after myocardial infarction by oxidizing CaMKII. *J. Clin. Invest.* **2013**, *123*, 1262–1274. [CrossRef]
43. Vazquez, E.J.; Berthiaume, J.M.; Kamath, V.; Achike, O.; Buchanan, E.; Montano, M.M.; Chandler, M.P.; Miyagi, M.; Rosca, M.G. Mitochondrial complex I defect and increased fatty acid oxidation enhance protein lysine acetylation in the diabetic heart. *Cardiovasc. Res.* **2015**, *107*, 453–465. [CrossRef]
44. Deeds, M.C.; Anderson, J.M.; Armstrong, A.S.; Gastineau, D.A.; Hiddinga, H.J.; Jahangir, A.; Eberhardt, N.L.; Kudva, Y.C. Single dose streptozotocin induced diabetes: considerations for study design in islet transplantation models. *Lab. Anim.* **2011**, *45*, 131–140. [CrossRef] [PubMed]
45. Turkieh, A.; Caubère, C.; Barutaut, M.; Desmoulin, F.; Harmancey, R.; Galinier, M.; Berry, M.; Dambrin, C.; Polidori, C.; Casteilla, L.; et al. Apolipoprotein O is mitochondrial and promotes lipotoxicity in heart. *J. Clin. Invest.* **2014**, *124*, 2277–2286. [CrossRef] [PubMed]
46. Wai, T.; García-Prieto, J.; Baker, M.J.; Merkwirth, C.; Benit, P.; Rustin, P.; Rupérez, F.J.; Barbas, C.; Ibañez, B.; Langer, T. Imbalanced OPA1 processing and mitochondrial fragmentation cause heart failure in mice. *Science* **2015**, *350*. [CrossRef]
47. Liu, W.; Ruiz-Velasco, A.; Wang, S.; Khan, S.; Zi, M.; Jungmann, A.; Dolores Camacho-Muñoz, M.; Guo, J.; Du, G.; Xie, L.; et al. Metabolic stress-induced cardiomyopathy is caused by mitochondrial dysfunction due to attenuated Erk5 signaling. *Nat. Commun.* **2017**, *8*, 494. [CrossRef] [PubMed]
48. Lehmann, L.H.; Jebessa, Z.H.; Kreusser, M.M.; Horsch, A.; He, T.; Kronlage, M.; Dewenter, M.; Sramek, V.; Oehl, U.; Krebs-Haupenthal, J.; et al. A proteolytic fragment of histone deacetylase 4 protects the heart from failure by regulating the hexosamine biosynthetic pathway. *Nat. Med.* **2017**, *24*, 62. [CrossRef] [PubMed]
49. Lopaschuk, G.D.; Ussher, J.R.; Folmes, C.D.L.; Jaswal, J.S.; Stanley, W.C. Myocardial fatty acid metabolism in health and disease. *Physiol. Rev.* **2010**, *90*, 207–258. [CrossRef] [PubMed]
50. Fillmore, N.; Mori, J.; Lopaschuk, G.D. Mitochondrial fatty acid oxidation alterations in heart failure, ischaemic heart disease and diabetic cardiomyopathy. *Br. J. Pharmacol.* **2014**, *171*, 2080–2090. [CrossRef]
51. Szczepaniak, L.S.; Dobbins, R.L.; Metzger, G.J.; Sartoni-D'Ambrosia, G.; Arbique, D.; Vongpatanasin, W.; Unger, R.; Victor, R.G. Myocardial triglycerides and systolic function in humans: In vivo evaluation by localized proton spectroscopy and cardiac imaging. *Magn. Reson. Med.* **2003**, *49*, 417–423. [CrossRef]
52. Peterson, L.; Herrero, P.; Schechtman, K.; Racette, S.; Waggoner, A.; Kisrieva-Ware, Z.; Dence, C.; Klein, S.; Marsala, J.; Meyer, T.; et al. Effect of obesity and insulin resistance on myocardial substrate metabolism and efficiency in young women. *Circulation* **2004**, *109*, 2191–2196. [CrossRef]
53. Mazumder, P.K.; O'Neill, B.T.; Roberts, M.W.; Buchanan, J.; Yun, U.J.; Cooksey, R.C.; Boudina, S.; Abel, E.D. Impaired cardiac efficiency and increased fatty acid oxidation in insulin-resistant ob/ob mouse hearts. *Diabetes* **2004**, *53*, 2366–2374. [CrossRef]

54. Belke, D.D.; Severson, D.L. Diabetes in mice with monogenic obesity: The db/db mouse and its use in the study of cardiac consequences. In *Animal Models in Diabetes Research*. Joost, H.-G.; Al-Hasani, H., Schürmann, A., Eds.; Humana Press: Totowa, NJ, USA, 2012; pp. 47–57.
55. Ansley, D.M.; Wang, B. Oxidative stress and myocardial injury in the diabetic heart. *J. Pathol.* **2013**, *229*, 232–241. [CrossRef] [PubMed]
56. Zhang, L.; Ussher, J.R.; Oka, T.; Cadete, V.J.J.; Wagg, C.; Lopaschuk, G.D. Cardiac diacylglycerol accumulation in high fat-fed mice is associated with impaired insulin-stimulated glucose oxidation. *Cardiovasc. Res.* **2011**, *89*, 148–156. [CrossRef] [PubMed]
57. Kuramoto, K.; Okamura, T.; Yamaguchi, T.; Nakamura, T.Y.; Wakabayashi, S.; Morinaga, H.; Nomura, M.; Yanase, T.; Otsu, K.; Usuda, N.; et al. Perilipin 5, a lipid droplet-binding protein, protects heart from oxidative burden by sequestering fatty acid from excessive oxidation. *J. Biol. Chem.* **2012**, *287*, 23852–23863. [CrossRef] [PubMed]
58. Chatham, J.C.; Forder, J.R. A 13C-NMR Study of glucose oxidation in the intact functioning rat heart following diabetes-induced cardiomyopathy. *J. Mol. Cell. Cardiol.* **1993**, *25*, 1203–1213. [CrossRef] [PubMed]
59. Chatham, J.C.; Forder, J.R. Relationship between cardiac function and substrate oxidation in hearts of diabetic rats. *Am. J. Physiol. Heart Circ. Physiol.* **1997**, *273*, H52–H58. [CrossRef] [PubMed]
60. Wu, C.-Y.; Satapati, S.; Gui, W.; Wynn, R.M.; Sharma, G.; Lou, M.; Qi, X.; Burgess, S.C.; Malloy, C.; Khemtong, C.; et al. A novel inhibitor of pyruvate dehydrogenase kinase stimulates myocardial carbohydrate oxidation in diet-induced obesity. *J. Biol. Chem.* **2018**, *293*, 9604–9613. [CrossRef]
61. Saeedi, R.; Grist, M.; Wambolt, R.B.; Bescond-Jacquet, A.; Lucien, A.; Allard, M.F. Trimetazidine normalizes postischemic function of hypertrophied rat hearts. *J. Pharmacol. Exp. Ther.* **2005**, *314*, 446–454. [CrossRef]
62. Dyck, J.R.; Cheng, J.F.; Stanley, W.C.; Barr, R.; Chandler, M.P.; Brown, S.; Wallace, D.; Arrhenius, T.; Harmon, C.; Yang, G.; et al. Malonyl coenzyme a decarboxylase inhibition protects the ischemic heart by inhibiting fatty acid oxidation and stimulating glucose oxidation. *Circ. Res.* **2004**, *94*, e78–e84.
63. Keung, W.; Ussher, J.R.; Jaswal, J.S.; Raubenheimer, M.; Lam, V.H.M.; Wagg, C.S.; Lopaschuk, G.D. Inhibition of carnitine palmitoyltransferase-1 activity alleviates insulin resistance in diet-induced obese mice. *Diabetes* **2013**, *62*, 711–720. [CrossRef]
64. Timmers, S.; Nabben, M.; Bosma, M.; van Bree, B.; Lenaers, E.; van Beurden, D.; Schaart, G.; Westerterp-Plantenga, M.S.; Langhans, W.; Hesselink, M.K.C.; et al. Augmenting muscle diacylglycerol and triacylglycerol content by blocking fatty acid oxidation does not impede insulin sensitivity. *Proc. Natl. Acad. Sci. USA* **2012**, *109*, 11711–11716. [CrossRef]
65. Henstridge, D.C.; Bruce, C.R.; Drew, B.G.; Tory, K.; Kolonics, A.; Estevez, E.; Chung, J.; Watson, N.; Gardner, T.; Lee-Young, R.S.; et al. Activating HSP72 in rodent skeletal muscle increases mitochondrial number and oxidative capacity and decreases insulin resistance. *Diabetes* **2014**, *63*, 1881–1894. [CrossRef]
66. Henstridge, D.C.; Estevez, E.; Allen, T.L.; Heywood, S.E.; Gardner, T.; Yang, C.; Mellett, N.A.; Kingwell, B.A.; Meikle, P.J.; Febbraio, M.A. Genetic manipulation of cardiac Hsp72 levels does not alter substrate metabolism but reveals insights into high-fat feeding-induced cardiac insulin resistance. *Cell Stress Chaperones* **2015**, *20*, 461–472. [CrossRef] [PubMed]
67. Holloway, G.P.; Chou, C.J.; Lally, J.; Stellingwerff, T.; Maher, A.C.; Gavrilova, O.; Haluzik, M.; Alkhateeb, H.; Reitman, M.L.; Bonen, A. Increasing skeletal muscle fatty acid transport protein 1 (FATP1) targets fatty acids to oxidation and does not predispose mice to diet-induced insulin resistance. *Diabetologia* **2011**, *54*, 1457. [CrossRef] [PubMed]
68. Snowling, N.J.; Hopkins, W.G. Effects of different modes of exercise training on glucose control and risk factors for complications in type 2 diabetic patients. *A Meta. Analysis* **2006**, *29*, 2518–2527. [CrossRef] [PubMed]
69. Perry, R.J.; Peng, L.; Cline, G.W.; Wang, Y.; Rabin-Court, A.; Song, J.D.; Zhang, D.; Zhang, X.-M.; Nozaki, Y.; Dufour, S.; et al. Mechanisms by which a Very-Low-Calorie Diet Reverses Hyperglycemia in a Rat Model of Type 2 Diabetes. *Cell Metab.* **2018**, *27*, 210–217.e213. [CrossRef] [PubMed]
70. Meex, R.C.R.; Schrauwen-Hinderling, V.B.; Moonen-Kornips, E.; Schaart, G.; Mensink, M.; Phielix, E.; van de Weijer, T.; Sels, J.-P.; Schrauwen, P.; Hesselink, M.K.C. Restoration of muscle mitochondrial function and metabolic flexibility in type 2 diabetes by exercise training is paralleled by increased myocellular fat storage and improved insulin sensitivity. *Diabetes* **2010**, *59*, 572–579. [CrossRef] [PubMed]

71. Finckenberg, P.; Eriksson, O.; Baumann, M.; Merasto, S.; Lalowski, M.M.; Levijoki, J.; Haasio, K.; Kytö, V.; Muller, D.N.; Luft, F.C.; et al. Caloric restriction ameliorates angiotensin II-induced mitochondrial remodeling and cardiac hypertrophy. *Hypertension* **2012**, *59*, 76–84. [CrossRef]
72. Holloszy, J.O. Regulation of mitochondrial biogenesis and GLUT4 expression by exercise. *Compr. Physiol.* **2011**, *1*, 921–940.
73. Johnson, M.L.; Distelmaier, K.; Lanza, I.R.; Irving, B.A.; Robinson, M.M.; Konopka, A.R.; Shulman, G.I.; Nair, K.S. Mechanism by which caloric restriction improves insulin sensitivity in sedentary obese adults. *Diabetes* **2016**, *65*, 74–84. [CrossRef]
74. Rabøl, R.; Svendsen, P.F.; Skovbro, M.; Boushel, R.; Haugaard, S.B.; Schjerling, P.; Schrauwen, P.; Hesselink, M.K.C.; Nilas, L.; Madsbad, S.; Dela, F. Reduced skeletal muscle mitochondrial respiration and improved glucose metabolism in nondiabetic obese women during a very low calorie dietary intervention leading to rapid weight loss. *Metab. Clin. Exp.* **2009**, *58*, 1145–1152. [CrossRef]
75. Veiga, F.M.S.; Graus-Nunes, F.; Rachid, T.L.; Barreto, A.B.; Mandarim-de-Lacerda, C.A.; Souza-Mello, V. Anti-obesogenic effects of WY14643 (PPAR-alpha agonist): Hepatic mitochondrial enhancement and suppressed lipogenic pathway in diet-induced obese mice. *Biochimie* **2017**, *140*, 106–116. [CrossRef] [PubMed]
76. Rasouli, N.; Kern, P.A.; Elbein, S.C.; Sharma, N.K.; Das, S.K. Improved insulin sensitivity after treatment with PPARγ and PPARα ligands is mediated by genetically modulated transcripts. *Pharmacogenet. Genomics* **2012**, *22*, 484–497. [CrossRef] [PubMed]
77. Coletta, D.K.; Fernandez, M.; Cersosimo, E.; Gastaldelli, A.; Musi, N.; DeFronzo, R.A. The effect of muraglitazar on adiponectin signalling, mitochondrial function and fat oxidation genes in human skeletal muscle in vivo. *Diabet. Med.* **2015**, *32*, 657–664. [CrossRef] [PubMed]
78. Tanaka, T.; Yamamoto, J.; Iwasaki, S.; Asaba, H.; Hamura, H.; Ikeda, Y.; Watanabe, M.; Magoori, K.; Ioka, R.X.; Tachibana, K.; et al. Activation of peroxisome proliferator-activated receptor δ induces fatty acid β-oxidation in skeletal muscle and attenuates metabolic syndrome. *Proc. Natl. Acad. Sci. USA* **2003**, *100*, 15924–15929. [CrossRef] [PubMed]
79. Yaribeygi, H.; Atkin, S.L.; Sahebkar, A. Mitochondrial dysfunction in diabetes and the regulatory roles of antidiabetic agents on the mitochondrial function. *J. Cell. Physiol.* **2018**. [CrossRef] [PubMed]
80. Nisr, R.B.; Affourtit, C. Insulin acutely improves mitochondrial function of rat and human skeletal muscle by increasing coupling efficiency of oxidative phosphorylation. *Biochim. Biophys. Acta* **2014**, *1837*, 270–276. [CrossRef]
81. Petersen, K.F.; Dufour, S.; Shulman, G.I. Decreased insulin-stimulated ATP synthesis and phosphate transport in muscle of insulin-resistant offspring of type 2 diabetic parents. *PLoS Med.* **2005**, *2*, e233. [CrossRef] [PubMed]
82. Mizuno, M.; Kuno, A.; Yano, T.; Miki, T.; Oshima, H.; Sato, T.; Nakata, K.; Kimura, Y.; Tanno, M.; Miura, T. Empagliflozin normalizes the size and number of mitochondria and prevents reduction in mitochondrial size after myocardial infarction in diabetic hearts. *Physiol. Rep.* **2018**, *6*, e13741. [CrossRef]
83. Vettor, R.; Inzucchi, S.E.; Fioretto, P. The cardiovascular benefits of empagliflozin: SGLT2-dependent and -independent effects. *Diabetologia* **2017**, *60*, 395–398. [CrossRef]
84. Zhou, H.; Wang, S.; Zhu, P.; Hu, S.; Chen, Y.; Ren, J. Empagliflozin rescues diabetic myocardial microvascular injury via AMPK-mediated inhibition of mitochondrial fission. *Redox Biol.* **2017**, *15*, 335–346. [CrossRef]
85. Sa-nguanmoo, P.; Tanajak, P.; Kerdphoo, S.; Jaiwongkam, T.; Pratchayasakul, W.; Chattipakorn, N.; Chattipakorn, S.C. SGLT2-inhibitor and DPP-4 inhibitor improve brain function via attenuating mitochondrial dysfunction, insulin resistance, inflammation, and apoptosis in HFD-induced obese rats. *Toxicol. Appl. Pharmacol.* **2017**, *333*, 43–50. [CrossRef] [PubMed]
86. Hawley, S.A.; Ford, R.J.; Smith, B.K.; Gowans, G.J.; Mancini, S.J.; Pitt, R.D.; Day, E.A.; Salt, I.P.; Steinberg, G.R.; Hardie, D.G. The Na$^+$/Glucose cotransporter inhibitor canagliflozin activates AMPK by inhibiting mitochondrial function and increasing cellular AMP levels. *Diabetes* **2016**, *65*, 2784–2794. [CrossRef] [PubMed]
87. Secker, P.F.; Beneke, S.; Schlichenmaier, N.; Delp, J.; Gutbier, S.; Leist, M.; Dietrich, D.R. Canagliflozin mediated dual inhibition of mitochondrial glutamate dehydrogenase and complex I: an off-target adverse effect. *Cell Death Dis.* **2018**, *9*, 226. [CrossRef] [PubMed]

88. Sun, D.; Yang, F. Metformin improves cardiac function in mice with heart failure after myocardial infarction by regulating mitochondrial energy metabolism. *Biochemical and Biophysical Research Communications* **2017**, *486*, 329–335. [CrossRef] [PubMed]
89. Wang, Q.; Zhang, M.; Torres, G.; Wu, S.; Ouyang, C.; Xie, Z.; Zou, M.-H. Metformin suppresses diabetes-accelerated atherosclerosis via the inhibition of Drp1-mediated mitochondrial fission. *Diabetes* **2017**, *66*, 193–205. [CrossRef] [PubMed]
90. Loubiere, C.; Clavel, S.; Gilleron, J.; Harisseh, R.; Fauconnier, J.; Ben-Sahra, I.; Kaminski, L.; Laurent, K.; Herkenne, S.; Lacas-Gervais, S.; et al. The energy disruptor metformin targets mitochondrial integrity via modification of calcium flux in cancer cells. *Sci. Rep.* **2017**, *7*, 5040. [CrossRef] [PubMed]
91. Wheaton, W.W.; Weinberg, S.E.; Hamanaka, R.B.; Soberanes, S.; Sullivan, L.B.; Anso, E.; Glasauer, A.; Dufour, E.; Mutlu, G.M.; Budigner, G.S.; et al. Metformin inhibits mitochondrial complex I of cancer cells to reduce tumorigenesis. *eLife* **2014**, *3*, e02242. [CrossRef] [PubMed]
92. Cameron, A.R.; Logie, L.; Patel, K.; Erhardt, S.; Bacon, S.; Middleton, P.; Harthill, J.; Forteath, C.; Coats, J.T.; Kerr, C.; et al. Metformin selectively targets redox control of complex I energy transduction. *Redox Biol.* **2017**, *14*, 187–197. [CrossRef]
93. Gui, D.Y.; Sullivan, L.B.; Luengo, A.; Hosios, A.M.; Bush, L.N.; Gitego, N.; Davidson, S.M.; Freinkman, E.; Thomas, C.J.; Vander Heiden, M.G. Environment dictates dependence on mitochondrial complex i for nad+ and aspartate production and determines cancer cell sensitivity to Metformin. *Cell Metab.* **2016**, *24*, 716–727. [CrossRef]
94. Ghosh, S.; Patel, N.; Rahn, D.; McAllister, J.; Sadeghi, S.; Horwitz, G.; Berry, D.; Wang, K.X.; Swerdlow, R.H. The thiazolidinedione pioglitazone alters mitochondrial function in human neuron-like cells. *Mol. Pharmacol.* **2007**, *71*, 1695–1702. [CrossRef]
95. Bogacka, I.; Xie, H.; Bray, G.A.; Smith, S.R. Pioglitazone induces mitochondrial biogenesis in human subcutaneous adipose tissue in vivo. *Diabetes* **2005**, *54*, 1392–1399. [CrossRef]
96. Wang, W.; Karamanlidis, G.; Tian, R. Novel targets for mitochondrial medicine. *Sci. Transl. Med.* **2016**, *8*, 326rv323. [CrossRef] [PubMed]
97. Nakahira, K.; Kyung, S.-Y.; Rogers, A.J.; Gazourian, L.; Youn, S.; Massaro, A.F.; Quintana, C.; Osorio, J.C.; Wang, Z.; Zhao, Y.; et al. Circulating mitochondrial dna in patients in the icu as a marker of mortality: derivation and validation. *PLoS Med.* **2014**, *10*, e1001577. [CrossRef] [PubMed]
98. Gu, X.; Yao, Y.; Wu, G.; Lv, T.; Luo, L.; Song, Y. The plasma mitochondrial dna is an independent predictor for post-traumatic systemic inflammatory response syndrome. *PLoS ONE* **2013**, *8*, e72834. [CrossRef] [PubMed]
99. Jeannin, P.; Chaze, T.; Giai Gianetto, Q.; Matondo, M.; Gout, O.; Gessain, A.; Afonso, P.V. Proteomic analysis of plasma extracellular vesicles reveals mitochondrial stress upon HTLV-1 infection. *Sci. Rep.* **2018**, *8*, 5170. [CrossRef] [PubMed]
100. Tretter, L.; Patocs, A.; Chinopoulos, C. Succinate, an intermediate in metabolism, signal transduction, ROS, hypoxia, and tumorigenesis. *Biochim. Biophys. Acta* **2016**, *1857*, 1086–1101. [CrossRef] [PubMed]
101. Moschoi, R.; Imbert, V.; Nebout, M.; Chiche, J.; Mary, D.; Prebet, T.; Saland, E.; Castellano, R.; Pouyet, L.; Collette, Y.; et al. Protective mitochondrial transfer from bone marrow stromal cells to acute myeloid leukemic cells during chemotherapy. *Blood* **2016**, *128*, 253–264. [CrossRef] [PubMed]
102. Marlein, C.R.; Zaitseva, L.; Piddock, R.E.; Robinson, S.D.; Edwards, D.R.; Shafat, M.S.; Zhou, Z.; Lawes, M.; Bowles, K.M.; Rushworth, S.A. NADPH oxidase-2 derived superoxide drives mitochondrial transfer from bone marrow stromal cells to leukemic blasts. *Blood* **2017**, *130*, 1649–1660. [CrossRef] [PubMed]
103. Jackson, M.V.; Morrison, T.J.; Doherty, D.F.; McAuley, D.F.; Matthay, M.A.; Kissenpfennig, A.; O'Kane, C.M.; Krasnodembskaya, A.D. Mitochondrial transfer via tunneling nanotubes is an important mechanism by which mesenchymal stem cells enhance macrophage phagocytosis in the in vitro and in vivo models of ARDS. *Stem Cells* **2016**, *34*, 2210–2223. [CrossRef]
104. Islam, M.N.; Das, S.R.; Emin, M.T.; Wei, M.; Sun, L.; Westphalen, K.; Rowlands, D.J.; Quadri, S.K.; Bhattacharya, S.; Bhattacharya, J. Mitochondrial transfer from bone-marrow–derived stromal cells to pulmonary alveoli protects against acute lung injury. *Nat. Med.* **2012**, *18*, 759. [CrossRef]
105. Ahmad, T.; Mukherjee, S.; Pattnaik, B.; Kumar, M.; Singh, S.; Kumar, M.; Rehman, R.; Tiwari, B.K.; Jha, K.A.; Barhanpurkar, A.P.; et al. Miro1 regulates intercellular mitochondrial transport & enhances mesenchymal stem cell rescue efficacy. *EMBO J.* **2014**, *33*, 994–1010. [PubMed]

106. Masuzawa, A.; Black, K.M.; Pacak, C.A.; Ericsson, M.; Barnett, R.J.; Drumm, C.; Seth, P.; Bloch, D.B.; Levitsky, S.; Cowan, D.B.; et al. Transplantation of autologously derived mitochondria protects the heart from ischemia-reperfusion injury. *Am. J. Physiol. Heart Circ. Physiol.* **2013**, *304*, H966–H982. [CrossRef] [PubMed]
107. Zhang, Q.; Raoof, M.; Chen, Y.; Sumi, Y.; Sursal, T.; Junger, W.; Brohi, K.; Itagaki, K.; Hauser, C.J. Circulating mitochondrial DAMPs cause inflammatory responses to injury. *Nature* **2010**, *464*, 104. [CrossRef] [PubMed]
108. Boudreau, L.H.; Duchez, A.-C.; Cloutier, N.; Soulet, D.; Martin, N.; Bollinger, J.; Paré, A.; Rousseau, M.; Naika, G.S.; Lévesque, T.; et al. Platelets release mitochondria serving as substrate for bactericidal group IIA-secreted phospholipase A2 to promote inflammation. *Blood* **2014**, *124*, 2173–2183. [CrossRef] [PubMed]
109. Pasquier, J.; Guerrouahen, B.S.; Al Thawadi, H.; Ghiabi, P.; Maleki, M.; Abu-Kaoud, N.; Jacob, A.; Mirshahi, M.; Galas, L.; Rafii, S.; et al. Preferential transfer of mitochondria from endothelial to cancer cells through tunneling nanotubes modulates chemoresistance. *J. Transl. Med.* **2013**, *11*, 94. [CrossRef] [PubMed]
110. Antanavičiūtė, I.; Rysevaitė, K.; Liutkevičius, V.; Marandykina, A.; Rimkutė, L.; Sveikatienė, R.; Uloza, V.; Skeberdis, V.A. Long-Distance communication between laryngeal carcinoma cells. *PLoS ONE* **2014**, *9*, e99196. [CrossRef]
111. Searle, J.; Lawson, T.A.; Abbott, P.J.; Harmon, B.; Kerr, J.F.R. An electron-microscope study of the mode of cell death induced by cancer-chemotherapeutic agents in populations of proliferating normal and neoplastic cells. *J. Pathol.* **1975**, *116*, 129–138. [CrossRef]
112. Zitvogel, L.; Regnault, A.; Lozier, A.; Wolfers, J.; Flament, C.; Tenza, D.; Ricciardi-Castagnoli, P.; Raposo, G.; Amigorena, S. Eradication of established murine tumors using a novel cell-free vaccine: dendritic cell derived exosomes. *Nat. Med.* **1998**, *4*, 594. [CrossRef]
113. Wolfers, J.; Lozier, A.; Raposo, G.; Regnault, A.; Théry, C.; Masurier, C.; Flament, C.; Pouzieux, S.; Faure, F.; Tursz, T.; et al. Tumor-derived exosomes are a source of shared tumor rejection antigens for CTL cross-priming. *Nat. Med.* **2001**, *7*, 297. [CrossRef]
114. Théry, C.; Ostrowski, M.; Segura, E. Membrane vesicles as conveyors of immune responses. *Nat. Rev. Immunol.* **2009**, *9*, 581. [CrossRef]
115. Hessvik, N.P.; Llorente, A. Current knowledge on exosome biogenesis and release. *Cell. Mol. Life Sci.* **2018**, *75*, 193–208. [CrossRef] [PubMed]
116. Choi, D.-S.; Kim, D.-K.; Kim, Y.-K.; Gho, Y.S. Proteomics, transcriptomics and lipidomics of exosomes and ectosomes. *Proteomics* **2013**, *13*, 1554–1571. [CrossRef] [PubMed]
117. Whitham, M.; Parker, B.L.; Friedrichsen, M.; Hingst, J.R.; Hjorth, M.; Hughes, W.E.; Egan, C.L.; Cron, L.; Watt, K.I.; Kuchel, R.P.; et al. Extracellular vesicles provide a means for tissue crosstalk during exercise. *Cell Metab.* **2018**, *27*, 237–251.e4. [CrossRef] [PubMed]
118. Hirsova, P.; Ibrahim, S.H.; Krishnan, A.; Verma, V.K.; Bronk, S.F.; Werneburg, N.W.; Charlton, M.R.; Shah, V.H.; Malhi, H.; Gores, G.J. Lipid-induced signaling causes release of inflammatory extracellular vesicles from hepatocytes. *Gastroenterology* **2016**, *150*, 956–967. [CrossRef] [PubMed]
119. Crewe, C.; Joffin, N.; Rutkowski, J.M.; Kim, M.; Zhang, F.; Towler, D.A.; Gordillo, R.; Scherer, P.E. An endothelial-to-adipocyte extracellular vesicle axis governed by metabolic state. *Cell* **2018**, *175*, 695–708.e13. [CrossRef] [PubMed]
120. Thomou, T.; Mori, M.A.; Dreyfuss, J.M.; Konishi, M.; Sakaguchi, M.; Wolfrum, C.; Rao, T.N.; Winnay, J.N.; Garcia-Martin, R.; Grinspoon, S.K.; et al. Adipose-derived circulating miRNAs regulate gene expression in other tissues. *Nature* **2017**, *542*, 450. [CrossRef] [PubMed]
121. Murakami, T.; Horigome, H.; Tanaka, K.; Nakata, Y.; Ohkawara, K.; Katayama, Y.; Matsui, A. Impact of weight reduction on production of platelet-derived microparticles and fibrinolytic parameters in obesity. *Thromb. Res.* **2007**, *119*, 45–53. [CrossRef]
122. Eguchi, A.; Lazic, M.; Armando, A.M.; Phillips, S.A.; Katebian, R.; Maraka, S.; Quehenberger, O.; Sears, D.D.; Feldstein, A.E. Circulating adipocyte-derived extracellular vesicles are novel markers of metabolic stress. *J. Cell Mol. Med.* **2016**, *94*, 1241–1253. [CrossRef]
123. Heinrich, L.F.; Andersen, D.K.; Cleasby, M.E.; Lawson, C. Long-term high fat feeding of rats results in increased numbers of circulating microvesicles with pro-inflammatory effects on endothelial cells. *Br. J. Nutr.* **2015**, *113*, 1704–1711. [CrossRef]

124. Freeman, D.W.; Hooten, N.N.; Eitan, E.; Green, J.; Mode, N.A.; Bodogai, M.; Zhang, Y.; Lehrmann, E.; Zonderman, A.B.; Biragyn, A.; et al. Altered extracellular vesicle concentration, cargo and function in diabetes mellitus. *Diabetes* **2018**, *67*, 2377–2388. [CrossRef]
125. Hubal, M.J.; Nadler, E.P.; Ferrante, S.C.; Barberio, M.D.; Suh, J.-H.; Wang, J.; Dohm, G.L.; Pories, W.J.; Mietus-Snyder, M.; Freishtat, R.J. Circulating adipocyte-derived exosomal MicroRNAs associated with decreased insulin resistance after gastric bypass. *Obesity* **2017**, *25*, 102–110. [CrossRef] [PubMed]
126. Ying, W.; Riopel, M.; Bandyopadhyay, G.; Dong, Y.; Birmingham, A.; Seo, J.B.; Ofrecio, J.M.; Wollam, J.; Hernandez-Carretero, A.; Fu, W.; et al. Adipose tissue macrophage-derived exosomal miRNAs can modulate In Vivo and In Vitro insulin sensitivity. *Cell* **2017**, *171*, 372–384.e12. [CrossRef] [PubMed]
127. Pardo, F.; Villalobos-Labra, R.; Sobrevia, B.; Toledo, F.; Sobrevia, L. Extracellular vesicles in obesity and diabetes mellitus. *Mol. Aspects Med.* **2018**, *60*, 81–91. [CrossRef] [PubMed]
128. Deng, Z.-b.; Poliakov, A.; Hardy, R.W.; Clements, R.; Liu, C.; Liu, Y.; Wang, J.; Xiang, X.; Zhang, S.; Zhuang, X.; et al. Adipose tissue exosome-like vesicles mediate activation of macrophage-induced insulin resistance. *Diabetes* **2009**, *58*, 2498–2505. [CrossRef] [PubMed]
129. Müller, G. Microvesicles/exosomes as potential novel biomarkers of metabolic diseases. *Diabetes Metab. Syndr. Obes.* **2012**, *5*, 247–282. [CrossRef] [PubMed]
130. Neuspiel, M.; Schauss, A.C.; Braschi, E.; Zunino, R.; Rippstein, P.; Rachubinski, R.A.; Andrade-Navarro, M.A.; McBride, H.M. Cargo-selected transport from the mitochondria to peroxisomes is mediated by vesicular carriers. *Curr. Biol.* **2008**, *18*, 102–108. [CrossRef] [PubMed]
131. Soubannier, V.; McLelland, G.-L.; Zunino, R.; Braschi, E.; Rippstein, P.; Fon, E.A.; McBride, H.M. A vesicular transport pathway shuttles cargo from mitochondria to lysosomes. *Curr. Biol.* **2012**, *22*, 135–141. [CrossRef]
132. Soubannier, V.; Rippstein, P.; Kaufman, B.A.; Shoubridge, E.A.; McBride, H.M. Reconstitution of mitochondria derived vesicle formation demonstrates selective enrichment of oxidized cargo. *PLoS ONE* **2012**, *7*, e52830. [CrossRef]
133. McLelland, G.-L.; Soubannier, V.; Chen, C.X.; McBride, H.M.; Fon, E.A. Parkin and PINK1 function in a vesicular trafficking pathway regulating mitochondrial quality control. *EMBO J.* **2014**, *33*, 282–295. [CrossRef]
134. Raposo, G.; Stoorvogel, W. Extracellular vesicles: exosomes, microvesicles, and friends. *J. Cell Biol.* **2013**, *200*, 373–383. [CrossRef]
135. Lee, J.; Moon, P.; Lee, I.; Baek, M. Proteomic analysis of extracellular vesicles released by adipocytes of otsuka long-evans tokushima fatty (OLETF) rats. *Protein J.* **2015**, *34*, 220–235. [PubMed]
136. Szczesny, B.; Marcatti, M.; Ahmad, A.; Montalbano, M.; Brunyánszki, A.; Bibli, S.-I.; Papapetropoulos, A.; Szabo, C. Mitochondrial DNA damage and subsequent activation of Z-DNA binding protein 1 links oxidative stress to inflammation in epithelial cells. *Sci. Rep.* **2018**, *8*, 914. [PubMed]
137. Kannan, A.; Wells, R.B.; Sivakumar, S.; Komatsu, S.; Singh, K.P.; Samten, B.; Philley, J.V.; Sauter, E.R.; Ikebe, M.; Idell, S.; et al. Mitochondrial reprogramming regulates breast cancer progression. *Clin. Cancer Res.* **2016**, *22*, 3348–3360. [PubMed]
138. Phinney, D.G.; Di Giuseppe, M.; Njah, J.; Sala, E.; Shiva, S.; St Croix, C.M.; Stolz, D.B.; Watkins, S.C.; Di, Y.P.; Leikauf, G.D.; et al. Mesenchymal stem cells use extracellular vesicles to outsource mitophagy and shuttle microRNAs. *Nat. Commun.* **2015**, *6*, 8472. [PubMed]
139. Baixauli, F.; López-Otín, C.; Mittelbrunn, M. Exosomes and autophagy: coordinated mechanisms for the maintenance of cellular fitness. *Front. Immunol.* **2014**, *5*, 403. [PubMed]
140. Wang, J.; Liu, X.; Qiu, Y.; Shi, Y.; Cai, J.; Wang, B.; Wei, X.; Ke, Q.; Sui, X.; Wang, Y.; et al. Cell adhesion-mediated mitochondria transfer contributes to mesenchymal stem cell-induced chemoresistance on T cell acute lymphoblastic leukemia cells. *J. Hematol. Oncol.* **2018**, *11*, 11. [PubMed]
141. Haraszti, R.A.; Didiot, M.-C.; Sapp, E.; Leszyk, J.; Shaffer, S.A.; Rockwell, H.E.; Gao, F.; Narain, N.R.; DiFiglia, M.; Kiebish, M.A.; et al. High-resolution proteomic and lipidomic analysis of exosomes and microvesicles from different cell sources. *J. Extracell. Vesicles* **2016**, *5*, 32570. [PubMed]
142. Croston, T.L.; Shepherd, D.L.; Thapa, D.; Nichols, C.E.; Lewis, S.E.; Dabkowski, E.R.; Jagannathan, R.; Baseler, W.A.; Hollander, J.M. Evaluation of the cardiolipin biosynthetic pathway and its interactions in the diabetic heart. *Life Sci.* **2013**, *93*, 313–322.

143. Han, X.; Yang, J.; Yang, K.; Zhao, Z.; Abendschein, D.R.; Gross, R.W. Alterations in myocardial cardiolipin content and composition occur at the very earliest stages of diabetes: a shotgun lipidomics study. *Biochemistry* **2007**, *46*, 6417–6428.
144. Han, X.; Yang, J.; Cheng, H.; Yang, K.; Abendschein, D.R.; Gross, R.W. Shotgun lipidomics identifies cardiolipin depletion in diabetic myocardium linking altered substrate utilization with mitochondrial dysfunction. *Biochemistry* **2005**, *44*, 16684–16694.
145. Dolinsky, V.W.; Cole, L.K.; Sparagna, G.C.; Hatch, G.M. Cardiac mitochondrial energy metabolism in heart failure: Role of cardiolipin and sirtuins. *Biochim. Biophys. Acta* **2016**, *1861*, 1544–1554. [CrossRef] [PubMed]
146. Menshikova, E.V.; Ritov, V.B.; Dube, J.J.; Amati, F.; Stefanovic-Racic, M.; Toledo, F.G.S.; Coen, P.M.; Goodpaster, B.H. Calorie restriction-induced weight loss and exercise have differential effects on skeletal muscle mitochondria despite similar effects on insulin sensitivity. *J. Gerontol.* **2018**, *73*, 81–87. [CrossRef] [PubMed]
147. Li, J.; Romestaing, C.; Han, X.L.; Li, Y.A.; Hao, X.B.; Wu, Y.Y.; Sun, C.; Liu, X.L.; Jefferson, L.S.; Xiong, J.W.; et al. Cardiolipin remodeling by alcat1 links oxidative stress and mitochondrial dysfunction to obesity. *Cell Metab.* **2010**, *12*, 154–165. [CrossRef]
148. Xiao, M.; Zhong, H.; Xia, L.; Tao, Y.; Yin, H. Pathophysiology of mitochondrial lipid oxidation: Role of 4-hydroxynonenal (4-HNE) and other bioactive lipids in mitochondria. *Free Radic. Biol. Med.* **2017**, *111*, 316–327. [CrossRef] [PubMed]
149. Zhao, Z.; Zhang, X.; Zhao, C.; Choi, J.; Shi, J.; Song, K.; Turk, J.; Ma, Z.A. Protection of pancreatic beta-cells by group VIA phospholipase A(2)-mediated repair of mitochondrial membrane peroxidation. *Endocrinology* **2010**, *151*, 3038–3048. [CrossRef]
150. Stealing from the Body: How Cancer Recharges Its Batteries. Available online: https://www.uea.ac.uk/about/-/stealing-from-the-body-how-cancer-recharges-its-batteries (accessed on 1 September 2018).
151. Sansone, P.; Savini, C.; Kurelac, I.; Chang, Q.; Amato, L.B.; Strillacci, A.; Stepanova, A.; Iommarini, L.; Mastroleo, C.; Daly, L.; et al. Packaging and transfer of mitochondrial DNA via exosomes regulate escape from dormancy in hormonal therapy-resistant breast cancer. *Proc. Natl. Acad. Sci. USA* **2017**, *114*, E9066–E9075. [CrossRef] [PubMed]
152. Zhou, M.C.; Zhu, L.; Cui, X.; Feng, L.; Zhao, X.; He, S.; Ping, F.; Li, W.; Li, Y. Reduced peripheral blood mtDNA content is associated with impaired glucose-stimulated islet β cell function in a Chinese population with different degrees of glucose tolerance. *Diabetes Metab. Res. Rev.* **2016**, *32*, 768–774. [CrossRef] [PubMed]
153. Malik, A.N.; Parsade, C.K.; Ajaz, S.; Crosby-Nwaobi, R.; Gnudi, L.; Czajka, A.; Sivaprasad, S. Altered circulating mitochondrial DNA and increased inflammation in patients with diabetic retinopathy. *Diabetes Res. Clin. Pract.* **2015**, *110*, 257–265. [CrossRef] [PubMed]
154. Xu, F.X.; Zhou, X.; Shen, F.; Pang, R.; Liu, S.M. Decreased peripheral blood mitochondrial DNA content is related to HbA1c, fasting plasma glucose level and age of onset in Type 2 diabetes mellitus. *Diabet. Med.* **2012**, *29*, e47–e54. [CrossRef]
155. Oexle, K.; Oberle, J.; Finckh, B.; Kohlschütter, A.; Nagy, M.; Seibel, P.; Seissler, J.; Hübner, C. Islet cell antibodies in diabetes mellitus associated with a mitochondrial tRNALeu(UUR) gene mutation. *Exp. Clin. Endocrinol. Diabetes* **1996**, *104*, 212–217. [CrossRef]
156. Li, J.-Y.; Kong, K.-W.; Chang, M.-H.; Cheung, S.-C.; Lee, H.-C.; Pang, C.-Y.; Wei, Y.-H. MELAS syndrome associated with a tandem duplication in the D-loop of mitochondrial DNA. *Acta Neurol. Scand.* **1996**, *93*, 450–455. [CrossRef] [PubMed]
157. Boesch, P.; Weber-Lotfi, F.; Ibrahim, N.; Tarasenko, V.; Cosset, A.; Paulus, F.; Lightowlers, R.N.; Dietrich, A. DNA repair in organelles: Pathways, organization, regulation, relevance in disease and aging. *Biochim. Biophys. Acta* **2011**, *1813*, 186–200. [CrossRef] [PubMed]
158. Gilkerson, R. Commentary: Mitochondrial DNA damage and loss in diabetes. *Diabetes Metab. Res. Rev.* **2016**, *32*, 672–674. [CrossRef] [PubMed]
159. Guescini, M.; Genedani, S.; Stocchi, V.; Agnati, L.F. Astrocytes and Glioblastoma cells release exosomes carrying mtDNA. *J. Neural. Transm.* **2009**, *117*, 1. [CrossRef] [PubMed]

© 2019 by the author. Licensee MDPI, Basel, Switzerland. This article is an open access article distributed under the terms and conditions of the Creative Commons Attribution (CC BY) license (http://creativecommons.org/licenses/by/4.0/).

Review

Cardiovascular Manifestations of Mitochondrial Disease

Jason Duran, Armando Martinez and Eric Adler *

Department of Cardiology, University of California San Diego Medical Center, La Jolla, CA 92037, USA; jaduran@ucsd.edu (J.D.); arm021@ucsd.edu (A.M.)
* Correspondence: eradler@ucsd.edu

Received: 27 February 2019; Accepted: 22 April 2019; Published: 11 May 2019

Abstract: Genetic mitochondrial cardiomyopathies are uncommon causes of heart failure that may not be seen by most physicians. However, the prevalence of mitochondrial DNA mutations and somatic mutations affecting mitochondrial function are more common than previously thought. In this review, the pathogenesis of genetic mitochondrial disorders causing cardiovascular disease is reviewed. Treatment options are presently limited to mostly symptomatic support, but preclinical research is starting to reveal novel approaches that may lead to better and more targeted therapies in the future. With better understanding and clinician education, we hope to improve clinician recognition and diagnosis of these rare disorders in order to improve ongoing care of patients with these diseases and advance research towards discovering new therapeutic strategies to help treat these diseases.

Keywords: mitochondrial; genetic mutations; cardiovascular disease; heart failure; cardiomyopathy

1. Introduction

Mitochondrial diseases affect nearly 1 in 5000–10,000 births [1–3] and genetic mutations in mitochondrial DNA (mtDNA) are even more commonly found, with studies on umbilical cord blood reporting mutations in as many as 1 in 200 samples [3,4]. Although relatively common, mitochondrial diseases have a wide array of clinical presentations and their dysfunction affects a wide variety of organs and tissues [5] including the heart [6]. As such the diagnosis of a mitochondrial disorder can be complex and easily overlooked. Thus, physician exposure to mitochondrial disorders is quite limited despite its relatively common incidence.

Normal mitochondria serve to supply energy in the form of adenosine triphosphate (ATP), generate and regulate radical oxygen species (ROS), buffer cytosolic calcium ions, and regulate cellular apoptosis via the mitochondrial permeability pore complex [1]. Mitochondria are more richly expressed in tissues with high energy demand, including the heart, brain, skeletal muscle and endocrine system, and their dysfunction will disproportionately affect these systems [7]. Tissues that are more metabolically active will typically have greater vulnerability to defects in mtDNA [8] and they will be affected earlier and more severely than less metabolic tissues [9]. Cardiomyopathy is common and usually increases mortality, with one study reporting that 17% of infants and children hospitalized with an array of mitochondrial diseases had cardiac manifestations, and patients with cardiomyopathy had markedly increased mortality (71%) compared to those without a cardiac phenotype (26%) [10]. Mitochondrial mutations can have variable expression in tissues, and cells can exhibit heteroplasmy in which a single cell can have a mixed population mitochondria with either wild-type or mutant-type mtDNA [8].

The variability of expression of a mutation to mtDNA coupled with the wide variety of symptoms and characteristics of each disorder can make these diseases difficult to identify and diagnose. The diagnostic gold standard of mitochondrial disorders continues to be muscle biopsy, however this is only achieved if physicians are aware of mitochondrial disorders and thinking about them in their differential when diagnosing a patient. In this review, we discuss the pathophysiology of mitochondrial disorders and seek to provide a comprehensive clinical resource on mitochondrial cardiomyopathies and their treatment options. We begin by discussing mitochondrial dysfunction in normal aging, followed by a review of primary (congenital) and secondary mitochondrial disorders with cardiovascular phenotypes. We seek to improve physician recognition and identification of patients with mitochondrial disorders so that we can improve future recognition, diagnosis and treatments of these diseases.

2. Mitochondrial Dysfunction in Normal Aging

The heart is one of the most metabolically active tissues in the human body, relying heavily on mitochondrial ATP production to maintain the high energy demand of individually contracting cardiac myocytes. Cardiac aging, or senescence, is coupled with a decline in mitochondrial function, increased production of reactive oxygen species (ROS) and suppressed mitophagy [11,12]. The mitochondrial free radical theory of aging supposes that, with time, free radicals produced as unwanted byproducts of normal mitochondrial metabolism from the respiratory chain on the inner mitochondrial membrane build up in the cell where they are converted to hydrogen peroxide by superoxide dismutase. These hydrogen peroxides are then converted by the Fenton reaction to damaging free radicals that are toxic to nearly all molecules in the cell [12,13]. On the microscopic level, it has been well-established that cells progressively accumulate damage to peptides and lipid molecules with time [14].

In addition to damaging proteins and lipids, free radicals also produces mutations in mitochondrial DNA. Over time, and especially in highly metabolic cells like cardiomyocytes that have high cellular concentrations of mitochondria, ROS-induced molecular damage leads to progressive accumulation of mitochondrial DNA mutations [15–17]. This ultimately leads to increasingly greater mitochondrial dysfunction, and, in damaged cells, dysfunctional mitochondria divide and produce more dysfunctional mitochondria that contain greater number of mtDNA mutations [11,17]. Over time, this leads to progressive cellular dysfunction at the organellar level. In metabolically active and mitochondria-rich cells like cardiomyocytes, the cumulation of damaged peptides, lipids and organelles is more pronounced and has a more pronounced contribution to cellular dysfunction and aging [11].

Maintaining protein homeostasis is also central to the normal function of any cell. When peptides get damaged or degraded, the ability to continue normal cellular function relies on the ability to break down and dispose of damage proteins. The cell must also then replace these damage peptides with normally functioning new proteins [11]. The limited ability to remove damaged proteins or replace them can be damaging to normal cell function [18]. Experimental mechanisms that enhance protein homeostasis include reducing insulin-like growth factor-1 (IGF-1) signaling [19,20], caloric restriction or inhibiting mammalian target of rapamycin (mTOR) signaling with rapamycin treatment [21,22]. As the cellular proteome ages, damaged proteins start to build up, protein degradation slows, and cells progressively accumulate more damaged molecules and become progressively more dysfunctional. This is consistent with experimental results that have demonstrated increased levels of protein ubiquitination in the aging heart, [21] but without increased rates of protein turnover [11,21]. On the cellular level, these microscopic changes lead to changes on the organellar and cellular level, with increasingly dysfunctional autophagy with age.

3. Primary Mitochondrial Disorders

3.1. Mitochondrial Encephalopathy with Lactic Acidosis and Stroke-Like Episodes (MELAS)

First described by Pavlakis et al. in 1984, [23] MELAS has a reported prevalence of 0.18 per 100,000 [24]. Typically presenting in childhood, with 76–80% presenting before 20 years old, the disease rapidly progresses after presentation [25,26]. It has a wide-ranging phenotype that includes stroke-like symptoms (weakness, aphasia, vision loss), encephalopathy (manifesting often as seizures or dementia), lactic acidosis and myopathy [24,26,27]. Additional symptoms include coma, vomiting, fever, headaches, ataxia, external opthalmoplegia, diabetes, hearing impairment, developmental delay, and short stature (Table 1) [26,27].

Cardiac dysfunction has been reported to occur in approximately 30–32% of cases, [24,28] with both hypertrophic and dilated cardiomyopathies having been reported [29]. Conduction abnormalities have also been reported, most notably Wolff–Parkinson–White syndrome [26,30]. The most common genetic mutation involved is A3243G, which has been reported in up to 80% of MELAS presentations [25]. Other mutations that have been identified include T3271C, A3252G, T9957C, 14787del4, G14453A, A13084T, A13045C, A12770G, A11084G, T3949C, G3946A, G3697A, G3376A, T3308C, A13514G, G13513A, G3697, and it is likely several others have yet to be identified (Table 2) [31,32].

Studies have shown treatment with nitric oxide (NO) precursors increase NO production and reportedly reduce stroke-like symptoms [33,34]. One study found plasma lactate decreased significantly after citrulline supplementation [35]. Though no studies have looked directly at the effects of NO precursor therapy on cardiac manifestations, some have theorized they may work by increasing NO production and decreasing cellular damage in cardiac tissue [36]. One case report found improvement in seizure control and reduction in stroke-like episodes after initiation of a ketogenic diet [37]. Reports have shown possible benefits from creatine, CoQ10, and lipoic acid therapy including improvement in symptoms of body composition, strength, and lactate level [38].

Table 1. Summary of mitochondrial disorders with cardiac phenotypes.

Disease	Incidence Age at Onset/Death	Primary Phenotype	Cardiac Manifestations	Genetic Mutations	Treatments
MELAS	0.18/100,000 <20 years Rapid progression to death after onset	• Stroke-like symptoms • Encephalopathy • Lactic Acidemia • Myopathy	30% of cases: • HCM, DCM • Conduction abnormalities (WPW)	A3243G (80% of cases)	• Nitric oxide precursors • Citrulline supplementation • Ketogenic diet • Creatine, CoQ10, lipoic acid therapy
Leigh Syndrome	1/32,000–40,000 <1–2 years Median age of death 2.4 years	• Encephalopathy with cognitive and behavioral dysfunction • Seizures • Hypotonia/Ataxia • Oculomotor dysfunction • Respiratory dysfunction	20% of cases: • HCM • Pericardial effusions • Conduction abnormalities	Mutations to SURF1 gene G1513A (WPW and HCM)	• High dose biotin • Thiamine for SLC19A3 • CoQ10
MERRF	0.9 or <1/100,000 10–20 years Progression to death within 2–15 years of onset (median 8.4 years)	• Myoclonus • Lactic acidosis • Cerebellar ataxia • Muscle weakness • Ragged red fibers	• DCM • HCM • Conduction abnormalities (WPW, SVT, RBBB)	A8344G (83–90% of cases and 53% of cases with cardiac involvement)	• Treatment of seizures with antiepileptic drugs
MIDD	6/100,000 (~1% of patients with diabetes) <35 years	• Diabetes (type I or II) • Bilateral neurosensory hearing loss	• LVH (55%) • HCM (15–30%) • Conduction abnormalities (WPW, SSS, Afib)	A3243G	• Treatment of diabetes • Cochlear implantation • CoQ10
NARP	1/12,000–40,000 3–12 months	• Neuropathy • Ataxia • Retinitis Pigmentosa • Deafness • Myoclonic epilepsy	• DCM • HCM • Conduction abnormalities (WPW)	Point mutations at 8993 MT-ATP6 gene (most commonly T8993G, then T8993C)	• Experimental: mitochondrially targeted obligate heterodimeric zinc finger nucleases to target mitochondrial DNA with the NARP T8993C mutation
GRACILE	1/47,000 (in Finland, may be lower worldwide) Onset in utero 50% die within first 4 months of life, remainder die by 4 years	• Growth restriction • Aminoaciduria • Cholestasis • Iron overload • Early death	• Prolonged QT • Reduced levels of complex III in myocardial tissues post mortem	homozygous point mutation A232G within the BCS1L gene	• Symptomatic/palliative care (usually lethal in first months of life).

Table 1. Cont.

Disease	Incidence Age at Onset/Death	Primary Phenotype	Cardiac Manifestations	Genetic Mutations	Treatments
MNGIE	* Only 100 cases ever reported • Mean age of onset 18 years • Mean age of death 35 years	• Chronic intestinal dysmotility • Leukoencephalopathy • Failure to thrive • Ptosis • Ophthalmoparesis • Peripheral neuropathy	• LVH • Conduction abnormalities (prolonged QT, SVT, sudden cardiac death)	Loss of function mutations to thymidine phosphorylase (TP) gene, chromosome 22q13.32-qter	• Allogeneic hematopoietic stem cell transplantation • Platelet transfusion • Hemodialysis or peritoneal dialysis • Experimental gene therapies aimed to restore thymidine phosphorylase activity
Barth Syndrome	1:300,000–400,000 • <1 year • Most die within first 4 years of life	• DCM • Skeletal myopathy (proximal) • Neutropenia • Growth retardation	• DCM • Cardiac abnormalities in utero • Endocardial fibroelastosis • LV noncompaction • Conduction abnormalities (prolonged WT, SVT, WPW, VT)	G4.5 gene (TAZ gene) on Xq28	• Treatment of heart failure • Cardiac transplantation
LHON	1/31,000–50,000 • 2–87 years	• Acute/subacute painless vision loss • Dystonia • Peripheral neuropathy	• LV hypertrebeculation • Prolonged QT • WPW	90% caused by G11778A (ND4 gene), C3460A (ND1 gene), and the T14484C (NG6 gene) which all cause dysfunction in complex I	• EPI-743: experimental drug targeting glutathione production • Idebenone (antioxidant that can slow vision loss)
Pearson Syndrome	1/1,000,000 • Presents in infancy • most deaths by 3 years of age	• Transfusion-dependent anemia (presenting finding) • Severe Infections • Liver, Kidney, Pancreas and CNS abnormalities	• Increased wall thickness • Depolarization abnormalities • Prolonged QT	Large deletions ranging from 4.9–14 kb	• Treatment of anemia with transfusions, EPO, GCSF • Treatment of pancreatic insufficiency with enzyme replacement
Kearns-Sayre Syndrome	1–3/100,000	• Progressive external ophthalmoplegia (ptosis, weakness of eye muscles) • Retinopathy • Ataxia • Elevated CSF proteins	• Conduction abnormalities (AV block requiring PPM) • Cardiac arrest	Large deletions ranging from 1000 to 10,000 base pairs	• ECG screening • PPM implantation for AV block
CPEO	1–3/100,000	• Loss of motor function of the eye and eyelid • Spectrum of disorders with PS and KSS	• HCM • Arrhythmias	Large mtDNA deletions similar to PS or KSS AD: mutations to nuclear-encoded genes ANT1, C10orf2 and POLG	• Symptomatic/supportive care

Table 1. Cont.

Disease	Incidence Age at Onset/Death	Primary Phenotype	Cardiac Manifestations	Genetic Mutations	Treatments
Freiderich's Ataxia	1-47:1,000,000 Presents in childhood Mean life expectancy 40 years	• Progressive ataxia/areflexia • Progressive and life-threatening cardiomyopathy (2/3 patients die from cardiovascular causes)	• HCM • DCM	Expansion of DNA triplet intron repeat GAA in the frataxin (FXN) gene	• Idebenone (antioxidant), reduces cardiac hypertrophy

Abbreviations: hypertrophic cardiomyopathy (HCM), dilated cardiomyopathy (DCM), Wolff-Parkinson-White syndrome (WPW), supraventricular tachycardia (SVT), right bundle branch block (RBBB), left ventricular hypertrophy (LVH), sick sinus syndrome (SSS), atrial fibrillation (Afib), ventricular tachyarrhythmias, permanent pacemaker (PPM), autosomal dominant (AD).

Table 2. Genetic mutations associated with mitochondrial disorders.

Disease	Mutations
MELAS	**A324G (80% of cases)**, T3271C, A3252G, T9957C, 14787del4, G1445...[?] A13084T, A13045C, A12770G, A11084G, T3949C, G3946A, C3697A, G3376A, T3308C, A13514G, G13513A, G3697
Leigh Syndrome	Mutations to **SURF1** gene, G13513A (WPW and HCM), 312del10/insAT, T8993C, T8993G, C68ST, 772delCC, 751C>T, 845delCT, 868insT, G385A, G618C, T751C, A8344G, A3243G, G13513A, and C1177A.
MERRF	**A8344G**, T8356C, G8363A
MIDD	A3243G, point mutations at 568 and 8281
NARP	Point mutations at **8993**, mutations to **MT-ATP6** gene (most commonly **T8993G**, then **T8993C**), C8839C, G8989C, 8618insT, T9185C, and a 2 bp microdeletion 9127e9128 del AT
GRACILE	Homozygous point mutation A232G within the BCS1L gene
MNGIE	Loss of function mutations to **thymidine phosphorylase (TP)** gene, chromosome 22q13.32-qter
Barth Syndrome	**G4.5 gene** (TAZ gene) on Xq28
LHON	90% caused by **G11778A** (ND4 gene), G3460A (ND1 gene), and the T14484C (NG6 gene) which all cause dysfunction in complex I
Pearson Syndrome	Large deletions ranging from 4.9-14 kb
Kearns-Sayre Syndrome	Large deletions ranging from 1000 to 10,000 base pairs
CPEO	Large mtDNA deletions similar to PS or KSS, AD form: mutations to nuclear-encoded genes *ANT1*, *C10orf2* and *POLG*
Freiderich's Ataxia	Expansion of DNA **triplet intron repeat GAA in the frataxin (FXN) gene**

*Principal mutations of each disorder are **bolded** (that most frequently attributed to each disease).

3.2. Leigh Syndrome

Leigh Syndrome (LS), or subacute necrotizing encephalomyelopathy, was first described by Dr. Denis Leigh, a British neuropsychiatrist, in 1951 [39]. It is a mitochondrial disorder that causes symmetric necrotic lesions in subcortical areas of the brain, particularly the brain stem and basal ganglia. Prevalence is thought to be between 1 per 32,000 and 1 per 40,000 [40,41]. Symptom onset occurs early in life, with initial presentation in infancy or adolescence, and with a majority of presentations before age 1 and 81% by age 2 [42,43]. Initial presentation is often some combination of signs and symptoms of progressive encephalopathy with cognitive and behavioral dysfunction, seizures, hypotonia, oculomotor abnormalities, ataxia, and respiratory dysfunction (Table 1) [40,42,44].

Patients often present with late-stage disease, with 39% of patients dying by the age of 21 years, at a median age of 2.4 years [42,43]. Atypical and slowly progressing presentations have also been described with more rare mutations. Lactate levels are often elevated and diagnosis is usually confirmed by radiologic findings of characteristic bilateral symmetric subcortical hypodensities on computed tomography (CT) or hyperintense signal abnormalities on T2-weighted magnetic resonance imaging (MRI) [42,45,46]. Cardiac manifestations can occur in up to 18–21% of cases, including cardiomyopathy, pericardial effusion, and conduction abnormalities [42,43,47,48]. Of those, hypertrophic cardiomyopathy was found in 50% of patients with cardiac manifestations (Table 1) [43]. One particular study associated the G13513A mutation with development of Wolff–Parkinson–White and hypertrophic cardiomyopathy [48]. One of the major identified causes of LS is Cytochrome c oxidase (COX) deficiency secondary to mutations in the SURF1 gene [49,50]. Over 35 other gene mutations have been identified in association with LS, some of the more studied mutations include 312del10/insAT, T8993C, T8993G, C688T, 772delCC, 751C>T, 845delCT, 868insT, G385A, G618C, T751C, A8344G, A3243G, G13513A, and C1177A (Table 2) [49–54].

As with other mitochondrial disorders, treatment options for LS are limited. One case series showed improvement in life-expectancy with Thiamine therapy in patients whose LS was secondary to a SLC19A3 mutation [31]. High dose biotin should be administered to every patient with suspected LS, given similarities with biotin-responsive basal ganglia disease. Some case reports have shown positive results for treating with Coenxyme Q10.

3.3. Myoclonic Epilepsy with Ragged Red Fibers (MERRF)

In 1921 Dr. Ramsey Hunt, an American neurologist, first published in the journal *Brain* a syndrome afflicting 6 patients that resembled Freidrich's ataxia which he described as "dyssynergia cereballaris myoclonica." It was not until 1973 that Tsairis et al. discovered mitochondrial abnormalities in a family with this disorder, and not until 1980 until Dr. Fukuhara discovered these symptoms associated with ragged red muscle fibers [55]. Finally in 1990 Schoffner et al. first identified a mitochondrial mutation associated with this disease [56]. Myoclonic epilepsy with ragged red fibers (MERRF) has a heterogeneous phenotype that most commonly includes myoclonus, lactic acidosis, cerebellar ataxia, muscle weakness, and ragged red fibers on muscle biopsy [55,57]. Less common symptoms include generalized seizures, short stature, hearing impairment, peripheral neuropathy, headaches, multiple lipomas, vomiting, cognitive dysfunction, and dementia [55,57,58]. Age of onset is typically the second decade of life, and the time of disease progression ranges from 2 to 15 years with a mean time of 8.5 years [58,59]. Reports of the exact incidence of MERRF vary widely, but according to the National Organization for Rare Disorders, the disease occurs in less than 0.9/100,000 patients. Patients often report frequent myoclonic seizures with rare episodes of generalized tonic–clonic seizures, often only partially controlled by anti-epileptic drugs [58]. Reported cardiac abnormalities include Wolff Parkinson White (WPW), supraventricular tachycardia (SVT), right bundle branch block (RBBB), and both dilated and hypertrophic cardiomyopathy (Table 1) [55,59–61].

The A8344G mutation is the most commonly identified mutation in patients with MERRF, occurring in 83–90% of cases [57,58]. One case series found cardiac abnormalities in 53% of MERRF patients with this mutation [61]. T8356C has been reported in one patient to be associated with MERRF

(Table 2) [62]. G8363A was reported in a case series of 9 patients with clinic presentations fitting MERRF and 4/9 patients were found to cardiac abnormalities [60]. This mutation involves coding for the same tRNA gene as other mutations associated with MERRF. Symptomatic treatment of MERRF is focused on symptomatic management of myoclonus and epilepsy with antiepileptic drugs [58].

3.4. Maternally Inherited Diabetes and Deafness (MIDD)

Maternally inherited diabetes and deafness (MIDD), is a mitochondrial disease first characterized in 1992 [63]. Approximately 0.5–2.8% of all diabetic patients have MIDD, with incidences varying by ethnic origin [64,65]. As the name suggests the defining clinical features are diabetes (both type 1 and 2) and bilateral neurosensory hearing loss, each one being equally likely to be the presenting symptom [66]. Typically diabetes develops at a relatively younger age (with half of all patients presenting before age 35), often in a patient with a low or normal BMI [65–68]. Majority of MIDD patients are non-insulin dependent at presentation however approximately half will have relatively rapid progression to inulin dependence within the first several years of diagnosis [65,66,69,70]. Studies have found macular pattern dystrophy in up to 85% of patients [66,71]. Other less frequent findings include myopathy, neuropsychiatric symptoms, short stature and renal disease [66,69]. Brain atrophy, ptosis, constipation, diarrhea, and pseudo-obstruction have also been reported [72–74]. Several case reports have shown a possible association with histories of spontaneous abortion, preterm pregnancy, and placenta accrete (Table 1) [74,75].

Other studies have found left ventricular hypertrophy in up to 55% of patients and cardiomyopathy in up to 15–30% [66,67,76,77]. One study found myocyte hypertrophy and vacuolization, an increased number of mitochondria and evidence of large mitochondria on endomyocardial biopsy [77]. Conduction disorders including WPW, sick sinus syndrome and atrial fibrillation have been found in up to 27% of cases (Table 1) [66–69]. Another study has shown diabetics with the A3243G mutation had more impairment in autonomic nervous function when compared to other diabetics [78]. The mitochondrial mutation A3243G causes a majority of the cases of MIDD, however other mutations have been identified in rare cases, including point mutations at 568 and 8281 (Table 2) [63,64,79].

Treatment of MIDD is primarily symptomatic. MIDD diabetes is managed similarly to other forms with the exception of avoidance of metformin and low threshold for insulin initiation [65,80]. Hearing impairment is closely followed and patients often benefit from cochlear implantation [65,81]. Case reports have shown treatment with Coenzyme Q10 appeared to improve left ventricular function in patients with MIDD [77,82]. One open trial study conducted on 28 patients in Japan found that treatment with CoQ10 reduced progression of insulin secretion defects, exercise intolerance, and hearing loss [83]. Visual symptoms, like acuity loss, are rare in MIDD [71].

3.5. Neuropathy, Ataxia and Retinitis Pigmentosa (NARP)

Classically the syndrome of neuropathy, ataxia, and retinitis pigmentosa (NARP) has been described as a combination of deafness, a myoclonic epilepsy, muscle weakness, retinal pigmentosa, and ataxia [84]. The disease was first described by Fryer et al. in 1994 in a family with seven children who presented with a heterogenous phenotype of Leigh's syndrome [85]. Further investigations however have shown NARP to be phenotypically variable ranging from mild or late-onset symptoms like retinal dystrophy to a more severe multisystemic phenotype that can include developmental delay, dementia, seizures, muscle weakness, sensory neuropathy, ataxia, and retinal pigmentosa and with typical age of onset in the first year of life [84,86]. NARP is estimated to effect about 1/12,000–40,000 according to the National Organization of Rare Disorders and the National Institutes of Health. Several studies have shown a correlation between mutation load and severity of symptoms, especially in the more common mutations at 8993 [87]. Individuals with 80% or more are more likely to have the classic presentation, those below 70% have milder symptoms, and below 50% is often asymptomatic or late onset [86,88–90]. Individuals with >90% mutation load appear to present with Leigh syndrome, indicating clinical presentations are on a NARP/Leigh spectrum (Table 1) [91,92].

Further studies have shown that factors other than mutation load affect a patient's phenotype. Proposed variables for further investigation include mtDNA background, environmental, autosomal, tissue-specific factors, and nuclear modifier genes likely play a role in determining clinical picture and disease severity [93–95]. One case report followed a patient with NARP who developed peripartium dilated cardiomyopathy, ventricular pre-excitation, and non-sustained ventricular tachycardia during 4 years of follow up [96]. However this patient had a A8344G, which has been associated with MERRF and not NARP. Though the 8993 mutation has been linked to cases of hypertrophic cardiomyopathy, cases have been in patients with the more severe Leigh syndrome phenotype [97]. More studies focused on long term follow up to determine late sequelae of NARP are needed to determine cardiac outcomes. NARP is caused by mutations in the MT-ATP6 gene, most commonly T8993G and occasionally T8993C [84,86,88]. Other reported mutations include G8839C, G8989C, 8618insT, T9185C, and a 2-bp microdeletion 9127e9128 del AT [89,91,98–100]. Other mutations in the MT-ATP6 gene have been documented but have been observed to cause the more severe presentations seen in Leigh syndrome [95,101]. A recent study demonstrated the promise of using mitochondrially targeted obligate heterodimeric zinc finger nucleases to target mitochondrial DNA with the NARP T8993G mutation in cell models [102]. Otherwise treatment has primarily been on symptom improvement (Table 2).

3.6. GRACILE

First described in 1998, GRACILE syndrome is a severe mitochondrial disease named after the significant clinical features: growth restriction, aminoaciduria, cholestasis, iron overload, and early death [103]. Incidence is approximated to be at least 1/47,000 based on a study conducted in Finland, however its global prevalence may actually be lower as Finland reports the highest number of cases of GRACILE worldwide [104]. Onset begins with growth restriction at the end of the first trimester, with patient's average birth weight at standard deviations below predicted normal [105]. A study of 29 GRACILE cases found a median lactate of 12.8 mmol/L, pH of 7.00, and all patients had elevated transaminases [106]. The liver develops macrovesicular steatosis, cholestasis with iron accumulation, rapidly progressive fibrosis and cirrhosis [107]. Lactic acidosis is present and often refractory to treatment, further contributing to failure to thrive. Cause of death appears to occur due to energy depletion, with approximately half of all patients dying within the first two weeks of life and the remainder die by 4 months of life [104,106,107]. One patient was found to have prolonged QT [105]. Autopsy-derived myocardial tissue samples showed higher respiratory chain enzyme activity when compared to autopsy controls [104]. Another study on isolated complex III from GRACILE patients found reduced levels and activity of complex III in myocardial tissues, however no cardiac dysfunction or histopathological abnormalities were found (Table 1) [108].

GRACILE is caused by the homozygous point mutation A232G within the BCS1L gene [104]. BCS1L is a mitochondrial chaperone protein that guides the assembly of complex III of the mitochondrial respiratory chain [104]. It is the most common and severe BCS1L disorder, however multiple mutations within the BCS1L gene have been identified and are linked to a variety of phenotypes [109,110]. Identified exclusively in Finland and included as part of the Finnish disease heritage, it is postulated to have accumulated in the Finnish population through founder effect and genetic drift [105,106]. Patients with the homozygous A232G have shown a strong consistency in the resulting phenotype [111]. BCS1L deficiency did not result in decreased complex III activity in tissues from Finnish GRACILE patients, but patients with BCSL1 deficiency did demonstrate reduced Rieske FeS levels [104]. While the pathophysiology is still largely unknown, a recent study showed tissue specific deficiencies in BCS1L and Rieske FeS proteins in the liver, kidney, and heart tissues were associated with decreased levels of assembled complex III (Table 2) [108]. This pathology was identified only in kidney and liver samples, so despite cardiac tissue also having decreased levels/activity of complex III, there appears to still be adequate respiratory chain enzyme activity [107,108]. Given rapid mortality in these patients, treatment is often symptomatic or palliative [112]. Different methods have been implemented to attempt to correct acidosis without clear success [112]. One study in a mouse model of GRACILE syndrome found decreased survival in mice fed a high carbohydrate diet, despite improved levels of several amino acids and urea cycle intermediates [113].

3.7. Mitochondrial Neurogastrointestinal Encephalopathy (MNGIE)

Mitochondrial neurogastrointestinal encephalopathy (MNGIE) is an autosomal recessive disease of multiple organ systems characterized by chronic intestinal dysmotility, leukoencephalopathy, failure to thrive, ptosis, ophthalmoparesis, and peripheral neuropathy [114–116]. Gastrointestinal (GI) symptoms are the most prominent manifestation and include recurrent nausea, vomiting, diarrhea, abdominal pain, gastroparesis, and intestinal pseudo-obstruction with dymotility [117,118]. Brain leukoencephalopathy is present in nearly all cases, but is largely asymptomatic [119]. Brain MRI typically shows symmetric T2 hyperintensities of cerebral white matter [119]. Reported in only ~100 cases, patients classically present between the second and fifth decades of life and die in early adulthood as symptoms progress, with mean age at onset around 18 years of age and mean age at death of about 35 years [117,120,121]. One study found abnormal electrocardiogram (ECG) findings, with 16% of patients in one study demonstrating left ventricular hypertrophy (LVH), and there are single case reports of prolonged QT, cardiac arrest, and SVT (Table 1) [120,121].

MNGIE is caused by loss of function mutations to the thymidine phosphorylase (TP) gene located on chromosome 22q13.32-qter (Table 2) [116]. TP catabolizes thymidine to thymine and loss of enzymatic activity causes accumulation of thymidine and deoxyuridine creating an imbalance in reservoir of cellular nucleosides and nucleotides [117]. This results in impaired mitochondrial DNA replication that can lead to further deletions and mutations [122]. TP is expressed in the GI system, brain, peripheral nerves, spleen and bladder, [120] a distribution pattern that accounts for most clinical findings. Despite low expression in muscles, muscle biopsies show mitochondrial dysfunction with abnormal mtDNA, COX deficiency, and ragged-red fibers [120,121,123]. Given the variability of symptom severity and progression without clear correlation between genotype and phenotype, it is hypothesized that environmental factors and genetic modifiers may play a role in determining a patient's clinical course [121].

Relative to other mitochondrial diseases, significant advancements have been made towards the treatment of MNGIE, but early diagnosis and treatment is crucial to avoid disease progression and accumulation of mutations [124]. Given significant GI manifestations and malnutrition, preventative treatment with nutritional therapy is vital but challenging [125]. Allogeneic hematopoietic stem cell transplant has shown to be a viable treatment option [124,126–128]. Stem cell transplantation can restore TP activity and normalizes the pool of nucleotide and nucleoside [126,127]. Despite these molecular improvements, there remains a high complication rate in transplanted patients, with one study showing only 37.5% of patients alive at time of study follow up [128]. Other treatments that have been studied include platelet infusions, peritoneal dialysis, hemodialysis, and carrier erythrocyte entrapped recombinant TP all of which were shown to partially and temporarily restore TP activity levels to varying degrees [129–132]. Liver transplantation has succeeded in producing lasting restoration of TP activity and symptom improvement in a single case report, though further investigation is needed [133]. Gene therapy-based models for restoration of normal TP enzyme expression has been studied in murine models with promising initial findings [134].

3.8. Barth Syndrome

Barth syndrome, first described in 1983 by Dutch neurologist Dr. Peter Barth, [135] is an X-linked mitochondrial disorder causing congenital dilated cardiomyopathy, skeletal myopathy (predominantly proximal), neutropenia (can be intermittent), and growth retardation in young male infants [135–138]. Prevalence has been estimated to be approximately 1:300,000–400,000 live births, though this is likely an underestimation due to relatively high numbers of spontaneous abortions and still births of male fetuses observed in known Barth syndrome families [139,140]. Disease symptoms typically present within the first year of life with failure to thrive and cardiac abnormalities with or without severe infections, and some studies have even shown cardiac abnormalities in utero [141–143]. Mortality is highest within the first four years of life with most deaths due to either cardiac failure secondary to dilated cardiomyopathy or severe sepsis in the setting of neutropenia [136,137,143]. Reported overall survival rate is 49% at age 5, but this has been steadily increasing in recent years, improving from 20% in patients born before 2000 to 70% in patients born after 2000, likely due to increasing awareness leading to earlier diagnosis and treatment [143]. Despite the high mortality rates in childhood, patients can live into their late 40s (Table 1) [141].

As many as 91–94% of patients develop structural cardiac manifestations including dilated cardiomyopathy, endocardial fibroelastosis, or left-ventricular non-compaction (hypertrabeculation) [137,143–148]. Conduction abnormalities are also common and include prolonged QT, WPW, SVT, and ventricular tachyarrhythmias [143,146,147]. Neutropenia occurs in 70–84% of patients with variable severity and can be persistent or intermittent [148,149]. Elevations in urine 3-methylglutaconic acid and abnormally low cholesterol levels are less common but still present in a significant percentage of patients (Table 1) [143,147,150].

Barth syndrome is an X-lined recessive disorder caused by mutations in the G4.5 gene (TAZ gene) on Xq28, which encodes a highly conserved ascyltransferase (Table 2) [138,151,152]. The TAZ gene encodes an enzyme important in remodeling of cardiolipin [153,154]. Cardiolipin is exclusive to the mitochondrial membrane and plays a key role in proper structure and function of the mitochondria [153,155]. Loss of this enzyme leads to a relative excess of monolysocardiolipin. Measurement of monolysocardilipin/cardiolipin ratio is an important diagnostic test with high sensitivity and specificity for Barth syndrome [156,157]. Though there is currently no curative treatment for Barth syndrome, early recognition and diagnosis can allow for aggressive and prophylactic management [140,158]. Treatment entails typical heart-failure medication regimens and antibiotics during periods of neutropenia or infection [158]. Despite medical management, approximately 12% of patients require cardiac transplantation, with mean age 3.8 years at time of transplant [148].

3.9. Leber's Hereditary Optic Neuropathy (LHON)

This disorder was first described by Dr. Albrecht von Graefe in 1858, but was named after Dr. Theodore Leber who published a case series in 1871 of 15 patients from 4 families afflicted with LHON [159]. LHON classically presents as acute or subacute, painless vision loss typically in young adulthood although symptom onset varies widely and has been reported as early as 2 and as late as 87 years [160–163]. One study showed 11.5% of cases occurring before age 10 [161]. Vision loss typically starts in one eye starting with central vision loss and then progresses to the second eye often days to months later [162,163]. Vision loss is due to degeneration of retinal ganglion cells [164]. Other non-visual, neurologic clinical findings have been reported including: dystonia, peripheral neuropathy, resting tremor, wide spread multiple sclerosis-like white matter lesions and Leigh-like encephalopathy [165–172]. European studies have estimated the prevalence to be between 1:31,000–50,000 with a 80–90% of cases occurring in males [173–175]. Case reports have found left ventricular hypertrabeculation in patients with LHON [176,177]. Another study showed a statistically significant longer QTc in LHON subjects from one American family when compared to controls [178,179]. Other cardiac associations reported include WPW (~9% of LHON subjects) and myocardial thickening (Table 1) [177,179–181].

Approximately 90% cases are caused by one of three mtDNA mutations: G11778A (ND4 gene), G3460A (ND1 gene), and the T14484C (NG6 gene) all of which cause dysfunction in complex I of the mitochondrial respiratory chain (Table 2) [182–185]. Other mutations have been reported in rare cases [186–189]. A majority of mutation carriers do not develop clinical LHON, with 50% of male and 10% of female carriers affected [190]. The reason for this pattern is unclear, but several hypotheses of additional determining factors have emerged including varying levels of oxidative stress, environmental factors, mtDNA haplogroups, and hormones [191–195]. Earlier onset of symptoms and the 14484 mutation have relatively higher rates of spontaneous recovery of vision and are associated with better prognosis [196,197].

Several promising therapies are under investigation. A small, open-label, prospective trial looking at drug EPI-743, which targets glutathione production, showed arrest and reversal of disease progression in 4 of 5 subjects [198]. Some case reports, one double-blinded RCT, and one retrospective study have pointed to beneficial effects on vision with a strong antioxidant, idebenone [199–202]. Given the observed gender bias in disease penetration, studies have begun to investigate the role of estrogen-like molecules on LHON cell-lines with promising preliminary results [203]. A handful of gene-therapy clinical trials are currently underway after promising findings in pre-clinical trials with adeno-associated viral vectors, [204–208] but no studies have evaluated the efficacies of these treatment on cardiac manifestations of this disease.

3.10. Pearson Syndrome

Pearson syndrome (PS), first described by pediatric hematologist-oncologist Dr. Howard Pearson in 1979, is a rare disease affecting the liver, kidney, pancreas, bone marrow and CNS, and more rarely can present with cardiovascular findings. This disease presents initially with transfusion-dependent anemia that may improve, but affected patients develop increasingly severe infections [209]. Incidence was estimated to be roughly 1/1,000,000 in a case series of 11 Italian children, [209] and mortality was high with most infants dying by 3 years of age, [210] although later studies have reported survival of one patient to 6.6 years of age [209]. Multiple cases have been reported of both phenotypic and genetic transformation of the disease from PS to Leigh syndrome (with development of dysphagia, ataxia, peripheral neuropathy and ophthalmoparesis) [211,212] or Kearn's sayre syndrome (with development of progressive external ophtalmoplegia, retinopathy and ataxia) [212–215]. Cardiac abnormalities, while not hallmark symptoms of the disease, have been reported including increased ventricular wall thickness, depolarization abnormalities and prolonged QT. Mortality likely results from failure of other organ symptoms, as most patients die before any severe cardiac abnormalities can result (Table 1) [209].

Genetic deletions of mtDNA in this disorder is widely variable, and may account for why PS can transform into other mitochondrial spectrum disorders like Kearns Sayre Syndrome (KSS), progressive external ophthalmoplegia (PEO) or Leigh Syndrome (LS) [214]. In a study of 15 patients diagnosed with PS, large mitochondrial DNA deletions were typical, with 9 patients having 4.9 kb deletions and 6 having deletions ranging from 9 to 14 kb (Table 2) [210]. Treatment of this disease primarily involves treatment of anemia with transfusions, erythropoietin (EPO) or granulocyte colony stimulating factor (GCSF), and replacement of pancreatic enzymes has been reported in patients with exocrine pancreatic insufficiency [209].

3.11. Kearns–Sayre Syndrome

Kearns–Sayre syndrome (KSS) is a rare mitochondrial disorder that primarily affects the eyes among many other organs including the heart, was first reported by Dr. Kearns with a case series involving nine patients in 1965, and has more recently been described on the spectrum of PEO and PS [212–215]. Most patients with KSS develop the hallmark symptom of progressive external ophthalmoplegia which causes ptosis and weakness or paralysis of the muscles of the eye and some patients may also develop retinopathy [216]. Cardiac conduction defects are common, as are ataxia and abnormally high cerebrospinal fluid (CSF) protein levels (Table 1). This disorder is very rare, effecting only 1–3 per 100,000 individuals and are caused by large deletions in mtDNA ranging from 1000 to 10,000 base pairs that results in loss of genes involved in oxidative phosphorylation (Table 2) [216]. For cardiac manifestations, AV block is a common complication, with case reports of cardiac arrest and conduction abnormalities including complete heart block requiring permanent cardiac pacemaker implantation (Table 1) [217–220]. All patients with KSS or KSS spectrum disorders should be regularly screened with ECGs to monitor for progressive AV block, and there should be low threshold for monitoring in any KSS patient presenting with syncope.

3.12. Chronic Progressive External Ophthalmoplegia

First reported by Zeviani et al. in 1989, [221] chronic progressive external opthalmoplegia (PEO) is an inherited mitochondrial disorder characterized by loss of motor function of the muscles of the eye an eyelid [222–224]. Other associated symptoms include skeletal and cardiac myopathy and dysphagia (Table 1) [223]. Cardiac arrhythmias may also develop. PEO results from large scale mtDNA deletions can occur as part of a spectrum of other large deletion mitochondrial disorders including Kearns–Sayre syndrome [222,223] and Pearson Syndrome, [225–227] as mentioned above. Disease incidence of large mtDNA deletions ranges from 1.2–2.9/100,000 [228,229]. Autosomal dominant PEO [230] has also been reported with mutations reported to 3 nuclear encoded genes mutations have been identified in three nuclear encoded genes: *ANT1, C10orf2* and *POLG* (Table 2). As with KSS and other large deletion spectrum disorders, symptomatic supportive care remains the only treatment option [224].

3.13. Friederich's Ataxia

Friederich's ataxia (FA), first described by German physician Dr. Nikolaus Freiderich in 1863, is a neurodegenerative disease that presents with progressive ataxia and areflexia, and progressive life-threatening cardiomyopathy, and both hypertrophic and dilated cardiomyopathies have been reported [231,232]. FA has an incidence of 1–47:1,000,000 [233] that typically presents in childhood, with a mean life expectancy is 40 years of age [234]. About two-thirds of patients will die from cardiac causes (Table 1) [234–236]. FA is an autosomal recessive disease caused by expansion of DNA triplet intron repeat GAA in the frataxin (FXN) gene (Table 2) [233,234,237]. Frataxin is a protein found on the inner mitochondrial membrane and is involved in oxidative phosphorylation, and its deficiency can lead to decreased cellular ATP production and disruption of iron chelation within mitochondria [237]. Treatment relies on antioxidant therapy to reduce buildup of reactive oxygen species and decrease oxidative damage. One novel drug, idebenone, an antioxidant that is similar to coenzyme Q, has been suggested for treatment of FA cardiomyopathy early in disease course based on preclinical studies, [238] and has been shown in randomized blinded trial to reduce measures of cardiac hypertrophy in FA-cardiomyopathy patients [239].

4. Secondary Mitochondrial Myopathies

4.1. Mitochondrial Dysfunction in Ischemia

During myocardial infarction, rupture of an atherosclerotic plaque leads to occlusion of a coronary artery and downstream myocardial ischemia. Endocardial myocytes are the most sensitive to cardiac ischemia, as this segment of the myocardium has the highest imbalance of energy, with cell death starting as early as 20 minutes after occlusion of the infarct related artery [240]. The ischemic insult then moves transmurally from the endocardium to the epicardium. In the era of cardiac catheterization, prompt revascularization has served to rapidly restore blood flow to these damaged tissues, however the consequence of this early restoration of blood flow is reperfusion injury, which produces an additional wave of cell death after the primary insult. The combination of ischemia and reperfusion leaves individual cells in three states of injury: dead cardiomyocytes, reversibly injured cells, and hypoperfused but viable cardiomyocytes [240]. Prompt reperfusion has been successful at saving hypoperfused viable cardiomyocytes, but studies have shown that few reversibly injured cells are rescued after reperfusion [240,241]. Additionally, reperfusion causes a second wave of cell death from altered metabolism of cells damaged by ischemia-reperfusion injury, which results in destruction of mitochondria and release of excessive ROS into damaged cells [240–242]. In animal studies, as many as 38% of cell death results solely from the reperfusion phase of injury, [242] and reperfusion injury accounts for up to 50% of the final infarct size [241].

Damaged mitochondria can continue to propagate myocyte cell death through activation of programmed cell death pathways. Mitochondrial electron transport chains are damaged during the ischemic phase of injury [243]. This results in excessive production and release of ROSs, dysregulation of calcium handling, and abnormal mitochondrial swelling leads to mitochondrial membrane disruption by the mitochondrial permeability transition pore in the inner mitochondrial membrane, [244–246] or via mitochondrial outer membrane permeabilization [246]. Formation of these pore complexes leads to release of toxic mitochondrial ROSs and proteins that activate programmed cell death mechanisms [240,242–244,246,247]. Thus, widespread cell death occurs in several stages following coronary artery occlusion, resulting from the ischemic phase, reperfusion phase, and mitochondrial dysfunction phase resulting in cell death both through necrosis and activation of programmed cell death pathways [240,242,244,247].

4.2. Mitochondrial Dysfunction in Diabetic Cardiomyopathy

There is growing clinical and preclinical evidence that mitochondrial dysfunction is central to cardiomyopathy observed in diabetic patients [248]. The pathogenesis of mitochondrial dysfunction in diabetes may be similar to that of ischemia-reperfusion, but the exact molecular mechanisms of this dysfunction remain poorly understood [248]. Clinically, diabetic cardiomyopathy was first described over 40 years ago in a case series of four patients with heart failure and diabetes, but with normal coronary arteries [249]. Subsequent studies have confirmed these observations [250,251] and more recent studies estimate the prevalence of diastolic heart failure in as many as 60% of type 2 diabetics [252]. While the exact molecular mechanisms of diabetic cardiomyopathy remains unknown, it is likely a multifactorial process involving disruption of normal respiration via the electron transport chain. This leads to increased ROS production and disrupted fatty acid oxidation.

As outlined by the mitochondrial free radial theory of aging, higher levels of ROS lead to hydroxylation and damage of cellular peptides and lipids [248]. These pathologic changes also result in impaired respiration. Normal hearts primarily generate ATP from mitochondrial fatty acid oxidation (60–70% of ATP generated) and less so from glucose, lactate or other substrates (30–40%) [248,253]. However, in the diabetic hearts, mitochondria are more reliant on fatty acid oxidation and suffer from impaired mitophagy, with greater inability to break down oxidative fatty acids, further perpetuating accumulation of oxygen free radical species. This disrupts mitochondrial calcium handling and myocardial E-C coupling, and elevated levels of cellular ROS can lead to disrupted mitochondrial biogenesis, increased mitochondrial permeability and ultimately increased cell death [248].

5. Conclusions

In this review, we seek to improve clinician awareness and recognition of mitochondrial disorders affecting the cardiovascular system. Because of the wide variability of presentations and their relative infrequency in the global population, recognition and diagnosis of these disorders can be challenging. Treatment options for primary mitochondrial diseases are limited and primarily involve support care, although some early studies suggest a role for antioxidant therapy. Future research will be needed to discover novel therapeutic approaches to advance our care of these diseases, including breakthroughs in gene therapy and improved understanding of the genetic expression and penetrance of these mutations. It is our hope that with better recognition of these disorders by clinicians that we can improve diagnosis of these disorders and advance long-term patient care and research into new and more viable treatment options.

Author Contributions: Conceptualization, E.A and J.D.; Writing-original draft preparation, A.M. and J.D.; Writing-review and editing, E.A. and J.D.; Supervision, E.A.

Funding: This research received no external funding.

Conflicts of Interest: The authors declare no conflict of interest.

References

1. Meyers, D.E.; Basha, H.I.; Koenig, M.K. Cardiac manifestations of mitochondrial disorders. *Tex. Heart Inst. J.* **2013**, *40*, 635–636. [PubMed]
2. Meyers, D.E.; Basha, H.I.; Koenig, M.K. Mitochondrial cardiomyopathy: Pathophysiology, diagnosis, and management. *Tex. Heart Inst. J.* **2013**, *40*, 385–394.
3. Chinnery, P.F.; Elliott, H.R.; Hudson, G.; Samuels, D.C.; Relton, C.L. Epigenetics, epidemiology and mitochondrial DNA diseases. *Int. J. Epidemiol.* **2012**, *41*, 177–187. [CrossRef] [PubMed]
4. Elliott, H.R.; Samuels, D.C.; Eden, J.A.; Relton, C.L.; Chinnery, P.F. Pathogenic mitochondrial DNA mutations are common in the general population. *Am. J. Hum. Genet.* **2008**, *83*, 254–260. [CrossRef] [PubMed]
5. Debray, F.G.; Lambert, M.; Chevalier, I.; Robitaille, Y.; Decarie, J.C.; Shoubridge, E.A.; Robinson, B.H.; Mitchell, G.A. Long-term outcome and clinical spectrum of 73 pediatric patients with mitochondrial diseases. *Pediatrics* **2007**, *119*, 722–733. [CrossRef] [PubMed]
6. Anan, R.; Nakagawa, M.; Miyata, M.; Higuchi, I.; Nakao, S.; Suehara, M.; Osame, M.; Tanaka, H. Cardiac involvement in mitochondrial diseases. A study on 17 patients with documented mitochondrial DNA defects. *Circulation* **1995**, *91*, 955–961. [CrossRef] [PubMed]
7. Lane, N. Evolution. The costs of breathing. *Science* **2011**, *334*, 184–185. [CrossRef] [PubMed]
8. Wallace, D.C. Why do we still have a maternally inherited mitochondrial DNA? Insights from evolutionary medicine. *Annu. Rev. Biochem.* **2007**, *76*, 781–821. [CrossRef]
9. Hirano, M.; Davidson, M.; DiMauro, S. Mitochondria and the heart. *Curr. Opin. Cardiol.* **2001**, *16*, 201–210. [CrossRef]
10. Holmgren, D.; Wahlander, H.; Eriksson, B.O.; Oldfors, A.; Holme, E.; Tulinius, M. Cardiomyopathy in children with mitochondrial disease; clinical course and cardiological findings. *Eur. Heart J.* **2003**, *24*, 280–288. [CrossRef]
11. Tocchi, A.; Quarles, E.K.; Basisty, N.; Gitari, L.; Rabinovitch, P.S. Mitochondrial dysfunction in cardiac aging. *Biochim. Biophys. Acta* **2015**, *1847*, 1424–1433. [CrossRef] [PubMed]
12. Bratic, A.; Larsson, N.G. The role of mitochondria in aging. *J. Clin. Investig.* **2013**, *123*, 951–957. [CrossRef]
13. Stadtman, E.R. Protein oxidation and aging. *Science* **1992**, *257*, 1220–1224. [CrossRef] [PubMed]
14. Balaban, R.S.; Nemoto, S.; Finkel, T. Mitochondria, oxidants, and aging. *Cell* **2005**, *120*, 483–495. [CrossRef] [PubMed]
15. Trifunovic, A.; Larsson, N.G. Mitochondrial dysfunction as a cause of ageing. *J. Intern. Med.* **2008**, *263*, 167–178. [CrossRef] [PubMed]
16. Trounce, I.; Byrne, E.; Marzuki, S. Decline in skeletal muscle mitochondrial respiratory chain function: Possible factor in ageing. *Lancet* **1989**, *1*, 637–639. [CrossRef]
17. Kujoth, G.C.; Hiona, A.; Pugh, T.D.; Someya, S.; Panzer, K.; Wohlgemuth, S.E.; Hofer, T.; Seo, A.Y.; Sullivan, R.; Jobling, W.A.; et al. Mitochondrial DNA mutations, oxidative stress, and apoptosis in mammalian aging. *Science* **2005**, *309*, 481–484. [CrossRef]
18. Koga, H.; Kaushik, S.; Cuervo, A.M. Protein homeostasis and aging: The importance of exquisite quality control. *Ageing Res. Rev.* **2011**, *10*, 205–215. [CrossRef]
19. Abbas, A.; Grant, P.J.; Kearney, M.T. Role of IGF-1 in glucose regulation and cardiovascular disease. *Expert Rev. Cardiovasc. Ther.* **2008**, *6*, 1135–1149. [CrossRef]
20. Puglielli, L. Aging of the brain, neurotrophin signaling, and Alzheimer's disease: Is IGF1-R the common culprit? *Neurobiol. Aging* **2008**, *29*, 795–811. [CrossRef]
21. Dai, D.F.; Karunadharma, P.P.; Chiao, Y.A.; Basisty, N.; Crispin, D.; Hsieh, E.J.; Chen, T.; Gu, H.; Djukovic, D.; Raftery, D.; et al. Altered proteome turnover and remodeling by short-term caloric restriction or rapamycin rejuvenate the aging heart. *Aging Cell* **2014**, *13*, 529–539. [CrossRef]
22. Johnson, S.C.; Rabinovitch, P.S.; Kaeberlein, M. mTOR is a key modulator of ageing and age-related disease. *Nature* **2013**, *493*, 338–345. [CrossRef]
23. Pavlakis, S.G.; Phillips, P.C.; DiMauro, S.; De Vivo, D.C.; Rowland, L.P. Mitochondrial myopathy, encephalopathy, lactic acidosis, and strokelike episodes: A distinctive clinical syndrome. *Ann. Neurol.* **1984**, *16*, 481–488. [CrossRef]

24. Yatsuga, S.; Povalko, N.; Nishioka, J.; Katayama, K.; Kakimoto, N.; Matsuishi, T.; Kakuma, T.; Koga, Y.; Taro Matsuoka for, M.S.G.i.J. MELAS: A nationwide prospective cohort study of 96 patients in Japan. *Biochim. Biophys. Acta* **2012**, *1820*, 619–624. [CrossRef]
25. Goto, Y.; Horai, S.; Matsuoka, T.; Koga, Y.; Nihei, K.; Kobayashi, M.; Nonaka, I. Mitochondrial myopathy, encephalopathy, lactic acidosis, and stroke-like episodes (MELAS): A correlative study of the clinical features and mitochondrial DNA mutation. *Neurology* **1992**, *42*, 545–550. [CrossRef]
26. Hirano, M.; Ricci, E.; Koenigsberger, M.R.; Defendini, R.; Pavlakis, S.G.; DeVivo, D.C.; DiMauro, S.; Rowland, L.P. Melas: An original case and clinical criteria for diagnosis. *Neuromuscul. Disord.* **1992**, *2*, 125–135. [CrossRef]
27. Zhang, J.; Guo, J.; Fang, W.; Jun, Q.; Shi, K. Clinical features of MELAS and its relation with A3243G gene point mutation. *Int. J. Clin. Exp. Pathol.* **2015**, *8*, 13411–13415.
28. Mancuso, M.; Orsucci, D.; Angelini, C.; Bertini, E.; Carelli, V.; Comi, G.P.; Donati, A.; Minetti, C.; Moggio, M.; Mongini, T.; et al. The m.3243A>G mitochondrial DNA mutation and related phenotypes. A matter of gender? *J. Neurol.* **2014**, *261*, 504–510. [CrossRef] [PubMed]
29. Okajima, Y.; Tanabe, Y.; Takayanagi, M.; Aotsuka, H. A follow up study of myocardial involvement in patients with mitochondrial encephalomyopathy, lactic acidosis, and stroke-like episodes (MELAS). *Heart* **1998**, *80*, 292–295. [CrossRef] [PubMed]
30. Sproule, D.M.; Kaufmann, P.; Engelstad, K.; Starc, T.J.; Hordof, A.J.; De Vivo, D.C. Wolff-Parkinson-White syndrome in Patients With MELAS. *Arch. Neurol.* **2007**, *64*, 1625–1627. [CrossRef] [PubMed]
31. El-Hattab, A.W.; Adesina, A.M.; Jones, J.; Scaglia, F. MELAS syndrome: Clinical manifestations, pathogenesis, and treatment options. *Mol. Genet. Metab.* **2015**, *116*, 4–12. [CrossRef]
32. Lott, M.T.; Leipzig, J.N.; Derbeneva, O.; Xie, H.M.; Chalkia, D.; Sarmady, M.; Procaccio, V.; Wallace, D.C. mtDNA Variation and Analysis Using Mitomap and Mitomaster. *Curr. Protoc. Bioinformat.* **2013**, *44*, 1–23.
33. El-Hattab, A.W.; Emrick, L.T.; Hsu, J.W.; Chanprasert, S.; Almannai, M.; Craigen, W.J.; Jahoor, F.; Scaglia, F. Impaired nitric oxide production in children with MELAS syndrome and the effect of arginine and citrulline supplementation. *Mol. Genet. Metab.* **2016**, *117*, 407–412. [CrossRef]
34. Ganetzky, R.D.; Falk, M.J. 8-year retrospective analysis of intravenous arginine therapy for acute metabolic strokes in pediatric mitochondrial disease. *Mol. Genet. Metab.* **2018**, *123*, 301–308. [CrossRef]
35. El-Hattab, A.W.; Emrick, L.T.; Williamson, K.C.; Craigen, W.J.; Scaglia, F. The effect of citrulline and arginine supplementation on lactic acidemia in MELAS syndrome. *Meta Gene* **2013**, *1*, 8–14. [CrossRef]
36. El-Hattab, A.W.; Zarante, A.M.; Almannai, M.; Scaglia, F. Therapies for mitochondrial diseases and current clinical trials. *Mol. Genet. Metab.* **2017**, *122*, 1–9. [CrossRef]
37. Steriade, C.; Andrade, D.M.; Faghfoury, H.; Tarnopolsky, M.A.; Tai, P. Mitochondrial encephalopathy with lactic acidosis and stroke-like episodes (MELAS) may respond to adjunctive ketogenic diet. *Pediatr. Neurol.* **2014**, *50*, 498–502. [CrossRef]
38. Rodriguez, M.C.; MacDonald, J.R.; Mahoney, D.J.; Parise, G.; Beal, M.F.; Tarnopolsky, M.A. Beneficial effects of creatine, CoQ10, and lipoic acid in mitochondrial disorders. *Muscle Nerve* **2007**, *35*, 235–242. [CrossRef]
39. Leigh, D. Subacute necrotizing encephalomyelopathy in an infant. *J. Neurol. Neurosurg. Psychiatry* **1951**, *14*, 216–221. [CrossRef]
40. Rahman, S.; Blok, R.B.; Dahl, H.H.; Danks, D.M.; Kirby, D.M.; Chow, C.W.; Christodoulou, J.; Thorburn, D.R. Leigh syndrome: Clinical features and biochemical and DNA abnormalities. *Ann. Neurol.* **1996**, *39*, 343–351. [CrossRef]
41. Darin, N.; Oldfors, A.; Moslemi, A.R.; Holme, E.; Tulinius, M. The incidence of mitochondrial encephalomyopathies in childhood: Clinical features and morphological, biochemical, and DNA abnormalities. *Ann. Neurol.* **2001**, *49*, 377–383. [CrossRef] [PubMed]
42. Lee, H.F.; Tsai, C.R.; Chi, C.S.; Lee, H.J.; Chen, C.C. Leigh syndrome: Clinical and neuroimaging follow-up. *Pediatr. Neurol.* **2009**, *40*, 88–93. [CrossRef] [PubMed]
43. Sofou, K.; De Coo, I.F.; Isohanni, P.; Ostergaard, E.; Naess, K.; De Meirleir, L.; Tzoulis, C.; Uusimaa, J.; De Angst, I.B.; Lonnqvist, T.; et al. A multicenter study on Leigh syndrome: Disease course and predictors of survival. *Orphanet J. Rare Dis.* **2014**, *9*, 52. [CrossRef] [PubMed]
44. Huntsman, R.J.; Sinclair, D.B.; Bhargava, R.; Chan, A. Atypical presentations of leigh syndrome: A case series and review. *Pediatr. Neurol.* **2005**, *32*, 334–340. [CrossRef] [PubMed]

45. Bonfante, E.; Koenig, M.K.; Adejumo, R.B.; Perinjelil, V.; Riascos, R.F. The neuroimaging of Leigh syndrome: Case series and review of the literature. *Pediatr. Radiol.* **2016**, *46*, 443–451. [CrossRef] [PubMed]
46. Sonam, K.; Khan, N.A.; Bindu, P.S.; Taly, A.B.; Gayathri, N.; Bharath, M.M.; Govindaraju, C.; Arvinda, H.R.; Nagappa, M.; Sinha, S.; et al. Clinical and magnetic resonance imaging findings in patients with Leigh syndrome and SURF1 mutations. *Brain Dev.* **2014**, *36*, 807–812. [CrossRef]
47. Hadzsiev, K.; Maasz, A.; Kisfali, P.; Kalman, E.; Gomori, E.; Pal, E.; Berenyi, E.; Komlosi, K.; Melegh, B. Mitochondrial DNA 11777C>A mutation associated Leigh syndrome: Case report with a review of the previously described pedigrees. *Neuromol. Med.* **2010**, *12*, 277–284. [CrossRef]
48. Wang, S.B.; Weng, W.C.; Lee, N.C.; Hwu, W.L.; Fan, P.C.; Lee, W.T. Mutation of mitochondrial DNA G13513A presenting with Leigh syndrome, Wolff-Parkinson-White syndrome and cardiomyopathy. *Pediatr. Neonatol.* **2008**, *49*, 145–149. [CrossRef]
49. Pequignot, M.O.; Dey, R.; Zeviani, M.; Tiranti, V.; Godinot, C.; Poyau, A.; Sue, C.; Di Mauro, S.; Abitbol, M.; Marsac, C. Mutations in the SURF1 gene associated with Leigh syndrome and cytochrome C oxidase deficiency. *Hum. Mutat.* **2001**, *17*, 374–381. [CrossRef]
50. Yang, Y.L.; Sun, F.; Zhang, Y.; Qian, N.; Yuan, Y.; Wang, Z.X.; Qi, Y.; Xiao, J.X.; Wang, X.Y.; Qi, Z.Y.; et al. Clinical and laboratory survey of 65 Chinese patients with Leigh syndrome. *Chin. Med. J.* **2006**, *119*, 373–377. [CrossRef]
51. Tiranti, V.; Hoertnagel, K.; Carrozzo, R.; Galimberti, C.; Munaro, M.; Granatiero, M.; Zelante, L.; Gasparini, P.; Marzella, R.; Rocchi, M.; et al. Mutations of SURF-1 in Leigh disease associated with cytochrome c oxidase deficiency. *Am. J. Hum. Genet.* **1998**, *63*, 1609–1621. [CrossRef]
52. Moslemi, A.R.; Tulinius, M.; Darin, N.; Aman, P.; Holme, E.; Oldfors, A. SURF1 gene mutations in three cases with Leigh syndrome and cytochrome c oxidase deficiency. *Neurology* **2003**, *61*, 991–993. [CrossRef]
53. Poyau, A.; Buchet, K.; Bouzidi, M.F.; Zabot, M.T.; Echenne, B.; Yao, J.; Shoubridge, E.A.; Godinot, C. Missense mutations in SURF1 associated with deficient cytochrome c oxidase assembly in Leigh syndrome patients. *Hum. Genet.* **2000**, *106*, 194–205. [CrossRef]
54. Gerards, M.; Kamps, R.; van Oevelen, J.; Boesten, I.; Jongen, E.; de Koning, B.; Scholte, H.R.; de Angst, I.; Schoonderwoerd, K.; Sefiani, A.; et al. Exome sequencing reveals a novel Moroccan founder mutation in SLC19A3 as a new cause of early-childhood fatal Leigh syndrome. *Brain* **2013**, *136*, 882–890. [CrossRef] [PubMed]
55. Fukuhara, N.; Tokiguchi, S.; Shirakawa, K.; Tsubaki, T. Myoclonus epilepsy associated with ragged-red fibres (mitochondrial abnormalities): Disease entity or a syndrome? Light-and electron-microscopic studies of two cases and review of literature. *J. Neurol. Sci.* **1980**, *47*, 117–133. [CrossRef]
56. Shoffner, J.M.; Lott, M.T.; Lezza, A.M.; Seibel, P.; Ballinger, S.W.; Wallace, D.C. Myoclonic epilepsy and ragged-red fiber disease (MERRF) is associated with a mitochondrial DNA tRNA(Lys) mutation. *Cell* **1990**, *61*, 931–937. [CrossRef]
57. Fang, W.; Huang, C.C.; Chu, N.S.; Lee, C.C.; Chen, R.S.; Pang, C.Y.; Shih, K.D.; Wei, Y.H. Myoclonic epilepsy with ragged-red fibers (MERRF) syndrome: Report of a Chinese family with mitochondrial DNA point mutation in tRNA(Lys) gene. *Muscle Nerve* **1994**, *17*, 52–57. [CrossRef]
58. Lorenzoni, P.J.; Scola, R.H.; Kay, C.S.; Arndt, R.C.; Silvado, C.E.; Werneck, L.C. MERRF: Clinical features, muscle biopsy and molecular genetics in Brazilian patients. *Mitochondrion* **2011**, *11*, 528–532. [CrossRef] [PubMed]
59. Ozawa, M.; Goto, Y.; Sakuta, R.; Tanno, Y.; Tsuji, S.; Nonaka, I. The 8,344 mutation in mitochondrial DNA: A comparison between the proportion of mutant DNA and clinico-pathologic findings. *Neuromuscul. Disord.* **1995**, *5*, 483–488. [CrossRef]
60. Santorelli, F.M.; Mak, S.C.; El-Schahawi, M.; Casali, C.; Shanske, S.; Baram, T.Z.; Madrid, R.E.; DiMauro, S. Maternally inherited cardiomyopathy and hearing loss associated with a novel mutation in the mitochondrial tRNA(Lys) gene (G8363A). *Am. J. Hum. Genet.* **1996**, *58*, 933–939.
61. Catteruccia, M.; Sauchelli, D.; Della Marca, G.; Primiano, G.; Cuccagna, C.; Bernardo, D.; Leo, M.; Camporeale, A.; Sanna, T.; Cianfoni, A.; et al. "Myo-cardiomyopathy" is commonly associated with the A8344G "MERRF" mutation. *J. Neurol.* **2015**, *262*, 701–710. [CrossRef]
62. Silvestri, G.; Moraes, C.T.; Shanske, S.; Oh, S.J.; DiMauro, S. A new mtDNA mutation in the tRNA(Lys) gene associated with myoclonic epilepsy and ragged-red fibers (MERRF). *Am. J. Hum. Genet.* **1992**, *51*, 1213–1217.

63. van den Ouweland, J.M.; Lemkes, H.H.; Ruitenbeek, W.; Sandkuijl, L.A.; de Vijlder, M.F.; Struyvenberg, P.A.; van de Kamp, J.J.; Maassen, J.A. Mutation in mitochondrial tRNA(Leu)(UUR) gene in a large pedigree with maternally transmitted type II diabetes mellitus and deafness. *Nat. Genet.* **1992**, *1*, 368–371. [CrossRef]
64. Maassen, J.A.; Janssen, G.M.; 't Hart, L.M. Molecular mechanisms of mitochondrial diabetes (MIDD). *Ann. Med.* **2005**, *37*, 213–221. [CrossRef] [PubMed]
65. Murphy, R.; Turnbull, D.M.; Walker, M.; Hattersley, A.T. Clinical features, diagnosis and management of maternally inherited diabetes and deafness (MIDD) associated with the 3243A>G mitochondrial point mutation. *Diabet. Med.* **2008**, *25*, 383–399. [CrossRef] [PubMed]
66. Guillausseau, P.J.; Massin, P.; Dubois-LaForgue, D.; Timsit, J.; Virally, M.; Gin, H.; Bertin, E.; Blickle, J.F.; Bouhanick, B.; Cahen, J.; et al. Maternally inherited diabetes and deafness: A multicenter study. *Ann. Intern. Med.* **2001**, *134*, 721–728. [CrossRef]
67. Suzuki, S.; Oka, Y.; Kadowaki, T.; Kanatsuka, A.; Kuzuya, T.; Kobayashi, M.; Sanke, T.; Seino, Y.; Nanjo, K.; Society, R.C.o.S.T.o.D.M.w.G.M.o.t.J.D. Clinical features of diabetes mellitus with the mitochondrial DNA 3243 (A-G) mutation in Japanese: Maternal inheritance and mitochondria-related complications. *Diabetes Res. Clin. Pract.* **2003**, *59*, 207–217. [CrossRef]
68. Zhu, J.; Yang, P.; Liu, X.; Yan, L.; Rampersad, S.; Li, F.; Li, H.; Sheng, C.; Cheng, X.; Zhang, M.; et al. The clinical characteristics of patients with mitochondrial tRNA Leu(UUR)m.3243A > G mutation: Compared with type 1 diabetes and early onset type 2 diabetes. *J. Diabetes Complicat.* **2017**, *31*, 1354–1359. [CrossRef]
69. Guillausseau, P.J.; Dubois-Laforgue, D.; Massin, P.; Laloi-Michelin, M.; Bellanné-Chantelot, C.; Gin, H.; Bertin, E.; Blickle, J.F.; Bauduceau, B.; Bouhanick, B.; et al. Heterogeneity of diabetes phenotype in patients with 3243 bp mutation of mitochondrial DNA (Maternally Inherited Diabetes and Deafness or MIDD). *Diabetes Metab.* **2004**, *30*, 181–186. [CrossRef]
70. Whittaker, R.G.; Schaefer, A.M.; McFarland, R.; Taylor, R.W.; Walker, M.; Turnbull, D.M. Prevalence and progression of diabetes in mitochondrial disease. *Diabetologia* **2007**, *50*, 2085–2089. [CrossRef] [PubMed]
71. Smith, P.R.; Bain, S.C.; Good, P.A.; Hattersley, A.T.; Barnett, A.H.; Gibson, J.M.; Dodson, P.M. Pigmentary retinal dystrophy and the syndrome of maternally inherited diabetes and deafness caused by the mitochondrial DNA 3243 tRNA(Leu) A to G mutation. *Ophthalmology* **1999**, *106*, 1101–1108. [CrossRef]
72. Narbonne, H.; Paquis-Fluckinger, V.; Valero, R.; Heyries, L.; Pellissier, J.F.; Vialettes, B. Gastrointestinal tract symptoms in Maternally Inherited Diabetes and Deafness (MIDD). *Diabetes Metab.* **2004**, *30*, 61–66. [CrossRef]
73. Robberecht, K.; Decock, C.; Stevens, A.; Seneca, S.; De Bleecker, J.; Leroy, B.P. Ptosis as an associated finding in maternally inherited diabetes and deafness. *Ophthalmic Genet.* **2010**, *31*, 240–243. [CrossRef] [PubMed]
74. Lien, L.M.; Lee, H.C.; Wang, K.L.; Chiu, J.C.; Chiu, H.C.; Wei, Y.H. Involvement of nervous system in maternally inherited diabetes and deafness (MIDD) with the A3243G mutation of mitochondrial DNA. *Acta Neurol. Scand.* **2001**, *103*, 159–165. [CrossRef]
75. Aggarwal, P.; Gill-Randall, R.; Wheatley, T.; Buchalter, M.B.; Metcalfe, J.; Alcolado, J.C. Identification of mtDNA mutation in a pedigree with gestational diabetes, deafness, Wolff-Parkinson-White syndrome and placenta accreta. *Hum. Hered.* **2001**, *51*, 114–116. [CrossRef]
76. Majamaa-Voltti, K.; Peuhkurinen, K.; Kortelainen, M.L.; Hassinen, I.E.; Majamaa, K. Cardiac abnormalities in patients with mitochondrial DNA mutation 3243A>G. *BMC Cardiovasc. Disord.* **2002**, *2*, 12. [CrossRef]
77. Azevedo, O.; Vilarinho, L.; Almeida, F.; Ferreira, F.; Guardado, J.; Ferreira, M.; Lourenço, A.; Medeiros, R.; Almeida, J. Cardiomyopathy and kidney disease in a patient with maternally inherited diabetes and deafness caused by the 3243A>G mutation of mitochondrial DNA. *Cardiology* **2010**, *115*, 71–74. [CrossRef] [PubMed]
78. Momiyama, Y.; Suzuki, Y.; Ohtomo, M.; Atsumi, Y.; Matsuoka, K.; Ohsuzu, F.; Kimura, M. Cardiac autonomic nervous dysfunction in diabetic patients with a mitochondrial DNA mutation: Assessment by heart rate variability. *Diabetes Care* **2002**, *25*, 2308–2313. [CrossRef]
79. Tabebi, M.; Charfi, N.; Kallabi, F.; Alila-Fersi, O.; Ben Mahmoud, A.; Tlili, A.; Keskes-Ammar, L.; Kamoun, H.; Abid, M.; Mnif, M.; et al. Whole mitochondrial genome screening of a family with maternally inherited diabetes and deafness (MIDD) associated with retinopathy: A putative haplotype associated to MIDD and a novel MT-CO2 m.8241T>G mutation. *J. Diabetes Complicat.* **2017**, *31*, 253–259. [CrossRef] [PubMed]
80. Maassen, J.A.; 'T Hart, L.M.; Van Essen, E.; Heine, R.J.; Nijpels, G.; Jahangir Tafrechi, R.S.; Raap, A.K.; Janssen, G.M.; Lemkes, H.H. Mitochondrial diabetes: Molecular mechanisms and clinical presentation. *Diabetes* **2004**, *53* (Suppl. 1), S103–S109. [CrossRef]

81. Sinnathuray, A.R.; Raut, V.; Awa, A.; Magee, A.; Toner, J.G. A review of cochlear implantation in mitochondrial sensorineural hearing loss. *Otol. Neurotol.* **2003**, *24*, 418–426. [CrossRef]
82. Salles, J.E.; Moisés, V.A.; Almeida, D.R.; Chacra, A.R.; Moisés, R.S. Myocardial dysfunction in mitochondrial diabetes treated with Coenzyme Q10. *Diabetes Res. Clin. Pract.* **2006**, *72*, 100–103. [CrossRef]
83. Suzuki, S.; Hinokio, Y.; Ohtomo, M.; Hirai, M.; Hirai, A.; Chiba, M.; Kasuga, S.; Satoh, Y.; Akai, H.; Toyota, T. The effects of coenzyme Q10 treatment on maternally inherited diabetes mellitus and deafness, and mitochondrial DNA 3243 (A to G) mutation. *Diabetologia* **1998**, *41*, 584–588. [CrossRef]
84. Holt, I.J.; Harding, A.E.; Petty, R.K.; Morgan-Hughes, J.A. A new mitochondrial disease associated with mitochondrial DNA heteroplasmy. *Am. J. Hum. Genet.* **1990**, *46*, 428–433.
85. Fryer, A.; Appleton, R.; Sweeney, M.G.; Rosenbloom, L.; Harding, A.E. Mitochondrial DNA 8993 (NARP) mutation presenting with a heterogeneous phenotype including 'cerebral palsy'. *Arch. Dis. Child.* **1994**, *71*, 419–422. [CrossRef]
86. Porto, F.B.; Mack, G.; Sterboul, M.J.; Lewin, P.; Flament, J.; Sahel, J.; Dollfus, H. Isolated late-onset cone-rod dystrophy revealing a familial neurogenic muscle weakness, ataxia, and retinitis pigmentosa syndrome with the T8993G mitochondrial mutation. *Am. J. Ophthalmol.* **2001**, *132*, 935–937. [CrossRef]
87. Jonckheere, A.I.; Smeitink, J.A.; Rodenburg, R.J. Mitochondrial ATP synthase: Architecture, function and pathology. *J. Inherit. Metab. Dis.* **2012**, *35*, 211–225. [CrossRef]
88. White, S.L.; Collins, V.R.; Wolfe, R.; Cleary, M.A.; Shanske, S.; DiMauro, S.; Dahl, H.H.; Thorburn, D.R. Genetic counseling and prenatal diagnosis for the mitochondrial DNA mutations at nucleotide 8993. *Am. J. Hum. Genet.* **1999**, *65*, 474–482. [CrossRef]
89. López-Gallardo, E.; Solano, A.; Herrero-Martín, M.D.; Martínez-Romero, I.; Castaño-Pérez, M.D.; Andreu, A.L.; Herrera, A.; López-Pérez, M.J.; Ruiz-Pesini, E.; Montoya, J. NARP syndrome in a patient harbouring an insertion in the MT-ATP6 gene that results in a truncated protein. *J. Med. Genet.* **2009**, *46*, 64–67. [CrossRef]
90. Sciacco, M.; Prelle, A.; D'Adda, E.; Lamperti, C.; Bordoni, A.; Rango, M.; Crimi, M.; Comi, G.P.; Bresolin, N.; Moggio, M. Familial mtDNA T8993C transition causing both the NARP and the MILS phenotype in the same generation. A morphological, genetic and spectroscopic study. *J. Neurol.* **2003**, *250*, 1498–1500. [CrossRef]
91. Childs, A.M.; Hutchin, T.; Pysden, K.; Highet, L.; Bamford, J.; Livingston, J.; Crow, Y.J. Variable phenotype including Leigh syndrome with a 9185T>C mutation in the MTATP6 gene. *Neuropediatrics* **2007**, *38*, 313–316. [CrossRef]
92. Tatuch, Y.; Christodoulou, J.; Feigenbaum, A.; Clarke, J.T.; Wherret, J.; Smith, C.; Rudd, N.; Petrova-Benedict, R.; Robinson, B.H. Heteroplasmic mtDNA mutation (T—G) at 8993 can cause Leigh disease when the percentage of abnormal mtDNA is high. *Am. J. Hum. Genet.* **1992**, *50*, 852–858.
93. D'Aurelio, M.; Vives-Bauza, C.; Davidson, M.M.; Manfredi, G. Mitochondrial DNA background modifies the bioenergetics of NARP/MILS ATP6 mutant cells. *Hum. Mol. Genet.* **2010**, *19*, 374–386. [CrossRef]
94. Claeys, K.G.; Abicht, A.; Häusler, M.; Kleinle, S.; Wiesmann, M.; Schulz, J.B.; Horvath, R.; Weis, J. Novel genetic and neuropathological insights in neurogenic muscle weakness, ataxia, and retinitis pigmentosa (NARP). *Muscle Nerve* **2016**, *54*, 328–333. [CrossRef]
95. Kara, B.; Arıkan, M.; Maraş, H.; Abacı, N.; Cakıris, A.; Ustek, D. Whole mitochondrial genome analysis of a family with NARP/MILS caused by m.8993T>C mutation in the MT-ATP6 gene. *Mol. Genet. Metab.* **2012**, *107*, 389–393. [CrossRef]
96. Limongelli, G.; Tome-Esteban, M.; Dejthevaporn, C.; Rahman, S.; Hanna, M.G.; Elliott, P.M. Prevalence and natural history of heart disease in adults with primary mitochondrial respiratory chain disease. *Eur. J. Heart Fail.* **2010**, *12*, 114–121. [CrossRef]
97. Pastores, G.M.; Santorelli, F.M.; Shanske, S.; Gelb, B.D.; Fyfe, B.; Wolfe, D.; Willner, J.P. Leigh syndrome and hypertrophic cardiomyopathy in an infant with a mitochondrial DNA point mutation (T8993G). *Am. J. Med. Genet.* **1994**, *50*, 265–271. [CrossRef]
98. Duno, M.; Wibrand, F.; Baggesen, K.; Rosenberg, T.; Kjaer, N.; Frederiksen, A.L. A novel mitochondrial mutation m.8989G>C associated with neuropathy, ataxia, retinitis pigmentosa—The NARP syndrome. *Gene* **2013**, *515*, 372–375. [CrossRef]
99. Mordel, P.; Schaeffer, S.; Dupas, Q.; Laville, M.A.; Gérard, M.; Chapon, F.; Allouche, S. A 2 bp deletion in the mitochondrial ATP 6 gene responsible for the NARP (neuropathy, ataxia, and retinitis pigmentosa) syndrome. *Biochem. Biophys. Res. Commun.* **2017**, *494*, 133–137. [CrossRef]

100. Blanco-Grau, A.; Bonaventura-Ibars, I.; Coll-Cantí, J.; Melià, M.J.; Martinez, R.; Martínez-Gallo, M.; Andreu, A.L.; Pinós, T.; García-Arumí, E. Identification and biochemical characterization of the novel mutation m.8839G>C in the mitochondrial ATP6 gene associated with NARP syndrome. *Genes Brain Behav.* **2013**, *12*, 812–820. [CrossRef]
101. Moslemi, A.R.; Darin, N.; Tulinius, M.; Oldfors, A.; Holme, E. Two new mutations in the MTATP6 gene associated with Leigh syndrome. *Neuropediatrics* **2005**, *36*, 314–318. [CrossRef]
102. Gammage, P.A.; Rorbach, J.; Vincent, A.I.; Rebar, E.J.; Minczuk, M. Mitochondrially targeted ZFNs for selective degradation of pathogenic mitochondrial genomes bearing large-scale deletions or point mutations. *EMBO Mol. Med.* **2014**, *6*, 458–466. [CrossRef]
103. Fellman, V.; Rapola, J.; Pihko, H.; Varilo, T.; Raivio, K.O. Iron-overload disease in infants involving fetal growth retardation, lactic acidosis, liver haemosiderosis, and aminoaciduria. *Lancet* **1998**, *351*, 490–493. [CrossRef]
104. Visapää, I.; Fellman, V.; Vesa, J.; Dasvarma, A.; Hutton, J.L.; Kumar, V.; Payne, G.S.; Makarow, M.; Van Coster, R.; Taylor, R.W.; et al. GRACILE syndrome, a lethal metabolic disorder with iron overload, is caused by a point mutation in BCS1L. *Am. J. Hum. Genet.* **2002**, *71*, 863–876. [CrossRef]
105. Fellman, V. The GRACILE syndrome, a neonatal lethal metabolic disorder with iron overload. *Blood Cells Mol. Dis.* **2002**, *29*, 444–450. [CrossRef]
106. Fellman, V.; Kotarsky, H. Mitochondrial hepatopathies in the newborn period. *Semin. Fetal Neonatal Med.* **2011**, *16*, 222–228. [CrossRef]
107. Rapola, J.; Heikkilä, P.; Fellman, V. Pathology of lethal fetal growth retardation syndrome with aminoaciduria, iron overload, and lactic acidosis (GRACILE). *Pediatr. Pathol. Mol. Med.* **2002**, *21*, 183–193. [CrossRef]
108. Kotarsky, H.; Karikoski, R.; Mörgelin, M.; Marjavaara, S.; Bergman, P.; Zhang, D.L.; Smet, J.; van Coster, R.; Fellman, V. Characterization of complex III deficiency and liver dysfunction in GRACILE syndrome caused by a BCS1L mutation. *Mitochondrion* **2010**, *10*, 497–509. [CrossRef]
109. de Lonlay, P.; Valnot, I.; Barrientos, A.; Gorbatyuk, M.; Tzagoloff, A.; Taanman, J.W.; Benayoun, E.; Chrétien, D.; Kadhom, N.; Lombès, A.; et al. A mutant mitochondrial respiratory chain assembly protein causes complex III deficiency in patients with tubulopathy, encephalopathy and liver failure. *Nat. Genet.* **2001**, *29*, 57–60. [CrossRef]
110. Serdaroğlu, E.; Takcı, Ş.; Kotarsky, H.; Çil, O.; Utine, E.; Yiğit, Ş.; Fellman, V. A Turkish BCS1L mutation causes GRACILE-like disorder. *Turk. J. Pediatr.* **2016**, *58*, 658–661. [CrossRef]
111. Fellman, V.; Lemmelä, S.; Sajantila, A.; Pihko, H.; Järvelä, I. Screening of BCS1L mutations in severe neonatal disorders suspicious for mitochondrial cause. *J. Hum. Genet.* **2008**, *53*, 554–558. [CrossRef]
112. Lee, W.S.; Sokol, R.J. Mitochondrial hepatopathies: Advances in genetics, therapeutic approaches, and outcomes. *J. Pediatr.* **2013**, *163*, 942–948. [CrossRef]
113. Rajendran, J.; Tomašić, N.; Kotarsky, H.; Hansson, E.; Velagapudi, V.; Kallijärvi, J.; Fellman, V. Effect of High-Carbohydrate Diet on Plasma Metabolome in Mice with Mitochondrial Respiratory Chain Complex III Deficiency. *Int. J. Mol. Sci.* **2016**, *17*, 1824. [CrossRef] [PubMed]
114. Bardosi, A.; Creutzfeldt, W.; DiMauro, S.; Felgenhauer, K.; Friede, R.L.; Goebel, H.H.; Kohlschütter, A.; Mayer, G.; Rahlf, G.; Servidei, S. Myo-, neuro-, gastrointestinal encephalopathy (MNGIE syndrome) due to partial deficiency of cytochrome-c-oxidase. A new mitochondrial multisystem disorder. *Acta Neuropathol.* **1987**, *74*, 248–258. [CrossRef] [PubMed]
115. Hirano, M.; Silvestri, G.; Blake, D.M.; Lombes, A.; Minetti, C.; Bonilla, E.; Hays, A.P.; Lovelace, R.E.; Butler, I.; Bertorini, T.E. Mitochondrial neurogastrointestinal encephalomyopathy (MNGIE): Clinical, biochemical, and genetic features of an autosomal recessive mitochondrial disorder. *Neurology* **1994**, *44*, 721–727. [CrossRef] [PubMed]
116. Nishino, I.; Spinazzola, A.; Hirano, M. Thymidine phosphorylase gene mutations in MNGIE, a human mitochondrial disorder. *Science* **1999**, *283*, 689–692. [CrossRef] [PubMed]
117. Nishino, I.; Spinazzola, A.; Papadimitriou, A.; Hammans, S.; Steiner, I.; Hahn, C.D.; Connolly, A.M.; Verloes, A.; Guimarães, J.; Maillard, I.; et al. Mitochondrial neurogastrointestinal encephalomyopathy: An autosomal recessive disorder due to thymidine phosphorylase mutations. *Ann. Neurol.* **2000**, *47*, 792–800. [CrossRef]

118. Blondon, H.; Polivka, M.; Joly, F.; Flourie, B.; Mikol, J.; Messing, B. Digestive smooth muscle mitochondrial myopathy in patients with mitochondrial-neuro-gastro-intestinal encephalomyopathy (MNGIE). *Gastroenterol. Clin. Biol.* **2005**, *29*, 773–778. [CrossRef]
119. Scarpelli, M.; Ricciardi, G.K.; Beltramello, A.; Zocca, I.; Calabria, F.; Russignan, A.; Zappini, F.; Cotelli, M.S.; Padovani, A.; Tomelleri, G.; et al. The role of brain MRI in mitochondrial neurogastrointestinal encephalomyopathy. *Neuroradiol. J.* **2013**, *26*, 520–530. [CrossRef]
120. Hirano, M.; Nishigaki, Y.; Martí, R. Mitochondrial neurogastrointestinal encephalomyopathy (MNGIE): A disease of two genomes. *Neurologist* **2004**, *10*, 8–17. [CrossRef]
121. Garone, C.; Tadesse, S.; Hirano, M. Clinical and genetic spectrum of mitochondrial neurogastrointestinal encephalomyopathy. *Brain* **2011**, *134*, 3326–3332. [CrossRef]
122. Viscomi, C.; Zeviani, M. MtDNA-maintenance defects: Syndromes and genes. *J. Inherit. Metab. Dis.* **2017**, *40*, 587–599. [CrossRef] [PubMed]
123. Filosto, M.; Tomelleri, G.; Tonin, P.; Scarpelli, M.; Vattemi, G.; Rizzuto, N.; Padovani, A.; Simonati, A. Neuropathology of mitochondrial diseases. *Biosci. Rep.* **2007**, *27*, 23–30. [CrossRef]
124. Halter, J.; Schüpbach, W.; Casali, C.; Elhasid, R.; Fay, K.; Hammans, S.; Illa, I.; Kappeler, L.; Krähenbühl, S.; Lehmann, T.; et al. Allogeneic hematopoietic SCT as treatment option for patients with mitochondrial neurogastrointestinal encephalomyopathy (MNGIE): A consensus conference proposal for a standardized approach. *Bone Marrow Transplant.* **2011**, *46*, 330–337. [CrossRef] [PubMed]
125. Wang, J.; Chen, W.; Wang, F.; Wu, D.; Qian, J.; Kang, J.; Li, H.; Ma, E. Nutrition Therapy for Mitochondrial Neurogastrointestinal Encephalopathy with Homozygous Mutation of the TYMP Gene. *Clin. Nutr. Res.* **2015**, *4*, 132–136. [CrossRef] [PubMed]
126. Hirano, M.; Martí, R.; Casali, C.; Tadesse, S.; Uldrick, T.; Fine, B.; Escolar, D.M.; Valentino, M.L.; Nishino, I.; Hesdorffer, C.; et al. Allogeneic stem cell transplantation corrects biochemical derangements in MNGIE. *Neurology* **2006**, *67*, 1458–1460. [CrossRef] [PubMed]
127. Filosto, M.; Scarpelli, M.; Tonin, P.; Lucchini, G.; Pavan, F.; Santus, F.; Parini, R.; Donati, M.A.; Cotelli, M.S.; Vielmi, V.; et al. Course and management of allogeneic stem cell transplantation in patients with mitochondrial neurogastrointestinal encephalomyopathy. *J. Neurol.* **2012**, *259*, 2699–2706. [CrossRef] [PubMed]
128. Halter, J.P.; Michael, W.; Schüpbach, M.; Mandel, H.; Casali, C.; Orchard, K.; Collin, M.; Valcarcel, D.; Rovelli, A.; Filosto, M.; et al. Allogeneic haematopoietic stem cell transplantation for mitochondrial neurogastrointestinal encephalomyopathy. *Brain* **2015**, *138*, 2847–2858. [CrossRef] [PubMed]
129. Lara, M.C.; Weiss, B.; Illa, I.; Madoz, P.; Massuet, L.; Andreu, A.L.; Valentino, M.L.; Anikster, Y.; Hirano, M.; Martí, R. Infusion of platelets transiently reduces nucleoside overload in MNGIE. *Neurology* **2006**, *67*, 1461–1463. [CrossRef]
130. Yavuz, H.; Ozel, A.; Christensen, M.; Christensen, E.; Schwartz, M.; Elmaci, M.; Vissing, J. Treatment of mitochondrial neurogastrointestinal encephalomyopathy with dialysis. *Arch. Neurol.* **2007**, *64*, 435–438. [CrossRef]
131. Bax, B.E.; Bain, M.D.; Scarpelli, M.; Filosto, M.; Tonin, P.; Moran, N. Clinical and biochemical improvements in a patient with MNGIE following enzyme replacement. *Neurology* **2013**, *81*, 1269–1271. [CrossRef]
132. Röeben, B.; Marquetand, J.; Bender, B.; Billing, H.; Haack, T.B.; Sanchez-Albisua, I.; Schöls, L.; Blom, H.J.; Synofzik, M. Hemodialysis in MNGIE transiently reduces serum and urine levels of thymidine and deoxyuridine, but not CSF levels and neurological function. *Orphanet J. Rare Dis.* **2017**, *12*, 135. [CrossRef] [PubMed]
133. De Giorgio, R.; Pironi, L.; Rinaldi, R.; Boschetti, E.; Caporali, L.; Capristo, M.; Casali, C.; Cenacchi, G.; Contin, M.; D'Angelo, R.; et al. Liver transplantation for mitochondrial neurogastrointestinal encephalomyopathy. *Ann. Neurol.* **2016**, *80*, 448–455. [CrossRef] [PubMed]
134. Torres-Torronteras, J.; Cabrera-Pérez, R.; Barba, I.; Costa, C.; de Luna, N.; Andreu, A.L.; Barquinero, J.; Hirano, M.; Cámara, Y.; Martí, R. Long-Term Restoration of Thymidine Phosphorylase Function and Nucleoside Homeostasis Using Hematopoietic Gene Therapy in a Murine Model of Mitochondrial Neurogastrointestinal Encephalopathy. *Hum. Gene Ther.* **2016**, *27*, 656–667. [CrossRef] [PubMed]
135. Barth, P.G.; Scholte, H.R.; Berden, J.A.; Van der Klei-Van Moorsel, J.M.; Luyt-Houwen, I.E.; Van 't Veer-Korthof, E.T.; Van der Harten, J.J.; Sobotka-Plojhar, M.A. An X-linked mitochondrial disease affecting cardiac muscle, skeletal muscle and neutrophil leucocytes. *J. Neurol. Sci* **1983**, *62*, 327–355. [CrossRef]

136. Lev, D.; Nissenkorn, A.; Leshinsky-Silver, E.; Sadeh, M.; Zeharia, A.; Garty, B.Z.; Blieden, L.; Barash, V.; Lerman-Sagie, T. Clinical presentations of mitochondrial cardiomyopathies. *Pediatr. Cardiol.* **2004**, *25*, 443–450. [CrossRef] [PubMed]

137. Adès, L.C.; Gedeon, A.K.; Wilson, M.J.; Latham, M.; Partington, M.W.; Mulley, J.C.; Nelson, J.; Lui, K.; Sillence, D.O. Barth syndrome: Clinical features and confirmation of gene localisation to distal Xq28. *Am. J. Med. Genet.* **1993**, *45*, 327–334. [CrossRef]

138. D'Adamo, P.; Fassone, L.; Gedeon, A.; Janssen, E.A.; Bione, S.; Bolhuis, P.A.; Barth, P.G.; Wilson, M.; Haan, E.; Orstavik, K.H.; et al. The X-linked gene G4.5 is responsible for different infantile dilated cardiomyopathies. *Am. J. Hum. Genet.* **1997**, *61*, 862–867. [CrossRef] [PubMed]

139. Steward, C.G.; Newbury-Ecob, R.A.; Hastings, R.; Smithson, S.F.; Tsai-Goodman, B.; Quarrell, O.W.; Kulik, W.; Wanders, R.; Pennock, M.; Williams, M.; et al. Barth syndrome: An X-linked cause of fetal cardiomyopathy and stillbirth. *Prenat. Diagn.* **2010**, *30*, 970–976. [CrossRef]

140. Clarke, S.L.; Bowron, A.; Gonzalez, I.L.; Groves, S.J.; Newbury-Ecob, R.; Clayton, N.; Martin, R.P.; Tsai-Goodman, B.; Garratt, V.; Ashworth, M.; et al. Barth syndrome. *Orphanet J. Rare Dis.* **2013**, *8*, 23. [CrossRef]

141. Barth, P.G.; Valianpour, F.; Bowen, V.M.; Lam, J.; Duran, M.; Vaz, F.M.; Wanders, R.J. X-linked cardioskeletal myopathy and neutropenia (Barth syndrome): An update. *Am. J. Med. Genet. A* **2004**, *126A*, 349–354. [CrossRef] [PubMed]

142. Brady, A.N.; Shehata, B.M.; Fernhoff, P.M. X-linked fetal cardiomyopathy caused by a novel mutation in the TAZ gene. *Prenat. Diagn.* **2006**, *26*, 462–465. [CrossRef] [PubMed]

143. Rigaud, C.; Lebre, A.S.; Touraine, R.; Beaupain, B.; Ottolenghi, C.; Chabli, A.; Ansquer, H.; Ozsahin, H.; Di Filippo, S.; De Lonlay, P.; et al. Natural history of Barth syndrome: A national cohort study of 22 patients. *Orphanet J. Rare Dis.* **2013**, *8*, 70. [CrossRef] [PubMed]

144. Bleyl, S.B.; Mumford, B.R.; Thompson, V.; Carey, J.C.; Pysher, T.J.; Chin, T.K.; Ward, K. Neonatal, lethal noncompaction of the left ventricular myocardium is allelic with Barth syndrome. *Am. J. Hum. Genet.* **1997**, *61*, 868–872. [CrossRef] [PubMed]

145. Chin, T.K.; Perloff, J.K.; Williams, R.G.; Jue, K.; Mohrmann, R. Isolated noncompaction of left ventricular myocardium. A study of eight cases. *Circulation* **1990**, *82*, 507–513. [CrossRef]

146. Pignatelli, R.H.; McMahon, C.J.; Dreyer, W.J.; Denfield, S.W.; Price, J.; Belmont, J.W.; Craigen, W.J.; Wu, J.; El Said, H.; Bezold, L.I.; et al. Clinical characterization of left ventricular noncompaction in children: A relatively common form of cardiomyopathy. *Circulation* **2003**, *108*, 2672–2678. [CrossRef]

147. Spencer, C.T.; Bryant, R.M.; Day, J.; Gonzalez, I.L.; Colan, S.D.; Thompson, W.R.; Berthy, J.; Redfearn, S.P.; Byrne, B.J. Cardiac and clinical phenotype in Barth syndrome. *Pediatrics* **2006**, *118*, e337–e346. [CrossRef]

148. Roberts, A.E.; Nixon, C.; Steward, C.G.; Gauvreau, K.; Maisenbacher, M.; Fletcher, M.; Geva, J.; Byrne, B.J.; Spencer, C.T. The Barth Syndrome Registry: Distinguishing disease characteristics and growth data from a longitudinal study. *Am. J. Med. Genet. A* **2012**, *158A*, 2726–2732. [CrossRef]

149. Steward, C.G.; Groves, S.J.; Taylor, C.T.; Maisenbacher, M.K.; Versluys, B.; Newbury-Ecob, R.A.; Ozsahin, H.; Damin, M.K.; Bowen, V.M.; McCurdy, K.R.; et al. Neutropenia in Barth syndrome: Characteristics, risks, and management. *Curr. Opin. Hematol.* **2019**, *26*, 6–15. [CrossRef]

150. Kelley, R.I.; Cheatham, J.P.; Clark, B.J.; Nigro, M.A.; Powell, B.R.; Sherwood, G.W.; Sladky, J.T.; Swisher, W.P. X-linked dilated cardiomyopathy with neutropenia, growth retardation, and 3-methylglutaconic aciduria. *J. Pediatr.* **1991**, *119*, 738–747. [CrossRef]

151. Bolhuis, P.A.; Hensels, G.W.; Hulsebos, T.J.; Baas, F.; Barth, P.G. Mapping of the locus for X-linked cardioskeletal myopathy with neutropenia and abnormal mitochondria (Barth syndrome) to Xq28. *Am. J. Hum. Genet.* **1991**, *48*, 481–485. [PubMed]

152. Bione, S.; D'Adamo, P.; Maestrini, E.; Gedeon, A.K.; Bolhuis, P.A.; Toniolo, D. A novel X-linked gene, G4.5. is responsible for Barth syndrome. *Nat. Genet.* **1996**, *12*, 385–389. [CrossRef] [PubMed]

153. Vreken, P.; Valianpour, F.; Nijtmans, L.G.; Grivell, L.A.; Plecko, B.; Wanders, R.J.; Barth, P.G. Defective remodeling of cardiolipin and phosphatidylglycerol in Barth syndrome. *Biochem. Biophys. Res. Commun.* **2000**, *279*, 378–382. [CrossRef] [PubMed]

154. Gaspard, G.J.; McMaster, C.R. Cardiolipin metabolism and its causal role in the etiology of the inherited cardiomyopathy Barth syndrome. *Chem. Phys. Lipids* **2015**, *193*, 1–10. [CrossRef] [PubMed]

155. Houtkooper, R.H.; Turkenburg, M.; Poll-The, B.T.; Karall, D.; Pérez-Cerdá, C.; Morrone, A.; Malvagia, S.; Wanders, R.J.; Kulik, W.; Vaz, F.M. The enigmatic role of tafazzin in cardiolipin metabolism. *Biochim. Biophys. Acta* **2009**, *1788*, 2003–2014. [CrossRef] [PubMed]
156. Valianpour, F.; Mitsakos, V.; Schlemmer, D.; Towbin, J.A.; Taylor, J.M.; Ekert, P.G.; Thorburn, D.R.; Munnich, A.; Wanders, R.J.; Barth, P.G.; et al. Monolysocardiolipins accumulate in Barth syndrome but do not lead to enhanced apoptosis. *J. Lipid Res.* **2005**, *46*, 1182–1195. [CrossRef] [PubMed]
157. Kulik, W.; van Lenthe, H.; Stet, F.S.; Houtkooper, R.H.; Kemp, H.; Stone, J.E.; Steward, C.G.; Wanders, R.J.; Vaz, F.M. Bloodspot assay using HPLC-tandem mass spectrometry for detection of Barth syndrome. *Clin. Chem* **2008**, *54*, 371–378. [CrossRef] [PubMed]
158. Jefferies, J.L. Barth syndrome. *Am. J. Med. Genet. C Semin. Med. Genet.* **2013**, *163C*, 198–205. [CrossRef]
159. Meyerson, C.; Van Stavern, G.; McClelland, C. Leber hereditary optic neuropathy: Current perspectives. *Clin. Ophthalmol* **2015**, *9*, 1165–1176.
160. Giraudet, S.; Lamirel, C.; Amati-Bonneau, P.; Reynier, P.; Bonneau, D.; Miléa, D.; Cochereau, I. Never too old to harbour a young man's disease? *Br. J. Ophthalmol.* **2011**, *95*, 887–896. [CrossRef]
161. Barboni, P.; Savini, G.; Valentino, M.L.; La Morgia, C.; Bellusci, C.; De Negri, A.M.; Sadun, F.; Carta, A.; Carbonelli, M.; Sadun, A.A.; et al. Leber's hereditary optic neuropathy with childhood onset. *Invest. Ophthalmol. Vis. Sci.* **2006**, *47*, 5303–5309. [CrossRef]
162. Carelli, V.; Ross-Cisneros, F.N.; Sadun, A.A. Mitochondrial dysfunction as a cause of optic neuropathies. *Prog. Retin. Eye Res.* **2004**, *23*, 53–89. [CrossRef]
163. Newman, N.J. Leber's hereditary optic neuropathy. New genetic considerations. *Arch. Neurol.* **1993**, *50*, 540–548. [CrossRef] [PubMed]
164. Chao de la Barca, J.M.; Simard, G.; Amati-Bonneau, P.; Safiedeen, Z.; Prunier-Mirebeau, D.; Chupin, S.; Gadras, C.; Tessier, L.; Gueguen, N.; Chevrollier, A.; et al. The metabolomic signature of Leber's hereditary optic neuropathy reveals endoplasmic reticulum stress. *Brain* **2016**, *139*, 2864–2876. [CrossRef]
165. Jun, A.S.; Brown, M.D.; Wallace, D.C. A mitochondrial DNA mutation at nucleotide pair 14459 of the NADH dehydrogenase subunit 6 gene associated with maternally inherited Leber hereditary optic neuropathy and dystonia. *Proc. Natl. Acad. Sci. USA* **1994**, *91*, 6206–6210. [CrossRef]
166. Harding, A.E.; Sweeney, M.G.; Miller, D.H.; Mumford, C.J.; Kellar-Wood, H.; Menard, D.; McDonald, W.I.; Compston, D.A. Occurrence of a multiple sclerosis-like illness in women who have a Leber's hereditary optic neuropathy mitochondrial DNA mutation. *Brain* **1992**, *115 (Pt. 4)*, 979–989. [CrossRef]
167. Funalot, B.; Reynier, P.; Vighetto, A.; Ranoux, D.; Bonnefont, J.P.; Godinot, C.; Malthièry, Y.; Mas, J.L. Leigh-like encephalopathy complicating Leber's hereditary optic neuropathy. *Ann. Neurol.* **2002**, *52*, 374–377. [CrossRef] [PubMed]
168. Nikoskelainen, E.K.; Marttila, R.J.; Huoponen, K.; Juvonen, V.; Lamminen, T.; Sonninen, P.; Savontaus, M.L. Leber's "plus": Neurological abnormalities in patients with Leber's hereditary optic neuropathy. *J. Neurol. Neurosurg. Psychiatry* **1995**, *59*, 160–164. [CrossRef] [PubMed]
169. Perez, F.; Anne, O.; Debruxelles, S.; Menegon, P.; Lambrecq, V.; Lacombe, D.; Martin-Negrier, M.L.; Brochet, B.; Goizet, C. Leber's optic neuropathy associated with disseminated white matter disease: A case report and review. *Clin. Neurol. Neurosurg.* **2009**, *111*, 83–86. [CrossRef]
170. Kellar-Wood, H.; Robertson, N.; Govan, G.G.; Compston, D.A.; Harding, A.E. Leber's hereditary optic neuropathy mitochondrial DNA mutations in multiple sclerosis. *Ann. Neurol.* **1994**, *36*, 109–112. [CrossRef]
171. Meire, F.M.; Van Coster, R.; Cochaux, P.; Obermaier-Kusser, B.; Candaele, C.; Martin, J.J. Neurological disorders in members of families with Leber's hereditary optic neuropathy (LHON) caused by different mitochondrial mutations. *Ophthalmic Genet.* **1995**, *16*, 119–126. [CrossRef] [PubMed]
172. Pfeffer, G.; Burke, A.; Yu-Wai-Man, P.; Compston, D.A.; Chinnery, P.F. Clinical features of MS associated with Leber hereditary optic neuropathy mtDNA mutations. *Neurology* **2013**, *81*, 2073–2081. [CrossRef] [PubMed]
173. Yu-Wai-Man, P.; Griffiths, P.G.; Brown, D.T.; Howell, N.; Turnbull, D.M.; Chinnery, P.F. The epidemiology of Leber hereditary optic neuropathy in the North East of England. *Am. J. Hum. Genet.* **2003**, *72*, 333–339. [CrossRef]
174. Puomila, A.; Hämäläinen, P.; Kivioja, S.; Savontaus, M.L.; Koivumäki, S.; Huoponen, K.; Nikoskelainen, E. Epidemiology and penetrance of Leber hereditary optic neuropathy in Finland. *Eur. J. Hum. Genet.* **2007**, *15*, 1079–1089. [CrossRef] [PubMed]

175. Mascialino, B.; Leinonen, M.; Meier, T. Meta-analysis of the prevalence of Leber hereditary optic neuropathy mtDNA mutations in Europe. *Eur. J. Ophthalmol.* **2012**, *22*, 461–465. [CrossRef] [PubMed]
176. Finsterer, J.; Stöllberger, C.; Prainer, C.; Hochwarter, A. Lone noncompaction in Leber's hereditary optic neuropathy. *Acta Cardiol.* **2004**, *59*, 187–190. [CrossRef]
177. Finsterer, J.; Stöllberger, C.; Michaela, J. Familial left ventricular hypertrabeculation in two blind brothers. *Cardiovasc. Pathol.* **2002**, *11*, 146–148. [CrossRef]
178. Ortiz, R.G.; Newman, N.J.; Manoukian, S.V.; Diesenhouse, M.C.; Lott, M.T.; Wallace, D.C. Optic disk cupping and electrocardiographic abnormalities in an American pedigree with Leber's hereditary optic neuropathy. *Am. J. Ophthalmol.* **1992**, *113*, 561–566. [CrossRef]
179. Finsterer, J.; Stollberger, C.; Gatterer, E. Wolff-Parkinson-White syndrome and noncompaction in Leber's hereditary optic neuropathy due to the variant m.3460G>A. *J. Int. Med. Res.* **2018**, *46*, 2054–2060. [CrossRef]
180. Nikoskelainen, E.K.; Savontaus, M.L.; Huoponen, K.; Antila, K.; Hartiala, J. Pre-excitation syndrome in Leber's hereditary optic neuropathy. *Lancet* **1994**, *344*, 857–858. [CrossRef]
181. Mashima, Y.; Kigasawa, K.; Hasegawa, H.; Tani, M.; Oguchi, Y. High incidence of pre-excitation syndrome in Japanese families with Leber's hereditary optic neuropathy. *Clin. Genet.* **1996**, *50*, 535–537. [CrossRef] [PubMed]
182. Wallace, D.C.; Singh, G.; Lott, M.T.; Hodge, J.A.; Schurr, T.G.; Lezza, A.M.; Elsas, L.J.; Nikoskelainen, E.K. Mitochondrial DNA mutation associated with Leber's hereditary optic neuropathy. *Science* **1988**, *242*, 1427–1430. [CrossRef] [PubMed]
183. Huoponen, K.; Vilkki, J.; Aula, P.; Nikoskelainen, E.K.; Savontaus, M.L. A new mtDNA mutation associated with Leber hereditary optic neuroretinopathy. *Am. J. Hum. Genet.* **1991**, *48*, 1147–1153.
184. Johns, D.R.; Neufeld, M.J.; Park, R.D. An ND-6 mitochondrial DNA mutation associated with Leber hereditary optic neuropathy. *Biochem. Biophys. Res. Commun.* **1992**, *187*, 1551–1557. [CrossRef]
185. Riordan-Eva, P.; Harding, A.E. Leber's hereditary optic neuropathy: The clinical relevance of different mitochondrial DNA mutations. *J. Med. Genet.* **1995**, *32*, 81–87. [CrossRef]
186. Valentino, M.L.; Barboni, P.; Ghelli, A.; Bucchi, L.; Rengo, C.; Achilli, A.; Torroni, A.; Lugaresi, A.; Lodi, R.; Barbiroli, B.; et al. The ND1 gene of complex I is a mutational hot spot for Leber's hereditary optic neuropathy. *Ann. Neurol.* **2004**, *56*, 631–641. [CrossRef]
187. Valentino, M.L.; Avoni, P.; Barboni, P.; Pallotti, F.; Rengo, C.; Torroni, A.; Bellan, M.; Baruzzi, A.; Carelli, V. Mitochondrial DNA nucleotide changes C14482G and C14482A in the ND6 gene are pathogenic for Leber's hereditary optic neuropathy. *Ann. Neurol.* **2002**, *51*, 774–778. [CrossRef]
188. Chinnery, P.F.; Brown, D.T.; Andrews, R.M.; Singh-Kler, R.; Riordan-Eva, P.; Lindley, J.; Applegarth, D.A.; Turnbull, D.M.; Howell, N. The mitochondrial ND6 gene is a hot spot for mutations that cause Leber's hereditary optic neuropathy. *Brain* **2001**, *124*, 209–218. [CrossRef] [PubMed]
189. Achilli, A.; Iommarini, L.; Olivieri, A.; Pala, M.; Hooshiar Kashani, B.; Reynier, P.; La Morgia, C.; Valentino, M.L.; Liguori, R.; Pizza, F.; et al. Rare primary mitochondrial DNA mutations and probable synergistic variants in Leber's hereditary optic neuropathy. *PLoS ONE* **2012**, *7*, e42242. [CrossRef]
190. Yu-Wai-Man, P.; Turnbull, D.M.; Chinnery, P.F. Leber hereditary optic neuropathy. *J. Med. Genet.* **2002**, *39*, 162–169.
191. Battisti, C.; Formichi, P.; Cardaioli, E.; Bianchi, S.; Mangiavacchi, P.; Tripodi, S.A.; Tosi, P.; Federico, A. Cell response to oxidative stress induced apoptosis in patients with Leber's hereditary optic neuropathy. *J. Neurol. Neurosurg. Psychiatry* **2004**, *75*, 1731–1736. [CrossRef] [PubMed]
192. Giordano, C.; Montopoli, M.; Perli, E.; Orlandi, M.; Fantin, M.; Ross-Cisneros, F.N.; Caprarotta, L.; Martinuzzi, A.; Ragazzi, E.; Ghelli, A.; et al. Oestrogens ameliorate mitochondrial dysfunction in Leber's hereditary optic neuropathy. *Brain* **2011**, *134*, 220–234. [CrossRef] [PubMed]
193. Sadun, A.A.; Carelli, V.; Salomao, S.R.; Berezovsky, A.; Quiros, P.A.; Sadun, F.; DeNegri, A.M.; Andrade, R.; Moraes, M.; Passos, A.; et al. Extensive investigation of a large Brazilian pedigree of 11778/haplogroup J Leber hereditary optic neuropathy. *Am. J. Ophthalmol.* **2003**, *136*, 231–238. [CrossRef]
194. Kirkman, M.A.; Yu-Wai-Man, P.; Korsten, A.; Leonhardt, M.; Dimitriadis, K.; De Coo, I.F.; Klopstock, T.; Chinnery, P.F. Gene-environment interactions in Leber hereditary optic neuropathy. *Brain* **2009**, *132*, 2317–2326. [CrossRef]
195. Khan, N.A.; Govindaraj, P.; Soumittra, N.; Sharma, S.; Srilekha, S.; Ambika, S.; Vanniarajan, A.; Meena, A.K.; Uppin, M.S.; Sundaram, C.; et al. Leber's Hereditary Optic Neuropathy-Specific Mutation m.11778G>A

Exists on Diverse Mitochondrial Haplogroups in India. *Invest. Ophthalmol. Vis. Sci.* **2017**, *58*, 3923–3930. [CrossRef]
196. Majander, A.; Bowman, R.; Poulton, J.; Antcliff, R.J.; Reddy, M.A.; Michaelides, M.; Webster, A.R.; Chinnery, P.F.; Votruba, M.; Moore, A.T.; et al. Childhood-onset Leber hereditary optic neuropathy. *Br. J. Ophthalmol.* **2017**, *101*, 1505–1509. [CrossRef]
197. Johns, D.R.; Heher, K.L.; Miller, N.R.; Smith, K.H. Leber's hereditary optic neuropathy. Clinical manifestations of the 14484 mutation. *Arch. Ophthalmol.* **1993**, *111*, 495–498. [CrossRef]
198. Sadun, A.A.; Chicani, C.F.; Ross-Cisneros, F.N.; Barboni, P.; Thoolen, M.; Shrader, W.D.; Kubis, K.; Carelli, V.; Miller, G. Effect of EPI-743 on the clinical course of the mitochondrial disease Leber hereditary optic neuropathy. *Arch. Neurol.* **2012**, *69*, 331–338. [CrossRef]
199. Klopstock, T.; Yu-Wai-Man, P.; Dimitriadis, K.; Rouleau, J.; Heck, S.; Bailie, M.; Atawan, A.; Chattopadhyay, S.; Schubert, M.; Garip, A.; et al. A randomized placebo-controlled trial of idebenone in Leber's hereditary optic neuropathy. *Brain* **2011**, *134*, 2677–2686. [CrossRef]
200. Mashima, Y.; Hiida, Y.; Oguchi, Y. Remission of Leber's hereditary optic neuropathy with idebenone. *Lancet* **1992**, *340*, 368–369. [CrossRef]
201. Mashima, Y.; Kigasawa, K.; Wakakura, M.; Oguchi, Y. Do idebenone and vitamin therapy shorten the time to achieve visual recovery in Leber hereditary optic neuropathy? *J. Neuroophthalmol.* **2000**, *20*, 166–170. [CrossRef]
202. Carelli, V.; La Morgia, C.; Valentino, M.L.; Rizzo, G.; Carbonelli, M.; De Negri, A.M.; Sadun, F.; Carta, A.; Guerriero, S.; Simonelli, F.; et al. Idebenone treatment in Leber's hereditary optic neuropathy. *Brain* **2011**, *134*, e188. [CrossRef]
203. Pisano, A.; Preziuso, C.; Iommarini, L.; Perli, E.; Grazioli, P.; Campese, A.F.; Maresca, A.; Montopoli, M.; Masuelli, L.; Sadun, A.A.; et al. Targeting estrogen receptor β as preventive therapeutic strategy for Leber's hereditary optic neuropathy. *Hum. Mol. Genet.* **2015**, *24*, 6921–6931. [CrossRef] [PubMed]
204. Guy, J.; Qi, X.; Koilkonda, R.D.; Arguello, T.; Chou, T.H.; Ruggeri, M.; Porciatti, V.; Lewin, A.S.; Hauswirth, W.W. Efficiency and safety of AAV-mediated gene delivery of the human ND4 complex I subunit in the mouse visual system. *Invest. Ophthalmol. Vis. Sci.* **2009**, *50*, 4205–4214. [CrossRef]
205. Ellouze, S.; Augustin, S.; Bouaita, A.; Bonnet, C.; Simonutti, M.; Forster, V.; Picaud, S.; Sahel, J.A.; Corral-Debrinski, M. Optimized allotopic expression of the human mitochondrial ND4 prevents blindness in a rat model of mitochondrial dysfunction. *Am. J. Hum. Genet.* **2008**, *83*, 373–387. [CrossRef] [PubMed]
206. Yang, S.; Ma, S.Q.; Wan, X.; He, H.; Pei, H.; Zhao, M.J.; Chen, C.; Wang, D.W.; Dong, X.Y.; Yuan, J.J.; et al. Long-term outcomes of gene therapy for the treatment of Leber's hereditary optic neuropathy. *EBioMedicine* **2016**, *10*, 258–268. [CrossRef] [PubMed]
207. Koilkonda, R.D.; Yu, H.; Chou, T.H.; Feuer, W.J.; Ruggeri, M.; Porciatti, V.; Tse, D.; Hauswirth, W.W.; Chiodo, V.; Boye, S.L.; et al. Safety and effects of the vector for the Leber hereditary optic neuropathy gene therapy clinical trial. *JAMA Ophthalmol.* **2014**, *132*, 409–420. [CrossRef] [PubMed]
208. Guy, J.; Feuer, W.J.; Davis, J.L.; Porciatti, V.; Gonzalez, P.J.; Koilkonda, R.D.; Yuan, H.; Hauswirth, W.W.; Lam, B.L. Gene Therapy for Leber Hereditary Optic Neuropathy: Low- and Medium-Dose Visual Results. *Ophthalmology* **2017**, *124*, 1621–1634. [CrossRef]
209. Farruggia, P.; Di Cataldo, A.; Pinto, R.M.; Palmisani, E.; Macaluso, A.; Valvo, L.L.; Cantarini, M.E.; Tornesello, A.; Corti, P.; Fioredda, F.; et al. Pearson Syndrome: A Retrospective Cohort Study from the Marrow Failure Study Group of A.I.E.O.P. (Associazione Italiana Emato-Oncologia Pediatrica). *JIMD Rep.* **2016**, *26*, 37–43. [PubMed]
210. Rotig, A.; Bourgeron, T.; Chretien, D.; Rustin, P.; Munnich, A. Spectrum of mitochondrial DNA rearrangements in the Pearson marrow-pancreas syndrome. *Hum. Mol. Genet.* **1995**, *4*, 1327–1330. [CrossRef] [PubMed]
211. Santorelli, F.M.; Barmada, M.A.; Pons, R.; Zhang, L.L.; DiMauro, S. Leigh-type neuropathology in Pearson syndrome associated with impaired ATP production and a novel mtDNA deletion. *Neurology* **1996**, *47*, 1320–1323. [CrossRef]
212. Lee, H.F.; Lee, H.J.; Chi, C.S.; Tsai, C.R.; Chang, T.K.; Wang, C.J. The neurological evolution of Pearson syndrome: Case report and literature review. *Eur. J. Paediatr. Neurol.* **2007**, *11*, 208–214. [CrossRef] [PubMed]
213. McShane, M.A.; Hammans, S.R.; Sweeney, M.; Holt, I.J.; Beattie, T.J.; Brett, E.M.; Harding, A.E. Pearson syndrome and mitochondrial encephalomyopathy in a patient with a deletion of mtDNA. *Am. J. Hum. Genet.* **1991**, *48*, 39–42. [PubMed]

214. Mancuso, M.; Orsucci, D.; Angelini, C.; Bertini, E.; Carelli, V.; Comi, G.P.; Donati, M.A.; Federico, A.; Minetti, C.; Moggio, M.; et al. Redefining phenotypes associated with mitochondrial DNA single deletion. *J. Neurol.* **2015**, *262*, 1301–1309. [CrossRef] [PubMed]
215. Crippa, B.L.; Leon, E.; Calhoun, A.; Lowichik, A.; Pasquali, M.; Longo, N. Biochemical abnormalities in Pearson syndrome. *Am. J. Med. Genet. A* **2015**, *167A*, 621–628. [CrossRef] [PubMed]
216. Yamashita, S.; Nishino, I.; Nonaka, I.; Goto, Y. Genotype and phenotype analyses in 136 patients with single large-scale mitochondrial DNA deletions. *J. Hum. Genet.* **2008**, *53*, 598–606. [CrossRef] [PubMed]
217. Puri, A.; Pradhan, A.; Chaudhary, G.; Singh, V.; Sethi, R.; Narain, V.S. Symptomatic complete heart block leading to a diagnosis of Kearns-Sayre syndrome. *Indian Heart J.* **2012**, *64*, 515–517. [CrossRef]
218. Gobu, P.; Karthikeyan, B.; Prasath, A.; Santhosh, S.; Balachander, J. Kearns Sayre Syndrome (KSS) - A Rare Cause For Cardiac Pacing. *Indian Pacing Electrophysiol. J.* **2011**, *10*, 547–550.
219. van Beynum, I.; Morava, E.; Taher, M.; Rodenburg, R.J.; Karteszi, J.; Toth, K.; Szabados, E. Cardiac arrest in kearns-sayre syndrome. *JIMD Rep.* **2012**, *2*, 7–10.
220. Kabunga, P.; Lau, A.K.; Phan, K.; Puranik, R.; Liang, C.; Davis, R.L.; Sue, C.M.; Sy, R.W. Systematic review of cardiac electrical disease in Kearns-Sayre syndrome and mitochondrial cytopathy. *Int. J. Cardiol.* **2015**, *181*, 303–310. [CrossRef]
221. Zeviani, M.; Servidei, S.; Gellera, C.; Bertini, E.; DiMauro, S.; DiDonato, S. An autosomal dominant disorder with multiple deletions of mitochondrial DNA starting at the D-loop region. *Nature* **1989**, *339*, 309–311. [CrossRef]
222. Kearns, T.P.; Sayre, G.P. Retinitis pigmentosa, external ophthalmophegia, and complete heart block: Unusual syndrome with histologic study in one of two cases. *AMA Arch. Ophthalmol.* **1958**, *60*, 280–289. [CrossRef]
223. Akaike, M.; Kawai, H.; Yokoi, K.; Kunishige, M.; Mine, H.; Nishida, Y.; Saito, S. Cardiac dysfunction in patients with chronic progressive external ophthalmoplegia. *Clin. Cardiol.* **1997**, *20*, 239–243. [CrossRef]
224. Aure, K.; Ogier de Baulny, H.; Laforet, P.; Jardel, C.; Eymard, B.; Lombes, A. Chronic progressive ophthalmoplegia with large-scale mtDNA rearrangement: Can we predict progression? *Brain* **2007**, *130*, 1516–1524. [CrossRef]
225. Poulton, J.; Deadman, M.E.; Ramacharan, S.; Gardiner, R.M. Germ-line deletions of mtDNA in mitochondrial myopathy. *Am. J. Hum. Genet.* **1991**, *48*, 649–653.
226. Bernes, S.M.; Bacino, C.; Prezant, T.R.; Pearson, M.A.; Wood, T.S.; Fournier, P.; Fischel-Ghodsian, N. Identical mitochondrial DNA deletion in mother with progressive external ophthalmoplegia and son with Pearson marrow-pancreas syndrome. *J. Pediatr.* **1993**, *123*, 598–602. [CrossRef]
227. Shanske, S.; Tang, Y.; Hirano, M.; Nishigaki, Y.; Tanji, K.; Bonilla, E.; Sue, C.; Krishna, S.; Carlo, J.R.; Willner, J.; et al. Identical mitochondrial DNA deletion in a woman with ocular myopathy and in her son with pearson syndrome. *Am. J. Hum. Genet.* **2002**, *71*, 679–683. [CrossRef]
228. Remes, A.M.; Majamaa-Voltti, K.; Karppa, M.; Moilanen, J.S.; Uimonen, S.; Helander, H.; Rusanen, H.; Salmela, P.I.; Sorri, M.; Hassinen, I.E.; et al. Prevalence of large-scale mitochondrial DNA deletions in an adult Finnish population. *Neurology* **2005**, *64*, 976–981. [CrossRef]
229. Chinnery, P.F.; Johnson, M.A.; Wardell, T.M.; Singh-Kler, R.; Hayes, C.; Brown, D.T.; Taylor, R.W.; Bindoff, L.A.; Turnbull, D.M. The epidemiology of pathogenic mitochondrial DNA mutations. *Ann. Neurol.* **2000**, *48*, 188–193. [CrossRef]
230. Kiechl, S.; Horvath, R.; Luoma, P.; Kiechl-Kohlendorfer, U.; Wallacher-Scholz, B.; Stucka, R.; Thaler, C.; Wanschitz, J.; Suomalainen, A.; Jaksch, M.; et al. Two families with autosomal dominant progressive external ophthalmoplegia. *J. Neurol. Neurosurg. Psychiatry* **2004**, *75*, 1125–1128. [CrossRef]
231. Harding, A.E. Friedreich's ataxia: A clinical and genetic study of 90 families with an analysis of early diagnostic criteria and intrafamilial clustering of clinical features. *Brain* **1981**, *104*, 589–620. [CrossRef]
232. Durr, A.; Cossee, M.; Agid, Y.; Campuzano, V.; Mignard, C.; Penet, C.; Mandel, J.L.; Brice, A.; Koenig, M. Clinical and genetic abnormalities in patients with Friedreich's ataxia. *N. Engl. J. Med.* **1996**, *335*, 1169–1175. [CrossRef] [PubMed]
233. Koeppen, A.H. Friedreich's ataxia: Pathology, pathogenesis, and molecular genetics. *J. Neurol. Sci.* **2011**, *303*, 1–12. [CrossRef] [PubMed]
234. Jensen, M.K.; Bundgaard, H. Cardiomyopathy in Friedreich ataxia: Exemplifying the challenges faced by cardiologists in the management of rare diseases. *Circulation* **2012**, *125*, 1591–1593. [CrossRef]

235. Tsou, A.Y.; Paulsen, E.K.; Lagedrost, S.J.; Perlman, S.L.; Mathews, K.D.; Wilmot, G.R.; Ravina, B.; Koeppen, A.H.; Lynch, D.R. Mortality in Friedreich ataxia. *J. Neurol. Sci.* **2011**, *307*, 46–49. [CrossRef]
236. Schultz, J.C.; Hilliard, A.A.; Cooper, L.T., Jr.; Rihal, C.S. Diagnosis and treatment of viral myocarditis. *Mayo Clin. Proc.* **2009**, *84*, 1001–1009. [CrossRef]
237. Drinkard, B.E.; Keyser, R.E.; Paul, S.M.; Arena, R.; Plehn, J.F.; Yanovski, J.A.; Di Prospero, N.A. Exercise capacity and idebenone intervention in children and adolescents with Friedreich ataxia. *Arch. Phys. Med. Rehabil.* **2010**, *91*, 1044–1050. [CrossRef]
238. Rustin, P.; von Kleist-Retzow, J.C.; Chantrel-Groussard, K.; Sidi, D.; Munnich, A.; Rotig, A. Effect of idebenone on cardiomyopathy in Friedreich's ataxia: A preliminary study. *Lancet* **1999**, *354*, 477–479. [CrossRef]
239. Mariotti, C.; Solari, A.; Torta, D.; Marano, L.; Fiorentini, C.; Di Donato, S. Idebenone treatment in Friedreich patients: One-year-long randomized placebo-controlled trial. *Neurology* **2003**, *60*, 1676–1679. [CrossRef]
240. Lesnefsky, E.J.; Chen, Q.; Tandler, B.; Hoppel, C.L. Mitochondrial Dysfunction and Myocardial Ischemia-Reperfusion: Implications for Novel Therapies. *Annu. Rev. Pharmacol. Toxicol.* **2017**, *57*, 535–565. [CrossRef]
241. Yellon, D.M.; Hausenloy, D.J. Myocardial reperfusion injury. *N. Engl. J. Med.* **2007**, *357*, 1121–1135. [CrossRef]
242. Murphy, E.; Steenbergen, C. Mechanisms underlying acute protection from cardiac ischemia-reperfusion injury. *Physiol. Rev.* **2008**, *88*, 581–609. [CrossRef] [PubMed]
243. Chen, Q.; Camara, A.K.; Stowe, D.F.; Hoppel, C.L.; Lesnefsky, E.J. Modulation of electron transport protects cardiac mitochondria and decreases myocardial injury during ischemia and reperfusion. *Am. J. Physiol. Cell Physiol.* **2007**, *292*, C137–C147. [CrossRef] [PubMed]
244. Lesnefsky, E.J.; Chen, Q.; Hoppel, C.L. Mitochondrial Metabolism in Aging Heart. *Circ. Res.* **2016**, *118*, 1593–1611. [CrossRef] [PubMed]
245. Halestrap, A.P.; Clarke, S.J.; Javadov, S.A. Mitochondrial permeability transition pore opening during myocardial reperfusion–a target for cardioprotection. *Cardiovasc. Res.* **2004**, *61*, 372–385. [CrossRef]
246. Kubli, D.A.; Gustafsson, A.B. Mitochondria and mitophagy: The yin and yang of cell death control. *Circ. Res.* **2012**, *111*, 1208–1221. [CrossRef]
247. Kung, G.; Konstantinidis, K.; Kitsis, R.N. Programmed necrosis, not apoptosis, in the heart. *Circ. Res.* **2011**, *108*, 1017–1036. [CrossRef]
248. Bugger, H.; Abel, E.D. Mitochondria in the diabetic heart. *Cardiovasc. Res.* **2010**, *88*, 229–240. [CrossRef] [PubMed]
249. Rubler, S.; Dlugash, J.; Yuceoglu, Y.Z.; Kumral, T.; Branwood, A.W.; Grishman, A. New type of cardiomyopathy associated with diabetic glomerulosclerosis. *Am. J. Cardiol.* **1972**, *30*, 595–602. [CrossRef]
250. Hamby, R.I.; Zoneraich, S.; Sherman, L. Diabetic cardiomyopathy. *JAMA* **1974**, *229*, 1749–1754. [CrossRef]
251. Regan, T.J.; Lyons, M.M.; Ahmed, S.S.; Levinson, G.E.; Oldewurtel, H.A.; Ahmad, M.R.; Haider, B. Evidence for cardiomyopathy in familial diabetes mellitus. *J. Clin. Investig.* **1977**, *60*, 884–899. [CrossRef] [PubMed]
252. Bell, D.S. Diabetic cardiomyopathy. *Diabetes Care* **2003**, *26*, 2949–2951. [CrossRef] [PubMed]
253. Duncan, J.G. Mitochondrial dysfunction in diabetic cardiomyopathy. *Biochim. Biophys. Acta* **2011**, *1813*, 1351–1359. [CrossRef] [PubMed]

© 2019 by the authors. Licensee MDPI, Basel, Switzerland. This article is an open access article distributed under the terms and conditions of the Creative Commons Attribution (CC BY) license (http://creativecommons.org/licenses/by/4.0/).

Review

From Powerhouse to Perpetrator—Mitochondria in Health and Disease

Nima B. Fakouri [1], Thomas Lau Hansen [2], Claus Desler [2], Sharath Anugula [2] and Lene Juel Rasmussen [2,*]

[1] Laboratory of Molecular Gerontology, National Institute on Aging, National Institutes of Health, Baltimore, MD 21224, USA; nima.borhanfakouri@nih.gov
[2] Center for Healthy Aging, Department of Cellular and Molecular Medicine, University of Copenhagen, 2200 Copenhagen, Denmark; tlhansen@sund.ku.dk (T.L.H.); cdesler@sund.ku.dk (C.D.); sharath@sund.ku.dk (S.A.)
* Correspondence: lenera@sund.ku.dk

Received: 2 January 2019; Accepted: 5 March 2019; Published: 11 May 2019

Abstract: In this review we discuss the interaction between metabolic stress, mitochondrial dysfunction, and genomic instability. Unrepaired DNA damage in the nucleus resulting from excess accumulation of DNA damages and stalled replication can initiate cellular signaling responses that negatively affect metabolism and mitochondrial function. On the other hand, mitochondrial pathologies can also lead to stress in the nucleus, and cause sensitivity to DNA-damaging agents. These are examples of how hallmarks of cancer and aging are connected and influenced by each other to protect humans from disease.

Keywords: mitochondria; cancer; nucleotide metabolism; DNA damage; NAD$^+$

1. Introduction

It has been almost two decades since Hanahan and Weinberg for the first time classified the hallmarks of cancer [1]. Ten years later, they updated that list and introduced genomic instability and dysregulation of cellular energetics and mitochondrial function as emerging hallmarks [2]. Recently, both genomic instability and mitochondrial dysfunction are considered as two of the key hallmarks of aging [3]. Both of these have been implicated in several pathologies, as reviewed in References [4–6]. There are other hallmarks that are common between cancer and aging, such as epigenetic changes and altered cellular communication. Moreover, other hallmarks are in opposition to each other in cancer and aging. These include the dysregulation of apoptosis and senescence, which are stimulated in aging cells and suppressed in cancer cells [2,3]. Whether we can consider cancer as the disease of aging is a topic that is beyond the scope of this review, but age is the largest risk factor in the development of cancer [7].

One of the key questions that remain to be answered is, how are these hallmarks are connected and influenced by each other [3]? As these pathologic cellular changes occur gradually, understanding the connection between them would help to develop more effective therapeutic strategies to treat cancer or rather prevent it.

It has long been known that byproducts of cellular metabolism such as reactive oxygen and nitrogen species (ROS and RNS) can damage cellular components and macromolecules including DNA [8,9]. Damage to DNA can have severe effects on cells by blocking replication, transcription, generation of DNA double and single strand breaks, as well as chromosome rearrangements [10,11]. Activation of the DNA damage response (DDR) following DNA damage is an energy-demanding process [12], and can deplete cells of substrates such as NAD$^+$ and ATP, which can in turn lead to additional metabolic stress and mitochondrial dysfunction (Figure 1) [13–15].

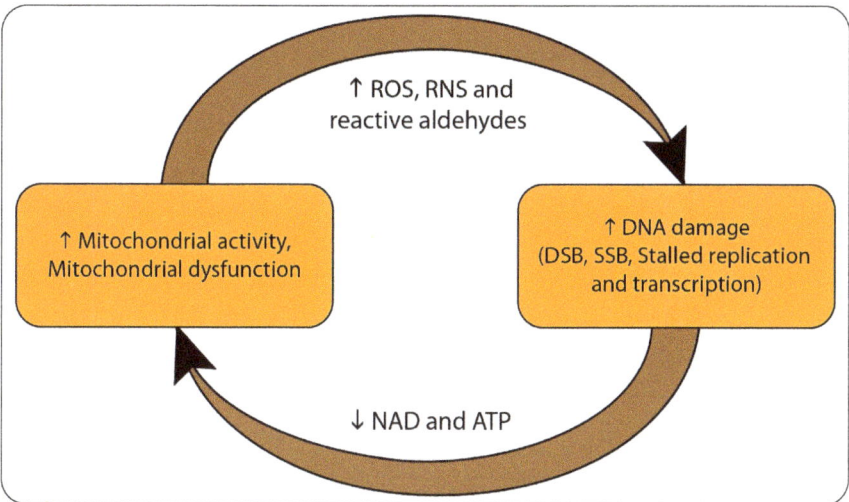

Figure 1. Illustration of the mitochondrial-nuclear interactions in aging or cancer. Abnormal metabolism and/or metabolic defects lead to metabolic stress and mitochondrial dysfunction. This is followed by the increased generation of reactive oxygen and nitrogen species (ROS and RNS) as well as reactive aldehydes. These reactive species can react and damage macromolecules such as proteins and DNA. Damage to DNA causes genomic instability via stalled replication and transcription, and the generation of double- and single-strand breaks (DSBs and SSBs, respectively) within the genome. Increased activities of the DNA damage response (DDR) deplete cells of key cellular substrates and cofactors, mainly ATP and NAD$^+$. This generates a positive feedback that enhances metabolic stress and mitochondrial dysfunction.

In this review, we aim to discuss the interaction between metabolic stress and mitochondrial dysfunction with genomic instability. Stress in the nucleus, such as the accumulation of DNA damage and stalled replication, negatively affects metabolism and mitochondrial function [13–17], while mitochondrial pathologies lead to stress in the nucleus [18] and cause sensitivity to DNA-damaging agents [19], as reviewed by Desler et al. in 2012 [20].

First, we explain how the accumulation of DNA damages and the activation of the DDR leads to mitochondrial dysfunction. Next, we explain how the dysregulation of mitochondrial function and metabolism contributes to the epigenetic changes, imbalanced dNTP pools, and genomic instability.

2. From DNA Damage to Mitochondrial Dysfunction

Activation of the main components of the DDR, including poly (ADP-ribose) polymerase (PARP) enzymes (mainly PARP1 and PARP2) as well as ataxia telangiectasia mutated (ATM) [21], DNA-dependent protein kinase (DNA-PK) [22,23] and P53 [24–27], is able to influence mitochondrial function and cellular metabolism. Chronic activation of PARP1 negatively affects cellular physiology and mitochondrial function [28]. Activation of ATM, DNA-PK, and P53 can influence mitochondrial and cellular metabolism to promote either cell survival or death (Figure 2). Here, we briefly describe how each of these enzymes are able to influence mitochondrial function.

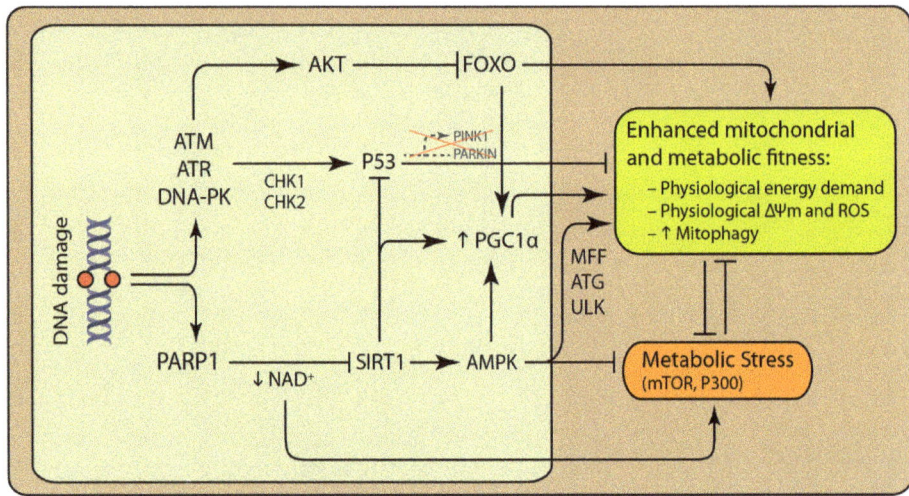

Figure 2. Core DNA damage response proteins are able to influence mitochondrial activity and quality control via multiple pathways. Abnormal DNA structure and certain genomic lesions, such as strand breaks, activate poly (ADP-ribose) polymerase (PARP) enzymes, mainly PARP1. Chronic activation of PARP1 can deplete the cell of NAD$^+$, which is a rate-limiting substrate for SIRT1. SIRT1 together with AMP-activated kinase (AMPK) can enhance metabolism and mitochondrial function by enhancing mitochondrial biogenesis and mitophagy. Activation of ataxia telangiectasia mutated (ATM), Rad3-related (ATR) and DNA-dependent protein kinase (DNA-PK) following DNA damage can promote the activation AKT and P53. Activation of AKT rewires cellular metabolism by inhibiting Forkhead box (FOXO) enzymes. Deacetylation of P53 by SIRT1 targets P53 for degradation. Decrease in SIRT1 activity stabilizes P53. P53 decreases mitophagy via inhibition of PTEN-induced kinase 1 (PINK1) and PARKIN transcription. SIRT1 promotes AMPK activity indirectly. Decrease in SIRT1 activity is followed by the decrease in activated AMPK.

3. PARP Modulates Mitochondrial Function and Cellular Metabolism

PARPs are a group of enzymes (16 in mice and 17 in humans) that are the main constituent of the cellular stress response [29]. PARPs cleave NAD$^+$ to nicotinamide (NAM) and ADP-ribose (ADPR), and the ADPR is subsequently transferred to certain amino acids within the target protein. The attachment of poly ADP-ribose (PAR) to the target proteins is referred to as PARylation, and it can affect protein–protein and protein–DNA interaction as well as protein localization [30]. PAR has a short half-life and is degraded almost directly after its formation by the activity of the PAR-degrading enzyme, poly(ADP-ribose) glycohydrolase (PARG) [31]. PARylation modulates several key cellular processes such as chromatin structure, transcription, translation, cell cycle, DNA repair, mitochondrial homeostasis, apoptosis, and metabolism [29,32]. PARP1 is activated by several mechanisms, including mono(ADP-ribosyl)ation, phosphorylation, and acetylation [29]. PARP1 possesses a DNA-binding domain that recognizes abnormal DNA structures such as gapped DNA, single- and double strand breaks, cruciform structures, and nucleosome linker DNA [33,34]. In the initial steps of the repair, PARP1 PARylates histones and facilitates the chromatin relaxation that provides more space for the recruitment of DNA repair proteins. Subsequent PAR generation recruits the DNA repair proteins via their PAR-binding domains [35,36]. Under mild genotoxic stress, PARP activation results in repair and survival. However, in response to DNA damage, PARP hyper-activation results in decrease in NAD$^+$ and ATP levels, mitochondrial dysfunction, and eventually cell death [32,35,37].

4. DNA Damage can Activate Both Pro-Survival and Pro-Death Pathways That Involve the Mitochondria

DNA damage and DDR can activate pathways that promote cell survival or death depending on the extent and type of DNA damages [38]. In addition to PARP1, other immediate sensors of DDR are enzymes belonging to the superfamily of phosphatidylinositol 3-kinase-related kinases (PIKKs), including ATM, ataxia telangiectasia and Rad3-related (ATR), and DNA-PK. Activated ATR and ATM, in turn, activate P53 through CHK1 and CHK2, respectively. Apart from preventing cell cycle progression and the recruitment of DNA repair proteins, these enzymes can modulate mitochondrial function and survival [23,39]. ATM, ATR, and DNA-PK are able to promote survival via the direct phosphorylation of AKT (also known as protein kinase B, PKB) independently of growth factor signaling [22,40–43]. However, the mechanism of this interaction is not fully understood [44]. This is particularly interesting, as in many cancer cells, the AKT is activated independently of growth factors [45]. Activated AKT stimulates glucose uptake and ATP production through glycolysis, one of the main hallmarks of cancer, also known as the Warburg effect [46]. In addition, activated AKT inhibits Forkhead box (FOXO) transcription factors [47,48]. FOXO proteins regulate the expression of key genes that are involved in mitochondrial biogenesis and homeostasis, such as peroxisome proliferator-activated receptor gamma co-activator 1alpha (PGC1a) and PTEN-induced kinase 1 (PINK1). Decrease in FOXO activity results in the decrease of mitochondrial biogenesis, mitophagy, autophagy, and lipolysis [49].

5. Mito-Nuclear Signaling in Aging and Cancer

For a long time, it was believed that mitochondria are regulated from the nucleus by the nuclear genome, and changes in mitochondria follow changes in the nucleus [50,51]. However, during recent years, accumulative evidence suggest that mitochondria and mitochondrial metabolites can influence nuclear processes and gene expression in response to various stimuli and environmental cues [52–54]. Mitochondrially generated ROS and intermediate metabolites are essential for several processes, including proliferation, epigenetic modifications, and post-translational modifications [54,55]. In addition, mitochondria contribute to genomic stability by replenishing dNTP pools for replication and repair of the genome (Figure 3) [56,57]. In this section we will discuss nuclear processes that are dependent on mitochondrial function and intermediate metabolites.

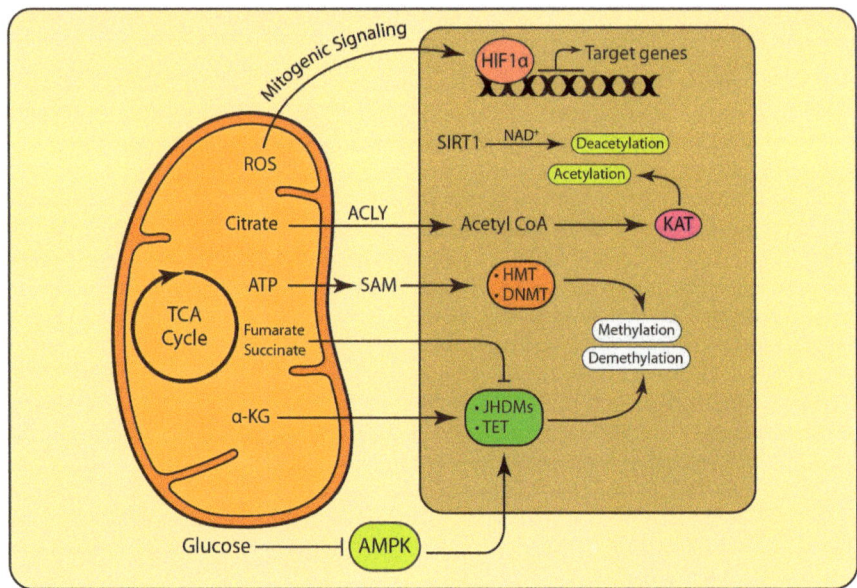

Figure 3. Mitochondrial function and metabolites influence nuclear processes. Mitochondrial ROS and secondary metabolites act as signaling molecules and cofactors that regulate fundamental nuclear processes. Mitochondrial ROS that are released into the cytosol stabilize hypoxia-inducible factor 1- α (HIF1α) and activate the transcription of genes that are involved in proliferation. Citrate generated through the TCA cycle is released into the cytosol and in the nucleus. It is further converted into acetyl-coenzyme A (acetyl CoA) that is used for the acetylation of target proteins. Through a reverse reaction, SIRT1 uses NAD$^+$ to deacetylate the target proteins. S-adenosylmethionine (SAM) is a methyl donor that is generated from methionine and ATP in the cytosol. Demethylases such as Jumonji C (JMJC) family members and the ten-eleven translocation (TET) methylcytosine hydroxylases use α-ketoglutarate (α-KG) as cofactor to remove methyl groups from proteins and DNA. AMPK stimulates the activity of TET enzymes, and thus the inhibition of AMPK by glucose impairs the function TET enzymes. The accumulation of fumarate and succinate due to impaired fumarate hydratase (FH) and succinate dehydrogenase (SDH) can inhibit α-KG-dependent demethylases and even cause defects in homologous recombination (HR) DNA repair.

6. Mitochondrial ROS Are Involved in Signaling and Determine Cell Fate

The mitochondrial electron transport chain (ETC) generates reactive oxygen species (ROS) as the byproduct of oxidative phosphorylation (OXPHOS) from different complexes, though mainly complexes I, II, and III. While complexes I and II exclusively create $O_2\cdot$ in the mitochondrial matrix, complex III produces $O_2\cdot$ in both the matrix and intermembrane space [58]. However, the ROS that are released outside of the matrix are converted into H_2O_2 by cytosolic superoxide dismutase 1 (SOD1) and participate in mitochondrial signaling through reversible cysteine oxidation [59]. Mitochondrially produced ROS can serve as second-messenger molecules. Mitochondrial ROS (mtROS) are required for the stabilization of HIFα (hypoxia-inducible factor 1-α) and the activation of downstream pathways that promote proliferation [60].

It is possible that the type of cell and energy demand determine the effect of ROS and mtDNA mutation over the cell. Cells that mostly rely on glycolysis will probably not be affected by mutation in mtDNA under physiological conditions. During the exposure to stress and stimuli, however, these cells might not be able to trigger an adaptive response to an increase in demand for ATP and NAD$^+$ [61]. During stress and increased energy demand, cells respond by boosting cellular respiration and mitochondrial activity to provide the cells with ATP and NAD$^+$ [12,62]. This is accompanied by increased mitochondrial membrane potential and ROS above the physiological level [61,63]. Increased ROS cause damage to macromolecules such as DNA, proteins, and lipids, which can cause cell death or malignancy, as reviewed by Sies et al. in 2017 [64]. The inability to enhance ATP production in response to stress and increased energy demand probably contributes to cellular deterioration during aging [65].

In contrast to increased ROS generation, a decrease in ROS in metabolically active or proliferative tissues interferes with cellular metabolism or proliferation that can promote senescence, as reviewed by Diebold and Chandel in 2016 [66].

Hematopoietic stem cells (HSCs) represent an excellent example for both situations [67]. HSCs are mainly quiescent, and they rely on glycolysis for ATP production [68]. While low levels of ROS prevent the proliferation of HSCs and maintain their quiescent state, stress and increased energy demand promote a shift toward ATP production by mitochondria and OXPHOS. This is accompanied by increased ROS and proliferation [67]. Chronic stress followed by enhanced mitochondrial activity and ROS generation leads to the depletion of HSCs, which is one of the hallmarks of aging [69–71].

7. Mitochondria Influence Post-Translational Modifications (PTMs) and Epigenetic Marks

Reversible acetylation and methylation are two frequently employed post-translational modifications that regulate a variety of protein functions, protein stability, gene expression [72], as well as DNA repair [73]. Epigenetic changes are also one of the main hallmarks of both aging and cancer [2,3]. The modifications include alterations in DNA methylation, modifications of histones, and chromatin remodeling. While DNA methylations show similar patterns in aging and cancer, the histone modifications show distinct patterns, as reviewed by Zane et al. in 2014 [74].

The multiple enzymatic systems assuring the generation and maintenance of epigenetic patterns include DNA methyltransferases, histone acetylases, deacetylases, methylases, and demethylases, as well as protein complexes implicated in chromatin remodeling. Most PTMs, such as phosphorylation, acetylation, methylation, and O-linked N-acetylglucosamine modification (O-GlcNAcylation), require metabolites as substrates [75]. If not all, the majority of substrates required for PTMs are intermediate metabolites generated by mitochondria [54]. Chromatin modifiers use metabolic intermediates as cofactors or substrates, but are also regulated by their availability. These metabolites include NAD$^+$ for deacetylation, acetyl-CoA for histone acetylation, S-adenosylmethionine (SAM) for histone as well as DNA methylation, and α-ketoglutarate (α-KG) for demethylation [54].

Lysine acetyltransferases (KATs) add acetyl groups to proteins while lysine deacetylases (KDACs) remove acetyl groups from proteins [76,77]. KATs such as GCN5, CBP/p300, and MYST use acetyl-coenzyme A (acetyl CoA) as an acetyl group donor for protein acetylation [78]. KDACs are classified into two groups with different catalytic mechanisms: Zn^{2+}-dependent histone deacetylases (HDAC1-11) and NAD$^+$-dependent deacetylases (SIRT1-7) [76,77]. Acetyl CoA is generated in mitochondria and converted to citrate through the TCA cycle. Citrate can then be exported from mitochondria. In the cytoplasm and nucleus, citrate is converted back into acetyl CoA via the function of ATP-citrate lyase (ACLY). Citrate is the major source of acetyl CoA in the cytoplasm and the nucleus. Depletion of mtDNA and the subsequent decrease in NAD negatively affect the TCA cycle and lead to a decrease in histone acetylation. This can be rescued by the restoration of electron flow and TCA cycle [60,79]. The degree of acetylation directly correlates with the availability of cofactors such as acetyl CoA and NAD$^+$. While increased nuclear acetyl CoA promotes increased acetylation and the formation of euchromatin, an increase in NAD$^+$ promotes deacetylation and the formation of heterochromatin, resulting in a decrease of gene expression [79]. MYC and AKT stimulate nutrient uptake and promote acetyl CoA production via ACLY. AKT can directly phosphorylate and activate ACLY to maintain the acetyl CoA levels, regardless of glucose concentrations [80].

Another key chromatin modification that is strongly interconnected with metabolism is methylation [54,81]. Methylation is regulated by S-adenosylmethionine (SAM) abundance, whereby

SAM serves as a universal methyl donor, synthesized from methionine and ATP by methionine adenosyltransferases (MATs) [54,81]. Histone and DNA methylation is removed by demethylases such as Jumonji C (JMJC) family members and the ten-eleven translocation (TET) methylcytosine hydroxylases, which use a dioxygenation reaction that requires Fe^{2+}, O_2, and α-ketoglutarate (α-KG) as cofactors [82]. α-KG is generated in the TCA cycle via the catabolism of glucose and glutamine. The function of lysine-specific histone demethylase 1A (LSD1; KDM1A) and LSD2 (KDM1B), which catalyze an amine oxidation reaction, is dependent on flavin adenine dinucleotide (FAD) [83,84].

The dysregulation of the TCA cycle and cellular metabolism affect PTMs, epigenetic changes, and choice of DNA repair pathway. Accumulation of succinate or fumarate, which occurs respectively in tumors deficient for succinate dehydrogenase (SDH) or fumarate hydratase (FH), similarly inhibits α-KG-dependent enzymes, leading to defects in homologous recombination (HR) DNA repair [53]. Changes in nutrient availability can directly affect chromatin modifications. The tumor suppressor ten-eleven translocation (TET) protein family of dioxygenases (TET1, TET2, and TET3) converts a 5mC DNA methylation to a hydroxymethylation, 5hmC [85]. AMP-activated kinase (AMPK) phosphorylates TET2 and stabilizes this tumor suppressor protein. However, increase in blood glucose levels impairs AMPK activity, which leads to destabilization of TET2 and the subsequent dysregulation of 5mCs and 5hmCs [86].

8. Regulation of dNTP Pools

In humans, nucleotide levels are maintained by the nucleotide salvage and/or de novo synthesis of ribo- and deoxyribonucleotide triphosphates (rNTPs and dNTPs).

It is generally accepted that the levels and especially the relative balance of the cytosolic dNTP pools have great influence on the replication, repair, and stability of the nuclear genome [87–92]. The efficiency and fidelity of most, if not all, polymerases and many repair enzymes are affected by the level of substrate nucleotides [87]. Too low a concentration of a nucleotide results in poor incorporation frequency, and a level which is too high risks the misincorporation of the high-concentration nucleotide [93–95]. Therefore, it is of no surprise that the process of synthesizing dNTPs in the right concentrations is governed by a generous amount of regulation, feedback loops, and redundancy [96,97].

Mitochondrial dysfunction, due to mutations in the mitochondrial genome or a decrease in the mitochondrial DNA (mtDNA) copy number, is associated with a poor prognosis of many types of cancer [98–102], also reviewed by Chatterjee et al. in 2006 [103]. The accumulation of mutations in mtDNA or impeded ETC have even been associated with tumor aggressiveness [99,101,104–106]. We have previously shown a relationship between mitochondrial respiration and the regulation of cytosolic dNTP pools, and demonstrated a co-occurring decrease of chromosomal stability [51].

The de novo synthesis of nucleotides is split up into purine and pyrimidine synthesis, which go through two distinct pathways.

Despite having separate synthesis pathways, both purine and pyrimidine-based ribonucleotides occupy a central role in cellular metabolism. In addition to being basal components of DNA and RNA, they function as phosphate donors in the transport of cellular energy and participate in enzymatic reactions as well as intracellular and extracellular signaling [107].

Both the salvage and the de novo synthesis pathways utilize an activated sugar intermediate: 5-phosphoribosyl-1-pyrophosphate (PRPP). PRPP is generated by the action of PRPP synthetase and is utilized in both purine and pyrimidine synthesis [108]. The purine nucleotides are synthesized from PRPP, through inosine 5-monophosphate (IMP), and further into AMP and GMP, through two separate but allosterically regulated pathways. Pyrimidines, on the other hand, originate from the precursor pyrimidine nucleotide, UMP, which is used to synthesize all the cellular ribosyl and deoxyribosyl pyrimidines including UTP, CTP, dUMP, and dTTP [108]. A key catalytic enzyme in this process is the dihydroorotate dehydrogenase (DHODH) [109], located in the inner mitochondrial membrane and functionally codependent with the OXPHOS [110]. The oxidation of dihydroorotate by DHODH, to form orotate, is a bottleneck reaction of the de novo synthesis of pyrimidines and is electrochemically coupled with the reduction of ubiquinone to ubiquinol [56,111–113]. Mitochondrial respiration can modulate the nucleotide synthesis at two separate steps. A decrease of the mitochondrial respiration is therefore linked to an inhibition of the DHODH enzyme, which in turn modulates the synthesis of pyrimidines [51,114]. Furthermore, the activity of the RNR complex is regulated by the binding of ATP to its active site and inhibited by dATP. By affecting the levels of cytosolic ATP, the mitochondria can influence the activity of RNR, and hence the levels of dNTPs. Furthermore, RNR is allosterically regulated by the relative levels of individual NTPs and dNTPs, and so the de novo synthesis of dNTPs and their relative balance is highly dependent on the mitochondrial supply of pyrimidines and ATP [114–116].

The cytosolic pool of dNTPs, which supplies the replication of the nuclear genome, is cell cycle-regulated. Synthesis is initiated at the beginning of the S-phase, and stopped upon reaching G2. In the de novo pathway, this phase-dependent synthesis of dNTPs is mediated through the RNR—specifically, through the cell cycle regulation of expression and degradation of the RNR subunit RNR-R2 [117]. In the salvage pathway, the phase-dependent synthesis is controlled by the translational regulation of constituents of the nucleotide salvage pathway [118]. Outside of S-phase, a dNTP hydrolase called SAMHD1 (SAM and HD domain-containing deoxynucleoside triphosphate triphosphohydrolase 1) depletes the dNTP pools by its hydrolase activity to block viral replication, amongst other things [119]. In response to DNA damage, however, dNTP levels can increase by up to 4-fold [96]. In this case, both the RNR and SAMHD1 are then recruited to the site of damage to tightly regulate the amount of dNTPs supplied to the DNA repair machinery [96,120].

Not only replication of the nuclear genome requires a balanced dNTP pool. Unlike the replication of the nuclear genome, the replication of mitochondrial DNA is not regulated by the cell cycle, but is carried out continuously in both mitotic and post-mitotic cells and tissue. Imbalance of the mitochondrial dNTP pools affects the replication of mtDNA, resulting in the accumulation of point mutations and deletions [121,122].

The dNTP pool of the mitochondrial compartment is somewhat separate from the much larger pool supplying the nuclear genome [123,124]. However, cytosolic de novo synthesis of dNTP is essential for mtDNA maintenance, even in post-mitotic cells and tissue [125–127].

P53R2 is a protein that substitutes RNR-R2 in post-mitotic cells in response to DNA damage. P53R2 is transcribed by P53 and results, when forming the complex, in the activation of RNR and the de novo synthesis of dNTPs intended as substrates for DNA repair mechanisms [128]. Mutations of the RRM2B gene encoding P53R2 have been shown to induce mtDNA replication and repair deficiency in post-mitotic, but not dividing, human fibroblasts [125]. In humans, mutations in the gene encoding the RRM2B subunit have been correlated with severe mtDNA depletion of muscle tissue, and $RRM2b^{-/-}$ mice further display a severe decrease of mtDNA content in liver, kidney, and muscle [126].

It is important to realize that balanced dNTP pools for mtDNA maintenance do not need to be of mitochondrial origin. Imbalances of the nuclear dNTP pools resulting from genetic predisposition, age, or even just diet [129] have the potential to start a vicious cycle, whereby failure to maintain mtDNA integrity results in the decreased synthesis of pyrimidines and further dNTP pool imbalance.

Mitochondrial dysfunction is therefore not only a risk to the mitotic cell, but also fully differentiated post-mitotic cells [130], and is involved in the etiology of a wide array of pathologies, including cancer and Alzheimer's disease [131–133].

9. Conclusions

According to recent advancements, hallmarks of cancer include genomic instability, dysregulation of cellular energetics, and mitochondrial dysfunction, which also are common pathways important for cellular aging. Mitochondrial dysfunction is associated with a poor prognosis of many types of cancer, which could very well be linked to an imbalance of the cytosolic dNTP pools, as both of these conditions are related to one of the hallmarks of cancer—chromosomal instability. A better understanding of these pathological cellular processes would advance the development of therapeutic modalities in the prevention of cancer and at the same time help the understanding of biological aging.

Author Contributions: N.B.F., T.L.H., C.D., S.A., L.J.R. wrote the review.

Funding: This research was funded by Nordea-fonden and Olav Thon Foundation.

Conflicts of Interest: The authors declare no conflict of interest.

References

1. Hanahan, D.; Weinberg, R.A. The hallmarks of cancer. *Cell* **2000**, *100*, 57–70. [CrossRef]
2. Hanahan, D.; Weinberg, R.A. Hallmarks of Cancer: The Next Generation. *Cell* **2011**, *144*, 646–674. [CrossRef] [PubMed]
3. López-Otín, C.; Blasco, M.A.; Partridge, L.; Serrano, M.; Kroemer, G. The hallmarks of aging. *Cell* **2013**, *153*, 1194–1217. [CrossRef]
4. Tubbs, A.; Nussenzweig, A. Endogenous DNA Damage as a Source of Genomic Instability in Cancer. *Cell* **2017**, *168*, 644–656. [CrossRef]
5. Sharma, P.; Sampath, H. Mitochondrial DNA Integrity: Role in Health and Disease. *Cells* **2019**, *8*, 100. [CrossRef] [PubMed]
6. Fan, P.; Xie, X.-H.; Chen, C.-H.; Peng, X.; Zhang, P.; Yang, C.; Wang, Y.-T. Molecular Regulation Mechanisms and Interactions Between Reactive Oxygen Species and Mitophagy. *DNA Cell Biol.* **2019**, *38*, 10–22. [CrossRef] [PubMed]
7. Aunan, J.R.; Cho, W.C.; Søreide, K. The Biology of Aging and Cancer: A Brief Overview of Shared and Divergent Molecular Hallmarks. *Aging Dis.* **2017**, *8*, 628. [CrossRef] [PubMed]
8. Burcham, P.C. Internal hazards: Baseline DNA damage by endogenous products of normal metabolism. *Mutat. Res./Genet. Toxicol. Environ. Mutagenesis* **1999**, *443*, 11–36. [CrossRef]
9. De Bont, R.; van Larebeke, N. Endogenous DNA damage in humans: A review of quantitative data. *Mutagenesis* **2004**, *19*, 169–185. [CrossRef]
10. Zeman, M.K.; Cimprich, K.A. Causes and consequences of replication stress. *Nat. Cell Biol.* **2014**, *16*, 2–9. [CrossRef] [PubMed]
11. White, R.R.; Vijg, J. Do DNA Double-Strand Breaks Drive Aging? *Mol. Cell* **2016**, *63*, 729–738. [CrossRef] [PubMed]
12. Qin, L.; Fan, M.; Candas, D.; Jiang, G.; Papadopoulos, S.; Tian, L.; Woloschak, G.; Grdina, D.J.; Li, J.J. CDK1 Enhances Mitochondrial Bioenergetics for Radiation-Induced DNA Repair. *Cell Rep.* **2015**, *13*, 2056–2063. [CrossRef]
13. Scheibye-Knudsen, M.; Mitchell, S.J.; Fang, E.F.; Iyama, T.; Ward, T.; Wang, J.; Dunn, C.A.; Singh, N.; Veith, S.; Hasan-Olive, M.M.; et al. A high-fat diet and NAD(+) activate Sirt1 to rescue premature aging in cockayne syndrome. *Cell Metab.* **2014**, *20*, 840–855. [CrossRef]
14. Fang, E.F.; Kassahun, H.; Croteau, D.L.; Scheibye-Knudsen, M.; Marosi, K.; Lu, H.; Shamanna, R.A.; Kalyanasundaram, S.; Bollineni, R.C.; Wilson, M.A.; et al. NAD(+) Replenishment Improves Lifespan and Healthspan in Ataxia Telangiectasia Models via Mitophagy and DNA Repair. *Cell Metab.* **2016**, *24*, 566–581. [CrossRef]
15. Fakouri, N.B.; Durhuus, J.A.; Regnell, C.E.; Angleys, M.; Desler, C.; Olive, M.H.; Martín-Pardillos, A.; Tsaalbi-Shtylik, A.; Thomsen, K.; Lauritzen, M.; et al. Rev1 contributes to proper mitochondrial function via the PARP-NAD+-SIRT1-PGC1α axis. *Sci. Rep.* **2017**, *7*, 12480. [CrossRef]
16. Rivera-Torres, J.; Acín-Perez, R.; Cabezas-Sánchez, P.; Osorio, F.G.; Gonzalez-Gómez, C.; Megias, D.; Cámara, C.; López-Otín, C.; Enríquez, J.A.; Luque-García, J.L.; et al. Identification of mitochondrial dysfunction in Hutchinson–Gilford progeria syndrome through use of stable isotope labeling with amino acids in cell culture. *J. Proteom.* **2013**, *91*, 466–477. [CrossRef] [PubMed]
17. Fang, E.F.; Scheibye-Knudsen, M.; Brace, L.E.; Kassahun, H.; SenGupta, T.; Nilsen, H.; Mitchell, J.R.; Croteau, D.L.; Bohr, V.A. Defective mitophagy in XPA via PARP-1 hyperactivation and NAD(+)/SIRT1 reduction. *Cell* **2014**, *157*, 882–896. [CrossRef]

18. Qian, W.; Choi, S.; Gibson, G.A.; Watkins, S.C.; Bakkenist, C.J.; Van Houten, B. Mitochondrial hyperfusion induced by loss of the fission protein Drp1 causes ATM-dependent G2/M arrest and aneuploidy through DNA replication stress. *J. Cell Sci.* **2012**, *125*, 5745–5757. [CrossRef] [PubMed]
19. Temelie, M.; Savu, D.I.; Moisoi, N. Intracellular and Intercellular Signalling Mechanisms following DNA Damage Are Modulated By PINK1. *Oxid. Med. Cell. Longev.* **2018**, *2018*, 1–15. [CrossRef]
20. Desler, C.; Hansen, T.L.; Frederiksen, J.B.; Marcker, M.L.; Singh, K.K.; Juel Rasmussen, L. Is There a Link between Mitochondrial Reserve Respiratory Capacity and Aging? *J. Aging Res.* **2012**, *2012*, 192503. [CrossRef]
21. Cosentino, C.; Grieco, D.; Costanzo, V. ATM activates the pentose phosphate pathway promoting anti-oxidant defence and DNA repair. *EMBO J.* **2011**, *30*, 546–555. [CrossRef] [PubMed]
22. Bozulic, L.; Surucu, B.; Hynx, D.; Hemmings, B.A. PKBα/Akt1 Acts Downstream of DNA-PK in the DNA Double-Strand Break Response and Promotes Survival. *Mol. Cell* **2008**, *30*, 203–213. [CrossRef]
23. Park, S.-J.; Gavrilova, O.; Brown, A.L.; Soto, J.E.; Bremner, S.; Kim, J.; Xu, X.; Yang, S.; Um, J.-H.; Koch, L.G.; et al. DNA-PK Promotes the Mitochondrial, Metabolic, and Physical Decline that Occurs During Aging. *Cell Metab.* **2017**, *25*, 1135–1146.e7. [CrossRef] [PubMed]
24. Matoba, S.; Kang, J.-G.; Patino, W.D.; Wragg, A.; Boehm, M.; Gavrilova, O.; Hurley, P.J.; Bunz, F.; Hwang, P.M. p53 regulates mitochondrial respiration. *Science* **2006**, *312*, 1650–1653. [CrossRef]
25. Bensaad, K.; Vousden, K.H. p53: New roles in metabolism. *Trends Cell Biol.* **2007**, *17*, 286–291. [CrossRef]
26. Hoshino, A.; Mita, Y.; Okawa, Y.; Ariyoshi, M.; Iwai-Kanai, E.; Ueyama, T.; Ikeda, K.; Ogata, T.; Matoba, S. Cytosolic p53 inhibits Parkin-mediated mitophagy and promotes mitochondrial dysfunction in the mouse heart. *Nat. Commun.* **2013**, *4*, 2308. [CrossRef] [PubMed]
27. Wang, D.B.; Kinoshita, C.; Kinoshita, Y.; Morrison, R.S. p53 and mitochondrial function in neurons. *Biochim. Biophys. Acta Mol. Basis Dis.* **2014**, *1842*, 1186–1197. [CrossRef]
28. Bai, P.; Nagy, L.; Fodor, T.; Liaudet, L.; Pacher, P. Poly(ADP-ribose) polymerases as modulators of mitochondrial activity. *Trends Endocrinol. Metab.* **2015**, *26*, 75–83. [CrossRef]
29. Luo, X.; Kraus, W.L. On PAR with PARP: Cellular stress signaling through poly(ADP-ribose) and PARP-1. *Genes Dev.* **2012**, *26*, 417–432. [CrossRef]
30. Krietsch, J.; Rouleau, M.; Pic, É.; Ethier, C.; Dawson, T.M.; Dawson, V.L.; Masson, J.-Y.; Poirier, G.G.; Gagné, J.-P. Reprogramming cellular events by poly(ADP-ribose)-binding proteins. *Mol. Asp. Med.* **2013**, *34*, 1066–1087. [CrossRef]
31. Davidovic, L.; Vodenicharov, M.; Affar, E.B.; Poirier, G.G. Importance of poly(ADP-ribose) glycohydrolase in the control of poly(ADP-ribose) metabolism. *Exp. Cell Res.* **2001**, *268*, 7–13. [CrossRef]
32. Bürkle, A.; Virag, L. Poly(ADP-ribose): PARadigms and PARadoxes. *Mol. Asp. Med.* **2013**, *34*, 1046–1065. [CrossRef]
33. Langelier, M.-F.; Planck, J.L.; Roy, S.; Pascal, J.M. Crystal structures of poly(ADP-ribose) polymerase-1 (PARP-1) zinc fingers bound to DNA: Structural and functional insights into DNA-dependent PARP-1 activity. *J. Biol. Chem.* **2011**, *286*, 10690–10701. [CrossRef] [PubMed]
34. Ali, A.A.E.; Timinszky, G.; Arribas-Bosacoma, R.; Kozlowski, M.; Hassa, P.O.; Hassler, M.; Ladurner, A.G.; Pearl, L.H.; Oliver, A.W. The zinc-finger domains of PARP1 cooperate to recognize DNA strand breaks. *Nat. Struct. Mol. Biol.* **2012**, *19*, 685–692. [CrossRef] [PubMed]
35. Gottschalk, A.J.; Timinszky, G.; Kong, S.E.; Jin, J.; Cai, Y.; Swanson, S.K.; Washburn, M.P.; Florens, L.; Ladurner, A.G.; Conaway, J.W.; et al. Poly(ADP-ribosyl)ation directs recruitment and activation of an ATP-dependent chromatin remodeler. *Proc. Natl. Acad. Sci. USA* **2009**, *106*, 13770–13774. [CrossRef] [PubMed]
36. Krishnakumar, R.; Kraus, W.L. The PARP side of the nucleus: Molecular actions, physiological outcomes, and clinical targets. *Mol. Cell* **2010**, *39*, 8–24. [CrossRef] [PubMed]
37. Fang, E.F.; Scheibye-Knudsen, M.; Chua, K.F.; Mattson, M.P.; Croteau, D.L.; Bohr, V.A. Nuclear DNA damage signalling to mitochondria in ageing. *Nat. Rev. Mol. Cell Biol.* **2016**, *17*, 308–321. [CrossRef] [PubMed]
38. Roos, W.P.; Thomas, A.D.; Kaina, B. DNA damage and the balance between survival and death in cancer biology. *Nat. Rev. Cancer* **2016**, *16*, 20–33. [CrossRef]
39. Maryanovich, M.; Zaltsman, Y.; Ruggiero, A.; Goldman, A.; Shachnai, L.; Zaidman, S.L.; Porat, Z.; Golan, K.; Lapidot, T.; Gross, A. An MTCH2 pathway repressing mitochondria metabolism regulates haematopoietic stem cell fate. *Nat. Commun.* **2015**, *6*, 7901. [CrossRef]

40. Feng, J.; Park, J.; Cron, P.; Hess, D.; Hemmings, B.A. Identification of a PKB/Akt hydrophobic motif Ser-473 kinase as DNA-dependent protein kinase. *J. Biol. Chem.* **2004**, *279*, 41189–41196. [CrossRef]
41. Matsuoka, S.; Ballif, B.A.; Smogorzewska, A.; McDonald, E.R.; Hurov, K.E.; Luo, J.; Bakalarski, C.E.; Zhao, Z.; Solimini, N.; Lerenthal, Y.; et al. ATM and ATR substrate analysis reveals extensive protein networks responsive to DNA damage. *Science* **2007**, *316*, 1160–1166. [CrossRef]
42. Park, J.; Feng, J.; Li, Y.; Hammarsten, O.; Brazil, D.P.; Hemmings, B.A. DNA-dependent protein kinase-mediated phosphorylation of protein kinase B requires a specific recognition sequence in the C-terminal hydrophobic motif. *J. Biol. Chem.* **2009**, *284*, 6169–6174. [CrossRef]
43. Toulany, M.; Maier, J.; Iida, M.; Rebholz, S.; Holler, M.; Grottke, A.; Jüker, M.; Wheeler, D.L.; Rothbauer, U.; Rodemann, H.P. Akt1 and Akt3 but not Akt2 through interaction with DNA-PKcs stimulate proliferation and post-irradiation cell survival of K-RAS-mutated cancer cells. *Cell Death Discov.* **2017**, *3*, 17072. [CrossRef] [PubMed]
44. Liu, Q.; Turner, K.M.; Yung, W.K.A.; Chen, K.; Zhang, W. Role of AKT signaling in DNA repair and clinical response to cancer therapy. *Neuro-Oncology* **2014**, *16*, 1313–1323. [CrossRef]
45. Stronach, E.A.; Chen, M.; Maginn, E.N.; Agarwal, R.; Mills, G.B.; Wasan, H.; Gabra, H. DNA-PK mediates AKT activation and apoptosis inhibition in clinically acquired platinum resistance. *Neoplasia* **2011**, *13*, 1069–1080. [CrossRef]
46. Robey, R.B.; Hay, N. Is Akt the "Warburg kinase"-Akt-energy metabolism interactions and oncogenesis. *Semin. Cancer Biol.* **2009**, *19*, 25–31. [CrossRef]
47. Brunet, A.; Bonni, A.; Zigmond, M.J.; Lin, M.Z.; Juo, P.; Hu, L.S.; Anderson, M.J.; Arden, K.C.; Blenis, J.; Greenberg, M.E. Akt promotes cell survival by phosphorylating and inhibiting a Forkhead transcription factor. *Cell* **1999**, *96*, 857–868. [CrossRef]
48. Kops, G.J.P.L.; de Ruiter, N.D.; De Vries-Smits, A.M.M.; Powell, D.R.; Bos, J.L.; Burgering, B.M.T. Direct control of the Forkhead transcription factor AFX by protein kinase B. *Nature* **1999**, *398*, 630–634. [CrossRef]
49. Webb, A.E.; Brunet, A. FOXO transcription factors: Key regulators of cellular quality control. *Trends Biochem. Sci.* **2014**, *39*, 159–169. [CrossRef] [PubMed]
50. Gray, M.W. Mitochondrial evolution. *Cold Spring Harb. Perspect. Biol.* **2012**, *4*, a011403. [CrossRef] [PubMed]
51. Lee, C.; Yen, K.; Cohen, P. Humanin: A harbinger of mitochondrial-derived peptides? *Trends Endocrinol. Metab.* **2013**, *24*, 222–228. [CrossRef] [PubMed]
52. Kim, K.H.; Son, J.M.; Benayoun, B.A.; Lee, C. The Mitochondrial-Encoded Peptide MOTS-c Translocates to the Nucleus to Regulate Nuclear Gene Expression in Response to Metabolic Stress. *Cell Metab.* **2018**, *28*, 516–524. [CrossRef]
53. Sulkowski, P.L.; Sundaram, R.K.; Oeck, S.; Corso, C.D.; Liu, Y.; Noorbakhsh, S.; Niger, M.; Boeke, M.; Ueno, D.; Kalathil, A.N.; et al. Krebs-cycle-deficient hereditary cancer syndromes are defined by defects in homologous-recombination DNA repair. *Nat. Genet.* **2018**, *50*, 1086–1092. [CrossRef]
54. Li, X.; Egervari, G.; Wang, Y.; Berger, S.L.; Lu, Z. Regulation of chromatin and gene expression by metabolic enzymes and metabolites. *Nat. Rev. Mol. Cell Biol.* **2018**, *19*, 563–578. [CrossRef] [PubMed]
55. Sivanand, S.; Rhoades, S.; Jiang, Q.; Lee, J.V.; Benci, J.; Zhang, J.; Yuan, S.; Viney, I.; Zhao, S.; Carrer, A.; et al. Nuclear Acetyl-CoA Production by ACLY Promotes Homologous Recombination. *Mol. Cell* **2017**, *67*, 252–265. [CrossRef] [PubMed]
56. Desler, C.; Munch-Petersen, B.; Stevnsner, T.; Matsui, S.-I.; Kulawiec, M.; Singh, K.K.; Rasmussen, L.J. Mitochondria as determinant of nucleotide pools and chromosomal stability. *Mutat. Res.* **2007**, *625*, 112–124. [CrossRef]
57. Rasmussen, A.K.; Chatterjee, A.; Rasmussen, L.J.; Singh, K.K. Mitochondria-mediated nuclear mutator phenotype in Saccharomyces cerevisiae. *Nucleic Acids Res.* **2003**, *31*, 3909–3917. [CrossRef]
58. Murphy, M.P. How mitochondria produce reactive oxygen species. *Biochem. J.* **2009**, *417*, 1–13. [CrossRef] [PubMed]
59. Sena, L.A.; Chandel, N.S. Physiological roles of mitochondrial reactive oxygen species. *Mol. Cell* **2012**, *48*, 158–167. [CrossRef] [PubMed]
60. Martínez-Reyes, I.; Diebold, L.P.; Kong, H.; Schieber, M.; Huang, H.; Hensley, C.T.; Mehta, M.M.; Wang, T.; Santos, J.H.; Woychik, R.; et al. TCA Cycle and Mitochondrial Membrane Potential Are Necessary for Diverse Biological Functions. *Mol. Cell* **2016**, *61*, 199–209. [CrossRef]

61. Westermann, B. Bioenergetic role of mitochondrial fusion and fission. *Biochim. Biophys. Acta Bioenerget.* **2012**, *1817*, 1833–1838. [CrossRef]
62. Brace, L.E.; Vose, S.C.; Stanya, K.; Gathungu, R.M.; Marur, V.R.; Longchamp, A.; Treviño-Villarreal, H.; Mejia, P.; Vargas, D.; Inouye, K.; et al. Increased oxidative phosphorylation in response to acute and chronic DNA damage. *NPJ Aging Mech. Dis.* **2016**, *2*, 16022. [CrossRef] [PubMed]
63. Youle, R.J.; van der Bliek, A.M. Mitochondrial fission, fusion, and stress. *Science* **2012**, *337*, 1062–1065. [CrossRef]
64. Sies, H.; Berndt, C.; Jones, D.P. Oxidative Stress. *Annu. Rev. Biochem.* **2017**, *86*, 715–748. [CrossRef]
65. Trifunovic, A.; Hansson, A.; Wredenberg, A.; Rovio, A.T.; Dufour, E.; Khvorostov, I.; Spelbrink, J.N.; Wibom, R.; Jacobs, H.T.; Larsson, N.-G. Somatic mtDNA mutations cause aging phenotypes without affecting reactive oxygen species production. *Proc. Natl. Acad. Sci. USA* **2005**, *102*, 17993–17998. [CrossRef]
66. Diebold, L.; Chandel, N.S. Mitochondrial ROS regulation of proliferating cells. *Free Radic. Biol. Med.* **2016**, *100*, 86–93. [CrossRef] [PubMed]
67. Maryanovich, M.; Gross, A. A ROS rheostat for cell fate regulation. *Trends Cell Biol.* **2013**, *23*, 129–134. [CrossRef] [PubMed]
68. Folmes, C.D.L.; Dzeja, P.P.; Nelson, T.J.; Terzic, A. Metabolic Plasticity in Stem Cell Homeostasis and Differentiation. *Cell Stem Cell* **2012**, *11*, 596–606. [CrossRef] [PubMed]
69. Nijnik, A.; Woodbine, L.; Marchetti, C.; Dawson, S.; Lambe, T.; Liu, C.; Rodrigues, N.P.; Crockford, T.L.; Cabuy, E.; Vindigni, A.; et al. DNA repair is limiting for haematopoietic stem cells during ageing. *Nature* **2007**, *447*, 686–690. [CrossRef]
70. Niedernhofer, L.J. DNA repair is crucial for maintaining hematopoietic stem cell function. *DNA Repair* **2008**, *7*, 523–529. [CrossRef] [PubMed]
71. Flach, J.; Bakker, S.T.; Mohrin, M.; Conroy, P.C.; Pietras, E.M.; Reynaud, D.; Alvarez, S.; Diolaiti, M.E.; Ugarte, F.; Forsberg, E.C.; et al. Replication stress is a potent driver of functional decline in ageing haematopoietic stem cells. *Nature* **2014**, *512*, 198–202. [CrossRef]
72. Drazic, A.; Myklebust, L.M.; Ree, R.; Arnesen, T. The world of protein acetylation. *Biochim. Biophys. Acta Proteins Proteom.* **2016**, *1864*, 1372–1401. [CrossRef] [PubMed]
73. Hauer, M.H.; Gasser, S.M. Chromatin and nucleosome dynamics in DNA damage and repair. *Genes Dev.* **2017**, *31*, 2204–2221. [CrossRef]
74. Zane, L.; Sharma, V.; Misteli, T. Common features of chromatin in aging and cancer: Cause or coincidence? *Trends Cell Biol.* **2014**, *24*, 686–694. [CrossRef] [PubMed]
75. Su, X.; Wellen, K.E.; Rabinowitz, J.D. Metabolic control of methylation and acetylation. *Curr. Opin. Chem. Biol.* **2016**, *30*, 52–60. [CrossRef] [PubMed]
76. Choudhary, C.; Weinert, B.T.; Nishida, Y.; Verdin, E.; Mann, M. The growing landscape of lysine acetylation links metabolism and cell signalling. *Nat. Rev. Mol. Cell Biol.* **2014**, *15*, 536–550. [CrossRef]
77. Ali, I.; Conrad, R.J.; Verdin, E.; Ott, M. Lysine Acetylation Goes Global: From Epigenetics to Metabolism and Therapeutics. *Chem. Rev.* **2018**, *118*, 1216–1252. [CrossRef] [PubMed]
78. Montgomery, D.C.; Sorum, A.W.; Guasch, L.; Nicklaus, M.C.; Meier, J.L. Metabolic Regulation of Histone Acetyltransferases by Endogenous Acyl-CoA Cofactors. *Chem. Biol.* **2015**, *22*, 1030–1039. [CrossRef]
79. Katada, S.; Imhof, A.; Sassone-Corsi, P. Connecting Threads: Epigenetics and Metabolism. *Cell* **2012**, *148*, 24–28. [CrossRef]
80. Lee, J.V.; Carrer, A.; Shah, S.; Snyder, N.W.; Wei, S.; Venneti, S.; Worth, A.J.; Yuan, Z.-F.; Lim, H.-W.; Liu, S.; et al. Akt-Dependent Metabolic Reprogramming Regulates Tumor Cell Histone Acetylation. *Cell Metab.* **2014**, *20*, 306–319. [CrossRef]
81. Kinnaird, A.; Zhao, S.; Wellen, K.E.; Michelakis, E.D. Metabolic control of epigenetics in cancer. *Nat. Rev. Cancer* **2016**, *16*, 694–707. [CrossRef]
82. Metzger, E.; Schüle, R. The expanding world of histone lysine demethylases. *Nat. Struct. Mol. Biol.* **2007**, *14*, 252–254. [CrossRef] [PubMed]
83. Forneris, F.; Binda, C.; Vanoni, M.A.; Mattevi, A.; Battaglioli, E. Histone demethylation catalysed by LSD1 is a flavin-dependent oxidative process. *FEBS Lett.* **2005**, *579*, 2203–2207. [CrossRef]
84. Metzger, E.; Imhof, A.; Patel, D.; Kahl, P.; Hoffmeyer, K.; Friedrichs, N.; Müller, J.M.; Greschik, H.; Kirfel, J.; Ji, S.; et al. Phosphorylation of histone H3T6 by PKCβI controls demethylation at histone H3K4. *Nature* **2010**, *464*, 792–796. [CrossRef]

85. He, Y.-F.; Li, B.-Z.; Li, Z.; Liu, P.; Wang, Y.; Tang, Q.; Ding, J.; Jia, Y.; Chen, Z.; Li, L.; et al. Tet-Mediated Formation of 5-Carboxylcytosine and Its Excision by TDG in Mammalian DNA. *Science* **2011**, *333*, 1303–1307. [CrossRef] [PubMed]
86. Wu, D.; Hu, D.; Chen, H.; Shi, G.; Fetahu, I.S.; Wu, F.; Rabidou, K.; Fang, R.; Tan, L.; Xu, S.; et al. Glucose-regulated phosphorylation of TET2 by AMPK reveals a pathway linking diabetes to cancer. *Nature* **2018**, *559*, 637–641. [CrossRef]
87. Pai, C.-C.; Kearsey, S.E. A Critical Balance: dNTPs and the Maintenance of Genome Stability. *Genes* **2017**, *8*, 57. [CrossRef]
88. Kunz, B.A. Mutagenesis and deoxyribonucleotide pool imbalance. *Mutat. Res.* **1988**, *200*, 133–147. [CrossRef]
89. Kunz, B.A.; Kohalmi, S.E.; Kunkel, T.A.; Mathews, C.K.; McIntosh, E.M.; Reidy, J.A. Deoxyribonucleoside triphosphate levels: A critical factor in the maintenance of genetic stability. *Mutat. Res. Toxicol.* **1994**, *318*, 1–64. [CrossRef]
90. Mathews, C.K. DNA precursor metabolism and genomic stability. *FASEB J.* **2006**, *20*, 1300–1314. [CrossRef] [PubMed]
91. Kumar, D.; Abdulovic, A.L.; Viberg, J.; Nilsson, A.K.; Kunkel, T.A.; Chabes, A. Mechanisms of mutagenesis in vivo due to imbalanced dNTP pools. *Nucleic Acids Res.* **2011**, *39*, 1360–1371. [CrossRef] [PubMed]
92. Reichard, P. Interactions between deoxyribonucleotide and DNA synthesis. *Annu. Rev. Biochem.* **1988**, *57*, 349–374. [CrossRef]
93. Meuth, M. The molecular basis of mutations induced by deoxyribonucleoside triphosphate pool imbalances in mammalian cells. *Exp. Cell Res.* **1989**, *181*, 305–316. [CrossRef]
94. Meuth, M. The genetic consequences of nucleotide precursor pool imbalance in mammalian cells. *Mutat. Res. Mol. Mech. Mutagen.* **1984**, *126*, 107–112. [CrossRef]
95. Haracska, L.; Prakash, S.; Prakash, L. Replication pastO6-Methylguanine by Yeast and Human DNA Polymerase η. *Mol. Cell. Biol.* **2000**, *20*, 8001–8007. [CrossRef]
96. Niida, H.; Shimada, M.; Murakami, H.; Nakanishi, M. Mechanisms of dNTP supply that play an essential role in maintaining genome integrity in eukaryotic cells. *Cancer Sci.* **2010**, *101*, 2505–2509. [CrossRef] [PubMed]
97. Niida, H.; Katsuno, Y.; Sengoku, M.; Shimada, M.; Yukawa, M.; Ikura, M.; Ikura, T.; Kohno, K.; Shima, H.; Suzuki, H.; et al. Essential role of Tip60-dependent recruitment of ribonucleotide reductase at DNA damage sites in DNA repair during G1 phase. *Genes Dev.* **2010**, *24*, 333–338. [CrossRef]
98. Tseng, L.-M.; Yin, P.-H.; Chi, C.-W.; Hsu, C.-Y.; Wu, C.-W.; Lee, L.-M.; Wei, Y.-H.; Lee, H.-C. Mitochondrial DNA mutations and mitochondrial DNA depletion in breast cancer. *Genes Chromosomes Cancer* **2006**, *45*, 629–638. [CrossRef]
99. Yu, M.; Zhou, Y.; Shi, Y.; Ning, L.; Yang, Y.; Wei, X.; Zhang, N.; Hao, X.; Niu, R. Reduced mitochondrial DNA copy number is correlated with tumor progression and prognosis in Chinese breast cancer patients. *IUBMB Life* **2007**, *59*, 450–457. [CrossRef] [PubMed]
100. Yamada, S.; Nomoto, S.; Fujii, T.; Kaneko, T.; Takeda, S.; Inoue, S.; Kanazumi, N.; Nakao, A. Correlation between copy number of mitochondrial DNA and clinico-pathologic parameters of hepatocellular carcinoma. *Eur. J. Surg. Oncol.* **2006**, *32*, 303–307. [CrossRef]
101. Matsuyama, W.; Nakagawa, M.; Wakimoto, J.; Hirotsu, Y.; Kawabata, M.; Osame, M. Mitochondrial DNA mutation correlates with stage progression and prognosis in non-small cell lung cancer. *Hum. Mutat.* **2003**, *21*, 441–443. [CrossRef] [PubMed]
102. Lièvre, A.; Chapusot, C.; Bouvier, A.-M.; Zinzindohoué, F.; Piard, F.; Roignot, P.; Arnould, L.; Beaune, P.; Faivre, J.; Laurent-Puig, P. Clinical value of mitochondrial mutations in colorectal cancer. *J. Clin. Oncol.* **2005**, *23*, 3517–3525. [CrossRef] [PubMed]
103. Chatterjee, A.; Mambo, E.; Sidransky, D. Mitochondrial DNA mutations in human cancer. *Oncogene* **2006**, *25*, 4663–4674. [CrossRef] [PubMed]
104. Simonnet, H. Low mitochondrial respiratory chain content correlates with tumor aggressiveness in renal cell carcinoma. *Carcinogenesis* **2002**, *23*, 759–768. [CrossRef]
105. Petros, J.A.; Baumann, A.K.; Ruiz-Pesini, E.; Amin, M.B.; Sun, C.Q.; Hall, J.; Lim, S.; Issa, M.M.; Flanders, W.D.; Hosseini, S.H.; et al. mtDNA mutations increase tumorigenicity in prostate cancer. *Proc. Natl. Acad. Sci. USA* **2005**, *102*, 719–724. [CrossRef] [PubMed]

106. Shidara, Y.; Yamagata, K.; Kanamori, T.; Nakano, K.; Kwong, J.Q.; Manfredi, G.; Oda, H.; Ohta, S. Positive contribution of pathogenic mutations in the mitochondrial genome to the promotion of cancer by prevention from apoptosis. *Cancer Res.* **2005**, *65*, 1655–1663. [CrossRef]
107. Ralevic, V.; Burnstock, G. Receptors for purines and pyrimidines. *Pharmacol. Rev.* **1998**, *50*, 413–492. [PubMed]
108. Smith, J.L. Enzymes of nucleotide synthesis. *Curr. Opin. Struct. Biol.* **1995**, *5*, 752–757. [CrossRef]
109. Fang, J.; Uchiumi, T.; Yagi, M.; Matsumoto, S.; Amamoto, R.; Takazaki, S.; Yamaza, H.; Nonaka, K.; Kang, D. Dihydro-orotate dehydrogenase is physically associated with the respiratory complex and its loss leads to mitochondrial dysfunction. *Biosci. Rep.* **2013**, *33*, e00021. [CrossRef]
110. Bader, B.; Knecht, W.; Fries, M.; Löffler, M. Expression, purification, and characterization of histidine-tagged rat and human flavoenzyme dihydroorotate dehydrogenase. *Protein Expr. Purif.* **1998**, *13*, 414–422. [CrossRef]
111. Löffler, M.; Jöckel, J.; Schuster, G.; Becker, C. Dihydroorotat-ubiquinone oxidoreductase links mitochondria in the biosynthesis of pyrimidine nucleotides. *Mol. Cell. Biochem.* **1997**, *174*, 125–129. [CrossRef]
112. Beuneu, C.; Auger, R.; Löffler, M.; Guissani, A.; Lemaire, G.; Lepoivre, M. Indirect inhibition of mitochondrial dihydroorotate dehydrogenase activity by nitric oxide. *Free Radic. Biol. Med.* **2000**, *28*, 1206–1213. [CrossRef]
113. Rawls, J.; Knecht, W.; Diekert, K.; Lill, R.; Löffler, M. Requirements for the mitochondrial import and localization of dihydroorotate dehydrogenase. *Eur. J. Biochem.* **2000**, *267*, 2079–2087. [CrossRef]
114. Uhlin, U.; Eklund, H. Structure of ribonucleotide reductase protein R1. *Nature* **1994**, *370*, 533–539. [CrossRef]
115. Jordan, A.; Reichard, P. Ribonucleotide reductases. *Annu. Rev. Biochem.* **1998**, *67*, 71–98. [CrossRef] [PubMed]
116. Kashlan, O.B.; Scott, C.P.; Lear, J.D.; Cooperman, B.S. A comprehensive model for the allosteric regulation of mammalian ribonucleotide reductase. Functional consequences of ATP- and dATP-induced oligomerization of the large subunit. *Biochemistry* **2002**, *41*, 462–474. [CrossRef]
117. Chabes, A.L.; Pfleger, C.M.; Kirschner, M.W.; Thelander, L. Mouse ribonucleotide reductase R2 protein: A new target for anaphase-promoting complex-Cdh1-mediated proteolysis. *Proc. Natl. Acad. Sci. USA* **2003**, *100*, 3925–3929. [CrossRef] [PubMed]
118. Munch-Petersen, B.; Cloos, L.; Jensen, H.K.; Tyrsted, G. Human thymidine kinase 1. Regulation in normal and malignant cells. *Adv. Enzyme Regul.* **1995**, *35*, 69–89. [CrossRef]
119. Goldstone, D.C.; Ennis-Adeniran, V.; Hedden, J.J.; Groom, H.C.T.; Rice, G.I.; Christodoulou, E.; Walker, P.A.; Kelly, G.; Haire, L.F.; Yap, M.W.; et al. HIV-1 restriction factor SAMHD1 is a deoxynucleoside triphosphate triphosphohydrolase. *Nature* **2011**, *480*, 379–382. [CrossRef]
120. Clifford, R.; Louis, T.; Robbe, P.; Ackroyd, S.; Burns, A.; Timbs, A.T.; Wright Colopy, G.; Dreau, H.; Sigaux, F.; Judde, J.G.; et al. SAMHD1 is mutated recurrently in chronic lymphocytic leukemia and is involved in response to DNA damage. *Blood* **2014**, *123*, 1021–1031. [CrossRef]
121. Song, S.; Wheeler, L.J.; Mathews, C.K. Deoxyribonucleotide pool imbalance stimulates deletions in HeLa cell mitochondrial DNA. *J. Biol. Chem.* **2003**, *278*, 43893–43896. [CrossRef]
122. López, L.C.; Akman, H.O.; García-Cazorla, A.; Dorado, B.; Martí, R.; Nishino, I.; Tadesse, S.; Pizzorno, G.; Shungu, D.; Bonilla, E.; et al. Unbalanced deoxynucleotide pools cause mitochondrial DNA instability in thymidine phosphorylase-deficient mice. *Hum. Mol. Genet.* **2009**, *18*, 714–722. [CrossRef] [PubMed]
123. Pontarin, G.; Gallinaro, L.; Ferraro, P.; Reichard, P.; Bianchi, V. Origins of mitochondrial thymidine triphosphate: Dynamic relations to cytosolic pools. *Proc. Natl. Acad. Sci. USA* **2003**, *100*, 12159–12164. [CrossRef] [PubMed]
124. Desler, C.; Munch-Petersen, B.; Rasmussen, L.J. The Role of Mitochondrial dNTP Levels in Cells with Reduced TK2 Activity. *Nucleosides Nucleotides Nucleic Acids* **2006**, *25*, 1171–1175. [CrossRef] [PubMed]
125. Pontarin, G.; Ferraro, P.; Bee, L.; Reichard, P.; Bianchi, V. Mammalian ribonucleotide reductase subunit p53R2 is required for mitochondrial DNA replication and DNA repair in quiescent cells. *Proc. Natl. Acad. Sci. USA* **2012**, *109*, 13302–13307. [CrossRef]
126. Bourdon, A.; Minai, L.; Serre, V.; Jais, J.-P.; Sarzi, E.; Aubert, S.; Chrétien, D.; de Lonlay, P.; Paquis-Flucklinger, V.; Arakawa, H.; et al. Mutation of RRM2B, encoding p53-controlled ribonucleotide reductase (p53R2), causes severe mitochondrial DNA depletion. *Nat. Genet.* **2007**, *39*, 776–780. [CrossRef] [PubMed]
127. Pontarin, G.; Ferraro, P.; Rampazzo, C.; Kollberg, G.; Holme, E.; Reichard, P.; Bianchi, V. Deoxyribonucleotide metabolism in cycling and resting human fibroblasts with a missense mutation in p53R2, a subunit of ribonucleotide reductase. *J. Biol. Chem.* **2011**, *286*, 11132–11140. [CrossRef] [PubMed]

128. Tanaka, H.; Arakawa, H.; Yamaguchi, T.; Shiraishi, K.; Fukuda, S.; Matsui, K.; Takei, Y.; Nakamura, Y. A ribonucleotide reductase gene involved in a p53-dependent cell-cycle checkpoint for DNA damage. *Nature* **2000**, *404*, 42–49. [CrossRef]
129. James, S.J.; Miller, B.J.; McGarrity, L.J.; Morris, S.M. The effect of folic acid and/or methionine deficiency on deoxyribonucleotide pools and cell cycle distribution in mitogen-stimulated rat lymphocytes. *Cell Prolif.* **1994**, *27*, 395–406. [CrossRef]
130. Micheli, V.; Camici, M.; G Tozzi, M.; L Ipata, P.; Sestini, S.; Bertelli, M.; Pompucci, G. Neurological Disorders of Purine and Pyrimidine Metabolism. *Curr. Top. Med. Chem.* **2011**, *11*, 923–947. [CrossRef]
131. Desler, C.; Marcker, M.L.; Singh, K.K.; Rasmussen, L.J. The importance of mitochondrial DNA in aging and cancer. *J. Aging Res.* **2011**, *2011*, 407536. [CrossRef] [PubMed]
132. Desler, C.; Frederiksen, J.H.; Angleys, M.; Maynard, S.; Keijzers, G.; Fagerlund, B.; Mortensen, E.L.; Osler, M.; Lauritzen, M.; Bohr, V.A.; et al. Increased deoxythymidine triphosphate levels is a feature of relative cognitive decline. *Mitochondrion* **2015**, *25*, 34–37. [CrossRef] [PubMed]
133. Maynard, S.; Hejl, A.-M.; Dinh, T.-S.T.; Keijzers, G.; Hansen, Å.M.; Desler, C.; Moreno-Villanueva, M.; Bürkle, A.; Rasmussen, L.J.; Waldemar, G.; et al. Defective mitochondrial respiration, altered dNTP pools and reduced AP endonuclease 1 activity in peripheral blood mononuclear cells of Alzheimer's disease patients. *Aging* **2015**, *7*, 793–815. [CrossRef] [PubMed]

 © 2019 by the authors. Licensee MDPI, Basel, Switzerland. This article is an open access article distributed under the terms and conditions of the Creative Commons Attribution (CC BY) license (http://creativecommons.org/licenses/by/4.0/).

Review

Mitochondrial Transport and Turnover in the Pathogenesis of Amyotrophic Lateral Sclerosis

Veronica Granatiero and Giovanni Manfredi *

Feil Family Brain and Mind Research Institute, Weill Cornell Medicine, 407 East 61st Street, New York, NY 10065, USA; veg2003@med.cornell.edu
* Correspondence: gim2004@med.cornell.edu; Tel.: +1-646-962-8172

Received: 16 December 2018; Accepted: 3 February 2019; Published: 11 May 2019

Abstract: Neurons are high-energy consuming cells, heavily dependent on mitochondria for ATP generation and calcium buffering. These mitochondrial functions are particularly critical at specific cellular sites, where ionic currents impose a large energetic burden, such as at synapses. The highly polarized nature of neurons, with extremely large axoplasm relative to the cell body, requires mitochondria to be efficiently transported along microtubules to reach distant sites. Furthermore, neurons are post-mitotic cells that need to maintain pools of healthy mitochondria throughout their lifespan. Hence, mitochondrial transport and turnover are essential processes for neuronal survival and function. In neurodegenerative diseases, the maintenance of a healthy mitochondrial network is often compromised. Numerous lines of evidence indicate that mitochondrial impairment contributes to neuronal demise in a variety of neurodegenerative diseases, including amyotrophic lateral sclerosis (ALS), where degeneration of motor neurons causes a fatal muscle paralysis. Dysfunctional mitochondria accumulate in motor neurons affected by genetic or sporadic forms of ALS, strongly suggesting that the inability to maintain a healthy pool of mitochondria plays a pathophysiological role in the disease. This article critically reviews current hypotheses on mitochondrial involvement in the pathogenesis of ALS, focusing on the alterations of mitochondrial axonal transport and turnover in motor neurons.

Keywords: mitochondria; ALS; axonal transport; mitophagy; SOD1; Miro1; PINK1; Parkin

1. Introduction

Amyotrophic lateral sclerosis (ALS) is a devastating neurodegenerative disorder that causes the death of both upper and lower motor neurons. It is the most common among the motor neuron diseases. Loss of motor neurons results in muscle denervation leading to progressive muscle weakness, causing respiratory failure, difficulty in speaking and swallowing, and eventually paralysis and death, typically between one and five years from the time of disease onset [1]. There are currently no effective treatments for ALS, with only two food and drug administration (FDA) approved drugs, which only extend survival by a few months [2]. Approximately, 10% of ALS cases are due to genetic causes, the remaining 90% are sporadic with unknown etiology [3]. The last decade has brought tremendous advances in the understanding of ALS genetics with over 20 different genes identified that account for almost 80% of all familial forms [4].

The nature of the genes involved in ALS indicates that this is an etiologically heterogeneous disease [5] with a multiplicity of initiating factors that trigger diverse pathogenic pathways, ultimately converging in motor neuron toxicity. In the majority of genetic forms of ALS, motor neuron degeneration is accompanied by the involvement of other neural systems, frequently resulting in frontotemporal dementia (FTD). Some of the most prominent alterations observed in ALS/FTD involve protein homeostasis, autophagy, RNA metabolism, axonal transport and mitochondrial abnormalities [5–12].

Mitochondrial abnormalities have often been implicated as secondary mechanisms of disease, because until recently no genetic forms of ALS have been ascribed to mutations in nuclear or mitochondrial DNA encoded proteins. Only in 2014, the first mutation in a mitochondrial gene, coiled–coiled helix containing domain 10 have been identified in pedigrees presenting with ALS, FTD, and myopathy [13]. Despite the scarcity of primary mitochondrial forms of ALS, mitochondrial alterations can be caused by mutant proteins that are not exclusively localized in mitochondria. Notable examples are mutations in Cu, Zn superoxide dismutase (SOD1), the first genetic form of ALS identified [14], and TAR DNA-binding protein (TDP43, [15]).

While motor neurons are the most affected cells in ALS, studies on the pathophysiology of the disease have highlighted the importance of non-cell autonomous mechanisms, which implicate other cell types in the central nervous system. Glial cells, including astrocytes [16–18], oligodendrocytes [19], and microglia [20,21] play toxic roles in mutant SOD1 mouse models of ALS. Particularly in astrocytes, mitochondria can play a significant role in causing toxicity to motor neurons. For example, failure of astrocytes to clear synaptic glutamate as a result of altered mitochondrial glutamate intermediary metabolism, can trigger neuronal excitotoxicity.

Importantly, in ALS neurons mitochondrial alterations are accompanied by defects of organelle dynamics, involving fusion, fission, and transport [12]. These impairments result not only in morphologically and functionally defective mitochondria, but also in mislocalized mitochondrial network [22]. In excitable cells with large axoplasm, such as motor neurons, loss of viable mitochondria at sites of high energy demand, such as the synapses that motor neurons form with muscle (neuromuscular junctions, NMJ), can have catastrophic consequences leading to muscle denervation.

This review article focuses on the alterations of mitochondrial transport in neurons and the role of mitochondrial quality control mechanisms in ALS.

2. Mitochondrial Axonal Transport in ALS

2.1. The Mitochondrial Transport Machinery

In neurons, mitochondria are highly dynamic organelles. They are characterized by fast transport along neuronal axons, in both anterograde and retrograde directions (i.e., from the soma to the periphery and vice versa). Mitochondria from the soma are anterogradely transported to sites where metabolic demand is high, such as the synapse [23,24], while retrograde transport provides essential information regarding the status and environment of distal sites. Thus, mitochondrial transport plays an essential role in maintaining healthy motor neurons, and alterations of rate-limiting components of the mitochondrial transport machinery may cause an imbalance of mitochondrial distribution in neurons. Furthermore, as active lysosomes are mostly localized in the soma, damaged mitochondria in axons are engulfed in autophagosomes and retro-transported to the soma to be degraded upon fusion with lysosomes [25]. Defective retrograde transport of autophagosomes could result in a delay of mitochondrial autophagy (mitophagy) fluxes and accumulation of damaged mitochondria. Thus, axonal transport and mitophagy are intimately interconnected processes.

For long-range axonal transport mitochondria move along microtubules. The kinesin superfamily of proteins and cytoplasmic dynein are the main microtubule-based motor proteins. They drive long distance transport of mitochondria and other membranous organelles through ATP-dependent mechanisms [26]. In axons, cytoplasmic dynein is responsible for the retrograde transport, moving mitochondria towards the soma, whereas kinesin drives anterograde mitochondrial transport from the soma to distal axonal regions and synaptic terminals. In dendritic spines, where microtubules exhibit mixed polarity in proximal regions, kinesin and dynein motors can transport mitochondria in either direction, depending on the microtubule polarity [26,27]. There are two main mechanisms by which molecular motors connect with their cargoes, direct linkage through cargo motor proteins or indirect linkage via linker/adaptor molecules. Several adaptor complexes have been identified, which ensure a precise regulation of mitochondrial motility. Among them, the best studied linkers are Milton and

Miro, first identified in Drosophila. Milton is a kinesin heavy chain-binding protein. In mammals, there are two Milton orthologues, TRAK1 and 2. Milton is linked indirectly with the mitochondrial outer membrane through interaction with an atypical Rho GTPase, Miro. Mammals have 2 orthologues of Miro (Miro1 and 2) [28,29]. Together Miro, Milton and kinesin provide mitochondria-specific axonal transport mechanisms.

The ATP/ADP ratio is part of the signaling involved in regulating mitochondrial transport. In regions with high ATP level, mitochondrial velocity increases, while upon ATP depletion mitochondrial velocity decreases [30]. Another important signal is provided by neuronal Ca^{2+} concentration. Miro contains two Ca^{2+}-sensitive helix-loop-helix structural domain, also called EF hand motifs, facing the cytosolic side, which sense cytosolic Ca^{2+} levels. Two different mechanisms have been proposed to explain how Miro regulates mitochondrial motility in a Ca^{2+}-dependent manner. According to the "motor-Miro binding" model, when Ca^{2+} around mitochondria is low, the C-terminal tail of kinesin is bound to the mitochondrion through its interaction with the Milton–Miro complex, thereby allowing for mitochondrial transport. When Ca^{2+} is high, it binds to Miro EF hands, causing a conformational change that results in the direct interaction of kinesin with Miro, which prevents mitochondrial movement [31]. According to the "motor-releasing model", upon Ca^{2+} increase, Miro remains attached to Milton and the mitochondrion, but dissociates from the kinesin, thereby arresting mitochondrial transport [32].

2.2. Alterations of the Mitochondrial Transport Machinery in ALS

There are two main hypotheses on the process of motor neuron degeneration in ALS, the "dying-forward" and the "dying-back" hypotheses. The former proposes that ALS is mainly a cortical motor neuron disorder, which mediates anterograde degeneration of anterior horn cells via glutamate excitotoxicity [33]. On the other hand, the "dying-back" hypothesis proposes that motor neurons degeneration in ALS starts distally at the nerve terminal or at the NMJ and progresses towards the soma [34]. In support to the latter, it was shown that early degeneration of the NMJ precedes the loss of neurons in the spinal cord of mutant SOD1 mice [35,36]. Mitochondrial transport abnormalities could significantly contribute to dying-back processes, because the distal regions of motor neurons may not be appropriately supplied with healthy, functional mitochondria, while damaged mitochondria may not be correctly turned over. A significant body of evidence points towards a causal relationship between deficits in axonal transport and degeneration of susceptible motor neurons in ALS [37,38]. For example, defects in neuronal mitochondrial morphology and axonal mitochondrial transport have been demonstrated in primary neuronal cultures from ALS mouse models [12,39,40]. Importantly, these abnormalities have also been observed in vivo in mutant SOD1 and TDP-43 ALS mouse models [41,42] and in Drosophila models [43]. Other studies in primary neuronal cultures and in vivo demonstrated that mutant FUS, a RNA-binding protein causative of familial ALS/FTD [44], induces motor neuron degeneration preceded by abnormalities in synaptic transmission [45] and mitochondrial abnormalities at the NMJ [46]. The common denominator of these studies was the finding that mitochondrial abnormalities and the impairment of mitochondrial axonal transport precede motor neuron degeneration. This evidence supports the "dying-back" hypothesis [34,47,48], in which mitochondrial transport abnormalities may play an instrumental role. Indeed, in mutant SOD1 mice deficits in bidirectional transport of mitochondria are described at the pre-symptomatic disease stage [49], suggesting that these alterations play an early causative role in NMJ degeneration.

The causes of mitochondrial transport abnormalities in ALS motor neurons are not fully understood. In some cases, alterations of the axonal transport machinery are directly involved. For example, studies suggest that SOD1 mutations impair dynein functions, as mutant SOD1 directly interacts with the dynein–dynactin complex, forming aggregates in the spinal cord and sciatic nerve of SOD1 transgenic mice [50,51]. Furthermore, a large body of evidence in models of familial ALS with mutations in SOD1 or TDP-43 indicates that mitochondrial damage and dysfunction is the result of the pathological accumulation of aggregated mutant proteins inside or on the surface of mitochondria [42,52–58].

Therefore, another possibility is that ALS mutant proteins damage mitochondria and impair their bioenergetics [59–61], thereby decreasing ATP availability for axonal transport. This is an attractive hypothesis, because it could provide a mechanism whereby unhealthy mitochondria are immobilized to facilitate their removal by autophagy, similar to the Ca^{2+} dependent mitochondrial arrest described above [31,32]. While it was shown that ATP levels affect mitochondrial motility [30], to our knowledge, the question of whether subpopulations of energy defective mitochondria have a selective decrease in transport, has not been addressed experimentally. This could be achieved, for example, by causing a mild impairment in ATP synthesis in a targeted subset of mitochondria in the neuronal soma and tracking their movement to neuronal processes over time, in comparison with healthy mitochondria.

Although long-range mitochondrial transport is microtubule-based, short range movement in presynaptic terminals and dendritic spines, where actin filaments form the cytoskeletal architecture, is mediated by actin-myosin motors. In cultured neurons, axonal mitochondria have been shown to travel along microtubules and actin microfilaments with different velocities and mechanisms [62]. The actin cytoskeleton is especially relevant to motor neuron diseases with altered actin dynamics [63]. Profilin1 (PFN1) is one of four isoforms of profilin, and the first actin-binding protein associated with familial ALS [64]. Initial studies suggested that the main function of profilin was to sequester actin monomers, thereby inhibiting F-actin formation [65,66]. However, later studies revealed that the amount of profilin present in cells is not sufficient to sequester abundant actin monomers, and a different function for profilin was proposed, as a catalytic converter for actin monomer recycling [67,68]. There are several hypotheses on the pathogenic mechanisms of mutant PFN1 in ALS. Both gain and loss of function have been proposed. The former is based on mutant PFN1 forming aggregates in motor neurons [69,70]. The latter on studies showing that mutant PFN1 causes the formation of cavities in its protein core structure, compromising protein stability and leading to misfolding and degradation [71]. Motor neurons of mutant PFN1 transgenic mice show aggregation of the protein and disruption of the actin cytoskeleton, accompanied by elevated ubiquitin and p62/SQSTM levels in motor neurons [72]. Another transgenic mouse model of mutant PFN1 revealed alterations of actin dynamics and reduced filamentous versus globular actin ratio [73]. Although mitochondrial transport and turnover have not yet been investigated in these mouse models, alterations in mitochondrial ultrastructure was reported in motor neuron axons [73], suggesting that cytoskeletal alterations in these mice affect mitochondria, possibly through impairment of their dynamics.

Myosins allow for cargo movement along actin cytoskeleton. Myo19 is the only myosin localized to mitochondria, and plays a physiological role in mitochondrial movement under conditions of glucose-starvation [74]. Recently, a new intersection point between microtubule-dependent and actin-dependent mitochondrial movement was described through Miro [75,76]. It was shown that, upon activation of pathways of mitochondrial degradation, Myo19 was digested together with Miro, thereby regulating mitochondrial movement and distribution. These findings raise the possibility that detachment of mitochondria from the actin cytoskeleton may be an important step in altering mitochondrial transport. Together these findings strongly suggest that microtubule- and actin-dependent mitochondrial transport mechanisms may be connected and that both mechanisms could be dysregulated in ALS.

3. Mitochondrial Turnover in ALS

3.1. Mitochondrial Quality Control Mechanisms

Mitochondrial quality control (MQC) is an important process in cellular homeostasis. Mitochondria with loss of membrane potential or subject to protein oxidation and misfolding become targets of MQC. There are three main known pathways of MQC: Protein degradation, vesicular degradation, and mitophagy. The first involves proteostatic selective elimination of damaged proteins. Mitochondria have internal proteases, such as the AAA-protease complex of the inner membrane [77] and the Lon protease of the matrix [78]. Mitochondria are also endowed with their own unfolded protein response, which is activated when misfolded proteins accumulate in the matrix [79] or in the intermembrane space [80]. Mitochondria rely on the cytosolic ubiquitin-proteasome system to eliminate damaged proteins destined to the outer membrane or in the case of the intermembrane space, before they engage in the mitochondrial import pathway [81,82]. Ubiquitin-ligases, such as Parkin, ubiquitinate oxidized or misfolded outer membrane proteins [83]. Parkin recruitment has been ascribed to the kinase PINK1, following its incomplete processing and import across the outer membrane of depolarized mitochondria [84–86]. Instead, maintenance of Parkin cytosolic localization has been attributed to its interaction with cleaved PINK1 released in the cytosol after processing by healthy mitochondria [87]. If its degradation fails in damaged mitochondria, PINK1 phosphorylates ubiquitin residues and Parkin, thereby activating a cascade of ubiquitination of outer membrane proteins [88]. Ubiquitination of outer membrane proteins is not exclusively performed by Parkin, as it can also be carried out by resident outer membrane ligases, such as MULAN [89] and MITOL/MARCH5, which ubiquitinate outer membrane proteins involved in mitochondrial fusion, Mfn1 and Mfn2, and fission, Drp1 [90,91]. Interestingly, MARCH5 ubiquitinates and increases the turnover of mutant SOD1 on the outer membrane [92].

Ubiquitination of outer membrane proteins is one of the best characterized signals for the activation of mitophagy and the regulation of mitochondrial motility through degradation of proteins involved in mitochondrial fusion/fission and transport (reviewed in [93]). Fragmentation of the network and immobilization facilitate the engulfment of damaged mitochondria in autophagic vesicles. The regulation of PINK1 by phosphorylation of specific amino acid residues [94] could provide an additional link between energetic defects in unhealthy mitochondria and the activation of MQC. What drives the switch from proteostasis to mitophagy is unclear, but the extent of mitochondrial damage is likely a discriminating factor: When proteostasis cannot repair mitochondria, mitophagy ensues. However, in some cases, a Parkin-dependent vesicular degradation of sections of mitochondrial membranes containing oxidized proteins, can be sufficient to repair the damage and prevent full-blown mitophagy [95]. Mitophagy involves ubiquitin-binding adaptors that recruit mitochondria to the autophagosome by binding to LC3 [88].

3.2. Mitochondrial Quality Control in ALS

The best characterized MQC pathway is mediated by the activation of PINK1 and Parkin [96]. Impairment of MQC is most commonly linked to Parkinson's disease, due to the discovery of inactivating mutations of these proteins in recessive forms of the disease [97–99]. However, MQC disturbances have also been associated with ALS, for example through the involvement of the mitophagy adaptor optineurin, which has been found to be mutated in familial forms of the disease [100]. Optineurin plays a role in PINK1-Parkin mediated mitophagy. After Parkin recruitment to the outer membrane, optineurin binds to ubiquitinated mitochondria, inducing autophagosome nucleation through LC3 recruitment [101]. ALS-associated optineurin mutations cause mitochondrial clearance impairment [102]. The link between optineurin and LC3, which finalizes autophagic clearance of damaged mitochondria, is Tank-binding kinase (TBK1). It was reported that TBK1 interacts with optineurin and by phosphorylating it at specific serine residues regulates its ability to bind to ubiquitinated mitochondrial proteins [103]. Therefore, TBK1-mediated phosphorylation of optineurin amplifies and reinforces mitophagy. As a consequence, inhibition or depletion of TBK1 delays mitophagy, resulting in accumulation of damaged mitochondria [104]. TBK1 mutations have been causally linked to familial forms of ALS [105], further reinforcing the relevance of MQC in ALS pathophysiology. Mutations in two additional genes involved in mitophagy were reported in familial ALS, p62 and VCP. p62/SQSTMQ1 is an adaptor for the binding of ubiquitin to LC3, recruiting mitochondria to the autophagosome [106,107]. Valosin-containing protein (VCP) is an ATPase with segregase activity, which can extract ubiquitinated proteins from organelle membranes and target them to proteasomal degradation [108]. Parkin-mediated outer membrane protein ubiquitination recruit VCP to mitochondria [109]. Mutations in VCP are associated with ALS [110]. VCP mutations cause mitochondrial structural changes in transgenic mice, and loss of VCP impairs the clearance of damaged mitochondria [109]. Overall, the genetics of familial ALS strongly emphasizes the relevance of MQC in disease pathogenesis.

Mitochondrial damage with loss of mitochondrial membrane potential may result in Parkin-mediated degradation of Miro [111]. Furthermore, it was reported that PINK1 directly interacts with Miro and Milton [112], and that it phosphorylates Miro in response to mitochondrial damage, thereby promoting its interaction with Parkin and its degradation [113]. Interestingly, Miro is decreased in ALS models [114,115], which may contribute to impairing mitochondrial axonal transport. Moreover, Miro overexpression in mutant SOD1 neurons restores mitochondrial axonal transport deficits [116]. Taken together, these data point to a role for Miro as a converging point between the mechanisms regulating MQC and axonal transport.

The actin cytoskeleton emerges as an additional converging point between mitochondrial turnover and mitochondrial motility. A recent study showed that, after Parkin recruitment to depolarized mitochondria, F-actin encapsulates damaged mitochondria through myosin VI (MYO6) complexing with Parkin [117]. Hence, damaged mitochondria are completely isolated from the rest of the mitochondrial network, and are prevented from fusing with healthy mitochondria. They also showed that lack of MYO6 induces accumulation of damaged mitochondria, because of a severe impairment in the clearance machinery, highlighting the importance of actin-related players in MQC and maintenance of mitochondrial homeostasis.

The therapeutic value of modulating autophagy and mitophagy in ALS is the object of debate. Stimulation of autophagy was attempted mostly in the SOD1 mouse model, using both pharmacological and genetic approaches [118–123]. There were discrepancies in the outcomes, with either beneficial or detrimental effects. This divergence of results could derive from the different approaches used and their effects on different central nervous system (CNS) cell types. A genetic approach to test the cell type specificity of autophagy modulation was implemented by deleting the critical autophagy gene Atg7 specifically in motor neurons of mutant SOD1 mice [124]. Autophagy inhibition accelerated early neuromuscular denervation and the onset of motor symptoms. Surprisingly, removal of Atg7 also extended the lifespan of the animals. The authors proposed that motor neuron autophagy contributes to maintaining neuromuscular innervation early on, but later causes a non-cell-autonomous effect that promotes disease progression. Although mitophagy was not investigated specifically, the findings raise the intriguing possibility of a phase-dependent involvement of MQC in ALS progression.

Mitophagy induction in ALS is supported by the finding of LC3 II increase in neurons of mutant SOD1 mice [125,126], accompanied by the accumulation of p62 and optineurin [115,127]. Furthermore, an increase in mitochondria-containing autophagosomes and autophagolysosomes was described in human ALS spinal cord [128]. Currently, there are no pharmaceutical approaches to selectively modulate MQC in ALS. However, it is possible to target genetically the main components of the machinery, such as Parkin. Interestingly, a progressive decrease in Parkin levels was documented in both cellular and animal models of ALS [115,129,130], suggesting that chronic activation of MQC secondary to mitochondrial damage causes Parkin depletion. In cultured ALS neurons, it was demonstrated that Parkin is responsible for Miro degradation and actively contributes to mitochondrial transport impairment [116]. Overexpression of Miro or ablation of PINK1 rescued the mitochondrial axonal transport deficits. However, genetic constitutive ablation of Parkin in mutant SOD1 mice delayed the decline of Miro and other components of the mitochondrial dynamics machinery, and resulted in a significant delay of neuromuscular degeneration and extension of lifespan [115]. Taken together, these findings support the notion that MQC is involved in ALS pathophysiology. Importantly, they also show that autophagy, and specifically MQC, may be a double-edge sword, with initial protective effects, which can become maladaptive and detrimental in the chronic phase of the disease.

4. Conclusions

Mitochondrial dynamics, axonal transport, and MQC are tightly intertwined processes that play a fundamental role in the homeostasis of neuronal mitochondrial network, in health and disease.

The genetics of familial ALS strongly suggest that these processes are affected during the course of the disease. Figure 1 summarizes the main molecular players of mitochondrial transport and turnover, whose alterations have been proposed to participate in ALS pathogenesis. More work is needed to achieve a detailed mechanistic understanding of the pathogenic pathways linking mitochondria transport and turnover with mitochondrial dysfunction and motor neuron degeneration in ALS. In particular, whether altered MQC in ALS motor neurons is the cause or the consequence of impaired mitochondrial function remains to be elucidated. Nevertheless, a wealth of clues indicate that alterations of these processes can be disease initiators or disease modifiers, with potentially interesting therapeutic implications. Studies focused on these important aspects of ALS pathophysiology will unveil novel disease mechanisms that could be addressed therapeutically by targeted approaches aimed at modulating mitochondrial transport and MQC through pharmacological intervention.

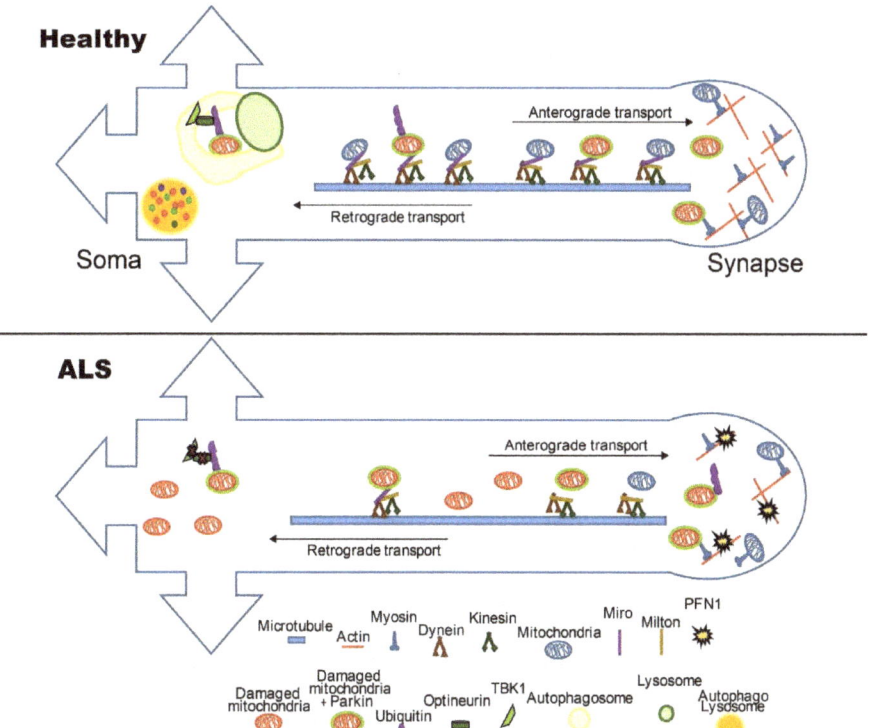

Figure 1. Schematic representation of key players in mitochondrial transport and turnover in healthy and amyotrophic lateral sclerosis (ALS) neurons. Mitochondria are transported in axons along microtubule tracks by dynein and kinesins, which are connected to mitochondria through cargo adaptors Milton and Miro. In ALS axons, the interactions between mitochondria and microtubules are disrupted, resulting in impaired transport. At synapses, mitochondria interact with the actin cytoskeleton, and mutations in proteins involved in actin dynamics, such as PFN1, can alter mitochondrial localization at this neuronal site. Ubiquitination of unhealthy mitochondria by Ub-ligases, such as Parkin, target mitochondria for degradation through the autophagy pathway. TBK1 and optineurin promote PINK1-Parkin ubiquitination of mitochondrial dynamics proteins, such as Miro. In ALS neurons, the quality control mechanisms are affected by dysfunction occurring at various steps of the mitophagy process.

Funding: This research was funded by grant NIH/NINDS R01NS062055.

Conflicts of Interest: The authors declare no conflict of interest.

References

1. Hardiman, O.; van den Berg, L.H.; Kiernan, M.C. Clinical diagnosis and management of amyotrophic lateral sclerosis. *Nat. Rev. Neurol.* **2011**, *7*, 639–649. [CrossRef]
2. De Vos, K.J.; Hafezparast, M. Neurobiology of axonal transport defects in motor neuron diseases: Opportunities for translational research? *Neurobiol. Dis.* **2017**, *105*, 283–299. [CrossRef]
3. Emery, A.E.; Holloway, S. Familial motor neuron diseases. *Adv. Neurol.* **1982**, *36*, 139–147. [PubMed]
4. Cirulli, E.T.; Lasseigne, B.N.; Petrovski, S.; Sapp, P.C.; Dion, P.A.; Leblond, C.S.; Couthouis, J.; Lu, Y.F.; Wang, Q.; Krueger, B.J.; et al. Exome sequencing in amyotrophic lateral sclerosis identifies risk genes and pathways. *Science* **2015**, *347*, 1436–1441. [CrossRef]
5. Taylor, J.P.; Brown, R.H., Jr.; Cleveland, D.W. Decoding ALS: From genes to mechanism. *Nature* **2016**, *539*, 197–206. [CrossRef]
6. Andrus, P.K.; Fleck, T.J.; Gurney, M.E.; Hall, E.D. Protein oxidative damage in a transgenic mouse model of familial amyotrophic lateral sclerosis. *J. Neurochem.* **1998**, *71*, 2041–2048. [CrossRef]
7. Wood, J.D.; Beaujeux, T.P.; Shaw, P.J. Protein aggregation in motor neurone disorders. *Neuropathol. Appl. Neurobiol.* **2003**, *29*, 529–545. [CrossRef]
8. Chang, Y.; Kong, Q.; Shan, X.; Tian, G.; Ilieva, H.; Cleveland, D.W.; Rothstein, J.D.; Borchelt, D.R.; Wong, P.C.; Lin, C.L. Messenger RNA oxidation occurs early in disease pathogenesis and promotes motor neuron degeneration in ALS. *PLoS ONE* **2008**, *3*, e2849. [CrossRef] [PubMed]
9. Shaw, P.J.; Ince, P.G.; Falkous, G.; Mantle, D. Oxidative damage to protein in sporadic motor neuron disease spinal cord. *Ann. Neurol.* **1995**, *38*, 691–695. [CrossRef] [PubMed]
10. Wiedemann, F.R.; Manfredi, G.; Mawrin, C.; Beal, M.F.; Schon, E.A. Mitochondrial DNA and respiratory chain function in spinal cords of ALS patients. *J. Neurochem.* **2002**, *80*, 616–625. [CrossRef]
11. Sasaki, S.; Iwata, M. Mitochondrial alterations in the spinal cord of patients with sporadic amyotrophic lateral sclerosis. *J. Neuropathol. Exp. Neurol.* **2007**, *66*, 10–16. [CrossRef] [PubMed]
12. De Vos, K.J.; Chapman, A.L.; Tennant, M.E.; Manser, C.; Tudor, E.L.; Lau, K.F.; Brownlees, J.; Ackerley, S.; Shaw, P.J.; McLoughlin, D.M.; et al. Familial amyotrophic lateral sclerosis-linked SOD1 mutants perturb fast axonal transport to reduce axonal mitochondria content. *Hum. Mol. Genet.* **2007**, *16*, 2720–2728. [CrossRef]
13. Bannwarth, S.; Ait-El-Mkadem, S.; Chaussenot, A.; Genin, E.C.; Lacas-Gervais, S.; Fragaki, K.; Berg-Alonso, L.; Kageyama, Y.; Serre, V.; Moore, D.G.; et al. A mitochondrial origin for frontotemporal dementia and amyotrophic lateral sclerosis through CHCHD10 involvement. *Brain J. Neurol.* **2014**, *137*, 2329–2345. [CrossRef]
14. Rosen, D.R.; Siddique, T.; Patterson, D.; Figlewicz, D.A.; Sapp, P.; Hentati, A.; Donaldson, D.; Goto, J.; O'Regan, J.P.; Deng, H.X.; et al. Mutations in Cu/Zn superoxide dismutase gene are associated with familial amyotrophic lateral sclerosis. *Nature* **1993**, *362*, 59–62. [CrossRef] [PubMed]
15. Sreedharan, J.; Blair, I.P.; Tripathi, V.B.; Hu, X.; Vance, C.; Rogelj, B.; Ackerley, S.; Durnall, J.C.; Williams, K.L.; Buratti, E.; et al. TDP-43 mutations in familial and sporadic amyotrophic lateral sclerosis. *Science* **2008**, *319*, 1668–1672. [CrossRef]
16. Yamanaka, K.; Chun, S.J.; Boillee, S.; Fujimori-Tonou, N.; Yamashita, H.; Gutmann, D.H.; Takahashi, R.; Misawa, H.; Cleveland, D.W. Astrocytes as determinants of disease progression in inherited amyotrophic lateral sclerosis. *Nat. Neurosci.* **2008**, *11*, 251–253. [CrossRef] [PubMed]
17. Nagai, M.; Re, D.B.; Nagata, T.; Chalazonitis, A.; Jessell, T.M.; Wichterle, H.; Przedborski, S. Astrocytes expressing ALS-linked mutated SOD1 release factors selectively toxic to motor neurons. *Nat. Neurosci.* **2007**, *10*, 615–622. [CrossRef] [PubMed]
18. Ferraiuolo, L.; Higginbottom, A.; Heath, P.R.; Barber, S.; Greenald, D.; Kirby, J.; Shaw, P.J. Dysregulation of astrocyte-motoneuron cross-talk in mutant superoxide dismutase 1-related amyotrophic lateral sclerosis. *Brain J. Neurol.* **2011**, *134*, 2627–2641. [CrossRef] [PubMed]
19. Yamanaka, K.; Boillee, S.; Roberts, E.A.; Garcia, M.L.; McAlonis-Downes, M.; Mikse, O.R.; Cleveland, D.W.; Goldstein, L.S. Mutant SOD1 in cell types other than motor neurons and oligodendrocytes accelerates onset of disease in ALS mice. *Proc. Natl. Acad. Sci. USA* **2008**, *105*, 7594–7599. [CrossRef] [PubMed]

20. Di Giorgio, F.P.; Carrasco, M.A.; Siao, M.C.; Maniatis, T.; Eggan, K. Non-cell autonomous effect of glia on motor neurons in an embryonic stem cell-based ALS model. *Nat. Neurosci.* **2007**, *10*, 608–614. [CrossRef]
21. Sargsyan, S.A.; Monk, P.N.; Shaw, P.J. Microglia as potential contributors to motor neuron injury in amyotrophic lateral sclerosis. *Glia* **2005**, *51*, 241–253. [CrossRef] [PubMed]
22. Vande Velde, C.; McDonald, K.K.; Boukhedimi, Y.; McAlonis-Downes, M.; Lobsiger, C.S.; Bel Hadj, S.; Zandona, A.; Julien, J.P.; Shah, S.B.; Cleveland, D.W. Misfolded SOD1 associated with motor neuron mitochondria alters mitochondrial shape and distribution prior to clinical onset. *PLoS ONE* **2011**, *6*, e22031. [CrossRef] [PubMed]
23. Hollenbeck, P.J.; Saxton, W.M. The axonal transport of mitochondria. *J. Cell Sci.* **2005**, *118*, 5411–5419. [CrossRef] [PubMed]
24. Kang, J.S.; Tian, J.H.; Pan, P.Y.; Zald, P.; Li, C.; Deng, C.; Sheng, Z.H. Docking of axonal mitochondria by syntaphilin controls their mobility and affects short-term facilitation. *Cell* **2008**, *132*, 137–148. [CrossRef] [PubMed]
25. Maday, S.; Holzbaur, E.L. Autophagosome assembly and cargo capture in the distal axon. *Autophagy* **2012**, *8*, 858–860. [CrossRef]
26. Hirokawa, N.; Niwa, S.; Tanaka, Y. Molecular motors in neurons: Transport mechanisms and roles in brain function, development, and disease. *Neuron* **2010**, *68*, 610–638. [CrossRef] [PubMed]
27. Sheng, Z.H.; Cai, Q. Mitochondrial transport in neurons: Impact on synaptic homeostasis and neurodegeneration. *Nat. Rev. Neurosci.* **2012**, *13*, 77–93. [CrossRef]
28. Fransson, A.; Ruusala, A.; Aspenstrom, P. Atypical Rho GTPases have roles in mitochondrial homeostasis and apoptosis. *J. Biol. Chem.* **2003**, *278*, 6495–6502. [CrossRef]
29. Fransson, S.; Ruusala, A.; Aspenstrom, P. The atypical Rho GTPases Miro-1 and Miro-2 have essential roles in mitochondrial trafficking. *Biochem. Biophys. Res. Commun.* **2006**, *344*, 500–510. [CrossRef]
30. Mironov, S.L. ADP regulates movements of mitochondria in neurons. *Biophys. J.* **2007**, *92*, 2944–2952. [CrossRef]
31. Wang, X.; Schwarz, T.L. The mechanism of Ca^{2+}-dependent regulation of kinesin-mediated mitochondrial motility. *Cell* **2009**, *136*, 163–174. [CrossRef]
32. Macaskill, A.F.; Rinholm, J.E.; Twelvetrees, A.E.; Arancibia-Carcamo, I.L.; Muir, J.; Fransson, A.; Aspenstrom, P.; Attwell, D.; Kittler, J.T. Miro1 is a calcium sensor for glutamate receptor-dependent localization of mitochondria at synapses. *Neuron* **2009**, *61*, 541–555. [CrossRef]
33. Eisen, A.; Weber, M. The motor cortex and amyotrophic lateral sclerosis. *Muscle Nerve* **2001**, *24*, 564–573. [CrossRef]
34. Kiernan, M.C.; Vucic, S.; Cheah, B.C.; Turner, M.R.; Eisen, A.; Hardiman, O.; Burrell, J.R.; Zoing, M.C. Amyotrophic lateral sclerosis. *Lancet* **2011**, *377*, 942–955. [CrossRef]
35. Fischer, L.R.; Culver, D.G.; Tennant, P.; Davis, A.A.; Wang, M.; Castellano-Sanchez, A.; Khan, J.; Polak, M.A.; Glass, J.D. Amyotrophic lateral sclerosis is a distal axonopathy: Evidence in mice and man. *Exp. Neurol.* **2004**, *185*, 232–240. [CrossRef]
36. Martineau, E.; Di Polo, A.; Vande Velde, C.; Robitaille, R. Dynamic neuromuscular remodeling precedes motor-unit loss in a mouse model of ALS. *eLife* **2018**, *7*. [CrossRef]
37. De Vos, K.J.; Grierson, A.J.; Ackerley, S.; Miller, C.C. Role of axonal transport in neurodegenerative diseases. *Annu. Rev. Neurosci.* **2008**, *31*, 151–173. [CrossRef]
38. Morfini, G.A.; Burns, M.; Binder, L.I.; Kanaan, N.M.; LaPointe, N.; Bosco, D.A.; Brown, R.H., Jr.; Brown, H.; Tiwari, A.; Hayward, L.; et al. Axonal transport defects in neurodegenerative diseases. *J. Neurosci. Off. J. Soc. Neurosci.* **2009**, *29*, 12776–12786. [CrossRef]
39. Wang, W.; Li, L.; Lin, W.L.; Dickson, D.W.; Petrucelli, L.; Zhang, T.; Wang, X. The ALS disease-associated mutant TDP-43 impairs mitochondrial dynamics and function in motor neurons. *Hum. Mol. Genet.* **2013**, *22*, 4706–4719. [CrossRef]
40. Magrane, J.; Sahawneh, M.A.; Przedborski, S.; Estevez, A.G.; Manfredi, G. Mitochondrial dynamics and bioenergetic dysfunction is associated with synaptic alterations in mutant SOD1 motor neurons. *J. Neurosci. Off. J. Soc. Neurosci.* **2012**, *32*, 229–242. [CrossRef]
41. Marinkovic, P.; Reuter, M.S.; Brill, M.S.; Godinho, L.; Kerschensteiner, M.; Misgeld, T. Axonal transport deficits and degeneration can evolve independently in mouse models of amyotrophic lateral sclerosis. *Proc. Natl. Acad. Sci. USA* **2012**, *109*, 4296–4301. [CrossRef] [PubMed]

42. Magrane, J.; Cortez, C.; Gan, W.B.; Manfredi, G. Abnormal mitochondrial transport and morphology are common pathological denominators in SOD1 and TDP43 ALS mouse models. *Hum. Mol. Genet.* **2014**, *23*, 1413–1424. [CrossRef] [PubMed]
43. Shahidullah, M.; Le Marchand, S.J.; Fei, H.; Zhang, J.; Pandey, U.B.; Dalva, M.B.; Pasinelli, P.; Levitan, I.B. Defects in synapse structure and function precede motor neuron degeneration in Drosophila models of FUS-related ALS. *J. Neurosci. Off. J. Soc. Neurosci.* **2013**, *33*, 19590–19598. [CrossRef] [PubMed]
44. Kwiatkowski, T.J., Jr.; Bosco, D.A.; Leclerc, A.L.; Tamrazian, E.; Vanderburg, C.R.; Russ, C.; Davis, A.; Gilchrist, J.; Kasarskis, E.J.; Munsat, T.; et al. Mutations in the FUS/TLS gene on chromosome 16 cause familial amyotrophic lateral sclerosis. *Science* **2009**, *323*, 1205–1208. [CrossRef] [PubMed]
45. Sharma, A.; Lyashchenko, A.K.; Lu, L.; Nasrabady, S.E.; Elmaleh, M.; Mendelsohn, M.; Nemes, A.; Tapia, J.C.; Mentis, G.Z.; Shneider, N.A. ALS-associated mutant FUS induces selective motor neuron degeneration through toxic gain of function. *Nat. Commun.* **2016**, *7*, 10465. [CrossRef] [PubMed]
46. So, E.; Mitchell, J.C.; Memmi, C.; Chennell, G.; Vizcay-Barrena, G.; Allison, L.; Shaw, C.E.; Vance, C. Mitochondrial abnormalities and disruption of the neuromuscular junction precede the clinical phenotype and motor neuron loss in hFUSWT transgenic mice. *Hum. Mol. Genet.* **2018**, *27*, 463–474. [CrossRef]
47. Dadon-Nachum, M.; Melamed, E.; Offen, D. The "dying-back" phenomenon of motor neurons in ALS. *J. Mol. Neurosci.* **2011**, *43*, 470–477. [CrossRef]
48. Krakora, D.; Macrander, C.; Suzuki, M. Neuromuscular junction protection for the potential treatment of amyotrophic lateral sclerosis. *Neurol. Res. Int.* **2012**, *2012*, 379657. [CrossRef]
49. Bilsland, L.G.; Sahai, E.; Kelly, G.; Golding, M.; Greensmith, L.; Schiavo, G. Deficits in axonal transport precede ALS symptoms in vivo. *Proc. Natl. Acad. Sci. USA* **2010**, *107*, 20523–20528. [CrossRef]
50. Ligon, L.A.; LaMonte, B.H.; Wallace, K.E.; Weber, N.; Kalb, R.G.; Holzbaur, E.L. Mutant superoxide dismutase disrupts cytoplasmic dynein in motor neurons. *Neuroreport* **2005**, *16*, 533–536. [CrossRef]
51. Zhang, F.; Strom, A.L.; Fukada, K.; Lee, S.; Hayward, L.J.; Zhu, H. Interaction between familial amyotrophic lateral sclerosis (ALS)-linked SOD1 mutants and the dynein complex. *J. Biol. Chem.* **2007**, *282*, 16691–16699. [CrossRef] [PubMed]
52. Igoudjil, A.; Magrane, J.; Fischer, L.R.; Kim, H.J.; Hervias, I.; Dumont, M.; Cortez, C.; Glass, J.D.; Starkov, A.A.; Manfredi, G. In vivo pathogenic role of mutant SOD1 localized in the mitochondrial intermembrane space. *J. Neurosci. Off. J. Soc. Neurosci.* **2011**, *31*, 15826–15837. [CrossRef]
53. Pickles, S.; Destroismaisons, L.; Peyrard, S.L.; Cadot, S.; Rouleau, G.A.; Brown, R.H., Jr.; Julien, J.P.; Arbour, N.; Vande Velde, C. Mitochondrial damage revealed by immunoselection for ALS-linked misfolded SOD1. *Hum. Mol. Genet.* **2013**, *22*, 3947–3959. [CrossRef] [PubMed]
54. Pickles, S.; Semmler, S.; Broom, H.R.; Destroismaisons, L.; Legroux, L.; Arbour, N.; Meiering, E.; Cashman, N.R.; Vande Velde, C. ALS-linked misfolded SOD1 species have divergent impacts on mitochondria. *Acta Neuropathol. Commun.* **2016**, *4*, 43. [CrossRef] [PubMed]
55. Li, Q.; Vande Velde, C.; Israelson, A.; Xie, J.; Bailey, A.O.; Dong, M.Q.; Chun, S.J.; Roy, T.; Winer, L.; Yates, J.R.; et al. ALS-linked mutant superoxide dismutase 1 (SOD1) alters mitochondrial protein composition and decreases protein import. *Proc. Natl. Acad. Sci. USA* **2010**, *107*, 21146–21151. [CrossRef]
56. Liu, J.; Lillo, C.; Jonsson, P.A.; Vande Velde, C.; Ward, C.M.; Miller, T.M.; Subramaniam, J.R.; Rothstein, J.D.; Marklund, S.; Andersen, P.M.; et al. Toxicity of familial ALS-linked SOD1 mutants from selective recruitment to spinal mitochondria. *Neuron* **2004**, *43*, 5–17. [CrossRef]
57. Magrane, J.; Hervias, I.; Henning, M.S.; Damiano, M.; Kawamata, H.; Manfredi, G. Mutant SOD1 in neuronal mitochondria causes toxicity and mitochondrial dynamics abnormalities. *Hum. Mol. Genet.* **2009**, *18*, 4552–4564. [CrossRef]
58. Cozzolino, M.; Pesaresi, M.G.; Amori, I.; Crosio, C.; Ferri, A.; Nencini, M.; Carri, M.T. Oligomerization of mutant SOD1 in mitochondria of motoneuronal cells drives mitochondrial damage and cell toxicity. *Antioxid. Redox Signal.* **2009**, *11*, 1547–1558. [CrossRef]
59. Mattiazzi, M.; D'Aurelio, M.; Gajewski, C.D.; Martushova, K.; Kiaei, M.; Beal, M.F.; Manfredi, G. Mutated human SOD1 causes dysfunction of oxidative phosphorylation in mitochondria of transgenic mice. *J. Biol. Chem.* **2002**, *277*, 29626–29633. [CrossRef]
60. Kawamata, H.; Manfredi, G. Different regulation of wild-type and mutant Cu,Zn superoxide dismutase localization in mammalian mitochondria. *Hum. Mol. Genet.* **2008**, *17*, 3303–3317. [CrossRef]

61. Ferri, A.; Cozzolino, M.; Crosio, C.; Nencini, M.; Casciati, A.; Gralla, E.B.; Rotilio, G.; Valentine, J.S.; Carri, M.T. Familial ALS-superoxide dismutases associate with mitochondria and shift their redox potentials. *Proc. Natl. Acad. Sci. USA* **2006**, *103*, 13860–13865. [CrossRef] [PubMed]
62. Morris, R.L.; Hollenbeck, P.J. Axonal transport of mitochondria along microtubules and F-actin in living vertebrate neurons. *J. Cell Biol.* **1995**, *131*, 1315–1326. [CrossRef] [PubMed]
63. Hensel, N.; Claus, P. The Actin Cytoskeleton in SMA and ALS: How Does It Contribute to Motoneuron Degeneration? *Neurosci. Rev. J. Bringing Neurobiol. Neurol. Psychiatry* **2018**, *24*, 54–72. [CrossRef]
64. Wu, C.H.; Fallini, C.; Ticozzi, N.; Keagle, P.J.; Sapp, P.C.; Piotrowska, K.; Lowe, P.; Koppers, M.; McKenna-Yasek, D.; Baron, D.M.; et al. Mutations in the profilin 1 gene cause familial amyotrophic lateral sclerosis. *Nature* **2012**, *488*, 499–503. [CrossRef] [PubMed]
65. Carlsson, L.; Nystrom, L.E.; Sundkvist, I.; Markey, F.; Lindberg, U. Actin polymerizability is influenced by profilin, a low molecular weight protein in non-muscle cells. *J. Mol. Biol.* **1977**, *115*, 465–483. [CrossRef]
66. Birbach, A. Profilin, a multi-modal regulator of neuronal plasticity. *Bioessays News Rev. Mol. Cell. Dev. Biol.* **2008**, *30*, 994–1002. [CrossRef]
67. Kang, F.; Purich, D.L.; Southwick, F.S. Profilin promotes barbed-end actin filament assembly without lowering the critical concentration. *J. Biol. Chem.* **1999**, *274*, 36963–36972. [CrossRef]
68. Yarmola, E.G.; Bubb, M.R. How depolymerization can promote polymerization: The case of actin and profilin. *Bioessays News Rev. Mol. Cell. Dev. Biol.* **2009**, *31*, 1150–1160. [CrossRef]
69. Smith, B.N.; Vance, C.; Scotter, E.L.; Troakes, C.; Wong, C.H.; Topp, S.; Maekawa, S.; King, A.; Mitchell, J.C.; Lund, K.; et al. Novel mutations support a role for Profilin 1 in the pathogenesis of ALS. *Neurobiol. Aging* **2015**, *36*, 1602.e17-27. [CrossRef]
70. Tanaka, Y.; Nonaka, T.; Suzuki, G.; Kametani, F.; Hasegawa, M. Gain-of-function profilin 1 mutations linked to familial amyotrophic lateral sclerosis cause seed-dependent intracellular TDP-43 aggregation. *Hum. Mol. Genet.* **2016**, *25*, 1420–1433. [CrossRef]
71. Boopathy, S.; Silvas, T.V.; Tischbein, M.; Jansen, S.; Shandilya, S.M.; Zitzewitz, J.A.; Landers, J.E.; Goode, B.L.; Schiffer, C.A.; Bosco, D.A. Structural basis for mutation-induced destabilization of profilin 1 in ALS. *Proc. Natl. Acad. Sci. USA* **2015**, *112*, 7984–7989. [CrossRef]
72. Yang, C.; Danielson, E.W.; Qiao, T.; Metterville, J.; Brown, R.H., Jr.; Landers, J.E.; Xu, Z. Mutant PFN1 causes ALS phenotypes and progressive motor neuron degeneration in mice by a gain of toxicity. *Proc. Natl. Acad. Sci. USA* **2016**, *113*, E6209–E6218. [CrossRef]
73. Fil, D.; DeLoach, A.; Yadav, S.; Alkam, D.; MacNicol, M.; Singh, A.; Compadre, C.M.; Goellner, J.J.; O'Brien, C.A.; Fahmi, T.; et al. Mutant Profilin1 transgenic mice recapitulate cardinal features of motor neuron disease. *Hum. Mol. Genet.* **2017**, *26*, 686–701. [CrossRef]
74. Shneyer, B.I.; Usaj, M.; Henn, A. Myo19 is an outer mitochondrial membrane motor and effector of starvation-induced filopodia. *J. Cell Sci.* **2016**, *129*, 543–556. [CrossRef]
75. Lopez-Domenech, G.; Covill-Cooke, C.; Ivankovic, D.; Halff, E.F.; Sheehan, D.F.; Norkett, R.; Birsa, N.; Kittler, J.T. Miro proteins coordinate microtubule- and actin-dependent mitochondrial transport and distribution. *Embo J.* **2018**, *37*, 321–336. [CrossRef]
76. Oeding, S.J.; Majstrowicz, K.; Hu, X.P.; Schwarz, V.; Freitag, A.; Honnert, U.; Nikolaus, P.; Bahler, M. Identification of Miro1 and Miro2 as mitochondrial receptors for myosin XIX. *J. Cell Sci.* **2018**, *131*, jcs219469. [CrossRef]
77. Gerdes, F.; Tatsuta, T.; Langer, T. Mitochondrial AAA proteases—Towards a molecular understanding of membrane-bound proteolytic machines. *Biochim. Biophys. Acta* **2012**, *1823*, 49–55. [CrossRef]
78. Matsushima, Y.; Kaguni, L.S. Matrix proteases in mitochondrial DNA function. *Biochim. Biophys. Acta* **2012**, *1819*, 1080–1087. [CrossRef]
79. Pellegrino, M.W.; Nargund, A.M.; Haynes, C.M. Signaling the mitochondrial unfolded protein response. *Biochim. Biophys. Acta* **2013**, *1833*, 410–416. [CrossRef]
80. Papa, L.; Germain, D. Estrogen receptor mediates a distinct mitochondrial unfolded protein response. *J. Cell Sci.* **2011**, *124*, 1396–1402. [CrossRef]
81. Radke, S.; Chander, H.; Schafer, P.; Meiss, G.; Kruger, R.; Schulz, J.B.; Germain, D. Mitochondrial protein quality control by the proteasome involves ubiquitination and the protease Omi. *J. Biol. Chem.* **2008**, *283*, 12681–12685. [CrossRef] [PubMed]

82. Karbowski, M.; Youle, R.J. Regulating mitochondrial outer membrane proteins by ubiquitination and proteasomal degradation. *Curr. Opin. Cell Biol.* **2011**, *23*, 476–482. [CrossRef]
83. Heo, J.M.; Rutter, J. Ubiquitin-dependent mitochondrial protein degradation. *Int. J. Biochem. Cell Biol.* **2011**, *43*, 1422–1426. [CrossRef]
84. Narendra, D.; Tanaka, A.; Suen, D.F.; Youle, R.J. Parkin is recruited selectively to impaired mitochondria and promotes their autophagy. *J. Cell Biol.* **2008**, *183*, 795–803. [CrossRef]
85. Narendra, D.P.; Jin, S.M.; Tanaka, A.; Suen, D.F.; Gautier, C.A.; Shen, J.; Cookson, M.R.; Youle, R.J. PINK1 is selectively stabilized on impaired mitochondria to activate Parkin. *PLoS Biol.* **2010**, *8*, e1000298. [CrossRef] [PubMed]
86. Deas, E.; Plun-Favreau, H.; Gandhi, S.; Desmond, H.; Kjaer, S.; Loh, S.H.; Renton, A.E.; Harvey, R.J.; Whitworth, A.J.; Martins, L.M.; et al. PINK1 cleavage at position A103 by the mitochondrial protease PARL. *Hum. Mol. Genet.* **2011**, *20*, 867–879. [CrossRef]
87. Vives-Bauza, C.; de Vries, R.L.; Tocilescu, M.; Przedborski, S. PINK1/Parkin direct mitochondria to autophagy. *Autophagy* **2010**, *6*, 315–316. [CrossRef]
88. Lazarou, M.; Sliter, D.A.; Kane, L.A.; Sarraf, S.A.; Wang, C.; Burman, J.L.; Sideris, D.P.; Fogel, A.I.; Youle, R.J. The ubiquitin kinase PINK1 recruits autophagy receptors to induce mitophagy. *Nature* **2015**, *524*, 309–314. [CrossRef]
89. Li, W.; Bengtson, M.H.; Ulbrich, A.; Matsuda, A.; Reddy, V.A.; Orth, A.; Chanda, S.K.; Batalov, S.; Joazeiro, C.A. Genome-wide and functional annotation of human E3 ubiquitin ligases identifies MULAN, a mitochondrial E3 that regulates the organelle's dynamics and signaling. *PLoS ONE* **2008**, *3*, e1487. [CrossRef] [PubMed]
90. Yonashiro, R.; Ishido, S.; Kyo, S.; Fukuda, T.; Goto, E.; Matsuki, Y.; Ohmura-Hoshino, M.; Sada, K.; Hotta, H.; Yamamura, H.; et al. A novel mitochondrial ubiquitin ligase plays a critical role in mitochondrial dynamics. *Embo J.* **2006**, *25*, 3618–3626. [CrossRef]
91. Karbowski, M.; Neutzner, A.; Youle, R.J. The mitochondrial E3 ubiquitin ligase MARCH5 is required for Drp1 dependent mitochondrial division. *J. Cell Biol.* **2007**, *178*, 71–84. [CrossRef] [PubMed]
92. Yonashiro, R.; Sugiura, A.; Miyachi, M.; Fukuda, T.; Matsushita, N.; Inatome, R.; Ogata, Y.; Suzuki, T.; Dohmae, N.; Yanagi, S. Mitochondrial ubiquitin ligase MITOL ubiquitinates mutant SOD1 and attenuates mutant SOD1-induced reactive oxygen species generation. *Mol. Biol. Cell* **2009**, *20*, 4524–4530. [CrossRef] [PubMed]
93. Ashrafi, G.; Schwarz, T.L. The pathways of mitophagy for quality control and clearance of mitochondria. *Cell Death Differ.* **2013**, *20*, 31–42. [CrossRef] [PubMed]
94. Aerts, L.; Craessaerts, K.; De Strooper, B.; Morais, V.A. PINK1 kinase catalytic activity is regulated by phosphorylation on serines 228 and 402. *J. Biol. Chem.* **2015**, *290*, 2798–2811. [CrossRef]
95. Soubannier, V.; McLelland, G.L.; Zunino, R.; Braschi, E.; Rippstein, P.; Fon, E.A.; McBride, H.M. A vesicular transport pathway shuttles cargo from mitochondria to lysosomes. *Curr. Biol.* **2012**, *22*, 135–141. [CrossRef]
96. Ravikumar, B.; Sarkar, S.; Davies, J.E.; Futter, M.; Garcia-Arencibia, M.; Green-Thompson, Z.W.; Jimenez-Sanchez, M.; Korolchuk, V.I.; Lichtenberg, M.; Luo, S.; et al. Regulation of mammalian autophagy in physiology and pathophysiology. *Physiol. Rev.* **2010**, *90*, 1383–1435. [CrossRef]
97. Kitada, T.; Asakawa, S.; Hattori, N.; Matsumine, H.; Yamamura, Y.; Minoshima, S.; Yokochi, M.; Mizuno, Y.; Shimizu, N. Mutations in the parkin gene cause autosomal recessive juvenile parkinsonism. *Nature* **1998**, *392*, 605–608. [CrossRef]
98. Valente, E.M.; Abou-Sleiman, P.M.; Caputo, V.; Muqit, M.M.; Harvey, K.; Gispert, S.; Ali, Z.; Del Turco, D.; Bentivoglio, A.R.; Healy, D.G.; et al. Hereditary early-onset Parkinson's disease caused by mutations in PINK1. *Science* **2004**, *304*, 1158–1160. [CrossRef]
99. Youle, R.J.; Narendra, D.P. Mechanisms of mitophagy. *Nat. Rev. Mol. Cell Biol.* **2011**, *12*, 9–14. [CrossRef]
100. Maruyama, H.; Morino, H.; Ito, H.; Izumi, Y.; Kato, H.; Watanabe, Y.; Kinoshita, Y.; Kamada, M.; Nodera, H.; Suzuki, H.; et al. Mutations of optineurin in amyotrophic lateral sclerosis. *Nature* **2010**, *465*, 223–226. [CrossRef]
101. Poole, A.C.; Thomas, R.E.; Yu, S.; Vincow, E.S.; Pallanck, L. The mitochondrial fusion-promoting factor mitofusin is a substrate of the PINK1/parkin pathway. *PLoS ONE* **2010**, *5*, e10054. [CrossRef]
102. Wong, Y.C.; Holzbaur, E.L. Optineurin is an autophagy receptor for damaged mitochondria in parkin-mediated mitophagy that is disrupted by an ALS-linked mutation. *Proc. Natl. Acad. Sci. USA* **2014**, *111*, E4439–E4448. [CrossRef]

103. Richter, B.; Sliter, D.A.; Herhaus, L.; Stolz, A.; Wang, C.; Beli, P.; Zaffagnini, G.; Wild, P.; Martens, S.; Wagner, S.A.; et al. Phosphorylation of OPTN by TBK1 enhances its binding to Ub chains and promotes selective autophagy of damaged mitochondria. *Proc. Natl. Acad. Sci. USA* **2016**, *113*, 4039–4044. [CrossRef]
104. Moore, A.S.; Holzbaur, E.L. Dynamic recruitment and activation of ALS-associated TBK1 with its target optineurin are required for efficient mitophagy. *Proc. Natl. Acad. Sci. USA* **2016**, *113*, E3349–E3358. [CrossRef]
105. Freischmidt, A.; Wieland, T.; Richter, B.; Ruf, W.; Schaeffer, V.; Muller, K.; Marroquin, N.; Nordin, F.; Hubers, A.; Weydt, P.; et al. Haploinsufficiency of TBK1 causes familial ALS and fronto-temporal dementia. *Nat. Neurosci.* **2015**, *18*, 631–636. [CrossRef]
106. Rubino, E.; Rainero, I.; Chio, A.; Rogaeva, E.; Galimberti, D.; Fenoglio, P.; Grinberg, Y.; Isaia, G.; Calvo, A.; Gentile, S.; et al. SQSTM1 mutations in frontotemporal lobar degeneration and amyotrophic lateral sclerosis. *Neurology* **2012**, *79*, 1556–1562. [CrossRef]
107. Goode, A.; Butler, K.; Long, J.; Cavey, J.; Scott, D.; Shaw, B.; Sollenberger, J.; Gell, C.; Johansen, T.; Oldham, N.J.; et al. Defective recognition of LC3B by mutant SQSTM1/p62 implicates impairment of autophagy as a pathogenic mechanism in ALS-FTLD. *Autophagy* **2016**, *12*, 1094–1104. [CrossRef]
108. Ye, Y.; Shibata, Y.; Kikkert, M.; van Voorden, S.; Wiertz, E.; Rapoport, T.A. Recruitment of the p97 ATPase and ubiquitin ligases to the site of retrotranslocation at the endoplasmic reticulum membrane. *Proc. Natl. Acad. Sci. USA* **2005**, *102*, 14132–14138. [CrossRef]
109. Kim, N.C.; Tresse, E.; Kolaitis, R.M.; Molliex, A.; Thomas, R.E.; Alami, N.H.; Wang, B.; Joshi, A.; Smith, R.B.; Ritson, G.P.; et al. VCP is essential for mitochondrial quality control by PINK1/Parkin and this function is impaired by VCP mutations. *Neuron* **2013**, *78*, 65–80. [CrossRef]
110. Johnson, J.O.; Mandrioli, J.; Benatar, M.; Abramzon, Y.; Van Deerlin, V.M.; Trojanowski, J.Q.; Gibbs, J.R.; Brunetti, M.; Gronka, S.; Wuu, J.; et al. Exome sequencing reveals VCP mutations as a cause of familial ALS. *Neuron* **2010**, *68*, 857–864. [CrossRef]
111. Wang, X.; Winter, D.; Ashrafi, G.; Schlehe, J.; Wong, Y.L.; Selkoe, D.; Rice, S.; Steen, J.; LaVoie, M.J.; Schwarz, T.L. PINK1 and Parkin target Miro for phosphorylation and degradation to arrest mitochondrial motility. *Cell* **2011**, *147*, 893–906. [CrossRef]
112. Weihofen, A.; Thomas, K.J.; Ostaszewski, B.L.; Cookson, M.R.; Selkoe, D.J. Pink1 forms a multiprotein complex with Miro and Milton, linking Pink1 function to mitochondrial trafficking. *Biochemistry* **2009**, *48*, 2045–2052. [CrossRef]
113. Shlevkov, E.; Kramer, T.; Schapansky, J.; LaVoie, M.J.; Schwarz, T.L. Miro phosphorylation sites regulate Parkin recruitment and mitochondrial motility. *Proc. Natl. Acad. Sci. USA* **2016**, *113*, E6097–E6106. [CrossRef]
114. Zhang, F.; Wang, W.; Siedlak, S.L.; Liu, Y.; Liu, J.; Jiang, K.; Perry, G.; Zhu, X.; Wang, X. Miro1 deficiency in amyotrophic lateral sclerosis. *Front. Aging Neurosci.* **2015**, *7*, 100. [CrossRef]
115. Palomo, G.M.; Granatiero, V.; Kawamata, H.; Konrad, C.; Kim, M.; Arreguin, A.J.; Zhao, D.; Milner, T.A.; Manfredi, G. Parkin is a disease modifier in the mutant SOD1 mouse model of ALS. *Embo Mol. Med.* **2018**, *10*. [CrossRef]
116. Moller, A.; Bauer, C.S.; Cohen, R.N.; Webster, C.P.; De Vos, K.J. Amyotrophic lateral sclerosis-associated mutant SOD1 inhibits anterograde axonal transport of mitochondria by reducing Miro1 levels. *Hum. Mol. Genet.* **2017**, *26*, 4668–4679. [CrossRef]
117. Kruppa, A.J.; Kishi-Itakura, C.; Masters, T.A.; Rorbach, J.E.; Grice, G.L.; Kendrick-Jones, J.; Nathan, J.A.; Minczuk, M.; Buss, F. Myosin VI-Dependent Actin Cages Encapsulate Parkin-Positive Damaged Mitochondria. *Dev. Cell* **2018**, *44*, 484–499 e486. [CrossRef]
118. Fornai, F.; Longone, P.; Cafaro, L.; Kastsiuchenka, O.; Ferrucci, M.; Manca, M.L.; Lazzeri, G.; Spalloni, A.; Bellio, N.; Lenzi, P.; et al. Lithium delays progression of amyotrophic lateral sclerosis. *Proc. Natl. Acad. Sci. USA* **2008**, *105*, 2052–2057. [CrossRef]
119. Pizzasegola, C.; Caron, I.; Daleno, C.; Ronchi, A.; Minoia, C.; Carri, M.T.; Bendotti, C. Treatment with lithium carbonate does not improve disease progression in two different strains of SOD1 mutant mice. *Amyotroph. Lateral Scler. Off. Publ. World Fed. Neurol. Res. Group Mot. Neuron Dis.* **2009**, *10*, 221–228. [CrossRef]
120. Zhang, X.; Li, L.; Chen, S.; Yang, D.; Wang, Y.; Zhang, X.; Wang, Z.; Le, W. Rapamycin treatment augments motor neuron degeneration in SOD1(G93A) mouse model of amyotrophic lateral sclerosis. *Autophagy* **2011**, *7*, 412–425. [CrossRef]

121. Zhang, K.; Shi, P.; An, T.; Wang, Q.; Wang, J.; Li, Z.; Duan, W.; Li, C.; Guo, Y. Food restriction-induced autophagy modulates degradation of mutant SOD1 in an amyotrophic lateral sclerosis mouse model. *Brain Res.* **2013**, *1519*, 112–119. [CrossRef]
122. Zhang, X.; Chen, S.; Song, L.; Tang, Y.; Shen, Y.; Jia, L.; Le, W. MTOR-independent, autophagic enhancer trehalose prolongs motor neuron survival and ameliorates the autophagic flux defect in a mouse model of amyotrophic lateral sclerosis. *Autophagy* **2014**, *10*, 588–602. [CrossRef]
123. Kim, J.; Kim, T.Y.; Cho, K.S.; Kim, H.N.; Koh, J.Y. Autophagy activation and neuroprotection by progesterone in the G93A-SOD1 transgenic mouse model of amyotrophic lateral sclerosis. *Neurobiol. Dis.* **2013**, *59*, 80–85. [CrossRef]
124. Rudnick, N.D.; Griffey, C.J.; Guarnieri, P.; Gerbino, V.; Wang, X.; Piersaint, J.A.; Tapia, J.C.; Rich, M.M.; Maniatis, T. Distinct roles for motor neuron autophagy early and late in the SOD1(G93A) mouse model of ALS. *Proc. Natl. Acad. Sci. USA* **2017**, *114*, E8294–E8303. [CrossRef]
125. Li, L.; Zhang, X.; Le, W. Altered macroautophagy in the spinal cord of SOD1 mutant mice. *Autophagy* **2008**, *4*, 290–293. [CrossRef]
126. Morimoto, N.; Nagai, M.; Ohta, Y.; Miyazaki, K.; Kurata, T.; Morimoto, M.; Murakami, T.; Takehisa, Y.; Ikeda, Y.; Kamiya, T.; et al. Increased autophagy in transgenic mice with a G93A mutant SOD1 gene. *Brain Res.* **2007**, *1167*, 112–117. [CrossRef]
127. Xie, Y.; Zhou, B.; Lin, M.Y.; Wang, S.; Foust, K.D.; Sheng, Z.H. Endolysosomal Deficits Augment Mitochondria Pathology in Spinal Motor Neurons of Asymptomatic fALS Mice. *Neuron* **2015**, *87*, 355–370. [CrossRef]
128. Sasaki, S. Autophagy in spinal cord motor neurons in sporadic amyotrophic lateral sclerosis. *J. Neuropathol. Exp. Neurol.* **2011**, *70*, 349–359. [CrossRef]
129. Lagier-Tourenne, C.; Polymenidou, M.; Hutt, K.R.; Vu, A.Q.; Baughn, M.; Huelga, S.C.; Clutario, K.M.; Ling, S.C.; Liang, T.Y.; Mazur, C.; et al. Divergent roles of ALS-linked proteins FUS/TLS and TDP-43 intersect in processing long pre-mRNAs. *Nat. Neurosci.* **2012**, *15*, 1488–1497. [CrossRef]
130. Polymenidou, M.; Lagier-Tourenne, C.; Hutt, K.R.; Huelga, S.C.; Moran, J.; Liang, T.Y.; Ling, S.C.; Sun, E.; Wancewicz, E.; Mazur, C.; et al. Long pre-mRNA depletion and RNA missplicing contribute to neuronal vulnerability from loss of TDP-43. *Nat. Neurosci.* **2011**, *14*, 459–468. [CrossRef]

© 2019 by the authors. Licensee MDPI, Basel, Switzerland. This article is an open access article distributed under the terms and conditions of the Creative Commons Attribution (CC BY) license (http://creativecommons.org/licenses/by/4.0/).

Review

Mitochondrial Dysfunction and Multiple Sclerosis

Isabella Peixoto de Barcelos, Regina M. Troxell and Jennifer S. Graves *

Department of Neurosciences, University of California San Diego, San Diego, CA 92093-0935, USA; ibarcelos@ucsd.edu (I.P.d.B.); rtroxell@ucsd.edu (R.M.T.)
* Correspondence: jgraves@ucsd.edu; Tel.: +1-(858)-246-2384

Received: 7 February 2019; Accepted: 30 April 2019; Published: 11 May 2019

Abstract: In recent years, several studies have examined the potential associations between mitochondrial dysfunction and neurodegenerative diseases such as multiple sclerosis (MS), Parkinson's disease and Alzheimer's disease. In MS, neurological disability results from inflammation, demyelination, and ultimately, axonal damage within the central nervous system. The sustained inflammatory phase of the disease leads to ion channel changes and chronic oxidative stress. Several independent investigations have demonstrated mitochondrial respiratory chain deficiency in MS, as well as abnormalities in mitochondrial transport. These processes create an energy imbalance and contribute to a parallel process of progressive neurodegeneration and irreversible disability. The potential roles of mitochondria in neurodegeneration are reviewed. An overview of mitochondrial diseases that may overlap with MS are also discussed, as well as possible therapeutic targets for the treatment of MS and other neurodegenerative conditions.

Keywords: multiple sclerosis; mitochondria; neuroinflammation; neurodegeneration

---

1. Introduction

Neurodegenerative diseases are characterized by neurologic dysfunction with a progressive course and consequent neuronal death [1]. Although these diseases, including multiple sclerosis, Alzheimer's disease, and Parkinson's disease, have different physiopathologies in their onset, they have a similar eventual course of gradual neurological decline and neuronal loss [2].

Multiple sclerosis (MS) is a leading cause of neurologic disability in young adults. MS is characterized by focal areas of demyelination in the white matter of the central nervous system (CNS) with secondary neuroaxonal degeneration [3,4]. The mean age of onset in females is approximately 30, compared to 33 years in males [5]. The sex ratio is 3:1, female to male, though men often progress more quickly and experience more rapid disability accumulation [3].

Among all patients with MS, about 85% present with a relapsing remitting form, which has alternate periods of acute demyelination (relapses) and periods of neurological recovery and stability (RRMS). For most patients, after 15–20 years the disease passes into a secondary progressive course (SPMS) which is characterized by an insidious progression of worsened neurological function with few or no acute relapses [3,4]. The remaining 10–15% of patients progress continuously from the first clinical manifestation of symptoms [4]; this is called the primary progressive form of multiple sclerosis (PPMS) and presents later in life, with a mean age of 45 years. The incidence of this form of the disease is approximately equivalent for men and women [6].

At the pathophysiological level, MS is characterized by two phases: At the initiation of a new lesion, there is a predominance of acute inflammation; subsequently, a state of chronic inflammation ensues with neurodegeneration. During the former, there is penetration of the blood brain barrier by activated immune cells against the myelin sheath. Inflammation in MS is due in part to components of both the innate and adaptive immune systems [7,8]. In brief, there is proliferation and dysregulation of pro-inflammatory T lymphocytes (Th1 and Th 17), as well as activation of B cells and secretion of inflammatory cytokines [9].

Pathognomonic inflammatory events in MS also activate neurodegenerative processes that lead to the destruction of oligodendrocytes, axons, and ultimately, neurons [7,8]. Brain, spinal cord and retinal atrophy are the result of the presence of neurodegeneration even at early stages of MS, meaning that both processes of acute inflammation and neurodegeneration co-exist since the first symptoms of the disease in the gray and white matter [3]. RRMS has a more prominent neuro-inflammatory phenotype, while the SPMS and PPMS forms are largely characterized by neurodegeneration [3].

The current available treatments for MS are directed against the acute episodes of neuroinflammation; this works well to prevent relapse events but is an approach with limited efficacy for protection against neurodegeneration, particularly in progressive forms of the disease [4]. The recent advancements in the understanding of mitochondrial dysfunction in neurodegenerative diseases, and in particular MS, bring new perspectives for future prevention of neuronal loss. We herein review the multifaceted role of mitochondria in MS pathology and the unique genetic factors that may contribute to the disease.

2. Mitochondria and Their Role in Neurodegeneration in Multiple Sclerosis

2.1. Mitochondria

Mitochondria uniquely have dual genomic expression of proteins that originate from both nuclear and mitochondrial DNA (mtDNA) [10]. The multi-copy nature of mitochondria gives rise to the concept of heteroplasmy (when both mutated and wild-type mtDNA molecules coexist in the same cell) and homoplasmy (when only mutant mtDNA molecules are present in the mitochondria of the cell). For a disease to manifest symptoms, the mutated mtDNA molecules in a tissue must increase to a pivotal threshold beyond which oxidative phosphorylation (OXPHOS) is impaired, thereby demonstrating a critical ratio of mutant to wild-type mtDNA [11–13]. The mitochondrial genome (5-μm circles, or 16.569 kilobases) is smaller than the nuclear genome, is highly compacted, and has only one DNA polymerase (polymerase γ) without any introns [13,14]. The lack of protective histones facilitates the accumulation of mtDNA mutations in an environment with a high concentration of reactive oxygen species (ROS) [15].

Mitochondria are organelles which are responsible for cellular bioenergetics via the Krebs cycle (with the production of NADH and FADH) and oxidative phosphorylation (OXPHOS), for cellular bioenergetics with secondary ATP production [16,17]. The mitochondria's main functions in bioenergetics include acting upon the electron transport chain (ETC) on the inner mitochondrial membrane, which is composed by four complexes (complex I, II, III and IV). They are also involved in the sequential reaction of reduction, OXPHOS, and electron flow (derived from NADH and FADH). This causes energy release which is used to transport protons from the matrix to the intermembrane space, creating an electrochemical gradient. ATP synthase (considered the complex V) uses this gradient to phosphorylate ADP to ATP [4].

Mitochondria participate in other crucial cells functions including calcium (Ca^{2+}) storage, cell signaling (proliferation, adaptation to different environments and stress response) and apoptosis [18,19]. Ca^{2+} storage in mitochondria is involved in the regulation of ion homeostasis, cell signaling, and apoptosis (when prolonged high levels of Ca^{2+} in plasma concentrations) [20].

Also, important to understanding the potential role of mitochondria in neuronal death is the regulation of mitochondrial outer membrane permeabilization (MOMP). This is well controlled by different mechanisms, and when significant permeation does occur there is an activation of caspases and a release of pro-apoptotic factors into the cytosol, initiating the apoptosis-cascade [21–26].

2.2. Inflammation and Glia in Multiple Sclerosis

During acute events of inflammation, immune cells (mainly CD4+ T helper lymphocytes and also CD8+ T cells) cross the blood–brain barrier (BBB) and B cells and monocytes are activated. The primary target in MS is the myelin sheath of CNS white matter, though in recent years there has been growing evidence of direct attack against cortical and deep gray matter [27]. The release of pro-inflammatory cytokines (e.g., IL-17, IL-4, IL-10, TNF-α), activation of microglia and macrophages with the release of toxic substances such as reactive oxygen species (ROS), tumor necrosis factor, reactive nitrogen species (RNS), and glutamate [28] further damage the myelin. Enzymes involved in this neuroinflammation include myeloperoxidase, xanthine, NADPH oxidases (responsible for neuronal injury) [29,30], excitotoxins, cytotoxic cytokines, proteases and, lipases [31].

The damaged BBB subsequently becomes increasingly permeable, allowing further migration of immune cells leading to the formation of plaques of focal demyelination [32,33]. Focal plaques may converge, forming confluent demyelinated areas in both the white and grey matter [34]. The lymphocytic neuroinflammation process that characterizes the acute phase of the disease leads not only to the damage of myelin fibers synthesized by oligodendrocytes, but also to the death of the oligodendrocytes. The combination of demyelination and loss of trophic stimuli of oligodendrocytes then progresses to axonal degeneration, axon and neuron death with permanent neurologic disability [35]. Remyelination, if it occurs, is often only partial and astrocytes form sclerotic glial scars in the damaged white matter [36]. Chronic inflammation is also responsible for cumulative oxidation of phospholipids and DNA strand breaks [37].

In multiple sclerosis, there is also production of intrathecal, oligoclonal IgG and IgM. Although investigated extensively, no clear antigenic pattern identifying a specific potential trigger for MS has been found in studying these CSF antibodies. In SPMS there are also "meningeal lymphoid-like structures" that correlate with the pathology of the gray matter [38].

2.3. Neurodegeneration in Multiple Sclerosis and Evidence for Mitochondrial Involvement

Historically, the neurodegeneration of MS was understood as a sequential process following chronic neuroinflammation, but some evidence suggests that the neurodegenerative component is already present during the initial clinical manifestations of the disease [3]. The number of relapses in RRMS does not correlate with the probability or latency of progression of SPMS [39]. Tissue atrophy is considered an imaging marker of neurodegeneration in MS, and cerebral, spinal, and retinal atrophy have been reported to be present at the first clinical manifestations of RRMS, affecting both white and gray matter [40–42]. The accepted explanation for this observation is that neuroinflammation is followed by a failure in the process of remyelination, axonal damage and Wallerian degeneration [42]. Normal appearing white matter (NAWM), with normal macroscopic appearance and microscopically normal myelination, has a decreased density of axons; this is, in part justified by Wallerian degeneration, but also indicates more widespread early damage than captured by routine MRI [43–48]. Neuropathological findings from brain tissue blocks of MS patients show evidence of gray matter lesions (axonal and dendritic transection, apoptotic neurons and demyelinated cortical plaques) [49] present from the time of initial disease onset; this is, particularly prominent in SPMS and PPMS.

The chronic neuro-inflammatory stimuli of MS disrupt neuro-axonal hemostasis, leading to a simultaneous increase in oxidative stress, marked by a rise in ROS, and secondary damage to mitochondria and macromolecules (mtDNA, proteins from ETC, lipids). Excitotoxicity and an imbalance of neurotrophic substances for neurons and oligodendrocytes occurs [3,50,51]. This damage impairs mitochondrial function (below described), which further increases ROS production in a vicious cycle [4]. The result is a reduction in the efficiency of energy production, creating an imbalance between energy generation and consumption. The final result is an environment with a failure to provide required levels of energy within the demyelinated axons, and after reduced ATP production reaches a critical point there is an imbalance in ionic homeostasis leading to activation of apoptosis mechanisms [52,53].

Of relevance to understanding the pathology of MS, the central nervous system (CNS) has increased susceptibility to oxidative damage because of the high metabolic rate (consumption of oxygen) of neurons and the rich composition of polyunsaturated fatty acids in CNS cells [54]. Furthermore, mitochondria influence the differentiation of oligodendroglial cells through overexpression of mitochondrial transcripts and mtDNA [55]. An environment of oxidative stress reduces the expression of these transcripts involved with oligodendrocyte differentiation [56]. Double strand breaks in mtDNA have been shown to cause an oligodendropathy and exaggerated injury responses in an animal model of MS [57]. Additional observations have demonstrated a potential direct link between mitochondrial dysfunction and oligodendrocyte myelination. N-acetyl aspartate (NAA), is a mitochondrial metabolite and also an indirect oligodendrocyte substrate for the production of myelin (after breakdown into acetate and aspartate). A lack in the availability of NAA, from damaged mitochondria, was associated with lower levels of acetate in the cortex (parietal and motor) in postmortem tissue from patients with MS [58].

Though they are often overridden and unable to counter the stress burden, there are compensatory mechanisms to counter these mitochondria-related degenerative processes. The body has intrinsic mechanisms of self-protection against ROS, including nuclear factor erythroid 2related factor 2 (NRF2) and antioxidant enzymes such as heme oxygenase 1 (HMOX1) which are activated during periods of hypoxic stress [59]. But after a critical point in the reduction in ATP production, an imbalance in ionic homeostasis occurs, leading to the activation of apoptosis mechanisms mediated by ions (Ca^{2+} dependent proteases) in those axons with chronic inflammation and demyelination [60].

2.3.1. Human Studies of Mitochondria Function in Multiple Sclerosis

Multiple human studies have demonstrated evidence of mitochondrial dysfunction in MS patients (Figure 1, Table 1). One publication compared 10 post mortem brains of patients with MS (n = 9 SPMS and n = 1 PPMS) to healthy controls paired for age and sex. The MS cortex exhibited distinctive levels of both mtDNA transcripts (488 decreased and 67 increased compared to controls), and nuclear mitochondrial DNA transcripts (26 decreased transcripts). In a study of function in the same samples there was a decrease in complex I and III activity from the neurons of motor cortex in MS patients, and a decrease in GABAergic synaptic components [52]. Another study of thirteen patients with MS (SPMS) identified large mtDNA deletions in neurons, with some showing specific deletions in the subunits of complex IV [59].

Additionally, reports of a decrease in PGC-1α levels (a transcriptional co-activator and regulator of mitochondrial function) in pyramidal neurons of MS patients (7 SPMS, 7 PPMS and 1 subtype not determined) was associated with reduced expression of mitochondrial machinery components (OXPHOS subunits, antioxidants and uncoupling proteins 4 and 5). This finding was confirmed in a functional model (with neuronal cells) showing association of these changes with more ROS production [60]. Another publication reported that increases in ROS affect the ability of NRF-2 (a transcription factor for ETC proteins) to bind promotors, even in apparently normal areas of gray matter cortex of SPMS patients [61]. Higher ROS production in the CNS of progressive MS patients (14 SPMS, 5 PPMS and 7 subtypes not determined) has also been associated with a rise in the number of mitochondria in axon and astrotcytes. Increased ROS is also associated with the translation of mitochondrial proteins in active and chronic inactive MS lesions, including elevated expression of proteins from the mitochondrial ETC complex IV and higher levels of a heat shock protein (mtHSP70) compared to the brain of controls. The mtHSP70 protein is a marker of mitochondrial stress [62]. A recent publication found in fronto-parietal areas decreased levels of the potent antioxidant, glutathione (GSH), in PPMS and SPMS compared with RRMS and controls, suggesting that oxidative stress affects the neurodegeneration phase more than the neuroinflammatory phase [63]. Another study compared the mitochondrial proteome from the brains of MS (eight SPMS) patients and controls. The findings showed different patterns by mass spectrometry in levels of human cytochrome c oxidase subunit 5b (COX5b), the brain-specific creatine kinase isoform, and the β-chain of hemoglobin between groups [64].

Table 1. Evidence of mitochondrial involvement in Progressive Forms of Multiple Sclerosis compared to Controls or RRMS.

MS Phenotype	Tissue	Cell Type	Mitochondria Pathology	Reference
1 PP 9 SP 8 C	Motor cortex	Neurons	—decreased expression of mitochondrial nuclear gene DNA —functionally reduced complex I and III activities	Dutta, R. et al. 2006 [52]
1 PP 9 SP 6 C	Chronic inactive lesions	Demyelinated axons	—increased total mitochondrial content and complex IV activity	Mahad, D.J. et al. 2009 [65]
8 SP 5 C	Grey matter in Cortex	NCD	—epigenetic changes affected by ROS, through the reduced capacity of NRF-2 (a transcription factor for ETC proteins)	Pandit, A. et al. 2009 [61]
5 PP 14 SP 7 ND 7 C	Active and chronic lesions	NCD	—increase in the levels of a heat shock protein (mtHSP70), a marker of mitochondrial stress —an increase in the number of mitochondria and in the translation of mitochondrial proteins	Witte, M.E. et al. 2009 [62]
13 SP 10 C	NCD	Neurons	—accumulation of large mtDNA deletions, with some showing specific deletion in the subunits of complex IV	Campbell, G.R. et al. 2011 [59]
2 PP 7 SP 1 RC	NCD	Acute and chronic demyelinated axons	—increased mitochondrial content and complex IV activity compared with remyelinating and myelinated axons	Zambonin, J.L. et al. 2011 [66]
8 SP 8 C	NCD	NCD	—different patterns of mass spectrometry in human cytochrome c oxidase subunit 5b (COX5b), the brain-specific creatine kinase isoform, and the β-chain of hemoglobin	Broadwater, L. et al. 2011 [64]
7 PP 7 SP 1 ND 9 C	NCD	Pyramidal neurons	—decrease in PGC-1α levels, OXPHOS subunits, antioxidants and uncoupling proteins 4 and 5	Witte, M.E. et al. 2013 [60]
20 PP 20 SP vs 21 RR	NCD	NCD	—decreased levels of glutathione (GSH), a potent antioxidant, signaling that oxidative stress more strongly affects the neurodegeneration phase than the neuroinflammation one	Choi, Y. et al. 2018 [63]

MS Type: PP = primary progressive; SP = secondary progressive; RR = relapsing progressive; C = controls; ND = not determined; NCD = Tissue or Cell Type not clearly defined.

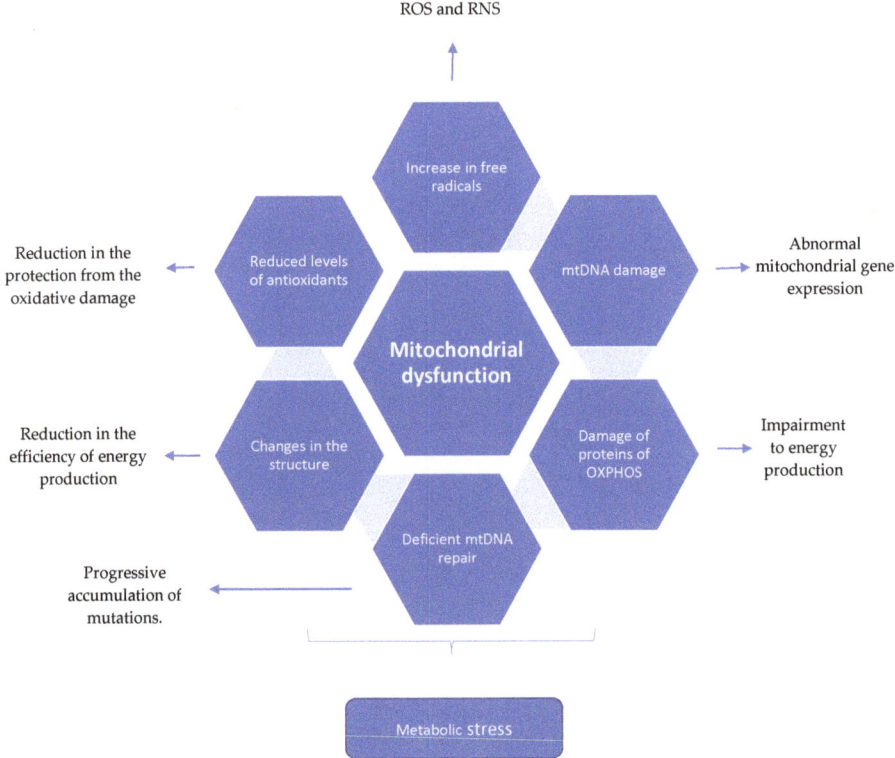

Figure 1. Mitochondrial dysfunction described in the literature associated with Multiple Sclerosis. Chronic neuroinflammation leading to mitochondrial dysfunction.

2.3.2. Neurodegeneration in Multiple Sclerosis Animal Models

There have been many attempts to reproduce the spectrum of inflammation (acute and chronic), demyelination/remyelination, and neurodegeneration that characterize the different clinical syndromes (PP, SP, RR) of the disease. There is no single experimental model that fully covers the spectrum of pathology in human MS. Each model available has strengths for certain questions, but without completely recapitulating all of the mitochondrial deficiencies in MS [67]. In trial design it is important to focus on the mechanism of the potential drug and choose the animal model in which it is possible to induce the disease process of interest [67,68].

In one of the most commonly used models for MS, experimental autoimmune encephalomyelitis (EAE), there are morphology changes in the mitochondria (swelling) [69], early mitochondrial dysfunction even in normal appearing white matter [70], and impairment of mitochondrial and axonal depolarization [71]. Some of the mitochondrial damage can be rescued with specific interventions such as gene therapy for expressing complex I ETC proteins [72,73] and antioxidant cocktails [74,75]. In a myelin basic protein (MBP) knockout, considered a model for the chronic demyelination of MS, there were increased numbers of mitochondria observed by electron microscopy. Additionally, there was a two-fold increase in the cytochrome c staining in the white matter, showing mitochondrial changes associated with cases of reduction in myelin [76]. A summary of previous animal models' findings regarding the association of mitochondrial involvement in multiple sclerosis is presented in Table 2.

Table 2. Mouse models to study multiple sclerosis [77].

MS Animal Model	Type of MS Modeled	Indication for Research	Mitochondrial Findings
EAE-SJL/J mice -C57BL/6J mice -Biozzi chronic EAE	-RR -PP and SP -RR -> SP	Understanding of the neuroinflammatory process after immunologic activation of the mice (SJL/J with PLP or MBP and C57BL/6J with MOG) [77,78]. Accumulative damage of neuroinflammation with secondary progression of the disease [68,79,80].	C57BL/6's mitochondria morphology changes (swelling) [69], early mitochondrial dysfunction in EAE disease [70] and impairment of mitochondrial and axonal depolarization [71]. C57Bl/6 model did not reproduce the cortex respiratory protein's alterations seen in MS patients [64].
TCR transgenic mice	-RR [78]	Understanding spontaneous neuroinflammatory process after immunologic activation [77].	-
TMEV	Demyelination and axonal damage	Infection mediated by Picornavirus inducing an encephalomyelitis (whole neuroaxis) [77].	-
Toxin-induced demyelination (Cuprizone, Lysolecithin, Ethidium bromide)	Demyelination and remyelination	Reproducible onset of demyelination and start of remyelination after interruption of toxic exposure. If chronic exposure of cuprizone also possible to see impairment of remyelination [81].	Cuprizone is a copper chelator an essential component of COX [82]. Mice's brain treated with cuprizone presented "giant" mitochondria in oligodendroglial cells [83]. Oligodendrocytes treated with cuprizone presented with decreased mitochondrial potential (in vitro) [84].

EAE = experimental autoimmune/allergic encephalomyelitis. TCR transgenic mice = T cell receptor (TCR) transgenic mouse models. TMEV = Theiler's murine encephalomyelitis virus. PLP = proteolipoprotein. MBP = myelin basic protein. MOG = myelin oligodendrocyte glycoprotein. COX = cytochrome oxidase.

2.4. Summary (Mechanism of Mitochondrial Dysfunction Perpetuating the CNS Injury in Multiple Sclerosis)

The brain has a high metabolic rate and consumes 20% of the total energy produced in the human body, which is mainly utilized in neurotransmission (more than half of that consumed to maintain the ionic equilibrium and the membrane potential) and the axoplasmic flow (to conduct nerve impulses); these functions depend substantially on mitochondria machinery [4,85]. The neurologic signal transmission is due to propagation of the membrane depolarization through the neuron, and the electrochemical gradient is created by the Na^+/K^+-ATPase, allocated in the nodes of Ranvier. Oligodendrocytes are not only responsible for the myelin sheath but also release lactate for the neuron as energy supply. With the chronic inflammation and myelin destruction, there is redistribution of the ion channels. Consequently, there is more ATP consumption by the increased number of Na^+/K^+-ATPase. With the purpose of balancing the ratio and demand for energy, mitochondria begin compensatory modifications (increasing in number and size, changing the localization in the neuron and its morphology). In parallel, the chronic inflammation creates an environment of oxidative stress secondary to ROS release by macrophages and the microglia and increases in glutamate released in response to neuronal damage. TNF-α damages the OXPHOS process through Ca^{++} regulated mechanisms [86]. With mitochondrial progressive accumulative damage (mtDNA alteration and increased heteroplasmy, OXPHOSP subunits dysfunction, alteration in proteins that regulates the migration of the organelle from neuron body to the axon) significant impairment in energy production develops. [51]. If ATP production is compromised, the Na^+/K^+-ATPase is not able to keep the gradient after an action potential, which leads to Na+ accumulation in the neuron cytoplasm. This forces the Na^+/Ca^{2+} channel to transfer Ca^{2+} inside the cell, activating the Ca^{2+} apoptosis-depend-cascade, which results in neuron death, Wallerian degeneration and irreversible neurologic dysfunction [34,53,65,66,86–89]. This process is represented in Figure 2. The progressive degenerative process initiated in the axon can continue to the neuron body and dendrites, also reaching presynaptic and postsynaptic neurons [34], chronic failure to provide energy to the tissue increases the oxidative stress in a vicious cycle that increases mitochondrial damage [51].

Important to mention is that mitochondrial DNA damage is amplified during the process of expansion of the clones (with deletions or mutations), changing the levels of heteroplasmy of the tissue [51,59]. This process increases the failure to provide appropriate energy supply for the tissue, contributing to the death of the cells [90].

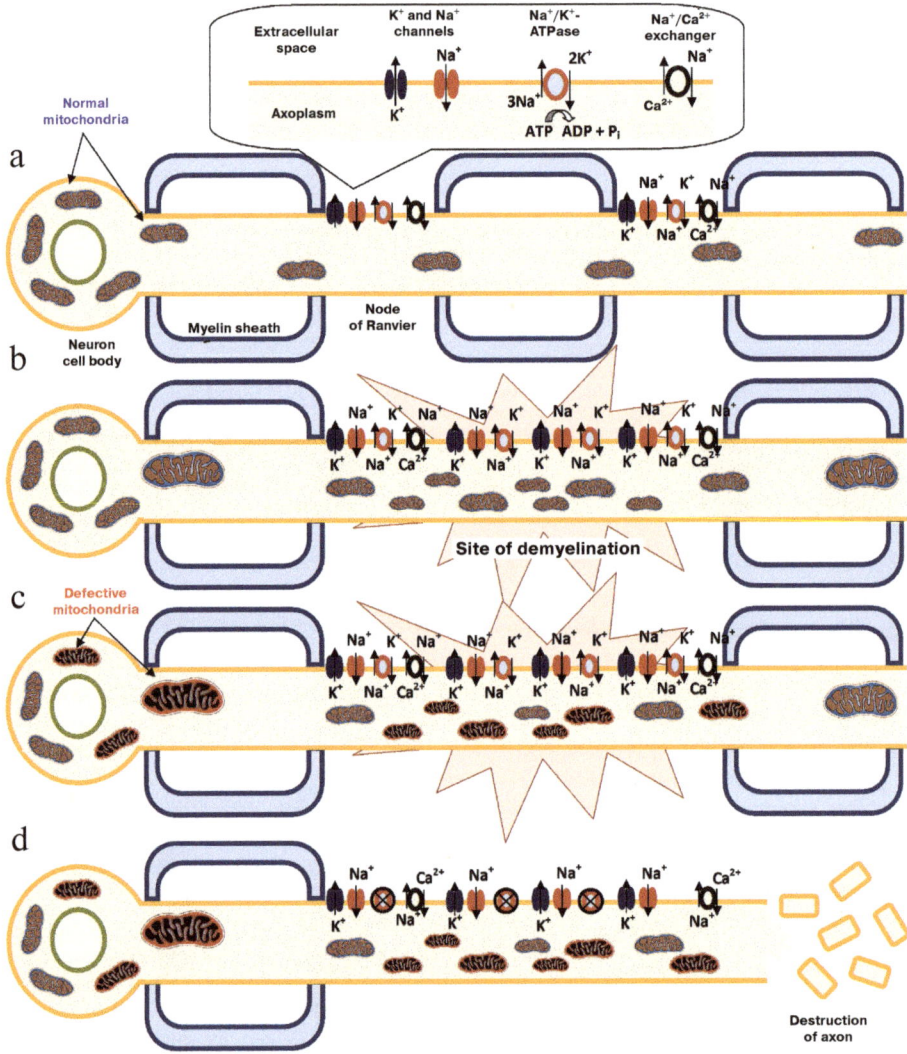

Figure 2. The role of mitochondria in the process of neurodegeneration. **a.** Normal nerve. **b.** Site of demyelination with secondary modification of the distribution of ion channels in the nerve. **c.** Structural and functional modification in mitochondria caused by oxidative stress. **d.** Cascade of apoptosis activated by Ca^{2+}. Figure reprinted with permission from the article "Involvement of Mitochondria in Neurodegeneration in Multiple Sclerosis", Kozin et al., Biochemistry (Moscow), 2018, Vol. 83, No. 7, pp. 813–830 [4].

3. Mitochondrial Mutations in Multiple Sclerosis and Overlapping Diseases

3.1. Mitochondrial Mutations and Multiple Sclerosis Risk

The current consensus is that MS is a multifactorial disease, with 25% of the risk related to heritable factors [91]. The important role of the class II region of the human leukocyte antigen (*HLA*) gene cluster has been well recognized for several decades. There are now over 100 loci identified in the HLA region found to be associated with susceptibility and over 200 in non-HLA loci [92]. Several single site mutations in mtDNA have been reported to increase the risk of MS, including the mtDNA nt13708A [93] and mtDNA T4216C [94] variants. A large consortium study by Tranah et al. examined mitochondrial DNA sequence variation and MS risk. In the discovery dataset they compared over 7000 MS cases and over 14,000 controls from seven countries. Haplotype group and more than 100 common mtDNA mutations were evaluated. While they reported an elevated risk of MS (OR 1.15, $p = 0002$) among haplotype JT carriers, they found no associations between common mtDNA mutations and MS risk [95].

3.2. Leber's Hereditary Optic Neuropathy

Leber's hereditary optic neuropathy (LHON) is a mitochondrial disease resulting in severe bilateral optic neuropathy, characterized by central vision loss and dyschromatopsia. There is degeneration of the retinal ganglion cells (RGC) and axonal tracts of the optic nerve. There are numerous mitochondrial mutations associated with LHON, but the vast majority of patients have one of three different mitochondrial mutations at nucleotides 3460, 11778 and 14484. The mutations are single amino acid substitutions in one of the mitochondrially encoded subunits of NADH: ubiquinone oxidoreductase, complex I of the electron transport chain (ETC). There is some evidence suggesting that the exposure to high nitric oxide concentrations could impair in vivo the ability to cope with the oxidative stress caused by the genetic defect, thereby driving the pathology in LHON. This was described by Flabella et al. in one patient carrying the 11788/ND4 mutation [96]. This same increase in ROS was described in MS as previously discussed [3,50,51].

Males are more frequently affected, and there is incomplete penetrance seen in LHON families [97]. There is a modest epidemiological overlap between MS and LHON, with a subset of patients developing both diseases (Table 3). Harding first described this association in 1992 in case studies of eight women with matrilineal relatives with LHON who presented with optic neuritis; six of the eight progressed to clinical MS with neurologic symptoms. Seven of the eight also had characteristic white matter lesions on MRI [98]. Since Harding's first report of the association of LHON and MS this relationship has continued to be observed with females being predominantly affected at a ratio of more than two to one [99]. In one review the incidence of demyelination among LHON affected persons was up to five percent, which is fifty times greater than the prevalence of MS in the general population [100]. RGC thinning is also noted in MS. While the exact pathophysiology may be different in LHON and MS, the mitochondrial dysfunction in LHON may be instructive to the understanding of mitochondria's role in MS.

3.3. Dominant Optic Atrophy and OPA1 Mutations

An additional example of potential overlap between mitochondrial genetic optic atrophy and MS has been described by Yu-Wai-Man et al. in a paper detailing three cases of MS-like disease associated with *OPA1* mutations (Table 3). *OPA1* mutations have previously been discussed in the literature in association with autosomal dominant optic atrophy (DOA), the most common inherited form of optic nerve visual loss. OPA1 has multiple roles in mitochondrial function as it encodes for an inner mitochondrial membrane protein, and is involved in respiratory chain complexes, cytochrome c molecules, and fusion/fission balance. There are over 90 known gene mutations (substitutions, deletions and insertions) associated with OPA1 mutations and thought to be due to a truncated protein [101]. Like LHON, DOA is a mitochondrial determined optic neuropathy preferentially affecting the ganglion cells within the inner retina. The exact relationship between *OPA1* proteins and MS has yet to be clearly elucidated, with only the above few cases being reported.

3.4. POLG1 Mutations

The mitochondrial gene *POLG1* is the larger catalytic subunit of polymerase gamma which is the only known DNA polymerase active in human mitochondria. *POLG1* mutations have been implicated in a number of mitochondrial disorders and more recently have also been identified in several cases of demyelination. Two cases of non-related individuals with novel *POLG1* mutations who had optic neuritis and white matter lesions consistent with clinical MS were reported in the literature. Of note both of these patients progressed into a more classic POLG1 phenotype with bilateral ophthalmoplegia, ptosis, myopathy, cardiomyopathy, ataxia, dysphagia, and hearing and cognitive impairment. These patients also had muscle biopsies showing red ragged fibers [99,102]. Therefore, their progression calls into question whether or not they truly had MS or if their initial presentations were instead MS mimics. Clearly more research is needed. Yet it is important to consider these cases as they may offer further evidence of the role of the mitochondria within MS and MS-like disease processes. Further research may lead to the discovery of more MS patients with mitochondrial mutations.

Table 3. Associations between MS and mitochondrial diseases.

Disease	Gene Mutation	MS Overlap	Overlap in Potential Mechanism
MS	mtDNA nt13708A mtDNA T4216C nt 11778 (G→A)	NA	NA
LHON	nt 3460 nt 11778 (G→A) nt 14484	5% LHON have evidence of demyelinating lesion	Degeneration of optic nerve
DOA	over 90 gene mutations	*OPA1* protein: known link to DOA, implicated in 3 patients with MS-like disease	*OPA1* mutation and truncated protein
POLG1	Not specified	Linked to cases of demyelination	Not specified

MS: multiple Sclerosis, LHON: Leber's hereditary optic neuropathy, DOA: autosomal dominant optic atrophy, *POLG1*: mitochondrial gene *POLG1*.

4. Potential Therapies and Targets

The treatment of most mitochondrial diseases is still largely supportive at this time, although some therapies have been tried such as vitamins, co-enzymes, creatine, free radical scavengers and hyperbaric oxygen treatments. Despite the widespread use of a multitude of co-enzymes and vitamin supplements there is currently limited evidence that these are effective in the treatment of primary mitochondrial disorders. For targeted treatment of MS the use of alpha lipoic acid and co-enzyme Q10 are being investigated. A randomized controlled phase 2 trial of alpha lipoic acid (ALA) versus placebo was studied in SPMS and found to slow whole brain atrophy [103]; further studies are ongoing. Other studies are examining a synthetic analogue of co-enzyme Q10, idebenone, which is being targeted for treatment of neurodegenerative disorders such as LHON [99]. Phase I/II trials of idebonone in PPMS demonstrated safety but initial data showed no change in progression of the disease (http://www.santhera.com/assets/files/press-releases/2018-03-05_PR_PPMS_e_final.pdf). An expansion study is ongoing with completion planned for later this year (https://www.clinicaltrials.gov/ct2/show/NCT01854359). In addition to these targets there are several other potential approaches for mitochondrial based therapy and limiting neurodegeneration in MS.

4.1. Mitochondrial Metabolism and Chronic Neuroinflammation

Neurodegeneration is in part driven by the activation of mononuclear phagocytes. When mononuclear phagocytes are persistently activated it can lead to a state of chronic neuroinflammation. Mitochondrial metabolism has a role in the inflammation cascade and targeting the metabolism of innate immune cells may be of benefit. Future studies may address this relationship to aid in the development of novel molecular and cellular therapies that could disrupt the state of chronic neuroinflammation as a way of preventing secondary neurologic damage [104]. Therapies that support cellular metabolism such as high dose biotin, iron and vitamin D have been proposed as possible treatment therapies in progressive MS, and studies looking at each of these treatments are ongoing (https://clinicaltrials.gov). The pilot studies of high dose biotin are encouraging and results suggest both a reduction in disease progression as well as decreased disability in PPMS [105]. Furthermore, these therapies may also have a role in preventing progression of RRMS to SPMS [50].

4.2. Gene Therapy

Gene therapies are being developed in mitochondrial disorders, though most are still in early phases of development. In vivo studies in mice using several vectors have been promising in some disease models such as LHON [106]. Gene Therapy GS010 was shown to be safe in LHON patients carrying the G11778A mutation in a phase 1/2 clinical trial (https://clinicaltrials.gov/ct2/show/NCT02064569). Although the results did not have sufficient power to definitively demonstrate efficacy, 6/14 patients who received GS010 had visual acuity improvements [107]. While these early results in LHON do not immediately translate to MS care, the suggestion of treatment effect is promising for the future of gene therapy in this field of mitochondrial dysfunction.

5. Conclusions

There is compelling data to suggest an important role for mitochondria in the pathophysiology of MS. Further work is needed to move from studies of association to understanding causal relationships between failure of mitochondrial function and MS phenotype. Targeting energy failure and mitochondrial dysfunction is a novel potential therapeutic approach for the challenging progressive phase of MS. Trials are already underway to begin exploring these pathways as treatment targets, including studies of biotin and alpha lipoic acid in progressive MS.

Funding: I.P.d.B. is recipient of a postdoctoral fellowship from the North American Mitochondrial Disease Consortium (NAMDC) supported by National Institute of Health (NIH) U54 grant NS078059.

Conflicts of Interest: The authors declare no conflict of interest related to the content of this manuscript.

References

1. Rafael, H. Omental transplantation for neurodegenerative diseases. *Am. J. Neurodegener. Dis.* **2014**, *3*, 50–63. [PubMed]
2. Burnside, S.W.; Hardingham, G.E. Transcriptional regulators of redox balance and other homeostatic processes with the potential to alter neurodegenerative disease trajectory. *Biochem. Soc. Trans.* **2017**, *45*, 1295–1303. [CrossRef]
3. Friese, M.A.; Schattling, B.; Fugger, L. Mechanisms of neurodegeneration and axonal dysfunction in multiple sclerosis. *Nat. Rev. Neurol.* **2014**, *10*, 225–238. [CrossRef] [PubMed]
4. Kozin, M.S.; Kulakova, O.G.; Favorova, O.O. Involvement of Mitochondria in Neurodegeneration in Multiple Sclerosis. *Biochemistry (Mosc)* **2018**, *83*, 813–830. [CrossRef]
5. Ahlgren, C.; Oden, A.; Lycke, J. High nationwide incidence of multiple sclerosis in Sweden. *PLoS ONE* **2014**, *9*, e108599. [CrossRef]
6. Koch, M.; Kingwell, E.; Rieckmann, P.; Tremlett, H. The natural history of primary progressive multiple sclerosis. *Neurology* **2009**, *73*, 1996–2002. [CrossRef] [PubMed]
7. Steinman, L. Multiple sclerosis: A two-stage disease. *Nat. Immunol.* **2001**, *2*, 762–764. [CrossRef]

8. Ellwardt, E.; Zipp, F. Molecular mechanisms linking neuroinflammation and neurodegeneration in MS. *Exp. Neurol.* **2014**, *262 Pt A*, 8–17. [CrossRef]
9. Sospedra, M.; Martin, R. Immunology of multiple sclerosis. *Annu. Rev. Immunol.* **2005**, *23*, 683–747. [CrossRef]
10. Haas, R.H.; Zolkipli, Z. Mitochondrial disorders affecting the nervous system. *Semin. Neurol.* **2014**, *34*, 321–340. [PubMed]
11. Pitceathly, R.D.; McFarland, R. Mitochondrial myopathies in adults and children: Management and therapy development. *Curr. Opin. Neurol.* **2014**, *27*, 576–582. [CrossRef] [PubMed]
12. Schaefer, A.M.; Taylor, R.W.; Turnbull, D.M.; Chinnery, P.F. The epidemiology of mitochondrial disorders–past, present and future. *Biochim. Biophys. Acta* **2004**, *1659*, 115–120. [CrossRef]
13. Mazunin, I.O.; Volod'ko, N.V.; Starikovskaia, E.B.; Sukernik, R.I. Mitochondrial genome and human mitochondrial diseases. *Mol. Biol. (Mosk)* **2010**, *44*, 755–772. [CrossRef]
14. Holt, I.J.; Reyes, A. Human mitochondrial DNA replication. *Cold Spring Harb. Perspect. Biol.* **2012**, *4*, 12. [CrossRef]
15. Wallace, D.C.; Fan, W.; Procaccio, V. Mitochondrial energetics and therapeutics. *Annu. Rev. Pathol.* **2010**, *5*, 297–348. [CrossRef]
16. Dyall, S.D.; Brown, M.T.; Johnson, P.J. Ancient invasions: From endosymbionts to organelles. *Science* **2004**, *304*, 253–257. [CrossRef]
17. Emes, R.D. The Logic of Evolution: Review of the Logic of Chance by Eugene V. Koonin. *Front. Genet.* **2012**, *3*, 135. [CrossRef]
18. Hunt, R.J.; Bateman, J.M. Mitochondrial retrograde signaling in the nervous system. *FEBS Lett.* **2017**, *592*, 663–678. [CrossRef]
19. Mitra, K. Mitochondrial fission-fusion as an emerging key regulator of cell proliferation and differentiation. *Bioessays* **2013**, *35*, 955–964. [CrossRef]
20. Glancy, B.; Balaban, R.S. Role of mitochondrial Ca2+ in the regulation of cellular energetics. *Biochemistry* **2012**, *51*, 2959–2973. [CrossRef]
21. Suhaili, S.H.; Karimian, H.; Stellato, M.; Lee, T.H.; Aguilar, M.I. Mitochondrial outer membrane permeabilization: A focus on the role of mitochondrial membrane structural organization. *Biophys. Rev.* **2017**, *9*, 443–457. [CrossRef] [PubMed]
22. Zamzami, N.; Kroemer, G. The mitochondrion in apoptosis: How Pandora's box opens. *Nat. Rev. Mol. Cell Biol.* **2001**, *2*, 67–71. [CrossRef]
23. Tait, S.W.; Green, D.R. Mitochondria and cell death: Outer membrane permeabilization and beyond. *Nat. Rev. Mol. Cell Biol.* **2010**, *11*, 621–632. [CrossRef]
24. Walle, L.V.; Lamkanfi, M.; Vandenabeele, P. The mitochondrial serine protease HtrA2/Omi: An overview. *Cell Death Differ.* **2008**, *15*, 453–460. [CrossRef]
25. Ding, Z.J.; Chen, X.; Tang, X.X.; Wang, X.; Song, Y.L.; Chen, X.D.; Wang, J.; Wang, R.F.; Mi, W.J.; Chen, F.Q.; et al. Apoptosis-inducing factor and calpain upregulation in glutamate-induced injury of rat spiral ganglion neurons. *Mol. Med. Rep.* **2015**, *12*, 1685–1692. [CrossRef]
26. Jang, D.S.; Penthala, N.R.; Apostolov, E.O.; Wang, X.; Crooks, P.A.; Basnakian, A.G. Novel cytoprotective inhibitors for apoptotic endonuclease G. *DNA Cell Biol.* **2015**, *34*, 92–100. [CrossRef]
27. Von Budingen, H.C.; Bar-Or, A.; Zamvil, S.S. B cells in multiple sclerosis: Connecting the dots. *Curr. Opin. Immunol.* **2011**, *23*, 713–720. [CrossRef]
28. Cunningham, C. Microglia and neurodegeneration: The role of systemic inflammation. *Glia* **2013**, *61*, 71–90. [CrossRef] [PubMed]
29. Gray, E.; Thomas, T.L.; Betmouni, S.; Scolding, N.; Love, S. Elevated activity and microglial expression of myeloperoxidase in demyelinated cerebral cortex in multiple sclerosis. *Brain Pathol.* **2008**, *18*, 86–95. [CrossRef]
30. Fischer, M.T.; Sharma, R.; Lim, J.L.; Haider, L.; Frischer, J.M.; Drexhage, J.; Mahad, D.; Bradl, M.; van Horssen, J.; Lassmann, H. NADPH oxidase expression in active multiple sclerosis lesions in relation to oxidative tissue damage and mitochondrial injury. *Brain* **2012**, *135*, 886–899. [CrossRef]
31. Lassmann, H. Models of multiple sclerosis: New insights into pathophysiology and repair. *Curr. Opin. Neurol.* **2008**, *21*, 242–247. [CrossRef] [PubMed]

32. Bjartmar, C.; Wujek, J.R.; Trapp, B.D. Axonal loss in the pathology of MS: Consequences for understanding the progressive phase of the disease. *J. Neurol. Sci.* **2003**, *206*, 165–171. [CrossRef]
33. Bruck, W. The pathology of multiple sclerosis is the result of focal inflammatory demyelination with axonal damage. *J. Neurol.* **2005**, *252* (Suppl. 5), v3–v9. [CrossRef] [PubMed]
34. Dendrou, C.A.; Fugger, L.; Friese, M.A. Immunopathology of multiple sclerosis. *Nat. Rev. Immunol.* **2015**, *15*, 545–558. [CrossRef] [PubMed]
35. Fünfschilling, U.; Supplie, L.M.; Mahad, D.; Boretius, S.; Saab, A.S.; Edgar, J.; Brinkmann, B.G.; Kassmann, C.M.; Tzvetanova, I.D.; Möbius, W.; et al. Glycolytic oligodendrocytes maintain myelin and long-term axonal integrity. *Nature* **2012**, *485*, 517–521. [CrossRef] [PubMed]
36. Popescu, B.F.; Lucchinetti, C.F. Pathology of demyelinating diseases. *Annu. Rev. Pathol.* **2012**, *7*, 185–217. [CrossRef] [PubMed]
37. Fischer, M.T.; Wimmer, I.; Höftberger, R.; Gerlach, S.; Haider, L.; Zrzavy, T.; Hametner, S.; Mahad, D.; Binder, C.J.; Krumbholz, M.; et al. Disease-specific molecular events in cortical multiple sclerosis lesions. *Brain* **2013**, *136*, 1799–1815. [CrossRef]
38. Howell, O.W.; Reeves, C.A.; Nicholas, R.; Carassiti, D.; Radotra, B.; Gentleman, S.M.; Serafini, B.; Aloisi, F.; Roncaroli, F.; Magliozzi, R.; et al. Meningeal inflammation is widespread and linked to cortical pathology in multiple sclerosis. *Brain* **2011**, *134*, 2755–2771. [CrossRef]
39. Scalfari, A.; Neuhaus, A.; Daumer, M.; Deluca, G.C.; Muraro, P.A.; Ebers, G.C. Early relapses, onset of progression, and late outcome in multiple sclerosis. *JAMA Neurol.* **2013**, *70*, 214–222. [CrossRef] [PubMed]
40. Barkhof, F.; Calabresi, P.A.; Miller, D.H.; Reingold, S.C. Imaging outcomes for neuroprotection and repair in multiple sclerosis trials. *Nat. Rev. Neurol.* **2009**, *5*, 256–266. [CrossRef]
41. Chard, D.T.; Griffin, C.M.; Parker, G.J.; Kapoor, R.; Thompson, A.J.; Miller, D.H. Brain atrophy in clinically early relapsing-remitting multiple sclerosis. *Brain* **2002**, *125*, 327–337. [CrossRef] [PubMed]
42. Miller, D.H.; Barkhof, F.; Frank, J.A.; Parker, G.J.; Thompson, A.J. Measurement of atrophy in multiple sclerosis: Pathological basis, methodological aspects and clinical relevance. *Brain* **2002**, *125*, 1676–1695. [CrossRef] [PubMed]
43. Frischer, J.M.; Bramow, S.; Dal-Bianco, A.; Lucchinetti, C.F.; Rauschka, H.; Schmidbauer, M.; Laursen, H.; Sorensen, P.S.; Lassmann, H. The relation between inflammation and neurodegeneration in multiple sclerosis brains. *Brain* **2009**, *132*, 1175–1189. [CrossRef] [PubMed]
44. Kornek, B.; Storch, M.K.; Weissert, R.; Wallstroem, E.; Stefferl, A.; Olsson, T.; Linington, C.; Schmidbauer, M.; Lassmann, H. Multiple sclerosis and chronic autoimmune encephalomyelitis: A comparative quantitative study of axonal injury in active, inactive, and remyelinated lesions. *Am. J. Pathol.* **2000**, *157*, 267–276. [CrossRef]
45. Kutzelnigg, A.; Lucchinetti, C.F.; Stadelmann, C.; Brück, W.; Rauschka, H.; Bergmann, M.; Schmidbauer, M.; Parisi, J.E.; Lassmann, H. Cortical demyelination and diffuse white matter injury in multiple sclerosis. *Brain* **2005**, *128*, 2705–2712. [CrossRef] [PubMed]
46. DeLuca, G.C.; Ebers, G.C.; Esiri, M.M. Axonal loss in multiple sclerosis: A pathological survey of the corticospinal and sensory tracts. *Brain* **2004**, *127*, 1009–1018. [CrossRef]
47. Bitsch, A.; Schuchardt, J.; Bunkowski, S.; Kuhlmann, T.; Bruck, W. Acute axonal injury in multiple sclerosis. Correlation with demyelination and inflammation. *Brain* **2000**, *123 Pt 6*, 1174–1183. [CrossRef]
48. DeLuca, G.C.; Williams, K.; Evangelou, N.; Ebers, G.C.; Esiri, M.M. The contribution of demyelination to axonal loss in multiple sclerosis. *Brain* **2006**, *129 Pt 6*, 1507–1516. [CrossRef]
49. Peterson, J.W.; Bo, L.; Mork, S.; Chang, A.; Trapp, B.D. Transected neurites, apoptotic neurons, and reduced inflammation in cortical multiple sclerosis lesions. *Ann. Neurol.* **2001**, *50*, 389–400. [CrossRef]
50. Heidker, R.M.; Emerson, M.R.; LeVine, S.M. Metabolic pathways as possible therapeutic targets for progressive multiple sclerosis. *Neural Regen. Res.* **2017**, *12*, 1262–1267.
51. Mahad, D.H.; Trapp, B.D.; Lassmann, H. Pathological mechanisms in progressive multiple sclerosis. *Lancet Neurol.* **2015**, *14*, 183–193. [CrossRef]
52. Dutta, R.; McDonough, J.; Yin, X.; Peterson, J.; Chang, A.; Torres, T.; Gudz, T.; Macklin, W.B.; Lewis, D.A.; Fox, R.J.; et al. Mitochondrial dysfunction as a cause of axonal degeneration in multiple sclerosis patients. *Ann. Neurol.* **2006**, *59*, 478–489. [CrossRef]
53. Trapp, B.D.; Stys, P.K. Virtual hypoxia and chronic necrosis of demyelinated axons in multiple sclerosis. *Lancet Neurol.* **2009**, *8*, 280–291. [CrossRef]

54. Patergnani, S.; Fossati, V.; Bonora, M.; Giorgi, C.; Marchi, S.; Missiroli, S.; Rusielewicz, T.; Wieckowski, M.R.; Pinton, P. Mitochondria in Multiple Sclerosis: Molecular Mechanisms of Pathogenesis. *Int. Rev. Cell Mol. Biol.* **2017**, *328*, 49–103. [PubMed]
55. Schoenfeld, R.; Wong, A.; Silva, J.; Li, M.; Itoh, A.; Horiuchi, M.; Itoh, T.; Pleasure, D.; Cortopassi, G. Oligodendroglial differentiation induces mitochondrial genes and inhibition of mitochondrial function represses oligodendroglial differentiation. *Mitochondrion* **2010**, *10*, 143–150. [CrossRef]
56. French, H.M.; Reid, M.; Mamontov, P.; Simmons, R.A.; Grinspan, J.B. Oxidative stress disrupts oligodendrocyte maturation. *J. Neurosci. Res.* **2009**, *87*, 3076–3087. [CrossRef]
57. Madsen, P.M.; Pinto, M.; Patel, S.; McCarthy, S.; Gao, H.; Taherian, M.; Karmally, S.; Pereira, C.V.; Dvoriantchikova, G.; Ivanov, D.; et al. Mitochondrial DNA Double-Strand Breaks in Oligodendrocytes Cause Demyelination, Axonal Injury, and CNS Inflammatio. *J. Neurosci., vol.* **2017**, *37*, 10185–10199. [CrossRef]
58. Li, S.; Clements, R.; Sulak, M.; Gregory, R.; Freeman, E.; McDonough, J. Decreased NAA in gray matter is correlated with decreased availability of acetate in white matter in postmortem multiple sclerosis cortex. *Neurochem. Res.* **2013**, *38*, 2385–2396. [CrossRef]
59. Campbell, G.R.; Ziabreva, I.; Reeve, A.K.; Krishnan, K.J.; Reynolds, R.; Howell, O.; Lassmann, H.; Turnbull, D.M.; Mahad, D.J. Mitochondrial DNA deletions and neurodegeneration in multiple sclerosis. *Ann. Neurol.* **2011**, *69*, 481–492. [CrossRef] [PubMed]
60. Witte, M.E.; Nijland, P.G.; Drexhage, J.A.; Gerritsen, W.; Geerts, D.; van Het Hof, B.; Reijerkerk, A.; de Vries, H.E.; van der Valk, P.; van Horssen, J. Reduced expression of PGC-1alpha partly underlies mitochondrial changes and correlates with neuronal loss in multiple sclerosis cortex. *Acta Neuropathol.* **2013**, *125*, 231–243. [CrossRef]
61. Pandit, A.; Vadnal, J.; Houston, S.; Freeman, E.; McDonough, J. Impaired regulation of electron transport chain subunit genes by nuclear respiratory factor 2 in multiple sclerosis. *J. Neurol. Sci.* **2009**, *279*, 14–20. [CrossRef] [PubMed]
62. Witte, M.E.; Bø, L.; Rodenburg, R.J.; Belien, J.A.; Musters, R.; Hazes, T.; Wintjes, L.T.; Smeitink, J.A.; Geurts, J.J.; De Vries, H.E.; et al. Enhanced number and activity of mitochondria in multiple sclerosis lesions. *J. Pathol.* **2009**, *219*, 193–204. [CrossRef]
63. Choi, I.Y.; Lee, P.; Adany, P.; Hughes, A.J.; Belliston, S.; Denney, D.R.; Lynch, S.G. In vivo evidence of oxidative stress in brains of patients with progressive multiple sclerosis. *Mult. Scler.* **2018**, *24*, 1029–1038. [CrossRef]
64. Broadwater, L.; Pandit, A.; Clements, R.; Azzam, S.; Vadnal, J.; Sulak, M.; Yong, V.W.; Freeman, E.J.; Gregory, R.B.; McDonough, J. Analysis of the mitochondrial proteome in multiple sclerosis cortex. *Biochim. Biophys. Acta* **2011**, *1812*, 630–641. [CrossRef]
65. Mahad, D.J.; Ziabreva, I.; Campbell, G.; Lax, N.; White, K.; Hanson, P.S.; Lassmann, H.; Turnbull, D.M. Mitochondrial changes within axons in multiple sclerosis. *Brain* **2009**, *132*, 1161–1174. [CrossRef] [PubMed]
66. Zambonin, J.L.; Zhao, C.; Ohno, N.; Campbell, G.R.; Engeham, S.; Ziabreva, I.; Schwarz, N.; Lee, S.E.; Frischer, J.M.; Turnbull, D.M.; et al. Increased mitochondrial content in remyelinated axons: Implications for multiple sclerosis. *Brain* **2011**, *134*, 1901–1913. [CrossRef] [PubMed]
67. Lassmann, H.; Bradl, M. Multiple sclerosis: Experimental models and reality. *Acta Neuropathol.* **2017**, *133*, 223–244. [CrossRef]
68. Hampton, D.W.; Serio, A.; Pryce, G.; Al-Izki, S.; Franklin, R.J.; Giovannoni, G.; Baker, D.; Chandran, S. Neurodegeneration progresses despite complete elimination of clinical relapses in a mouse model of multiple sclerosis. *Acta Neuropathol. Commun.* **2013**, *1*, 84. [CrossRef] [PubMed]
69. Recks, M.S.; Stormanns, E.R.; Bader, J.; Arnhold, S.; Addicks, K.; Kuerten, S. Early axonal damage and progressive myelin pathology define the kinetics of CNS histopathology in a mouse model of multiple sclerosis. *Clin. Immunol.* **2013**, *149*, 32–45. [CrossRef] [PubMed]
70. Feng, J.; Tao, T.; Yan, W.; Chen, C.S.; Qin, X. Curcumin inhibits mitochondrial injury and apoptosis from the early stage in EAE mice. *Oxid. Med. Cell. Longev.* **2014**, *2014*, 728751. [CrossRef]
71. Sadeghian, M.; Mastrolia, V.; Rezaei Haddad, A.; Mosley, A.; Mullali, G.; Schiza, D.; Sajic, M.; Hargreaves, I.; Heales, S.; Duchen, M.R.; et al. Mitochondrial dysfunction is an important cause of neurological deficits in an inflammatory model of multiple sclerosis. *Sci. Rep.* **2016**, *6*, 33249. [CrossRef]
72. Talla, V.; Yu, H.; Chou, T.H.; Porciatti, V.; Chiodo, V.; Boye, S.L.; Hauswirth, W.W.; Lewin, A.S.; Guy, J. NADH-dehydrogenase type-2 suppresses irreversible visual loss and neurodegeneration in the EAE animal model of MS. *Mol. Ther.* **2013**, *21*, 1876–1888. [CrossRef]

73. Talla, V.; Koilkonda, R.; Porciatti, V.; Chiodo, V.; Boye, S.L.; Hauswirth, W.W.; Guy, J. Complex I subunit gene therapy with NDUFA6 ameliorates neurodegeneration in EAE. *Investig. Ophthalmol. Vis. Sci.* **2015**, *56*, 1129–1140. [CrossRef]
74. Fetisova, E.; Chernyak, B.; Korshunova, G.; Muntyan, M.; Skulachev, V. Mitochondria-targeted Antioxidants as a Prospective Therapeutic Strategy for Multiple Sclerosis. *Curr. Med. Chem.* **2017**, *24*, 2086–2114. [CrossRef] [PubMed]
75. Mao, P.; Manczak, M.; Shirendeb, U.P.; Reddy, P.H. MitoQ, a mitochondria-targeted antioxidant, delays disease progression and alleviates pathogenesis in an experimental autoimmune encephalomyelitis mouse model of multiple sclerosis. *Biochim. Biophys. Acta* **2013**, *1832*, 2322–2331. [CrossRef]
76. Andrews, H.; White, K.; Thomson, C.; Edgar, J.; Bates, D.; Griffiths, I.; Turnbull, D.; Nichols, P. Increased axonal mitochondrial activity as an adaptation to myelin deficiency in the Shiverer mouse. *J. Neurosci. Res.* **2006**, *83*, 1533–1539. [CrossRef]
77. Procaccini, C.; de Rosa, V.; Pucino, V.; Formisano, L.; Matarese, G. Animal models of Multiple Sclerosis. *Eur. J. Pharmacol.* **2015**, *759*, 182–191. [CrossRef]
78. Warford, J.; Robertson, G.S. New methods for multiple sclerosis drug discovery. *Expert Opin. Drug Discov.* **2011**, *6*, 689–699. [CrossRef]
79. Furlan, R.; Poliani, P.L.; Marconi, P.C.; Bergami, A.; Ruffini, F.; Adorini, L.; Glorioso, J.C.; Comi, G.; Martino, G. Central nervous system gene therapy with interleukin-4 inhibits progression of ongoing relapsing-remitting autoimmune encephalomyelitis in Biozzi AB/H mice. *Gene Ther.* **2001**, *8*, 13–19. [CrossRef]
80. Al-Izki, S.; Pryce, G.; O'Neill, J.K.; Butter, C.; Giovannoni, G.; Amor, S.; Baker, D. Practical guide to the induction of relapsing progressive experimental autoimmune encephalomyelitis in the Biozzi ABH mouse. *Mult. Scler. Relat. Disord.* **2012**, *1*, 29–38. [CrossRef]
81. Praet, J.; Guglielmetti, C.; Berneman, Z.; van der Linden, A.; Ponsaerts, P. Cellular and molecular neuropathology of the cuprizone mouse model: Clinical relevance for multiple sclerosis. *Neurosci. Biobehav. Rev.* **2014**, *47*, 485–505. [CrossRef] [PubMed]
82. Schreiner, B.; Heppner, F.L.; Becher, B. Modeling multiple sclerosis in laboratory animals. *Semin. Immunopathol.* **2009**, *31*, 479–495. [CrossRef] [PubMed]
83. Hemm, R.D.; Carlton, W.W.; Welser, J.R. Ultrastructural changes of cuprizone encephalopathy in mice. *Toxicol. Appl. Pharmacol.* **1971**, *18*, 869–882. [CrossRef]
84. Bénardais, K.; Kotsiari, A.; Skuljec, J.; Koutsoudaki, P.N.; Gudi, V.; Singh, V.; Vulinović, F.; Skripuletz, T.; Stangel, M. Cuprizone [bis(cyclohexylidenehydrazide)] is selectively toxic for mature oligodendrocytes. *Neurotox. Res.* **2013**, *24*, 244–250. [CrossRef] [PubMed]
85. Howarth, C.; Gleeson, P.; Attwell, D. Updated energy budgets for neural computation in the neocortex and cerebellum. *J. Cereb. Blood Flow Metab.* **2012**, *32*, 1222–1232. [CrossRef] [PubMed]
86. Dziedzic, T.; Metz, I.; Dallenga, T.; König, F.B.; Müller, S.; Stadelmann, C.; Brück, W. Wallerian degeneration: A major component of early axonal pathology in multiple sclerosis. *Brain Pathol.* **2010**, *20*, 976–985. [CrossRef]
87. Campbell, G.; Mahad, D.J. Mitochondrial dysfunction and axon degeneration in progressive multiple sclerosis. *FEBS Lett.* **2018**, *592*, 1113–1121. [CrossRef] [PubMed]
88. Kiryu-Seo, S.; Ohno, N.; Kidd, G.J.; Komuro, H.; Trapp, B.D. Demyelination increases axonal stationary mitochondrial size and the speed of axonal mitochondrial transport. *J. Neurosci.* **2010**, *30*, 6658–6666. [CrossRef] [PubMed]
89. Tsutsui, S.; Stys, P.K. Metabolic injury to axons and myelin. *Exp. Neurol.* **2013**, *246*, 26–34. [CrossRef]
90. Witte, M.E.; Mahad, D.J.; Lassmann, H.; van Horssen, J. Mitochondrial dysfunction contributes to neurodegeneration in multiple sclerosis. *Trends Mol. Med.* **2014**, *20*, 179–187. [CrossRef]
91. Willer, C.J.; Dyment, D.A.; Risch, N.J.; Sadovnick, A.D.; Ebers, G.C. Twin concordance and sibling recurrence rates in multiple sclerosis. *Proc. Natl. Acad. Sci. USA* **2003**, *100*, 12877–12882. [CrossRef] [PubMed]
92. Hollenbach, J.A.; Oksenberg, J.R. The immunogenetics of multiple sclerosis: A comprehensive review. *J. Autoimmun.* **2015**, *64*, 13–25. [CrossRef]
93. Yu, X.; Koczan, D.; Sulonen, A.M.; Akkad, D.A.; Kroner, A.; Comabella, M.; Costa, G.; Corongiu, D.; Goertsches, R.; Camina-Tato, M.; et al. mtDNA nt13708A variant increases the risk of multiple sclerosis. *PLoS ONE* **2008**, *3*, e1530. [CrossRef]

94. Andalib, S.; Emamhadi, M.; Yousefzadeh-Chabok, S.; Salari, A.; Sigaroudi, A.E.; Vafaee, M.S. MtDNA T4216C variation in multiple sclerosis: A systematic review and meta-analysis. *Acta Neurol. Belg.* **2016**, *116*, 439–443. [CrossRef]
95. Tranah, G.J.; Santaniello, A.; Caillier, S.J.; D'Alfonso, S.; Martinelli Boneschi, F.; Hauser, S.L.; Oksenberg, J.R. Mitochondrial DNA sequence variation in multiple sclerosis. *Neurology* **2015**, *85*, 325–330. [CrossRef]
96. Falabella, M.; Forte, E.; Magnifico, M.C.; Santini, P.; Arese, M.; Giuffrè, A.; Radić, K.; Chessa, L.; Coarelli, G.; Buscarinu, M.C.; et al. Evidence for Detrimental Cross Interactions between Reactive Oxygen and Nitrogen Species in Leber's Hereditary Optic Neuropathy Cells. *Oxid. Med. Cell. Longev.* **2016**, *2016*, 3187560. [CrossRef]
97. Howell, N. Leber hereditary optic neuropathy: Mitochondrial mutations and degeneration of the optic nerve. *Vision Res.* **1997**, *37*, 3495–3507. [CrossRef]
98. Harding, A.E.; Sweeney, M.G.; Miller, D.H.; Mumford, C.J.; Kellar-Wood, H.; Menard, D.; McDonald, W.I.; Compston, D.A. Occurrence of a multiple sclerosis-like illness in women who have a Leber's hereditary optic neuropathy mitochondrial DNA mutation. *Brain* **1992**, *115 Pt 4*, 979–989. [CrossRef]
99. Yu-Wai-Man, P.; Griffiths, P.G.; Chinnery, P.F. Mitochondrial optic neuropathies - disease mechanisms and therapeutic strategies. *Prog. Retin. Eye Res.* **2011**, *30*, 81–114. [CrossRef]
100. Palace, J. Multiple sclerosis associated with Leber's Hereditary Optic Neuropathy. *J. Neurol. Sci.* **2009**, *286*, 24–27. [CrossRef]
101. Zanna, C.; Ghelli, A.; Porcelli, A.M.; Karbowski, M.; Youle, R.J.; Schimpf, S.; Wissinger, B.; Pinti, M.; Cossarizza, A.; Vidoni, S.; et al. OPA1 mutations associated with dominant optic atrophy impair oxidative phosphorylation and mitochondrial fusion. *Brain* **2008**, *131*, 352–367. [CrossRef]
102. Echaniz-Laguna, A.; Chassagne, M.; de Sèze, J.; Mohr, M.; Clerc-Renaud, P.; Tranchant, C.; Mousson de Camaret, B. POLG1 variations presenting as multiple sclerosis. *Arch. Neurol.* **2010**, *67*, 1140–1143. [CrossRef]
103. Spain, R.; Powers, K.; Murchison, C.; Heriza, E.; Winges, K.; Yadav, V.; Cameron, M.; Kim, E.; Horak, F.; Simon, J.; et al. Lipoic acid in secondary progressive MS: A randomized controlled pilot trial. *Neurol. Neuroimmunol. Neuroinflamm.* **2017**, *4*, e374. [CrossRef]
104. Peruzzotti-Jametti, L.; Pluchino, S. Targeting Mitochondrial Metabolism in Neuroinflammation: Towards a Therapy for Progressive Multiple Sclerosis. *Trends Mol. Med.* **2018**, *24*, 838–855. [CrossRef] [PubMed]
105. Tourbah, A.; Lebrun-Frenay, C.; Edan, G.; Clanet, M.; Papeix, C.; Vukusic, S.; De Sèze, J.; Debouverie, M.; Gout, O.; Clavelou, P.; et al. MD1003 (high-dose biotin) for the treatment of progressive multiple sclerosis: A randomised, double-blind, placebo-controlled study. *Mult. Scler.* **2016**, *22*, 1719–1731. [CrossRef]
106. Jang, Y.-H.; Lim, K.-I. Recent Advances in Mitochondria-Targeted Gene Delivery. *Molecules* **2018**, *23*, 2316. [CrossRef] [PubMed]
107. Finsterer, J.; Zarrouk-Mahjoub, S. Re: Guy et al.: Gene therapy for Leber hereditary optic neuropathy: Low-and medium-dose visual results (Ophthalmology. 2017; 124:1621–1634). *Ophthalmology* **2018**, *125*, e14–e15. [CrossRef] [PubMed]

© 2019 by the authors. Licensee MDPI, Basel, Switzerland. This article is an open access article distributed under the terms and conditions of the Creative Commons Attribution (CC BY) license (http://creativecommons.org/licenses/by/4.0/).

Review

Mitochondrial Dysfunction in Parkinson's Disease—Cause or Consequence?

Chun Chen [1,2], Doug M. Turnbull [1,2] and Amy K. Reeve [1,2,]*

1. Wellcome Centre for Mitochondrial Research, Institute for Neuroscience, Newcastle University Institute for Ageing, Newcastle University, Newcastle upon Tyne, NE2 4HH, UK; c.chen20@ncl.ac.uk (C.C.); doug.turnbull@newcastle.ac.uk (D.M.T.)
2. Newcastle University LLHW Centre for Ageing and Vitality, Newcastle University, Newcastle upon Tyne, NE2 4HH, UK
* Correspondence: amy.reeve@newcastle.ac.uk; Tel.: +44-(0)191-208-3074

Received: 21 December 2018; Accepted: 5 February 2019; Published: 11 May 2019

Abstract: James Parkinson first described the motor symptoms of the disease that took his name over 200 years ago. While our knowledge of many of the changes that occur in this condition has increased, it is still unknown what causes this neurodegeneration and why it only affects some individuals with advancing age. Here we review current literature to discuss whether the mitochondrial dysfunction we have detected in Parkinson's disease is a pathogenic cause of neuronal loss or whether it is itself a consequence of dysfunction in other pathways. We examine research data from cases of idiopathic Parkinson's with that from model systems and individuals with familial forms of the disease. Furthermore, we include data from healthy aged individuals to highlight that many of the changes described are also present with advancing age, though not normally in the presence of severe neurodegeneration. While a definitive answer to this question may still be just out of reach, it is clear that mitochondrial dysfunction sits prominently at the centre of the disease pathway that leads to catastrophic neuronal loss in those affected by this disease.

Keywords: Parkinson's disease; mitochondria; ageing; neurodegenerative disease

1. Introduction

Mitochondrial dysfunction was proposed to be an integral player in the development of Parkinson's disease (PD) nearly 40 years ago, and since those initial discoveries, evidence of the role it may play in this neurodegeneration continues to increase. There is now evidence to suggest a role not only for a loss of mitochondrial function in terms of ATP provision and calcium buffering capacity, but also the degradation of these organelles through mitophagy and the interaction of mitochondria with other organelles and proteins in this disease. Furthermore, over the last 30 years, our understanding of the molecular pathways involved in the development of PD has grown immensely. These developments have been driven in part by improved genetic techniques allowing large-scale screens to be performed, but also by the speed with which disease-causing genes can be identified within affected families. In light of this, the number of genes associated with both early onset familial PD and the sporadic form of the disease continues to increase. This growing body of work continues to highlight pathways likely to be important for the development of PD, many of which centre around the function of mitochondria.

2. Mitochondrial Function

2.1. Mitochondrial Respiratory Chain

The first link between mitochondrial dysfunction and PD became evident in 1983, when 1-methyl-4-phenyl-1,2,3,6-tetrahydropyridine (MPTP) was found to cause parkinsonian-like symptoms in intravenous drug users [1,2]. Once MPTP penetrates the blood-brain barrier, this lipophilic compound is bio-transformed into its toxic form 1-methyl-4-phenylpyridinium (MPP+) by glial monoamine oxidase (MAO) [3]. MPP+ specifically interferes with the activity of respiratory chain (RC) complex I (NADH: Ubiquinone oxidoreductase) in dopaminergic (DA) neurons, causing selective neurodegeneration in both human and mouse substantia nigra (SN) [2,4,5]. Since this original study, several environmental toxins which also are capable of potent inhibition of mitochondrial complex I have been implicated in the epidemiology of sporadic PD. This raised awareness of the association of environmental exposure with the risk of developing PD later in life, via damage of RC function and increased oxidative stress (reviewed in [6,7]). These neurotoxins include rotenone, a wide spectrum insecticide [8]; organochlorine pesticides, such as chloripyrifos and permethrin, and macromolecular solvents such as 1-trichloromethyl-1,2,3,4-tetrahydro-β-carboline (TaClo) and trichloroethylene (TCE) (reviewed in [6]). However, the toxicity of these molecules may not simply lie in their inhibition of Complex I, but may also be related to exposure level, and their effect on other cellular processes, for example intracellular dopamine oxidation [9–12].

In post-mortem studies, varying degrees of Complex I (Figure 1) and complex II (succinate dehydrogenase, SDH) deficiency have been found in individual SN neurons from PD patients (~60% Complex I and ~65% Complex II deficiency [13]). A widespread decrease in complex I expression has also been observed in multiple brain regions in PD, including hippocampus, putamen, and pedunculopontine nucleus [14,15]. While immunofluorescent (protein expression) [13] and COX (Complex IV, cytochrome *c* oxidase)/SDH (enzyme activity) [16,17] assays have identified that 25–30% of SN neurons show a deficiency for Complex IV (Figure 1). This deficiency commonly presents alongside the aforementioned Complex I or Complex II deficiency in both PD and normal ageing [13,18]. It is possible that the predisposition for Complex I deficiency in SN neurons is due to a low reserve for Complex I function [19]. Furthermore, the high number of mitochondrial DNA (mtDNA) encoded subunits required for Complex I assembly confers that this complex is more likely to be affected by pathogenic mtDNA mutations in the susceptible neuronal populations. However, the causal role of Complex I deficiency in SN neuronal loss remains unclear. The loss of Complex I seems to be better tolerated than that of other RC subunits, given that defects in the expression of this complex are commonly detected in surviving SN neurons in PD cases and aged matched individuals, as well as in neurons from multiple other regions [13,14,18]. Mutations in the gene encoding the mitochondrial polymerase, *POLG*, in mitochondrial disease patients cause accumulated mitochondrial dysfunction similar to that detected in PD and ageing [20]. Complex I-negative neurons are found in infant *POLG* patients, before the prominent loss of neurons, which could further support a less detrimental effect of Complex I deficiency on neuronal survival, compared to the loss of other RC complexes [21].

Single neuron studies have attempted to interrogate the relationship between mitochondrial dysfunction and the formation of alpha-synuclein pathology [14,22]. Significantly higher expression of Complex I and Complex IV was found in SN neurons with alpha-synuclein pathology [22], whilst the presence of alpha-synuclein aggregation was less frequent in Complex I deficient SN neurons [14,22]. Both post-mortem studies suggested that alpha-synuclein pathology and abnormal RC function may be two independent factors in PD pathogenesis.

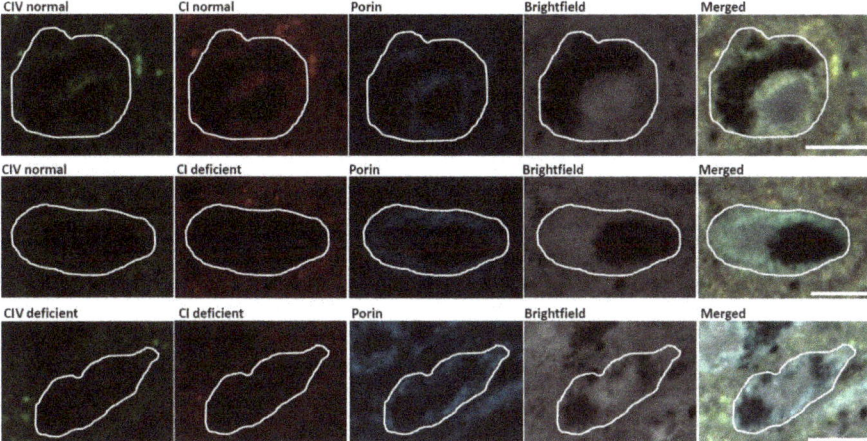

Figure 1. Mitochondrial respiratory chain (RC) deficiency in substantia nigra (SN) neurons of an individual with Parkinson's disease (PD). Immunofluorescent images demonstrating SN neurons with normal Complex I (CI) and Complex IV (CIV) expression, Complex I deficiency and normal Complex IV expression, and deficiency of both Complex I and Complex IV. Scale bar, 20 µm.

2.2. Mitochondrial DNA Defects

The potential consequences of clonally expanded mtDNA deletions and related biochemical defects on the survival of SN neurons have been investigated in several studies. Indeed, somatic mtDNA deletions reaching 50% have been described at the single SN neuron level in both PD and healthy ageing, with much higher levels in COX-deficient cells [16,17]. These deletions were found to be of various sizes (~2000 to ~9500bp), located within the major arc of mtDNA, and involve both tRNA and mitochondrial RC protein genes [20]. It is therefore evident that these multiple mtDNA deletions are somatically acquired, with the clonal expansion of different-sized deletions within each individual SN neuron, thus yielding a wide spectrum of mtDNA deletion breakpoints within an individual [20]. A recent post-mortem study revealed an accumulation of mtDNA deletions with advancing age in individual SN neurons, whilst such a trend did not occur in cortical or cerebellar neurons and the deletion load was much lower in these neuronal populations [23].

Although the mechanism for the formation of mtDNA deletions is still under debate, the necessity of mtDNA replication activity in deletion formation [24] was challenged by the absence of 3'-repeat retention in some of the detected mtDNA deletion species in SN neurons [20]. According to the "slipped-replication" theory, a single stranded loop occurs between 3'5' direct repeats which is then exposed and degraded during mtDNA replication, leading to the occurrence of deletions that are often located within the major arc of mtDNA [24]. Krishnan et al. proposed another possibility that mtDNA deletion formation may initiate from 3'–5' exonuclease activity during the repair of damage to mtDNA in the SN neuronal population [25]. mtDNA damage leads to the formation of double strand breaks (DSB), and misrepair of such breaks could cause the loss of several kilobases of mtDNA on both strands. Single strand mtDNA generated from this process would then anneal with homologous sequences, resulting in the double-stranded deleted molecule with both a 5' and 3' repeat [25]. The relationship between DSB repair and the generation of mtDNA deletions has been supported by several studies using inducible mitochondrial-targeted restriction endonuclease, *Pst*I, to trigger mtDNA DSB. Multiple mtDNA deletions were found in cultured *Pst*I neurons [26] and in DA neurons of a *Pst*I mouse model [27]. These mice also developed nigrostriatal degeneration and PD-related behavioral phenotypes [27]. Later, Moraes' team reported an acceleration effect of mtDNA DSB on neuronal ageing through reactive oxygen species (ROS) induced damage [28]. The nature of deletion formation through repair could therefore be regarded as an inevitable cost of mtDNA self-rescue/regulation against oxidative damage. This fits well with the strikingly high deletion load of mtDNA detected in aged SN neurons of both healthy individuals and PD patients [20,23], as ROS-induced mtDNA damage advances with ageing.

In addition to the higher mtDNA deletion load in the SN than other aged brain regions, such as the basal ganglia, cortical areas and cerebellum [23,29,30], Dolle et al. also revealed a age-related increase of mtDNA copy number within individual SN neurons [23]. More importantly, the upregulation of mtDNA copy number also correlates with the increase in mtDNA deletion load within the SN neuronal population. The corresponding increase of mtDNA copy number with mtDNA deletion level, together with an enhancement of mitochondrial function, were mirrored in a study of DA neurons from a premature ageing model (POLG^{D257A} mice) [31]. These findings suggested an ability of age SN neurons to adapt to mitochondrial dysfunction via the maintenance of their mtDNA population. Failure of such regulatory mechanisms of mtDNA copy number is described in PD patients in several studies. Grunewald et al. suggested depletion of mtDNA copy number and low expression of an essential mtDNA nucleoid protein, mitochondrial transcription factor A (TFAM) in the SN neurons of PD patients [13]. However, another study showed a decrease in mtDNA wildtype copy number but not in total copy number, alongside a decreased correlation between mtDNA copy number and deletion level in PD neurons compared to healthy aged neurons [23]. Reduction of mtDNA copy number in the peripheral blood [32] and cerebrospinal fluid [33,34] has also been detected in patients with PD and neurodegenerative disease.

These studies all suggest that the formation and accumulation of high levels of mtDNA deletions leads to mitochondrial dysfunction that will have an impact on the health and survival of SN neurons. However, many of the reported findings are also found in normal ageing in the SN. Therefore, mtDNA alterations are certainly one cause of the mitochondrial deficiency detected within these neurons and will undoubtedly increase the vulnerability of these neurons to loss in PD, but the question still remains whether these defects are causative of PD itself.

2.3. ROS Production and Oxidative Stress

The accumulation of somatic mtDNA deletions and the subsequent development of RC deficiency within aged SN neurons may lead to impaired mitochondrial membrane potential ($\Delta\Psi m$), reduced synthesis of ATP, and an increase in release of ROS leading to oxidative stress [35]. A higher rate of electron flux due to an increase in respiratory capacity to adapt to high metabolic demands in SN neurons has also been thought to cause increased ROS generation. The increase in numbers of electrons being transferred would increase the probability of electron capture by O_2 to generate superoxide before they reach COX [36]. In addition to the production and release of ROS from mitochondria, the process of dopamine metabolism by monoamine oxidase could also account for the accumulation of toxic oxidative species such as hydroxyl radicals [37]. Oxidative damage to mtDNA results in further deficiency in RC expression and creates a vicious circle of oxidative stress and bioenergetics failure. This "mitochondrial theory of ageing" has long been regarded as a rationale for mitochondrial changes with advancing age [38], which could contribute to the increase in SN neuronal vulnerability. However, this theory was challenged by more recent studies which manipulated the antioxidant capacity of several mouse models, finding impaired integrity of the mitochondrial genome which was not necessarily associated with an age-related phenotype or shortened lifespan (reviewed in [39]). Evidence from a premature ageing mouse model (POLG^{D257A} mutant) found that the generation of mitochondrial ROS or mtDNA damage did not increase with a high heteroplasmic level of mtDNA point mutation, and that there seems to be a threshold for mtDNA deletions to cause accelerated ageing [40].

In addition to mitochondrial ROS, other factors that may contribute to increased oxidative damage include iron accumulation and an increase in lipid peroxidation burden [41,42], while accumulation of hydroxyl and superoxide radicals due to a decline in glutathione content [43] have also been reported in the brain of PD patients. Recent studies also highlighted a direct toxicity of alpha-synuclein oligomers on mitochondria via induction of mitochondrial lipid peroxidation and oxidation of ATP synthase [44,45]. It was shown that alpha-synuclein leads to increased permeability of mitochondrial membranes and ROS production, ultimately leading to neuronal death [45].

2.4. Calcium Handling

Emerging evidence has suggested the importance of cytosolic calcium ions (Ca^{2+}) in the regionally selective nature of neuronal loss in PD pathogenesis. Dopaminergic neurons in the SN exhibit autonomous pacemaking activity in order to maintain regular release of dopamine to the striatum, facilitated by CaV1.3 L-type calcium channels. Action potentials generated from this process are broad, with a relatively slow rate (2–10 Hz), promoting a slow rhythmic activity accompanied with a slow oscillation in cytosolic Ca^{2+}. A relationship between Ca^{2+} oscillations and PD neuronal loss was initially proposed by experiments showing a decline in the number of calbindin (a Ca^{2+} binding protein), positive neurons in PD patients [46,47]. These slow Ca^{2+} oscillations promote the import of Ca^{2+} into the mitochondrial intermembrane space and matrix. Intramitochondrial Ca^{2+} is required for the activation of pyruvate dehydrogenase, isocitrate dehydrogenase and α-ketoglutarate (important enzymes in the tricarboxylic acid (TCA) cycle) and promotes RC function and ATP production. Mitochondrial sequestration of Ca^{2+} benefits synaptic neurotransmission in terms of assisting in the stabilization of the cytosolic Ca^{2+} concentration during vesicle exocytosis [48] and promoting the recovery of synaptic excitation after exocytosis [48,49]. The unique feature of CaV1.3 channels is that they open at relatively hyperpolarized $\Delta\Psi m$. Even so, the lack of powerful intracellular buffering of Ca^{2+} from the endoplasmic reticulum (ER) in DA neurons drives a sustained Ca^{2+} flux to enter mitochondria. This is able to promote ATP synthesis in the absence of a strong energy demand [50,51]. Ca^{2+} overload may force the opening of mitochondrial permeability transition pore (mPTP), leading to retrograde electron flux through the electron transport chain, resulting in increased ROS production, release of cytochrome *c*, and activation of apoptosis [52,53]. The increase in mitochondrial oxidative stress has also been identified in other neuronal populations, including the dorsal motor nucleus of the vagus (DMV) and the locus coeruleus (LC), which also express CaV1.3 channels. DA neurons in the ventral tegmental area (VTA) which have modest CaV1.3 channel currents however, demonstrate low mitochondrial oxidative stress and an improved ability to buffer cytosolic Ca^{2+} with ageing [54,55]. All of these indicate that CaV1.3 channels cause extra oxidative burden on mitochondria that may contribute to the selectively vulnerability of SN neurons (reviewed in Reference [56]). A recent report showed depletion of alpha-synuclein prevents the elevation of cytosolic Ca^{2+} concentration induced by MPP+ in mouse SN neurons, suggesting synergistic detrimental effects of these two pathological aspects in PD neurodegeneration [57].

2.5. Two Novel Familial PD Genes Associated with Mitochondrial Function

2.5.1. VPS35

A novel autosomal dominant PD related pathogenic mutation was identified in the vascular protein sorting associated protein 35 (*VPS35*, PARK17, OMIM 614203) gene [58,59], and has been linked to impaired RC complex I and II subunit assembly and activity [60]. The VPS35 protein is a core subunit of the cargo recognition subcomplex that mediates the sorting, trafficking, and endocytosis of synaptic vesicles [35]. The *VPS35* mutation-induced mitochondrial dysfunction has been suggested to be a consequence of impaired mitochondrial fusion, tipping the balance of the fission and fusion cycle toward excessive fission. VPS35 promotes the degradation of mitochondrial E3 ubiquitin ligase1 (MUL1), and stabilizes mitofusin 2 (MFN2) preventing its degradation by MUL1 [61]. VPS35 deficiency impairs the regulatory machinery of mitochondrial dynamics via degradation of MFN2 by MUL1 in DA neurons, leading to mitochondrial dysfunction and fragmentation, which may underlie related PD pathogenesis [61]. Furthermore, a recent report has shown prominent changes in dopaminergic synaptic function, including increased dopamine turnover, loss of the dopamine transporter (DAT) and increased vesicular monoamine transporter 2 (VMAT2) expression in a *VPS35* knock-in mouse model [62].

2.5.2. CHCHD2

Mutations in the coiled-coil-helix-coiled-coil-helix domain containing 2 (*CHCHD2*, PARK22, OMIM 616244) gene have recently been associated with autosomal dominant PD in Japanese families [63]. The role of *CHCHD2* as a PD-related gene was confirmed by genetic screening of affected Chinese families [64], however, studies in Caucasians, south Italians [65,66], and Brazilians [67] did not support its causative role in PD. It has been proposed that the localization of the CHCHD2 protein in the mitochondrial intermembrane space is beneficial to the maintenance of mitochondrial cristae structure and integrity of the mitochondrial RC [68,69]. Inhibition of CHCHD2 was found to impair mitochondrial Complex III (cytochrome bc_1 complex) and Complex IV activity, leading to decreases in OXPHOS activity and increased oxidative damage [69,70]. These mitochondrial structural and biochemical defects have also been shown in fibroblasts generated from a PD patient who carried a homozygous variant in *CHCHD2* [71]. In addition to the mitochondrial structural and biochemical defects, *Drosophila* carrying the *CHCHD2* mutations identified in patients also manifest motor symptoms and DA neurodegeneration [69].

2.6. Does Mitochondrial Dysfunction Drive the Development of Neurodegeneration in PD?

It is evident from other conditions, for example, mitochondrial disorders associated with mutations in the polymerase gamma gene, *POLG*, that a primary mitochondrial defect is sufficient to cause loss of SN neurons (and other neuronal populations) [18,72]. Furthermore, such mutations are often associated with the development of PD-like symptoms (reviewed in Reference [73]). However, attempting to disentangle "cause and effect" in the contribution of mitochondrial dysfunction to the pathogenesis of PD is difficult, since many of the changes we detect in those with PD are also present in healthy aged individuals. The mitochondrial defects present in these individuals often exist in the absence of cell loss or parkinsonian symptoms. For example, equivalent levels of mtDNA deletions are detected in SN neurons from individuals with PD and healthy controls [16,17] and neurons showing deficiencies for both Complex I and Complex IV are also detected in both instances [18]. Many of the processes described above initiate changes in other pathways and thus activate well defined responses which exist to mitigate to these changes, for example, ROS and antioxidants. Therefore, it might be suggested that in PD it is the ability of neurons to respond to mitochondrial functional changes that is impaired. Thus, neurons become more sensitive to changes in other pathways, which ultimately leads to their degeneration and loss.

3. Mitochondrial Turnover

3.1. Mitophagy

Mitophagy contributes to mitochondrial quality control by the selective clearance of damaged mitochondria, targeting an entire mitochondrion or fission fragmented mitochondria for autophagy. As a specialized form of autophagy, mitophagy comprises three necessary stages: the recognition of impaired mitochondria, the formation of autophagic membranes around the target, and the fusion of the mitoautophagosome with a lysosome (reviewed in Reference [74] and Reference [75]). Two familial PD-associated proteins, phosphatase and tensin homolog (PTEN)-induced putative kinase 1 (PINK1, PARK6, OMIM 608309) and E3 ubiquitin ligase, Parkin (PARK2, OMIM 602544), have been strongly implicated in the identification of damaged mitochondria for degradation via mitophagy [76]. PINK1 is a serine/threonine kinase localized to both mitochondria and the cytosol [77]. It accumulates on the outer membrane of damaged mitochondria, following a reduction in $\Delta\Psi m$, and recruits Parkin to trigger the engulfment of mitochondria by autophagosomes [78]. The recruitment of Parkin requires PINK1-mediated phosphorylation of both Parkin and ubiquitin, which activates a ubiquitin phosphorylation cascade for mitophagy adapters. This feed-forward amplification loop leads to further recruitment and ubiquitylation-activation of Parkin so that the engulfment can proceed [79]. However, several studies have described a Parkin-independent mitophagy pathway associated with other E3 ubiquitin ligases including Glycoprotein 78 (Gp78) [80] and Ariadne RBR Ubiquitin E3 ligase 1 (ARIH1) [81].

3.2. Mitochondrial Biogenesis

The constant generation of new mitochondria is a key process for the maintenance of a healthy mitochondrial population with advancing age. The fact that the mitochondrial genome has limited protein coding capacity determines the dependence of mitochondrial biogenesis on synchronized transcription activity from both the mitochondrial and nuclear genomes. This provides building blocks which aid the replacement of mitochondria or can be inserted into remaining functional mitochondria [82]. Central to the regulation of mitochondrial biogenesis is the nuclear transcription cascade, mediated by peroxisome proliferator-activated receptor γ (PPARγ) coactivator 1α (PGC-1α) [83,84]. PGC-1α and its downstream cofactors nuclear respiratory factors (NRF1, 2) largely contribute to the regulation of transcription initiation of all nuclear-encoded RC proteins that are expressed by the nuclear genome. TFAM and dimethyladenosine transferase 2 (TFB2M) can then act in response to intracellular oxidative stress [85,86]. Reduction in the mRNA level of nuclear-encoded RC genes that are responsive to PGC-1α has been identified in individual DA neurons of PD patients, and in neurons in regions with subclinical PD-related Lewy body neuropathology [87]. The transcriptional decline in genes that control mitochondrial biogenesis and oxidative energy metabolism, including RC related genes and some cytosolic ribosomal genes (related to protein synthesis), are altered with advancing age [88,89]. This is supported by the discovery of downregulated AMP- activated protein kinase (AMPK) signalling, a key player in the control of PGC-1α expression, alongside decreased mitochondrial biogenesis in aged individuals [90]. These data strongly indicate the contribution of impaired biogenesis to the accumulation of mitochondrial dysfunction during ageing and the neurodegenerative process.

PGC-1α and its transcriptional control is critical for maintaining neuron function. Loss of dopaminergic neurons and striatal degeneration have been demonstrated in animal models where the PGC-1α gene is silenced or knockout [57,91,92]. Several ex vivo and in vivo studies have demonstrated the beneficial effects of PGC-1α overexpression and the corresponding increase in transcription activities in the rescue of dopaminergic neuronal loss induced by MPTP [93], rotenone, or alpha-synuclein mediated toxicity [94–96]. In addition to enhancing mitochondrial biogenesis, PGC-1α is also capable of activating the expression of antioxidant enzymes in response to oxidative stress [97,98] and protecting DA neurons against neuroinflammation [99]. However, there is still a lack of consensus on the value of upregulating PGC-1α expression as a PD therapeutic intervention. An increase in PGC-1α expression is not necessarily associated with an activation of its target cofactors that facilitate mitochondrial biogenesis [40]. Major alterations in energy metabolism that impair normal neuronal function have been found in a study of long term overexpression of PGC-1α in an alpha-synuclein mutant mouse, carrying the A53T mutation [33]. However, mitochondrial hyperactivity and increased ROS production were also described in this model, and may also contribute to the severe DA neuronal loss.

Shin et al. first identified a link between Parkin and PGC-1α in the regulation of PGC-1α expression via a Parkin interacting substrate, PARIS (also known as zinc finger protein 746, ZNF746). Parkin mediates the proteasomal degradation of PARIS. In the absence of Parkin, PARIS is bound to the promoter of PGC-1α gene and suppresses its expression [100]. This causes declines in mitochondrial respiratory capacity and mitochondrial mass, ultimately leading to dopaminergic neuronal death [33,34,101]. The parkin-mediated PARIS-dependent control of mitochondrial biogenesis, adds further evidence that impaired mitochondrial turnover contributes to the neuronal loss in PD.

It is hypothesized that mitochondrial turnover is slowed in aged and PD neurons due to defective nuclear transcriptional control of mitochondrial proteins. Alongside this, there is a decline in the clearance of dysfunctional mitochondria that occurs in coordination with a decrease in biogenic activity and, as a result, mutant mitochondria accumulate in the neurodegenerative process. Evidence for a decline in mitophagy during ageing has been highlighted in several recent reviews [102–105]. The abnormal accumulation of alpha-synuclein in PD neurons may put an additional burden on protein degradation pathways, increasing the vulnerability of SN neurons to mitochondrial dysfunction and pushing these neurons toward early cell death (reviewed in [106]).

3.3. Familial PD Associated Genes Related to Mitochondrial Turnover

3.3.1. PINK1 and Parkin

Among many genes that are associated with familiar PD, *PINK1* and *Parkin* are highlighted due to their substantial involvement in mitochondrial maintenance. Although the PINK1/Parkin pathway is the key mediator of the process of mitophagy, recently more attention has been given to their regulatory roles in neuronal mitochondrial dynamics via their interactions with the Mitochondrial Rho GTPase1 (MIRO)/Milton complex (see below). In addition, the association of PINK1 and Parkin with the removal of mitochondrial proteins and impaired mitochondrial biogenesis has been suggested (reviewed in Reference [107]).

PINK1 has been suggested to also be important for the mitochondrial unfolded protein response (UPR^{mt}). The UPR^{mt} was first reported in cell lines with a mutant form of ornithine transcarbamylase (OTC) [108], which led to accumulation of a misfolded form of the protein which subsequently drove expression of mitochondrial quality control genes. Believed to be a stress response to damaged mitochondrial proteins and hence dysfunction, alterations in the UPR^{mt} pathway may be of pathogenic importance in PD. For example, PINK1 has been proposed to be involved in this pathway through its interaction with HTRA serine peptidase 2 (HTRA2, PARK13, OMIM 606441) [109] and tumor necrosis factor receptor-associated protein 1 (TRAP1) [110]. Importantly, mutations in the genes encoding both these proteins have been associated with the development of PD [111,112]. In addition, the brains of those individuals carrying *PINK1* mutations show decreased levels of HTRA2 phosphorylation [109], while the brains of mice with a knockout of *HTRA2* show an accumulation of misfolded proteins within the mitochondria leading to reduced respiratory function and neurodegeneration [113]. TRAP1 is a mitochondrial chaperone protein, whose knockout in *Drosophila* causes an age-related decrease in climbing ability and dopamine levels in the brain, increased sensitivity to mitochondrial toxins including rotenone and a decrease in ATP levels [114]. It is also a PINK1 effector whose expression in PINK1 deficient *Drosophila* rescues morphological and mitochondrial defects and is capable of reducing dopamine deficits [114,115]. Given the importance of the UPR^{mt} for the maintenance of mitochondrial function, this pathway may prove to be an important therapeutic target. Upregulation of the UPR^{mt} for example, may allow mitochondrial function to be improved through either degradation of unfolded mitochondrial proteins or increases in the upstream signalling cascade.

3.3.2. DJ-1 and LRRK2

The functions of DJ-1 (PARK7, OMIM 606324) and leucine-rich repeat kinase 2 (LRRK2, PARK8, OMIM 609007) are less well characterized in terms of mitochondrial health. Downregulation of DJ-1 causes multiple mitochondrial abnormalities, including increased mitochondrial fragmentation and reduced ΔΨm and connectivity [116,117]. Interestingly, this effect can be blocked by overexpression of Parkin and PINK1 [118]. A recent study using induced pluripotent stem cells (iPSC)-derived neurons reports the beneficial effect of LRRK2 in the removal of MIRO, and thus enhanced arrest of damaged mitochondria. Pathogenic *LRRK2* mutations disrupt this function, speeding up mitochondrial mobility and consequentially slowing the initiation of mitophagy [119].

3.3.3. ATP13A2 and GBA

Novel mutations in *ATP13A2* (ATPase type 13A2, PARK9, OMIM 610513) have been identified as a genetic cause of a rare juvenile-onset form of PD, named Kufor–Rakeb syndrome [120,121]. *ATP13A2* encodes an isoform of type P5B ATPase, which functions as a lysosomal ATPase transporter. It is believed to play critical roles in the regulation of vesicular transportation [122], endolysosomal activity [123,124], and glycolysis [125]. Deficiency of ATP13A2 causes lysosomal defects, and thus affects cation homeostasis, for example Zn^{2+}, leading to impaired mitochondrial respiratory function and defective mitophagy [89,122]. A recent study also found gliosis in multiple brain regions of heterozygous *ATP13A2* knockout mice, suggestive of the neuroinflammation which occurs in the early stages of PD development [123]. Another heterozygous mutation in the gene that encodes the lysosomal enzyme glucocerebrosidase (*GBA*, OMIM 606463) has also been identified as a powerful genetic risk factor for PD [126,127]. Those PD patients who carried a *GBA* mutation tended to have an earlier age of onset and an increased risk of developing dementia compared to idiopathic PD patients [128]. Together, this evidence highlights that lysosomal alterations may be crucial to the pathogenesis PD, exacerbating the accumulation of mitochondrial damage and abnormal protein aggregation [129].

3.4. Does Mitochondrial Turnover Drive the Development of Neurodegeneration in PD?

Although the process is still not clearly defined in neurons, data suggests that the degradation of dysfunctional mitochondria occurs through the autophagy related pathway, mitophagy. A number of studies have suggested that there is a decline in mitophagy with age (reviewed in References [102,103]). Furthermore, many of the genes that control the degradation of mitochondria through this pathway have been linked to ageing and lifespan in model organisms (reviewed in [130]). A decline in lysosome function has also been shown to occur with advancing age, mainly through the accumulation of lipofuscin within these organelles [131]. However, the fact that many of the genes responsible for familial PD have been shown to have important roles in autophagy or lysosomal pathways suggests that this is an important driver of the pathogenesis of PD. Alterations in the efficiency of such pathways will clearly exacerbate mitochondrial dysfunction, allowing defective mitochondria to accumulate. Mitochondrial dysfunction itself may also increase oxidative damage to proteins and organelles which may overwhelm these systems in PD, leading to neuronal loss. Again, we are faced with a vicious circle of damage, whereby mitochondrial dysfunction may be caused by defects in genes such as *ATP13A2*. However, lysosomal dysfunction may increase oxidative stress, which would then cause further mitochondrial and cellular damage.

4. Mitochondrial Dynamics, Transport, and Distribution

The precise distribution of mitochondria within neurons is a fundamental aspect of the correct function of these cells. Neuronal communication through electrical impulse and chemical synapses is fundamental for brain function, and, in PD, a loss of synapses has been shown to precede the loss of cell bodies from the substantia nigra. This decline in synaptic terminal density may begin decades before the first symptoms appear in affected individuals (reviewed in Reference [132]). Therefore, understanding the causes of this synaptic loss may be key to unlocking new therapies for PD.

As described above, mitochondrial production relies on signalling between the nucleus and the mitochondria, while the degradation of dysfunctional mitochondria requires the interaction of a number of proteins and signalling pathways that we are only just beginning to understand. Mitochondrial biogenesis generally occurs at the soma, in close proximity to the nucleus, although evidence has suggested that mtDNA replication may occur within axons at sites distant from the cell body [133–135]. The degradation of damaged mitochondria has generally been thought to occur within the cell body, since this was thought to be the only location of lysosomes, though recently this convention has also been challenged [136]. Data has also suggested that mitochondria may be donated to and removed from neurons by astrocytes, with such processes occurring in response to cellular stress [137,138]. Therefore, neurons require processes which allow mitochondrial movements to be responsive to changes within the local and intracellular environments.

Mitochondria perform a variety of roles within neurons to support their function. They provide localized ATP to regions of the neuron which are particularly energy demanding (including the nodes of Ranvier and pre-synaptic terminals) and buffer cytosolic calcium. These two functions of mitochondria support the release and recycling of synaptic vesicles and the maintenance of electrochemical gradients, which are intimately dependent on each other since they both rely on calcium signalling. Therefore, as the main organelles which support these processes, there is a need for mitochondria to be dynamic and responsive in their distribution. The movement and localization of mitochondria can be affected by their function. A loss of mitochondrial membrane potential, for example, has been proposed to drive directional mitochondrial movement. Vice versa, incorrect positioning of mitochondria within neurons may have a detrimental effect on the health and function of neurons (reviewed in [139,140]).

4.1. Fission and Fusion

The processes of fission and fusion control the network connectivity of mitochondria within neurons. These processes allow the creation or destruction of a mitochondrial syncytium, in which mitochondrial contents may be shared, or dysfunctional mitochondria partitioned for destruction. The degradation of dysfunctional mitochondria through mitophagy and the manipulation of this process has been suggested to be an important therapeutic avenue for many degenerative diseases [141]. While many of the specifics of exactly how mitochondria are detected as being targets for degradation remain unclear, it is likely that the fission of such organelles from the network will be a key process [142]. Fragmentation of the mitochondrial network occurs rapidly following the loss of mitochondrial membrane potential, and if this loss cannot be restored, fragmented mitochondria lose their "fusion" capabilities and are degraded [142]. Many of the proteins that control these two process have been characterized and alterations in these processes have been detected in a number of models of familial PD, yielding differences in the size of mitochondria and their ultrastructure [143]. Moreover, a loss of dynamin-1-like protein (DRP1), a key protein in the fission machinery within dopaminergic neurons, causes a Parkinson's like phenotype in affected mice due to degeneration of SN neurons. A staggering 85% of SN neurons had been lost by the age of 1 month in these animals and this loss was due to depletion of the axonal mitochondrial population [144]. Furthermore, mutations in *OPA1* (OMIM 605290) have been linked with parkinsonism, including alterations upon DAT-SPECT scanning [145]. The two heterozygous mutations reported in this paper also caused alterations to the structure and function of mitochondria [145]. In SHSY5Y cells, the overexpression of OPA1 has been shown to protect against the ultrastructural abnormalities induced by treatment with MPP$^+$, although interestingly this was not related to a reduction of fission [146]. The fusion of mitochondria has been further linked with PD by the finding that mitofusin 1 (MFN1) and MFN2 are ubiquitinated in a Parkin/PINK1 dependent manner [147,148]. The ubiquitination of mitochondrial proteins is a key step in their identification for degradation through mitophagy. Further details of this process have emerged showing that PINK1 phosphorylates MFN2, and this phosphorylation then recruits Parkin, which then ubiquitinates the protein [149]. This study also showed that in cardiomyocytes, a deficiency of MFN2 impedes mitophagy, leading to the accumulation of abnormal mitochondria and respiratory dysfunction [149]. The ubiquitination of the mitofusin proteins is impeded in fibroblasts from individuals with pathogenic *PINK1* or *Parkin* mutations [150]. Interestingly, other proteins important for the fission and fusion of mitochondria are not affected by mutations within these genes [150]. The ubiquitination of the mitofusins will act not only to identify mitochondria for degradation, but will also ensure that such mitochondria are unable to rejoin the mitochondrial network. In addition, as described above, many of the neurotoxins used to generate experimental models of Parkinson's Disease, including rotenone and MPTP, cause the fragmentation of the mitochondrial network, most likely due to their inhibitory effects on mitochondrial complex I [151–153]. In fact, their toxicity might be directly related to their effect on the balance of mitochondrial biogenesis and fission/fusion [154].

4.2. Mitochondrial Distribution

The correct positioning of mitochondria within neurons is required for localized supply of ATP and calcium buffering. By performing these essential tasks mitochondria become integral to the functioning and survival of neurons. The provision of ATP is crucial for the support of neuronal electrical activity, and in line with this, mitochondria are found clustered at sites with high energy demands, for example, at the Nodes of Ranvier in myelinated axons. Their location at such sites supports the function of ion channels such as the Na^+/K^+ ATPase, which is responsible for the restoration of resting membrane potential following an action potential. Furthermore, their distribution changes to mirror alterations in channel distribution, for example, following demyelination [155]. In addition, the positional holding of mitochondria at these locations changes in response to electrical activity. A decline or arrest in electrical activity will increase mitochondrial mobility away from nodes, while conversely, depolarization leads to a movement of mitochondria into these areas to provide ATP to repolarize the membrane (reviewed in Reference [156]).

Localized reliable sources of ATP are also essential for synaptic function. Within the presynaptic terminal, mitochondrial ATP is required for an array of processes, including vesicle recycling, exo and endocytosis of vesicles, and the loading of recycled vesicles with neurotransmitter (reviewed in Reference [157]). As crucial as this energy provision is the ability of mitochondria to buffer calcium (see above). The release of neurotransmitter from presynaptic terminals relies on a number of processes which are modulated by calcium signalling. As such, the concentration of calcium within the presynaptic terminal must be tightly regulated, to control the quantal release of neurotransmitter, the priming of terminals for the next depolarization, and changes in synaptic strength. Calcium buffering might be particularly important for the neurons of the substantia nigra, given their pace-making activity. Blockage of the calcium channel, which governs this activity, forces the neurons to utilize sodium channels to maintain this low frequency activity, which also protects neurons against mitochondrial dysfunction caused by rotenone treatment [158].

Mitochondria can be recruited to synapses, though not all synapses contain mitochondria. Synapses devoid of mitochondria have been found in a number of cell types, including hippocampal neurons, primary cortical neurons, and the DA neurons of the nigrostriatal pathway [159–161]. The number of these "empty" synapses was recently shown to decline in the presence of dopaminergic neuronal degeneration in PD [162]. This suggests two interesting hypotheses, either synapses which lack mitochondria show an increased vulnerability in PD and are the first to be lost, or that in the presence of degenerating neighbors, neurons can populate such synapses with mitochondria and the replenishment with ATP activates them as part of an adaptation to maintain transmission to those post-synaptic sites. In neurons with complex architecture, such as the neurons of the nigrostriatal pathway which have been suggested to have a huge number of pre-synaptic terminals, it may seem obvious that they would not utilize the full synaptic complement but keep some "in reserve" in case of loss of a proportion.

4.3. Mitochondrial Transport

The correct distribution of mitochondria and the maintenance of a healthy functional population within DA neurons relies on efficient transport from sites of genesis or to sites of degradation. The movement of mitochondria along axons relies on motor proteins and microtubules with different proteins mediating retrograde (towards the cell body) and anterograde (towards the terminus) transport. The directionality of mitochondrial movements is controlled by different sets of motors with dynein motors controlling retrograde transport, while anterograde movements are driven by kinesin motors (reviewed in Reference [157]). The transport of mitochondria over long distances within neurons is much more complicated than this and additional protein interactions are required namely between trafficking kinesin binding proteins (TRAKs), kinesin heavy chain isoform 5 (KIF5) and dynein, and the mitochondrial proteins MIRO1 and 2. These outer mitochondrial membrane proteins are responsible for the interaction of mitochondria with these molecular motors. An in-depth discussion of the mechanisms of mitochondrial transport is out of the realm of this review, however, there are a number of excellent reviews which cover this in detail [139,156,157,163].

Previous work has suggested that mitochondria are transported towards the cell body when they have lost their membrane potential, a key signal for their sequestration in autophagosomes, although other studies have been unable to replicate this [164,165]. Many of the interactions involved in these processes have been well characterized, as have the patterns of movement exhibited by mitochondria under normal and disease conditions. In line with the fact that mitochondria are predominantly held at sites where they are required to provide localized ATP or calcium buffering, the majority of mitochondria within neurons are stationary, with only 10–20% of mitochondria being motile in hippocampal neurons [160,166], with over 90% reported as being stationary in cortical and spinal neurons [167–169]. Those mitochondria move through showing a variety of velocities and movements, with some moving in one direction with others seemingly wobbling back and forth. These mitochondrial movements also change with maturation of neurons, with more mitochondria becoming immobile and enriched at sites such as synapses [168].

The docking of mitochondria at discrete sites of the neuron relies on calcium. Localized elevated calcium levels cause mitochondrial stalling. In cultured neurons, mitochondrial mobility is decreased by increasing calcium influx [159,170]. The ability of calcium to cause mitochondrial immobility relies on MIRO and the binding of calcium to the proteins two EF hands. Mutations within these domains leading to impaired calcium binding have been shown to prevent the ability of mitochondria to dock (reviewed in Reference [171]). The importance of calcium for the transport and mobilization of mitochondria is particularly interesting given the importance of calcium for the pace-making activity exhibited by dopaminergic SN neurons (reviewed in Reference [56]). Many studies have investigated mitochondrial trafficking in mouse and cell culture-based models of PD [172,173]. For example, the complex I inhibitor MPP$^+$ alters mitochondrial directionality, increasing the speed of retrograde movement and decreasing the opposing movements. Furthermore, this study also showed that of the mitochondria that were moving, MPP$^+$ caused the stalling of over 50% of these. Interestingly, this study also uncovered that mitochondria within DA neurons were much smaller than those within non-DA axons. This reduction in size in DA neurons may suggest that there are inherent differences in the levels of fission and fusion between DA neurons and other neuronal populations. Differences in transport in these cells may also impede the delivery of mitochondria to synapses and other discrete sites within the neurons [172]. Furthermore, in the DA neurons of the MitoPark mouse (conditional TFAM knockout) the supply of mitochondria to axonal terminals is impaired due to reduced anterograde transport [173].

4.4. Familial PD Genes and Their Association with Mitochondrial Dynamics

The importance of many of the genes associated with PD for mitochondrial dynamics lies in the ability of the proteins they encode to affect the transport of mitochondria. Of all the proteins which have been associated with familial PD, the ones most associated with mitochondrial dynamics, obviously given their roles in mitophagy, are PINK1 and Parkin. Initially thought of as key mediators of mitochondrial degradation through mitophagy, the impact and responsibility of Parkin and PINK1 on the maintenance of mitochondrial integrity is now believed to be more complex. Recent studies have also proposed roles for these two proteins in the mitochondrial unfolded protein response, in the modulation of mitochondrial fission and fusion, and finally in mitochondrial transport and trafficking (reviewed in depth in Reference [174]).

Early studies showed that PINK1, a mitochondrial outer membrane protein, was able to recruit Parkin to mitochondria upon mitochondrial depolarization with the uncoupler carbonyl cyanide m-chlorophenyl hydrazine (CCCP) [78,175]. We now know that PINK1 actively recruits Parkin to mitochondria and phosphorylates it, Parkin is then able to ubiquitinate outer mitochondrial membrane proteins signalling the mitochondrion for destruction. The mitochondrion is then enveloped by an autophagosome and trafficked to the lysosome (reviewed in Reference [176]). *PINK1* and *Parkin* mutations in animal models have been shown to lead to both mitochondrial structural and functional changes, which are often causative of the DA neuron degeneration in these models [177,178]. Parkin overexpression has the ability to rescue many of the phenotypic changes detected in *PINK1* mutants, including the mitochondrial defects and DA loss [177]. Furthermore, the expression of Parkin in cell lines has been shown to drive expression of a number of mitochondria associated proteins, including a number important for mitochondrial dynamics, trafficking, and the ubiquitin proteasome [179].

PINK1 and Parkin have also been linked to mitochondrial dynamics and trafficking through their interaction with a number of proteins essential to these processes. For example, overexpression of PINK1 or Parkin in neurons has been shown to decrease bidirectional mitochondrial movements. PINK1 phosphorylates MIRO, which then causes Parkin mediated degradation and kinesin release [180,181]. The knockdown of PINK1 has been shown to promote anterograde mitochondrial transport in *Drosophila* motor neurons, while conversely, a loss of MIRO in Hela cells was linked to perinuclear mitochondrial clustering and mitochondrial degradation by mitophagy [180]. PINK1 expression in mitochondria has been found to be linked to expression of both MIRO and Milton, with which it forms a complex. Furthermore, the interaction with these proteins can drive the association of PINK1 with mitochondria even in the absence of a PINK1 mitochondrial targeting sequence [182]. Finally, Parkin expression has been found to decrease the amount of mitochondrial movement in hippocampal neurons. This change in mitochondrial movement was also found to be associated with a reduction in axonal mitochondrial density [181].

4.5. Do Mitochondrial Dynamics Drive the Development of Neurodegeneration in PD?

Together these data suggest that many of the proteins that are known to be integral to the development of PD have now been shown to have crucial roles in maintaining the dynamics, distribution, and transport of mitochondria within neurons. We are now beginning to understand more about these dynamic aspects of mitochondria due to improved imaging techniques and models. It is clear, however, that the correct distribution of mitochondria within neurons is as important to their function as the interconnectivity of these organelles. Furthermore, it is not only the movements of functional mitochondria which need to be monitored and controlled, but also the localisation and transport of those which are dysfunctional and require degradation. Disruption of these processes in many models has been linked with impaired neuronal function or even neurodegeneration. With advancing age and in PD the neurons of the SN do accumulate mitochondria which are dysfunctional and may therefore be unable to provide the required level of ATP. Failure to remove these mitochondria from important sites such as the synapse will then lead to further detrimental changes (as described above) and neuronal loss.

5. Mitochondria and Protein Aggregation

Protein aggregation in Parkinson's disease is predominantly driven by alpha-synuclein, a small protein with a propensity to aggregate into oligomeric and fibrillar forms upon damage or mutation. This small protein has been proposed to function to support synaptic transmission, with proposed roles in vesicular packaging, synaptic vesicle trafficking and synaptic plasticity (reviewed in Reference [183]). The interaction of alpha-synuclein with mitochondria has been studied in depth over recent years, particularly since it has been suggested that it may be imported into mitochondria and cause inhibition of complex I [184]. The specific interaction of alpha-synuclein with mitochondrial complex I causes a reduction in its activity. This inhibition can be driven by both wild-type and mutant alpha-synuclein, and by oligomeric and fibrillar forms of the protein [45,185,186]. This functional disruption of mitochondria has most recently been shown to be also driven by an interaction of alpha-synuclein with complex V (ATP synthase) causing alterations in mitochondrial morphology, accompanied by an opening of the permeability transition pore, an increase in oxidative stress, and ultimately, neuronal death [45]. Recent work has highlighted that this capacity to interact with mitochondria may also impact on the ability of mitochondria to import essential proteins from the cytosol [184,187,188]. Alpha-synuclein has been shown to interact with the TOM (translocase of the outer membrane) complex in both tissue and culture-based systems. A blockage of import in this way has previously been shown to be sufficient to drive nigrostriatal degeneration [189].

Alpha-synuclein has been shown to cause fragmentation of the mitochondrial network in a number of models [190–192]. This fragmentation is DRP1 independent, occurs in the presence of both wild-type and mutant forms of the protein, and has been shown to precede the loss of mitochondrial function and neuron death. Interestingly, it has been suggested that the protein has no effect on the structure of the endoplasmic reticulum, the Golgi, or their interactions with the mitochondria. However, a recent study has shown that alpha-synuclein does interact with VAPB, an integral ER protein. This interaction then weakens the association of mitochondria with the ER, causing alteration in calcium buffering and the production of ATP [193]. The effect of alpha-synuclein on the structure of the mitochondria extends beyond merely affecting their fission and fusion with alterations to the ultrastructure of mitochondria also reported. Kamp et al. showed that PINK1, Parkin, and DJ-1 over expression was sufficient to be able to rescue the effect of alpha-synuclein on mitochondrial dynamics [191].

Alpha-synuclein has been suggested to be important for synaptic vesicle trafficking for many years, but recently, mounting evidence has suggested that it also affects the movement of mitochondria around neurons. In neurons exposed to alpha-synuclein, the balance between retrograde and anterograde transport is shifted. Wild type neurons showed a skewed preference for anterograde movement to maintain the axonal mitochondrial populations, and this preference was lost when alpha-synuclein expressed [194]. Kymograph analysis in a zebrafish model confirmed that alpha-synuclein reduced anterograde mitochondrial movements along axons and that the mitochondria within these axons, spent more time in a paused/stationary state [96]. The ability of alpha-synuclein to interrupt mitochondrial trafficking in this manner has been studied in depth, and is driven by its interaction with microtubules and cytoskeletal elements. Alpha-synuclein is capable of binding to a number of these proteins, including kinesin family member 5 (KIF5A), microtubule-associated protein 2 (MAP2), and Tau, with wild type synuclein showing the strongest binding to these proteins, with oligomers and fibrils showing weaker interactions [195]. The effect of alpha-synuclein on microtubule stability affects the distribution of mitochondria to neurites which would have detrimental effects on neuronal health and survival.

Does the Interaction of Alpha-Synuclein with Mitochondria Drive the Development of Neurodegeneration in PD?

Alpha-synuclein, of all the proposed contributors to Parkinson's pathogenesis, perhaps affects the most aspects of mitochondrial function. Alpha-synuclein has been shown to impact mitochondrial function, dynamics, transport, and protein import. Considering that alpha-synuclein aggregation is still one of the pathological hallmarks of PD, this would certainly suggest that in the presence of mitochondrial defects, it may lead to neurodegeneration. However, it is important to consider that studies have reported that Lewy bodies accumulate in neurons which do not have mitochondrial deficiencies, and that not all alpha-synuclein models recapitulate all of the characteristics of PD. Mitochondrial dysfunction, caused by alpha-synuclein or mtDNA mutations, may cause oxidative stress which will further damage alpha-synuclein, causing it to misfold into oligomeric forms. These alpha-synuclein aggregates are then capable of exacerbating mitochondrial dysfunction, as well as interfering with mitochondrial turnover and transport. This would lead to incorrect distribution of mitochondria, impaired calcium buffering, and ATP production increasing the sensitivity of these neurons to loss in Parkinson's.

6. Conclusions

Mitochondrial dysfunction, initially detected as Complex I deficiency, is found within the neurons of the SN, the population of neurons whose loss leads to the development of the motor symptoms associated with this disease. This deficiency may be caused by mtDNA deletions, by alpha-synuclein oligomers, or environmental toxins. A loss of mitochondrial function impedes ATP production and calcium buffering, key processes in the function of these neurons, and exacerbates oxidative stress. Such mitochondria should be targeted for degradation through the mitophagy pathway and replenished with new functional organelles through the PGC1α biogenesis pathway, both of which are affected by advancing age and have been shown to be linked to Parkinson's disease. Furthermore, dysfunctional mitochondria require transport to lysosomes and away from energy demanding sites. Recent evidence has also hinted that the dynamic processes that control the distribution and structure of mitochondria are also linked with PD. All these processes have the ability to cause further accumulation of defective mitochondria if they are themselves impaired.

Therefore, it is clear that alterations in mitochondrial function are important for neuronal survival and are a key driver of neuronal loss in Parkinson's disease (Figure 2). Since the initial description of Complex I deficiency in this disorder, our understanding not only of the complexities of mitochondria but also of the pathways and proteins whose function has an impact on, or is impacted by mitochondrial function, has dramatically increased. However, despite overwhelming evidence to suggest that the function of mitochondria is important for PD pathogenesis, the ability to identify the initial event in the cascade of changes that leads to neurodegeneration remains elusive. While here we have tried to categorize a number of proteins by their primary role (for example PINK1/Parkin as mitophagy proteins), it is becoming clear that many of these proteins have a role to play in a host of other pathways, while many of these also have differential consequences for the mitochondria (for example, the contribution of PINK1/Parkin to mitochondrial dynamics). While current techniques and models are driving this search forward, it is clear that the complex nature of the mitochondria and the role they play in a host of cellular pathways means that they have a large impact on a number of processes (energy provision, calcium buffering, ROS production, etc.). However, since their function is also impacted by changes in many of these pathways, it would seem that for every link between mitochondrial function and a pathway (for example, mitochondrial dynamics), a contrary relationship will then be found between the two processes. Understanding more about these interactions will uncover novel drugs which can modulate them not only to preserve mitochondrial function, but also to protect these organelles against damage and damaging moieties, driving and supporting the hunt for neuroprotective treatments.

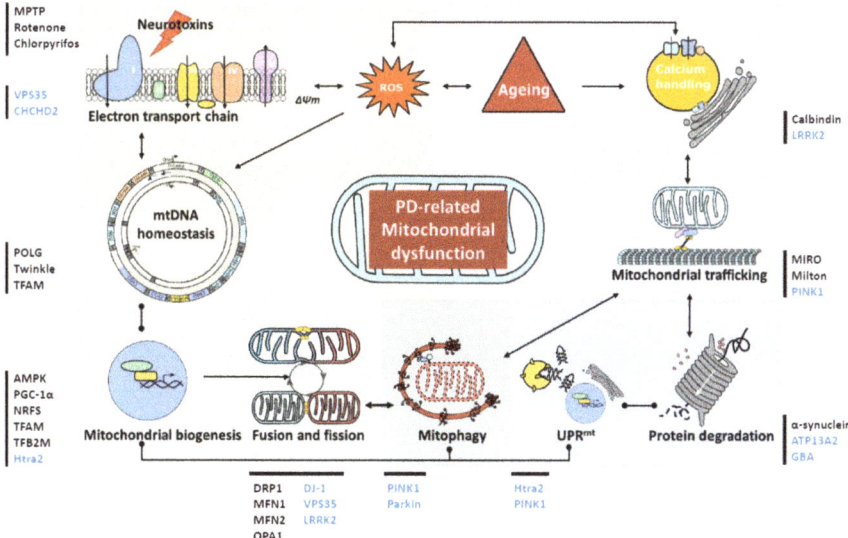

Figure 2. Schematic presentation of mitochondrial involvement in the pathogenesis of Parkinson's disease (PD). This diagram serves to highlight the complex links between the changes in mitochondrial

homeostasis, turnover, quality control and trafficking in cases of PD. These mitochondrial alternations are also intricately associated with the ageing process and impairments of the ubiquitin protease system that are attributed to Lewy body pathology. Lines with dots represent interactive effects, lines with arrows represent regulatory effects. Genes associated with familial PD are shown in blue, while we have also highlighted other proteins and toxins, recently been associated with PD, which impact on mitochondrial function. MPTP: 1-methyl-4-phenyl-1,2,3,6-tetrahydropyridine; VPS35: vacuolar protein sorting 35; CHCHD2: coiled-coil-helix-coiled-coil-helix domain containing 2; TFAM: mitochondrial transcription factor A; AMPK: AMP-activated protein kinase; PGC-1α: peroxisome proliferator-activated receptor γ (PPARγ) coactivator 1α; NRF: nuclear respiratory factors; TFB2M: dimethyladenosine transferase 2; DRP1: dynamin-1-like protein; MFN: mitofusin; LRRK2: Leucine-rich repeat kinase 2; PINK1: phosphatase and tensin homolog (PTEN)-induced putative kinase 1; GBA: lysosomal enzyme glucocerebrosidase; MIRO: mitochondrial Rho GTPase1; ΔΨm: mitochondrial membrane potential; ROS: reactive oxygen species; mtDNA: mitochondrial DNA; UPR^{mt}: the mitochondrial unfolded protein response.

Author Contributions: Writing—original draft preparation, C.C. and A.K.R.; writing—review and editing, A.K.R., C.C. and D.M.T.; supervision, A.K.R. and D.M.T.

Funding: All authors were supported by the Wellcome Centre of Mitochondrial Research (203105/Z16/Z) and MRC as part of the Newcastle Centre for Ageing and Vitality (MR/L016354/1). A.K.R. is also supported by a senior Parkinson's UK fellowship (F-1401).

Conflicts of Interest: The authors declare no conflict of interest.

References

1. Langston, J.W.; Ballard, P.; Tetrud, J.W.; Irwin, I. Chronic Parkinsonism in humans due to a product of meperidine-analog synthesis. *Science* **1983**, *219*, 979–980. [CrossRef] [PubMed]
2. Langston, J.W.; Ballard, P.A., Jr. Parkinson's disease in a chemist working with 1-methyl-4-phenyl-1,2,5,6-tetrahydropyridine. *N. Engl. J. Med.* **1983**, *309*, 310. [PubMed]
3. Levitt, P.; Pintar, J.E.; Breakefield, X.O. Immunocytochemical demonstration of monoamine oxidase B in brain astrocytes and serotonergic neurons. *Proc. Natl. Acad. Sci. USA* **1982**, *79*, 6385–6389. [CrossRef] [PubMed]
4. Schapira, A.H.; Cooper, J.M.; Dexter, D.; Jenner, P.; Clark, J.B.; Marsden, C.D. Mitochondrial complex I deficiency in Parkinson's disease. *Lancet* **1989**, *1*, 1269. [CrossRef]
5. Heikkila, R.E.; Nicklas, W.J.; Vyas, I.; Duvoisin, R.C. Dopaminergic toxicity of rotenone and the 1-methyl-4-phenylpyridinium ion after their stereotaxic administration to rats: Implication for the mechanism of 1-methyl-4-phenyl-1,2,3,6-tetrahydropyridine toxicity. *Neurosci. Lett.* **1985**, *62*, 389–394. [CrossRef]
6. Goldman, S.M. Environmental toxins and Parkinson's disease. *Annu. Rev. Pharmacol. Toxicol.* **2014**, *54*, 141–164. [CrossRef] [PubMed]
7. Nandipati, S.; Litvan, I. Environmental Exposures and Parkinson's Disease. *Int. J. Environ. Res. Public Health* **2016**, *13*, 881. [CrossRef] [PubMed]
8. Betarbet, R.; Sherer, T.B.; MacKenzie, G.; Garcia-Osuna, M.; Panov, A.V.; Greenamyre, J.T. Chronic systemic pesticide exposure reproduces features of Parkinson's disease. *Nat. Neurosci.* **2000**, *3*, 1301–1306. [CrossRef] [PubMed]
9. Nakamura, K.; Bindokas, V.P.; Marks, J.D.; Wright, D.A.; Frim, D.M.; Miller, R.J.; Kang, U.J. The selective toxicity of 1-methyl-4-phenylpyridinium to dopaminergic neurons: The role of mitochondrial complex I and reactive oxygen species revisited. *Mol. Pharmacol.* **2000**, *58*, 271–278. [CrossRef] [PubMed]
10. Richardson, J.R.; Quan, Y.; Sherer, T.B.; Greenamyre, J.T.; Miller, G.W. Paraquat neurotoxicity is distinct from that of MPTP and rotenone. *Toxicol. Sci.* **2005**, *88*, 193–201. [CrossRef] [PubMed]
11. Lotharius, J.; O'Malley, K.L. The parkinsonism-inducing drug 1-methyl-4-phenylpyridinium triggers intracellular dopamine oxidation. A novel mechanism of toxicity. *J. Biol. Chem.* **2000**, *275*, 38581–38588. [CrossRef] [PubMed]

12. Giordano, S.; Lee, J.; Darley-Usmar, V.M.; Zhang, J. Distinct effects of rotenone, 1-methyl-4-phenylpyridinium and 6-hydroxydopamine on cellular bioenergetics and cell death. *PLoS ONE* **2012**, *7*, e44610. [CrossRef] [PubMed]
13. Grunewald, A.; Rygiel, K.A.; Hepplewhite, P.D.; Morris, C.M.; Picard, M.; Turnbull, D.M. Mitochondrial DNA depletion in respiratory chain-deficient Parkinson disease neurons. *Ann. Neurol.* **2016**, *79*, 366–378. [CrossRef] [PubMed]
14. Flones, I.H.; Fernandez-Vizarra, E.; Lykouri, M.; Brakedal, B.; Skeie, G.O.; Miletic, H.; Lilleng, P.K.; Alves, G.; Tysnes, O.B.; Haugarvoll, K.; et al. Neuronal complex I deficiency occurs throughout the Parkinson's disease brain, but is not associated with neurodegeneration or mitochondrial DNA damage. *Acta Neuropathol.* **2018**, *135*, 409–425. [CrossRef] [PubMed]
15. Bury, A.G.; Pyle, A.; Elson, J.L.; Greaves, L.; Morris, C.M.; Hudson, G.; Pienaar, I.S. Mitochondrial DNA changes in pedunculopontine cholinergic neurons in Parkinson disease. *Ann. Neurol.* **2017**, *82*, 1016–1021. [CrossRef] [PubMed]
16. Bender, A.; Krishnan, K.J.; Morris, C.M.; Taylor, G.A.; Reeve, A.K.; Perry, R.H.; Jaros, E.; Hersheson, J.S.; Betts, J.; Klopstock, T.; et al. High levels of mitochondrial DNA deletions in substantia nigra neurons in aging and Parkinson disease. *Nat. Genet.* **2006**, *38*, 515–517. [CrossRef] [PubMed]
17. Kraytsberg, Y.; Kudryavtseva, E.; McKee, A.C.; Geula, C.; Kowall, N.W.; Khrapko, K. Mitochondrial DNA deletions are abundant and cause functional impairment in aged human substantia nigra neurons. *Nat. Genet.* **2006**, *38*, 518–520. [CrossRef] [PubMed]
18. Reeve, A.; Meagher, M.; Lax, N.; Simcox, E.; Hepplewhite, P.; Jaros, E.; Turnbull, D. The impact of pathogenic mitochondrial DNA mutations on substantia nigra neurons. *J. Neurosci.* **2013**, *33*, 10790–10801. [CrossRef] [PubMed]
19. Kann, O.; Huchzermeyer, C.; Kovács, R.; Wirtz, S.; Schuelke, M. Gamma oscillations in the hippocampus require high complex I gene expression and strong functional performance of mitochondria. *Brain* **2010**, *134*, 345–358. [CrossRef] [PubMed]
20. Reeve, A.K.; Krishnan, K.J.; Elson, J.L.; Morris, C.M.; Bender, A.; Lightowlers, R.N.; Turnbull, D.M. Nature of mitochondrial DNA deletions in substantia nigra neurons. *Am. J. Hum. Genet.* **2008**, *82*, 228–235. [CrossRef] [PubMed]
21. Tzoulis, C.; Tran, G.T.; Coxhead, J.; Bertelsen, B.; Lilleng, P.K.; Balafkan, N.; Payne, B.; Miletic, H.; Chinnery, P.F.; Bindoff, L.A. Molecular pathogenesis of polymerase gamma-related neurodegeneration. *Ann. Neurol.* **2014**, *76*, 66–81. [CrossRef] [PubMed]
22. Reeve, A.K.; Park, T.K.; Jaros, E.; Campbell, G.R.; Lax, N.Z.; Hepplewhite, P.D.; Krishnan, K.J.; Elson, J.L.; Morris, C.M.; McKeith, I.G.; Turnbull, D.M. Relationship between mitochondria and α-synuclein: A study of single substantia nigra neurons. *Arch. Neurol.* **2012**, *69*, 385–393. [CrossRef] [PubMed]
23. Dolle, C.; Flones, I.; Nido, G.S.; Miletic, H.; Osuagwu, N.; Kristoffersen, S.; Lilleng, P.K.; Larsen, J.P.; Tysnes, O.B.; Haugarvoll, K.; et al. Defective mitochondrial DNA homeostasis in the substantia nigra in Parkinson disease. *Nat. Commun.* **2016**, *7*, 13548. [CrossRef] [PubMed]
24. Clayton, D.A. Replication of animal mitochondrial DNA. *Cell* **1982**, *28*, 693–705. [CrossRef]
25. Krishnan, K.J.; Reeve, A.K.; Samuels, D.C.; Chinnery, P.F.; Blackwood, J.K.; Taylor, R.W.; Wanrooij, S.; Spelbrink, J.N.; Lightowlers, R.N.; Turnbull, D.M. What causes mitochondrial DNA deletions in human cells? *Nat. Genet.* **2008**, *40*, 275–279. [CrossRef] [PubMed]
26. Fukui, H.; Moraes, C.T. Mechanisms of formation and accumulation of mitochondrial DNA deletions in aging neurons. *Hum. Mol. Genet.* **2009**, *18*, 1028–1036. [CrossRef] [PubMed]
27. Pickrell, A.M.; Pinto, M.; Hida, A.; Moraes, C.T. Striatal dysfunctions associated with mitochondrial DNA damage in dopaminergic neurons in a mouse model of Parkinson's disease. *J. Neurosci.* **2011**, *31*, 17649–17658. [CrossRef] [PubMed]
28. Pinto, M.; Pickrell, A.M.; Wang, X.; Bacman, S.R.; Yu, A.; Hida, A.; Dillon, L.M.; Morton, P.D.; Malek, T.R.; Williams, S.L.; et al. Transient mitochondrial DNA double strand breaks in mice cause accelerated aging phenotypes in a ROS-dependent but p53/p21-independent manner. *Cell Death Differ.* **2017**, *24*, 288–299. [CrossRef] [PubMed]
29. Corral-Debrinski, M.; Horton, T.; Lott, M.T.; Shoffner, J.M.; Beal, M.F.; Wallace, D.C. Mitochondrial DNA deletions in human brain: Regional variability and increase with advanced age. *Nat. Genet.* **1992**, *2*, 324–329. [CrossRef] [PubMed]

30. Cortopassi, G.A.; Shibata, D.; Soong, N.W.; Arnheim, N. A pattern of accumulation of a somatic deletion of mitochondrial DNA in aging human tissues. *Proc. Natl. Acad. Sci. USA* **1992**, *89*, 7370–7374. [CrossRef] [PubMed]
31. Perier, C.; Bender, A.; Garcia-Arumi, E.; Melia, M.J.; Bove, J.; Laub, C.; Klopstock, T.; Elstner, M.; Mounsey, R.B.; Teismann, P.; et al. Accumulation of mitochondrial DNA deletions within dopaminergic neurons triggers neuroprotective mechanisms. *Brain* **2013**, *136*, 2369–2378. [CrossRef] [PubMed]
32. Pyle, A.; Anugrha, H.; Kurzawa-Akanbi, M.; Yarnall, A.; Burn, D.; Hudson, G. Reduced mitochondrial DNA copy number is a biomarker of Parkinson's disease. *Neurobiol. Aging* **2016**, *38*, 216.e7–216.e10. [CrossRef] [PubMed]
33. Pyle, A.; Brennan, R.; Kurzawa-Akanbi, M.; Yarnall, A.; Thouin, A.; Mollenhauer, B.; Burn, D.; Chinnery, P.F.; Hudson, G. Reduced cerebrospinal fluid mitochondrial DNA is a biomarker for early-stage Parkinson's disease. *Ann. Neurol.* **2015**, *78*, 1000–1004. [CrossRef] [PubMed]
34. Lowes, H.; Pyle, A.; Duddy, M.; Hudson, G. Cell-free mitochondrial DNA in progressive multiple sclerosis. *Mitochondrion* **2018**. [CrossRef] [PubMed]
35. Suski, J.M.; Lebiedzinska, M.; Bonora, M.; Pinton, P.; Duszynski, J.; Wieckowski, M.R. Relation between mitochondrial membrane potential and ROS formation. *Methods Mol. Biol.* **2012**, *810*, 183–205. [PubMed]
36. Rigoulet, M.; Yoboue, E.D.; Devin, A. Mitochondrial ROS generation and its regulation: Mechanisms involved in H(2)O(2) signaling. *Antioxid. Redox Signal.* **2011**, *14*, 459–468. [CrossRef] [PubMed]
37. Delcambre, S.; Nonnenmacher, Y.; Hiller, K. Dopamine metabolism and reactive oxygen species production. In *Mitochondrial Mechanisms of Degeneration and Repair in Parkinson's Disease*; Buhlman, L.M., Ed.; Springer: Berlin, Germany, 2016; pp. 25–47.
38. Linnane, A.W.; Marzuki, S.; Ozawa, T.; Tanaka, M. Mitochondrial DNA mutations as an important contributor to ageing and degenerative diseases. *Lancet* **1989**, *1*, 642–645. [CrossRef]
39. Surmeier, D.J.; Guzman, J.N.; Sanchez, J.; Schumacker, P.T. Physiological phenotype and vulnerability in Parkinson's disease. *Cold Spring Harb. Perspect. Med.* **2012**, *2*, a009290. [CrossRef] [PubMed]
40. Vermulst, M.; Bielas, J.H.; Kujoth, G.C.; Ladiges, W.C.; Rabinovitch, P.S.; Prolla, T.A.; Loeb, L.A. Mitochondrial point mutations do not limit the natural lifespan of mice. *Nat. Genet.* **2007**, *39*, 540–543. [CrossRef] [PubMed]
41. Faucheux, B.A.; Martin, M.E.; Beaumont, C.; Hauw, J.J.; Agid, Y.; Hirsch, E.C. Neuromelanin associated redox-active iron is increased in the substantia nigra of patients with Parkinson's disease. *J. Neurochem.* **2003**, *86*, 1142–1148. [CrossRef] [PubMed]
42. Wang, J.-Y.; Zhuang, Q.-Q.; Zhu, L.-B.; Zhu, H.; Li, T.; Li, R.; Chen, S.-F.; Huang, C.-P.; Zhang, X.; Zhu, J.-H. Meta-analysis of brain iron levels of Parkinson's disease patients determined by postmortem and MRI measurements. *Sci. Rep.* **2016**, *6*, 36669. [CrossRef] [PubMed]
43. Smeyne, M.; Smeyne, R.J. Glutathione metabolism and Parkinson's disease. *Free Radic. Biol. Med.* **2013**, *62*, 13–25. [CrossRef] [PubMed]
44. Angelova, P.R.; Horrocks, M.H.; Klenerman, D.; Gandhi, S.; Abramov, A.Y.; Shchepinov, M.S. Lipid peroxidation is essential for alpha-synuclein-induced cell death. *J. Neurochem.* **2015**, *133*, 582–589. [CrossRef] [PubMed]
45. Ludtmann, M.H.R.; Angelova, P.R.; Horrocks, M.H.; Choi, M.L.; Rodrigues, M.; Baev, A.Y.; Berezhnov, A.V.; Yao, Z.; Little, D.; Banushi, B.; et al. α-synuclein oligomers interact with ATP synthase and open the permeability transition pore in Parkinson's disease. *Nat. Commun.* **2018**, *9*, 2293. [CrossRef] [PubMed]
46. Yamada, T.; McGeer, P.L.; Baimbridge, K.G.; McGeer, E.G. Relative sparing in Parkinson's disease of substantia nigra dopamine neurons containing calbindin-D28K. *Brain Res.* **1990**, *526*, 303–307. [CrossRef]
47. German, D.C.; Manaye, K.F.; Sonsalla, P.K.; Brooks, B.A. Midbrain dopaminergic cell loss in Parkinson's disease and MPTP-induced parkinsonism: sparing of calbindin-D28k-containing cells. *Ann. NY Acad. Sci.* **1992**, *648*, 42–62. [CrossRef] [PubMed]
48. David, G.; Barrett, E.F. Stimulation-evoked increases in cytosolic [Ca(2+)] in mouse motor nerve terminals are limited by mitochondrial uptake and are temperature-dependent. *J. Neurosci.* **2000**, *20*, 7290–7296. [CrossRef] [PubMed]
49. Billups, B.; Forsythe, I.D. Presynaptic mitochondrial calcium sequestration influences transmission at mammalian central synapses. *J. Neurosci.* **2002**, *22*, 5840–5847. [CrossRef] [PubMed]
50. Wilson, C.J.; Callaway, J.C. Coupled oscillator model of the dopaminergic neuron of the substantia nigra. *J. Neurophysiol.* **2000**, *83*, 3084–3100. [CrossRef] [PubMed]

51. Wiel, C.; Lallet-Daher, H.; Gitenay, D.; Gras, B.; Le Calve, B.; Augert, A.; Ferrand, M.; Prevarskaya, N.; Simonnet, H.; Vindrieux, D.; et al. Endoplasmic reticulum calcium release through ITPR2 channels leads to mitochondrial calcium accumulation and senescence. *Nat. Commun.* **2014**, *5*, 3792. [CrossRef] [PubMed]
52. Mattson, M.P.; Chan, S.L. Calcium orchestrates apoptosis. *Nat. Cell Biol.* **2003**, *5*, 1041. [CrossRef] [PubMed]
53. Ludtmann, M.H.R.; Abramov, A.Y. Mitochondrial calcium imbalance in Parkinson's disease. *Neurosci. Lett.* **2018**, *663*, 86–90. [CrossRef] [PubMed]
54. Liu, Y.; Harding, M.; Pittman, A.; Dore, J.; Striessnig, J.; Rajadhyaksha, A.; Chen, X. Cav1.2 and Cav1.3 L-type calcium channels regulate dopaminergic firing activity in the mouse ventral tegmental area. *J. Neurophysiol.* **2014**, *112*, 1119–1130. [CrossRef] [PubMed]
55. Khaliq, Z.M.; Bean, B.P. Pacemaking in dopaminergic ventral tegmental area neurons: Depolarizing drive from background and voltage-dependent sodium conductances. *J. Neurosci.* **2010**, *30*, 7401–7413. [CrossRef] [PubMed]
56. Surmeier, D.J.; Schumacker, P.T.; Guzman, J.D.; Ilijic, E.; Yang, B.; Zampese, E. Calcium and Parkinson's disease. *Biochem. Biophys. Res. Commun.* **2017**, *483*, 1013–1019. [CrossRef] [PubMed]
57. Lieberman, O.J.; Choi, S.J.; Kanter, E.; Saverchenko, A.; Frier, M.D.; Fiore, G.M.; Wu, M.; Kondapalli, J.; Zampese, E.; Surmeier, D.J.; et al. α-synuclein-dependent calcium entry underlies differential sensitivity of cultured SN and VTA dopaminergic neurons to a parkinsonian neurotoxin. *eNeuro* **2017**, *4*. [CrossRef] [PubMed]
58. Wider, C.; Skipper, L.; Solida, A.; Brown, L.; Farrer, M.; Dickson, D.; Wszolek, Z.K.; Vingerhoets, F.J. Autosomal dominant dopa-responsive parkinsonism in a multigenerational Swiss family. *Parkinsonism Relat. Disord.* **2008**, *14*, 465–470. [CrossRef] [PubMed]
59. Vilarino-Guell, C.; Wider, C.; Ross, O.A.; Dachsel, J.C.; Kachergus, J.M.; Lincoln, S.J.; Soto-Ortolaza, A.I.; Cobb, S.A.; Wilhoite, G.J.; Bacon, J.A.; et al. VPS35 mutations in Parkinson disease. *Am. J. Hum. Genet.* **2011**, *89*, 162–167. [CrossRef] [PubMed]
60. Zhou, L.; Wang, W.; Hoppel, C.; Liu, J.; Zhu, X. Parkinson's disease-associated pathogenic VPS35 mutation causes complex I deficits. *Biochim. Biophys. Acta Mol. Basis Dis.* **2017**, *1863*, 2791–2795. [CrossRef] [PubMed]
61. Tang, F.L.; Liu, W.; Hu, J.X.; Erion, J.R.; Ye, J.; Mei, L.; Xiong, W.C. VPS35 deficiency or mutation causes dopaminergic neuronal loss by impairing mitochondrial fusion and function. *Cell Rep.* **2015**, *12*, 1631–1643. [CrossRef] [PubMed]
62. Cataldi, S.; Follett, J.; Fox, J.D.; Tatarnikov, I.; Kadgien, C.; Gustavsson, E.K. Altered dopamine release and monoamine transporters in Vps35 p.D620N knock-in mice. *NPJ Parkinson's Dis.* **2018**, *4*, 27. [CrossRef] [PubMed]
63. Funayama, M.; Ohe, K.; Amo, T.; Furuya, N.; Yamaguchi, J.; Saiki, S.; Li, Y.; Ogaki, K.; Ando, M.; Yoshino, H.; et al. CHCHD2 mutations in autosomal dominant late-onset Parkinson's disease: A genome-wide linkage and sequencing study. *Lancet Neurol.* **2015**, *14*, 274–282. [CrossRef]
64. Shi, C.H.; Mao, C.Y.; Zhang, S.Y.; Yang, J.; Song, B.; Wu, P.; Zuo, C.T.; Liu, Y.T.; Ji, Y.; Yang, Z.H.; et al. CHCHD2 gene mutations in familial and sporadic Parkinson's disease. *Neurobiol. Aging* **2016**, *38*, 217.e9–217.e13. [CrossRef] [PubMed]
65. Jansen, I.E.; Bras, J.M.; Lesage, S.; Schulte, C.; Gibbs, J.R.; Nalls, M.A.; Brice, A.; Wood, N.W.; Morris, H.; Hardy, J.A.; et al. CHCHD2 and Parkinson's disease. *Lancet Neurol.* **2015**, *14*, 678–679. [CrossRef]
66. Gagliardi, M.; Iannello, G.; Colica, C.; Annesi, G.; Quattrone, A. Analysis of CHCHD2 gene in familial Parkinson's disease from Calabria. *Neurobiol. Aging* **2017**, *50*, 169.e5–169.e6. [CrossRef] [PubMed]
67. Voigt, D.D.; Nascimento, C.M.; de Souza, R.B.; Cabello Acero, P.H.; Campos Junior, M.; da Silva, C.P.; Pereira, J.S.; Rosso, A.L.; Araujo Leite, M.A.; Vasconcellos, L.F.R.; et al. CHCHD2 mutational screening in Brazilian patients with familial Parkinson's disease. *Neurobiol. Aging* **2018**, *74*, 236.e7–236.e8. [CrossRef] [PubMed]
68. Bannwarth, S.; Ait-El-Mkadem, S.; Chaussenot, A.; Genin, E.C.; Lacas-Gervais, S.; Fragaki, K.; Berg-Alonso, L.; Kageyama, Y.; Serre, V.; Moore, D.G.; et al. A mitochondrial origin for frontotemporal dementia and amyotrophic lateral sclerosis through CHCHD10 involvement. *Brain* **2014**, *137*, 2329–2345. [CrossRef] [PubMed]
69. Meng, H.; Yamashita, C.; Shiba-Fukushima, K.; Inoshita, T.; Funayama, M.; Sato, S.; Hatta, T.; Natsume, T.; Umitsu, M.; Takagi, J.; et al. Loss of Parkinson's disease-associated protein CHCHD2 affects mitochondrial crista structure and destabilizes cytochrome c. *Nat. Commun.* **2017**, *8*, 15500. [CrossRef] [PubMed]

70. Longen, S.; Bien, M.; Bihlmaier, K.; Kloeppel, C.; Kauff, F.; Hammermeister, M.; Westermann, B.; Herrmann, J.M.; Riemer, J. Systematic analysis of the twin cx(9)c protein family. *J. Mol. Biol.* **2009**, *393*, 356–368. [CrossRef] [PubMed]
71. Lee, R.G.; Sedghi, M.; Salari, M.; Shearwood, A.J.; Stentenbach, M.; Kariminejad, A.; Goullee, H.; Rackham, O.; Laing, N.G.; Tajsharghi, H.; et al. Early-onset Parkinson disease caused by a mutation in CHCHD2 and mitochondrial dysfunction. *Neurol. Genet.* **2018**, *4*, e276. [CrossRef] [PubMed]
72. Betts-Henderson, J.; Jaros, E.; Krishnan, K.J.; Perry, R.H.; Reeve, A.K.; Schaefer, A.M.; Taylor, R.W.; Turnbull, D.M. Alpha-synuclein pathology and Parkinsonism associated with POLG1 mutations and multiple mitochondrial DNA deletions. *Neuropathol. Appl. Neurobiol.* **2009**, *35*, 120–124. [PubMed]
73. Hudson, G.; Chinnery, P.F. Mitochondrial DNA polymerase-gamma and human disease. *Hum. Mol. Genet.* **2006**, *15*, R244–R252. [CrossRef] [PubMed]
74. Yin, Z.; Pascual, C.; Klionsky, D.J. Autophagy: Machinery and regulation. *Microb. Cell* **2016**, *3*, 588–596. [CrossRef] [PubMed]
75. Andres, A.M.; Stotland, A.; Queliconi, B.B.; Gottlieb, R.A. A time to reap, a time to sow: Mitophagy and biogenesis in cardiac pathophysiology. *J. Mol. Cell Cardiol.* **2015**, *78*, 62–72. [CrossRef] [PubMed]
76. Poole, A.C.; Thomas, R.E.; Andrews, L.A.; McBride, H.M.; Whitworth, A.J.; Pallanck, L.J. The PINK1/Parkin pathway regulates mitochondrial morphology. *Proc. Natl. Acad. Sci. USA* **2008**, *105*, 1638–1643. [CrossRef] [PubMed]
77. Kitada, T.; Asakawa, S.; Hattori, N.; Matsumine, H.; Yamamura, Y.; Minoshima, S.; Yokochi, M.; Mizuno, Y.; Shimiz, N. Mutations in the parkin gene cause autosomal recessive juvenile parkinsonism. *Nature* **1998**, *392*, 605–608. [CrossRef] [PubMed]
78. Vives-Bauza, C.; Zhou, C.; Huang, Y.; Cui, M.; de Vries, R.L.A.; Kim, J.; May, J.; Tocilescu, M.A.; Liu, W.; Ko, H.S.; et al. PINK1-dependent recruitment of Parkin to mitochondria in mitophagy. *Proc. Natl. Acad. Sci. USA* **2010**, *107*, 378–383. [CrossRef] [PubMed]
79. Durcan, T.M.; Fon, E.A. The three 'P's of mitophagy: PARKIN, PINK1, and post-translational modifications. *Genes Dev.* **2015**, *29*, 989–999. [CrossRef] [PubMed]
80. Fu, M.; St-Pierre, P.; Shankar, J.; Wang, P.T.; Joshi, B.; Nabi, I.R. Regulation of mitophagy by the Gp78 E3 ubiquitin ligase. *Mol. Biol. Cell* **2013**, *24*, 1153–1162. [CrossRef] [PubMed]
81. Villa, E.; Proics, E.; Rubio-Patino, C.; Obba, S.; Zunino, B.; Bossowski, J.P.; Rozier, R.M.; Chiche, J.; Mondragon, L.; Riley, J.S.; et al. Parkin-independent mitophagy controls chemotherapeutic response in cancer cells. *Cell Rep.* **2017**, *20*, 2846–2859. [CrossRef] [PubMed]
82. Anderson, S.; Bankier, A.T.; Barrell, B.G.; de Bruijn, M.H.; Coulson, A.R.; Drouin, J.; Eperon, I.C.; Nierlich, D.P.; Roe, B.A.; Sanger, F.; et al. Sequence and organization of the human mitochondrial genome. *Nature* **1981**, *290*, 457–465. [CrossRef] [PubMed]
83. Ventura-Clapier, R.; Garnier, A.; Veksler, V. Transcriptional control of mitochondrial biogenesis: The central role of PGC-1alpha. *Cardiovasc. Res.* **2008**, *79*, 208–217. [CrossRef] [PubMed]
84. Scarpulla, R.C. Metabolic control of mitochondrial biogenesis through the PGC-1 family regulatory network. *Biochim. Biophys. Acta* **2011**, *1813*, 1269–1278. [CrossRef] [PubMed]
85. Gleyzer, N.; Vercauteren, K.; Scarpulla, R.C. Control of mitochondrial transcription specificity factors (TFB1M and TFB2M) by nuclear respiratory factors (NRF-1 and NRF-2) and PGC-1 family coactivators. *Mol. Cell Biol.* **2005**, *25*, 1354–1366. [CrossRef] [PubMed]
86. Scarpulla, R.C. Transcriptional paradigms in mammalian mitochondrial biogenesis and function. *Physiol. Rev.* **2008**, *88*, 611–638. [CrossRef] [PubMed]
87. Zheng, B.; Liao, Z.; Locascio, J.J.; Lesniak, K.A.; Roderick, S.S.; Watt, M.L.; Eklund, A.C.; Zhang-James, Y.; Kim, P.D.; Hauser, M.A.; et al. PGC-1alpha, a potential therapeutic target for early intervention in Parkinson's disease. *Sci. Transl. Med.* **2010**, *2*, 52ra73. [CrossRef] [PubMed]
88. Zahn, J.M.; Sonu, R.; Vogel, H.; Crane, E.; Mazan-Mamczarz, K.; Rabkin, R.; Davis, R.W.; Becker, K.G.; Owen, A.B.; Kim, S.K. Transcriptional profiling of aging in human muscle reveals a common aging signature. *PLoS Genet.* **2006**, *2*, e115. [CrossRef]
89. Kett, L.R.; Dauer, W.T. Endolysosomal dysfunction in Parkinson's disease: Recent developments and future challenges. *Mov. Disord.* **2016**, *31*, 1433–1443. [CrossRef] [PubMed]

90. Reznick, R.M.; Zong, H.; Li, J.; Morino, K.; Moore, I.K.; Yu, H.J.; Liu, Z.X.; Dong, J.; Mustard, K.J.; Hawley, S.A.; et al. Aging-associated reductions in AMP-activated protein kinase activity and mitochondrial biogenesis. *Cell Metab.* **2007**, *5*, 151–156. [CrossRef] [PubMed]
91. Jiang, H.; Kang, S.-U.; Zhang, S.; Karuppagounder, S.; Xu, J.; Lee, Y.-K.; Kang, B.-G.; Lee, Y.; Zhang, J.; Pletnikova, O.; et al. Adult conditional knockout of PGC-1α leads to loss of dopamine neurons. *eNeuro* **2016**, *3*. [CrossRef] [PubMed]
92. Ng, C.H.; Basil, A.H.; Hang, L.; Tan, R.; Goh, K.L.; O'Neill, S.; Zhang, X.; Yu, F.; Lim, K.L. Genetic or pharmacological activation of the Drosophila PGC-1alpha ortholog spargel rescues the disease phenotypes of genetic models of Parkinson's disease. *Neurobiol. Aging* **2017**, *55*, 33–37. [CrossRef] [PubMed]
93. Ye, Q.; Huang, W.; Li, D.; Si, E.; Wang, J.; Wang, Y.; Chen, C.; Chen, X. Overexpression of PGC-1alpha influences mitochondrial signal transduction of dopaminergic neurons. *Mol. Neurobiol.* **2016**, *53*, 3756–3770. [CrossRef] [PubMed]
94. Eschbach, J.; von Einem, B.; Muller, K.; Bayer, H.; Scheffold, A.; Morrison, B.E.; Rudolph, K.L.; Thal, D.R.; Witting, A.; Weydt, P.; et al. Mutual exacerbation of peroxisome proliferator-activated receptor gamma coactivator 1alpha deregulation and alpha-synuclein oligomerization. *Ann. Neurol.* **2015**, *77*, 15–32. [CrossRef] [PubMed]
95. Ciron, C.; Zheng, L.; Bobela, W.; Knott, G.W.; Leone, T.C.; Kelly, D.P.; Schneider, B.L. PGC-1alpha activity in nigral dopamine neurons determines vulnerability to alpha-synuclein. *Acta Neuropathol. Commun.* **2015**, *3*, 16. [CrossRef] [PubMed]
96. O'Donnell, K.C.; Lulla, A.; Stahl, M.C.; Wheat, N.D.; Bronstein, J.M.; Sagasti, A. Axon degeneration and PGC-1alpha-mediated protection in a zebrafish model of alpha-synuclein toxicity. *Dis. Model Mech.* **2014**, *7*, 571–582. [CrossRef] [PubMed]
97. Uldry, M.; Yang, W.; St-Pierre, J.; Lin, J.; Seale, P.; Spiegelman, B.M. Complementary action of the PGC-1 coactivators in mitochondrial biogenesis and brown fat differentiation. *Cell Metab.* **2006**, *3*, 333–341. [CrossRef] [PubMed]
98. St-Pierre, J.; Drori, S.; Uldry, M.; Silvaggi, J.M.; Rhee, J.; Jager, S.; Handschin, C.; Zheng, K.; Lin, J.; Yang, W.; et al. Suppression of reactive oxygen species and neurodegeneration by the PGC-1 transcriptional coactivators. *Cell* **2006**, *127*, 397–408. [CrossRef] [PubMed]
99. Di Giacomo, E.; Benedetti, E.; Cristiano, L.; Antonosante, A.; d'Angelo, M.; Fidoamore, A.; Barone, D.; Moreno, S.; Ippoliti, R.; Ceru, M.P.; et al. Roles of PPAR transcription factors in the energetic metabolic switch occurring during adult neurogenesis. *Cell Cycle* **2017**, *16*, 59–72. [CrossRef] [PubMed]
100. Shin, J.H.; Ko, H.S.; Kang, H.; Lee, Y.; Lee, Y.I.; Pletinkova, O.; Troconso, J.C.; Dawson, V.L.; Dawson, T.M. PARIS (ZNF746) repression of PGC-1alpha contributes to neurodegeneration in Parkinson's disease. *Cell* **2011**, *144*, 689–702. [CrossRef] [PubMed]
101. Peng, K.; Xiao, J.; Yang, L.; Ye, F.; Cao, J.; Sai, Y. Mutual Antagonism of PINK1/Parkin and PGC-1alpha Contributes to Maintenance of Mitochondrial Homeostasis in Rotenone-Induced Neurotoxicity. *Neurotox. Res.* **2018**, *35*, 331–343. [CrossRef] [PubMed]
102. Diot, A.; Morten, K.; Poulton, J. Mitophagy plays a central role in mitochondrial ageing. *Mamm. Genome* **2016**, *27*, 381–395. [CrossRef] [PubMed]
103. Fivenson, E.M.; Lautrup, S.; Sun, N.; Scheibye-Knudsen, M.; Stevnsner, T.; Nilsen, H.; Bohr, V.A.; Fang, E.F. Mitophagy in neurodegeneration and aging. *Neurochem. Int.* **2017**, *109*, 202–209. [CrossRef] [PubMed]
104. Shi, R.; Guberman, M.; Kirshenbaum, L.A. Mitochondrial quality control: The role of mitophagy in aging. *Trends Cardiovasc. Med.* **2018**, *28*, 246–260. [CrossRef] [PubMed]
105. Moreira, O.C.; Estébanez, B.; Martínez-Florez, S.; de Paz, J.A.; Cuevas, M.J.; González-Gallego, H. Mitochondrial Function and mitophagy in the elderly: Effects of exercise. *Oxid. Med. Cell. Longev.* **2017**, *2017*, 13. [CrossRef] [PubMed]
106. Reeve, A.; Simcox, E.; Turnbull, D. Ageing and Parkinson's disease: Why is advancing age the biggest risk factor? *Ageing Res. Rev.* **2014**, *14*, 19–30. [CrossRef] [PubMed]
107. Scarffe, L.A.; Stevens, D.A.; Dawson, V.L.; Dawson, T.M. Parkin and PINK1: Much more than mitophagy. *Trends Neurosci.* **2014**, *37*, 315–324. [CrossRef] [PubMed]
108. Zhao, Q.; Wang, J.; Levichkin, I.V.; Stasinopoulos, S.; Ryan, M.T.; Hoogenraad, N.J. A mitochondrial specific stress response in mammalian cells. *EMBO J.* **2002**, *21*, 4411–4419. [CrossRef] [PubMed]

109. Plun-Favreau, H.; Klupsch, K.; Moisoi, N.; Gandhi, S.; Kjaer, S.; Frith, D.; Harvey, K.; Deas, E.; Harvey, R.J.; McDonald, N.; et al. The mitochondrial protease HtrA2 is regulated by Parkinson's disease-associated kinase PINK1. *Nat. Cell Biol.* **2007**, *9*, 1243–1252. [CrossRef] [PubMed]
110. Pridgeon, J.W.; Olzmann, J.A.; Chin, L.-S.; Li, L. PINK1 protects against oxidative stress by phosphorylating mitochondrial chaperone TRAP1. *PLoS Biol.* **2007**, *5*, e172. [CrossRef] [PubMed]
111. Strauss, K.M.; Martins, L.M.; Plun-Favreau, H.; Marx, F.P.; Kautzmann, S.; Berg, D.; Gasser, T.; Wszolek, Z.; Muller, T.; Bornemann, A.; et al. Loss of function mutations in the gene encoding Omi/HtrA2 in Parkinson's disease. *Hum. Mol. Genet.* **2005**, *14*, 2099–2111. [CrossRef] [PubMed]
112. Fitzgerald, J.C.; Zimprich, A.; Carvajal Berrio, D.A.; Schindler, K.M.; Maurer, B.; Schulte, C.; Bus, C.; Hauser, A.K.; Kubler, M.; Lewin, R.; et al. Metformin reverses TRAP1 mutation-associated alterations in mitochondrial function in Parkinson's disease. *Brain* **2017**, *140*, 2444–2459. [CrossRef] [PubMed]
113. Moisoi, N.; Klupsch, K.; Fedele, V.; East, P.; Sharma, S.; Renton, A.; Plun-Favreau, H.; Edwards, R.E.; Teismann, P.; Esposti, M.D.; et al. Mitochondrial dysfunction triggered by loss of HtrA2 results in the activation of a brain-specific transcriptional stress response. *Cell Death Differ.* **2009**, *16*, 449–464. [CrossRef] [PubMed]
114. Costa, A.C.; Loh, S.H.; Martins, L.M. Drosophila Trap1 protects against mitochondrial dysfunction in a PINK1/parkin model of Parkinson's disease. *Cell Death Dis.* **2013**, *4*, e467. [CrossRef] [PubMed]
115. Zhang, L.; Karsten, P.; Hamm, S.; Pogson, J.H.; Müller-Rischart, A.K.; Exner, N.; Haass, C.; Whitworth, A.J.; Winklhofer, K.F.; Schulz, J.B.; et al. TRAP1 rescues PINK1 loss-of-function phenotypes. *Hum. Mol. Genet.* **2013**, *22*, 2829–2841. [CrossRef] [PubMed]
116. Thomas, K.J.; McCoy, M.K.; Blackinton, J.; Beilina, A.; van der Brug, M.; Sandebring, A.; Miller, D.; Maric, D.; Cedazo-Minguez, A.; Cookson, M.R. DJ-1 acts in parallel to the PINK1/parkin pathway to control mitochondrial function and autophagy. *Hum. Mol. Genet.* **2011**, *20*, 40–50. [CrossRef] [PubMed]
117. Wang, X.; Petrie, T.G.; Liu, Y.; Liu, J.; Fujioka, H.; Zhu, X. Parkinson's disease-associated DJ-1 mutations impair mitochondrial dynamics and cause mitochondrial dysfunction. *J. Neurochem.* **2012**, *121*, 830–839. [CrossRef] [PubMed]
118. McCoy, M.K.; Cookson, M.R. DJ-1 regulation of mitochondrial function and autophagy through oxidative stress. *Autophagy* **2011**, *7*, 531–532. [CrossRef] [PubMed]
119. Hsieh, C.H.; Shaltouki, A.; Gonzalez, A.E.; Bettencourt da Cruz, A.; Burbulla, L.F.; St Lawrence, E.; Schule, B.; Krainc, D.; Palmer, T.D.; Wang, X. Functional impairment in Miro degradation and mitophagy is a shared feature in familial and sporadic Parkinson's disease. *Cell Stem Cell* **2016**, *19*, 709–724. [CrossRef] [PubMed]
120. Park, J.-S.; Blair, N.F.; Sue, C.M. The role of ATP13A2 in Parkinson's disease: Clinical phenotypes and molecular mechanisms. *Mov. Disord.* **2015**, *30*, 770–779. [CrossRef] [PubMed]
121. Suleiman, J.; Hamwi, N.; El-Hattab, A.W. ATP13A2 novel mutations causing a rare form of juvenile-onset Parkinson disease. *Brain Dev.* **2018**, *40*, 824–826. [CrossRef] [PubMed]
122. Park, J.S.; Koentjoro, B.; Veivers, D.; Mackay-Sim, A.; Sue, C.M. Parkinson's disease-associated human ATP13A2 (PARK9) deficiency causes zinc dyshomeostasis and mitochondrial dysfunction. *Hum. Mol. Genet.* **2014**, *23*, 2802–2815. [CrossRef] [PubMed]
123. Rayaprolu, S.; Seven, Y.B.; Howard, J.; Duffy, C.; Altshuler, M.; Moloney, C.; Giasson, B.I.; Lewis, J. Partial loss of ATP13A2 causes selective gliosis independent of robust lipofuscinosis. *Mol. Cell Neurosci.* **2018**, *92*, 17–26. [CrossRef] [PubMed]
124. Sato, S.; Koike, M.; Funayama, M.; Ezaki, J.; Fukuda, T.; Ueno, T.; Uchiyama, Y.; Hattori, N. Lysosomal storage of subunit c of mitochondrial ATP Synthase in brain-specific Atp13a2-deficient mice. *Am. J. Pathol.* **2016**, *186*, 3074–3082. [CrossRef] [PubMed]
125. Park, J.S.; Koentjoro, B.; Davis, R.L.; Sue, C.M. Loss of ATP13A2 impairs glycolytic function in Kufor-Rakeb syndrome patient-derived cell models. *Parkinsonism Relat. Disord.* **2016**, *27*, 67–73. [CrossRef] [PubMed]
126. Goker-Alpan, O.; Lopez, G.; Vithayathil, J.; Davis, J.; Hallett, M.; Sidransky, E. The spectrum of parkinsonian manifestations associated with glucocerebrosidase mutations. *Arch. Neurol.* **2008**, *65*, 1353–1357. [CrossRef] [PubMed]
127. Sidransky, E.; Lopez, G. The link between the GBA gene and parkinsonism. *Lancet Neurol.* **2012**, *11*, 986–998. [CrossRef]

128. Gan-Or, Z.; Amshalom, I.; Kilarski, L.L.; Bar-Shira, A.; Gana-Weisz, M.; Mirelman, A.; Marder, K.; Bressman, S.; Giladi, N.; Orr-Urtreger, A. Differential effects of severe vs mild GBA mutations on Parkinson disease. *Neurology* **2015**, *84*, 880–887. [CrossRef] [PubMed]
129. Sato, S.; Li, Y.; Hattori, N. Lysosomal defects in ATP13A2 and GBA associated familial Parkinson's disease. *J. Neural Transm.* **2017**, *124*, 1395–1400. [CrossRef] [PubMed]
130. Sun, N.; Youle, R.J.; Finkel, T. The mitochondrial basis of aging. *Mol. Cell* **2016**, *61*, 654–666. [CrossRef] [PubMed]
131. Brunk, U.T.; Terman, A. Lipofuscin: Mechanisms of age-related accumulation and influence on cell function. *Free Radic. Biol. Med.* **2002**, *33*, 611–619. [CrossRef]
132. Dauer, W.; Przedborski, S. Parkinson's disease: Mechanisms and models. *Neuron* **2003**, *39*, 889–909. [CrossRef]
133. Amiri, M.; Hollenbeck, P.J. Mitochondrial biogenesis in the axons of vertebrate peripheral neurons. *Dev. Neurobiol.* **2008**, *68*, 1348–1361. [CrossRef] [PubMed]
134. Lentz, S.I.; Edwards, J.L.; Backus, C.; McLean, L.L.; Haines, K.M.; Feldman, E.L. Mitochondrial DNA (mtDNA) biogenesis: visualization and duel incorporation of BrdU and EdU into newly synthesized mtDNA in vitro. *J. Histochem. Cytochem.* **2010**, *58*, 207–218. [CrossRef] [PubMed]
135. Calkins, M.J.; Reddy, P.H. Assessment of newly synthesized mitochondrial DNA using BrdU labeling in primary neurons from Alzheimer's disease mice: Implications for impaired mitochondrial biogenesis and synaptic damage. *Biochim. Biophys. Acta* **2011**, *1812*, 1182–1189. [CrossRef] [PubMed]
136. Ferguson, S.M. Axonal transport and maturation of lysosomes. *Curr. Opin. Neurobiol.* **2018**, *51*, 45–51. [CrossRef] [PubMed]
137. Hayakawa, K.; Esposito, E.; Wang, X.; Terasaki, Y.; Liu, Y.; Xing, C.; Ji, X.; Lo, E.H. Transfer of mitochondria from astrocytes to neurons after stroke. *Nature* **2016**, *535*, 551. [CrossRef] [PubMed]
138. Davis, C.-h.O.; Kim, K.-Y.; Bushong, E.A.; Mills, E.A.; Boassa, D.; Shih, T.; Kinebuchi, M.; Phan, S.; Zhou, Y.; Bihlmeyer, N.A.; et al. Transcellular degradation of axonal mitochondria. *Proc. Natl. Acad. Sci. USA* **2014**, *111*, 9633–9638. [CrossRef] [PubMed]
139. Lin, M.Y.; Sheng, Z.H. Regulation of mitochondrial transport in neurons. *Exp. Cell Res.* **2015**, *334*, 35–44. [CrossRef] [PubMed]
140. Misgeld, T.; Schwarz, T.L. Mitostasis in neurons: Maintaining mitochondria in an extended cellular architecture. *Neuron* **2017**, *96*, 651–666. [CrossRef] [PubMed]
141. Menzies, F.M.; Fleming, A.; Caricasole, A.; Bento, C.F.; Andrews, S.P.; Ashkenazi, A.; Fullgrabe, J.; Jackson, A.; Jimenez Sanchez, M.; Karabiyik, C.; et al. Autophagy and neurodegeneration: Pathogenic mechanisms and therapeutic opportunities. *Neuron* **2017**, *93*, 1015–1034. [CrossRef] [PubMed]
142. Twig, G.; Elorza, A.; Molina, A.J.; Mohamed, H.; Wikstrom, J.D.; Walzer, G.; Stiles, L.; Haigh, S.E.; Katz, S.; Las, G.; et al. Fission and selective fusion govern mitochondrial segregation and elimination by autophagy. *EMBO J.* **2008**, *27*, 433–446. [CrossRef] [PubMed]
143. Van Laar, V.S.; Berman, S.B. The interplay of neuronal mitochondrial dynamics and bioenergetics: Implications for Parkinson's disease. *Neurobiol. Dis.* **2013**, *51*, 43–55. [CrossRef] [PubMed]
144. Berthet, A.; Margolis, E.B.; Zhang, J.; Hsieh, I.; Zhang, J.; Hnasko, T.S.; Ahmad, J.; Edwards, R.H.; Sesaki, H.; Huang, E.J.; et al. Loss of mitochondrial fission depletes axonal mitochondria in midbrain dopamine neurons. *J Neurosci.* **2014**, *34*, 14304–14317. [CrossRef] [PubMed]
145. Carelli, V.; Musumeci, O.; Caporali, L.; Zanna, C.; La Morgia, C.; Del Dotto, V.; Porcelli, A.M.; Rugolo, M.; Valentino, M.L.; Iommarini, L.; et al. Syndromic parkinsonism and dementia associated with OPA1 missense mutations. *Ann. Neurol.* **2015**, *78*, 21–38. [CrossRef] [PubMed]
146. Ramonet, D.; Perier, C.; Recasens, A.; Dehay, B.; Bove, J.; Costa, V.; Scorrano, L.; Vila, M. Optic atrophy 1 mediates mitochondria remodeling and dopaminergic neurodegeneration linked to complex I deficiency. *Cell Death Differ.* **2013**, *20*, 77–85. [CrossRef] [PubMed]
147. Glauser, L.; Sonnay, S.; Stafa, K.; Moore, D.J. Parkin promotes the ubiquitination and degradation of the mitochondrial fusion factor mitofusin 1. *J. Neurochem.* **2011**, *118*, 636–645. [CrossRef] [PubMed]
148. Gegg, M.E.; Cooper, J.M.; Chau, K.Y.; Rojo, M.; Schapira, A.H.; Taanman, J.W. Mitofusin 1 and mitofusin 2 are ubiquitinated in a PINK1/parkin-dependent manner upon induction of mitophagy. *Hum. Mol. Genet.* **2010**, *19*, 4861–4870. [CrossRef] [PubMed]

149. Chen, Y.; Dorn, G.W., 2nd. PINK1-phosphorylated mitofusin 2 is a Parkin receptor for culling damaged mitochondria. *Science* **2013**, *340*, 471–475. [CrossRef] [PubMed]
150. Rakovic, A.; Grünewald, A.; Kottwitz, J.; Brüggemann, N.; Pramstaller, P.P.; Lohmann, K.; Klein, C. Mutations in PINK1 and parkin impair ubiquitination of mitofusins in human fibroblasts. *PLoS ONE* **2011**, *6*, e16746. [CrossRef] [PubMed]
151. Barsoum, M.J.; Yuan, H.; Gerencser, A.A.; Liot, G.; Kushnareva, Y.; Graber, S.; Kovacs, I.; Lee, W.D.; Waggoner, J.; Cui, J.; et al. Nitric oxide-induced mitochondrial fission is regulated by dynamin-related GTPases in neurons. *EMBO J.* **2006**, *25*, 3900–3911. [CrossRef] [PubMed]
152. Plecita-Hlavata, L.; Lessard, M.; Santorova, J.; Bewersdorf, J.; Jezek, P. Mitochondrial oxidative phosphorylation and energetic status are reflected by morphology of mitochondrial network in INS-1E and HEP-G2 cells viewed by 4Pi microscopy. *Biochim. Biophys. Acta* **2008**, *1777*, 834–846. [CrossRef] [PubMed]
153. Wang, X.; Su, B.; Liu, W.; He, X.; Gao, Y.; Castellani, R.J.; Perry, G.; Smith, M.A.; Zhu, X. DLP1-dependent mitochondrial fragmentation mediates 1-methyl-4-phenylpyridinium toxicity in neurons: Implications for Parkinson's disease. *Aging Cell* **2011**, *10*, 807–823. [CrossRef] [PubMed]
154. Peng, K.; Yang, L.; Wang, J.; Ye, F.; Dan, G.; Zhao, Y.; Cai, Y.; Cui, Z.; Ao, L.; Liu, J.; et al. The interaction of mitochondrial biogenesis and fission/fusion mediated by PGC-1alpha regulates rotenone-induced dopaminergic neurotoxicity. *Mol. Neurobiol.* **2017**, *54*, 3783–3797. [CrossRef] [PubMed]
155. Zambonin, J.L.; Zhao, C.; Ohno, N.; Campbell, G.R.; Engeham, S.; Ziabreva, I.; Schwarz, N.; Lee, S.E.; Frischer, J.M.; Turnbull, D.M.; et al. Increased mitochondrial content in remyelinated axons: Implications for multiple sclerosis. *Brain* **2011**, *134*, 1901–1913. [CrossRef] [PubMed]
156. Schwarz, T.L. Mitochondrial trafficking in neurons. *Cold Spring Harb. Perspect. Biol.* **2013**, *5*. [CrossRef] [PubMed]
157. Devine, M.J.; Kittler, J.T. Mitochondria at the neuronal presynapse in health and disease. *Nat. Rev. Neurosci.* **2018**, *19*, 63–80. [CrossRef] [PubMed]
158. Chan, C.S.; Guzman, J.N.; Ilijic, E.; Mercer, J.N.; Rick, C.; Tkatch, T.; Meredith, G.E.; Surmeier, D.J. 'Rejuvenation' protects neurons in mouse models of Parkinson's disease. *Nature* **2007**, *447*, 1081–1086. [CrossRef] [PubMed]
159. Chang, D.T.; Honick, A.S.; Reynolds, I.J. Mitochondrial trafficking to synapses in cultured primary cortical neurons. *J. Neurosci.* **2006**, *26*, 7035–7045. [CrossRef] [PubMed]
160. Kang, J.S.; Tian, J.H.; Pan, P.Y.; Zald, P.; Li, C.; Deng, C.; Sheng, Z.H. Docking of axonal mitochondria by syntaphilin controls their mobility and affects short-term facilitation. *Cell* **2008**, *132*, 137–148. [CrossRef] [PubMed]
161. Obashi, K.; Okabe, S. Regulation of mitochondrial dynamics and distribution by synapse position and neuronal activity in the axon. *Eur. J. Neurosci.* **2013**, *38*, 2350–2363. [CrossRef] [PubMed]
162. Reeve, A.K.; Grady, J.P.; Cosgrave, E.M.; Bennison, E.; Chen, C.; Hepplewhite, P.D.; Morris, C.M. Mitochondrial dysfunction within the synapses of substantia nigra neurons in Parkinson's disease. *NPJ Parkinson's Dis.* **2018**, *4*, 9. [CrossRef] [PubMed]
163. Barnhart, E.L. Mechanics of mitochondrial motility in neurons. *Curr. Opin. Cell Biol.* **2016**, *38*, 90–99. [CrossRef] [PubMed]
164. Miller, K.E.; Sheetz, M.P. Axonal mitochondrial transport and potential are correlated. *J. Cell Sci.* **2004**, *117*, 2791–2804. [CrossRef] [PubMed]
165. Verburg, J.; Hollenbeck, P.J. Mitochondrial membrane potential in axons increases with local nerve growth factor or semaphorin signaling. *J. Neurosci.* **2008**, *28*, 8306–8315. [CrossRef] [PubMed]
166. Chen, Y.; Sheng, Z.H. Kinesin-1-syntaphilin coupling mediates activity-dependent regulation of axonal mitochondrial transport. *J. Cell Biol.* **2013**, *202*, 351–364. [CrossRef] [PubMed]
167. Sorbara, C.D.; Wagner, N.E.; Ladwig, A.; Nikic, I.; Merkler, D.; Kleele, T.; Marinkovic, P.; Naumann, R.; Godinho, L.; Bareyre, F.M.; et al. Pervasive axonal transport deficits in multiple sclerosis models. *Neuron* **2014**, *84*, 1183–1190. [CrossRef] [PubMed]
168. Lewis, T.L., Jr.; Turi, G.F.; Kwon, S.K.; Losonczy, A.; Polleux, F. Progressive decrease of mitochondrial motility during maturation of cortical axons in vitro and in vivo. *Curr. Biol.* **2016**, *26*, 2602–2608. [CrossRef] [PubMed]

169. Smit-Rigter, L.; Rajendran, R.; Silva, C.A.; Spierenburg, L.; Groeneweg, F.; Ruimschotel, E.M.; van Versendaal, D.; van der Togt, C.; Eysel, U.T.; Heimel, J.A.; et al. Mitochondrial dynamics in visual cortex are limited in vivo and not affected by axonal structural plasticity. *Curr. Biol.* **2016**, *26*, 2609–2616. [CrossRef] [PubMed]
170. Rintoul, G.L.; Filiano, A.J.; Brocard, J.B.; Kress, G.J.; Reynolds, I.J. Glutamate decreases mitochondrial size and movement in primary forebrain neurons. *J. Neurosci.* **2003**, *23*, 7881–7888. [CrossRef] [PubMed]
171. Sheng, Z.H.; Cai, Q. Mitochondrial transport in neurons: Impact on synaptic homeostasis and neurodegeneration. *Nat. Rev. Neurosci.* **2012**, *13*, 77–93. [CrossRef] [PubMed]
172. Kim-Han, J.S.; Antenor-Dorsey, J.A.; O'Malley, K.L. The parkinsonian mimetic, MPP+, specifically impairs mitochondrial transport in dopamine axons. *J. Neurosci.* **2011**, *31*, 7212–7221. [CrossRef] [PubMed]
173. Sterky, F.H.; Lee, S.; Wibom, R.; Olson, L.; Larsson, N.G. Impaired mitochondrial transport and Parkin-independent degeneration of respiratory chain-deficient dopamine neurons in vivo. *Proc. Natl. Acad. Sci. USA* **2011**, *108*, 12937–12942. [CrossRef] [PubMed]
174. Pickrell, A.M.; Youle, R.J. The roles of PINK1, parkin, and mitochondrial fidelity in Parkinson's disease. *Neuron* **2015**, *85*, 257–273. [CrossRef] [PubMed]
175. Narendra, D.; Tanaka, A.; Suen, D.F.; Youle, R.J. Parkin is recruited selectively to impaired mitochondria and promotes their autophagy. *J. Cell Biol.* **2008**, *183*, 795–803. [CrossRef] [PubMed]
176. McWilliams, T.G.; Muqit, M.M. PINK1 and Parkin: Emerging themes in mitochondrial homeostasis. *Curr. Opin. Cell Biol.* **2017**, *45*, 83–91. [CrossRef] [PubMed]
177. Park, J.; Lee, S.B.; Lee, S.; Kim, Y.; Song, S.; Kim, S.; Bae, E.; Kim, J.; Shong, M.; Kim, J.M.; et al. Mitochondrial dysfunction in Drosophila PINK1 mutants is complemented by parkin. *Nature* **2006**, *441*, 1157–1161. [CrossRef] [PubMed]
178. Gispert, S.; Ricciardi, F.; Kurz, A.; Azizov, M.; Hoepken, H.H.; Becker, D.; Voos, W.; Leuner, K.; Muller, W.E.; Kudin, A.P.; et al. Parkinson phenotype in aged PINK1-deficient mice is accompanied by progressive mitochondrial dysfunction in absence of neurodegeneration. *PLoS ONE* **2009**, *4*, e5777. [CrossRef] [PubMed]
179. Chan, N.C.; Salazar, A.M.; Pham, A.H.; Sweredoski, M.J.; Kolawa, N.J.; Graham, R.L.; Hess, S.; Chan, D.C. Broad activation of the ubiquitin-proteasome system by Parkin is critical for mitophagy. *Hum. Mol. Genet.* **2011**, *20*, 1726–1737. [CrossRef] [PubMed]
180. Liu, S.; Sawada, T.; Lee, S.; Yu, W.; Silverio, G.; Alapatt, P.; Millan, I.; Shen, A.; Saxton, W.; Kanao, T.; et al. Parkinson's disease–associated kinase PINK1 regulates miro protein level and axonal transport of mitochondria. *PLoS Genet.* **2012**, *8*, e1002537. [CrossRef] [PubMed]
181. Wang, X.; Winter, D.; Ashrafi, G.; Schlehe, J.; Wong, Y.L.; Selkoe, D.; Rice, S.; Steen, J.; LaVoie, M.J.; Schwarz, T.L. PINK1 and Parkin target Miro for phosphorylation and degradation to arrest mitochondrial motility. *Cell* **2011**, *147*, 893–906. [CrossRef] [PubMed]
182. Weihofen, A.; Thomas, K.J.; Ostaszewski, B.L.; Cookson, M.R.; Selkoe, D.J. Pink1 forms a multiprotein complex with Miro and Milton, linking Pink1 function to mitochondrial trafficking. *Biochemistry* **2009**, *48*, 2045–2052. [CrossRef] [PubMed]
183. Villar-Pique, A.; Lopes da Fonseca, T.; Outeiro, T.F. Structure, function and toxicity of alpha-synuclein: The Bermuda triangle in synucleinopathies. *J. Neurochem.* **2016**, *139*, 240–255. [CrossRef] [PubMed]
184. Devi, L.; Raghavendran, V.; Prabhu, B.M.; Avadhani, N.G.; Anandatheerthavarada, H.K. Mitochondrial import and accumulation of alpha-synuclein impair complex I in human dopaminergic neuronal cultures and Parkinson disease brain. *J. Biol. Chem.* **2008**, *283*, 9089–9100. [CrossRef] [PubMed]
185. Reeve, A.K.; Ludtmann, M.H.; Angelova, P.R.; Simcox, E.M.; Horrocks, M.H.; Klenerman, D.; Gandhi, S.; Turnbull, D.M.; Abramov, A.Y. Aggregated alpha-synuclein and complex I deficiency: Exploration of their relationship in differentiated neurons. *Cell Death Dis.* **2015**, *6*, e1820. [CrossRef] [PubMed]
186. Chinta, S.J.; Mallajosyula, J.K.; Rane, A.; Andersen, J.K. Mitochondrial alpha-synuclein accumulation impairs complex I function in dopaminergic neurons and results in increased mitophagy in vivo. *Neurosci. Lett.* **2010**, *486*, 235–239. [CrossRef] [PubMed]
187. Bender, A.; Desplats, P.; Spencer, B.; Rockenstein, E.; Adame, A.; Elstner, M.; Laub, C.; Mueller, S.; Koob, A.O.; Mante, M.; et al. TOM40 mediates mitochondrial dysfunction induced by alpha-synuclein accumulation in Parkinson's disease. *PLoS ONE* **2013**, *8*, e62277. [CrossRef] [PubMed]

188. Di Maio, R.; Barrett, P.J.; Hoffman, E.K.; Barrett, C.W.; Zharikov, A.; Borah, A.; Hu, X.; McCoy, J.; Chu, C.T.; Burton, E.A.; et al. α-Synuclein binds to TOM20 and inhibits mitochondrial protein import in Parkinson's disease. *Sci. Transl. Med.* **2016**, *8*, 342ra78. [CrossRef] [PubMed]
189. Yano, H.; Baranov, S.V.; Baranova, O.V.; Kim, J.; Pan, Y.; Yablonska, S.; Carlisle, D.L.; Ferrante, R.J.; Kim, A.H.; Friedlander, R.M. Inhibition of mitochondrial protein import by mutant huntingtin. *Nat. Neurosci.* **2014**, *17*, 822–831. [CrossRef] [PubMed]
190. Nakamura, K.; Nemani, V.M.; Azarbal, F.; Skibinski, G.; Levy, J.M.; Egami, K.; Munishkina, L.; Zhang, J.; Gardner, B.; Wakabayashi, J.; et al. Direct membrane association drives mitochondrial fission by the Parkinson disease-associated protein alpha-synuclein. *J. Biol. Chem.* **2011**, *286*, 20710–20726. [CrossRef] [PubMed]
191. Kamp, F.; Exner, N.; Lutz, A.K.; Wender, N.; Hegermann, J.; Brunner, B.; Nuscher, B.; Bartels, T.; Giese, A.; Beyer, K.; et al. Inhibition of mitochondrial fusion by alpha-synuclein is rescued by PINK1, Parkin and DJ-1. *EMBO J.* **2010**, *29*, 3571–3589. [CrossRef] [PubMed]
192. Guardia-Laguarta, C.; Area-Gomez, E.; Rub, C.; Liu, Y.; Magrane, J.; Becker, D.; Voos, W.; Schon, E.A.; Przedborski, S. Alpha-Synuclein is localized to mitochondria-associated ER membranes. *J. Neurosci.* **2014**, *34*, 249–259. [CrossRef] [PubMed]
193. Paillusson, S.; Gomez-Suaga, P.; Stoica, R.; Little, D.; Gissen, P.; Devine, M.J.; Noble, W.; Hanger, D.P.; Miller, C.C.J. Alpha-Synuclein binds to the ER-mitochondria tethering protein VAPB to disrupt Ca(2+) homeostasis and mitochondrial ATP production. *Acta Neuropathol.* **2017**, *134*, 129–149. [CrossRef] [PubMed]
194. Pozo Devoto, V.M.; Dimopoulos, N.; Alloatti, M.; Pardi, M.B.; Saez, T.M.; Otero, M.G.; Cromberg, L.E.; Marín-Burgin, A.; Scassa, M.E.; Stokin, G.B.; et al. αSynuclein control of mitochondrial homeostasis in human-derived neurons is disrupted by mutations associated with Parkinson's disease. *Sci. Rep.* **2017**, *7*, 5042. [CrossRef] [PubMed]
195. Prots, I.; Veber, V.; Brey, S.; Campioni, S.; Buder, K.; Riek, R.; Bohm, K.J.; Winner, B. Alpha-Synuclein oligomers impair neuronal microtubule-kinesin interplay. *J. Biol. Chem.* **2013**, *288*, 21742–21754. [CrossRef] [PubMed]

© 2019 by the authors. Licensee MDPI, Basel, Switzerland. This article is an open access article distributed under the terms and conditions of the Creative Commons Attribution (CC BY) license (http://creativecommons.org/licenses/by/4.0/).

Review

Mitochondrial Dysfunction and Stress Responses in Alzheimer's Disease

Ian Weidling [1,2] and Russell H. Swerdlow [1,2,3,4,*]

1. University of Kansas Alzheimer's Disease Center, Fairway, KS 66205, USA; iweidling2@kumc.edu
2. Department of Integrated and Molecular Physiology, University of Kansas Medical Center, Kansas City, KS 66160, USA
3. Department of Neurology, University of Kansas Medical Center, Kansas City, KS 66160, USA
4. Department of Biochemistry and Molecular Biology, University of Kansas Medical Center, Kansas City, KS 66160, USA
* Correspondence: rswerdlow@kumc.edu; Tel.: +01-913-588-0555

Received: 4 December 2018; Accepted: 16 January 2019; Published: 11 May 2019

Abstract: Alzheimer's disease (AD) patients display widespread mitochondrial defects. Brain hypometabolism occurs alongside mitochondrial defects, and correlates well with cognitive decline. Numerous theories attempt to explain AD mitochondrial dysfunction. Groups propose AD mitochondrial defects stem from: (1) mitochondrial-nuclear DNA interactions/variations; (2) amyloid and neurofibrillary tangle interactions with mitochondria, and (3) mitochondrial quality control defects and oxidative damage. Cells respond to mitochondrial dysfunction through numerous retrograde responses including the Integrated Stress Response (ISR) involving eukaryotic initiation factor 2α (eIF2α), activating transcription factor 4 (ATF4) and C/EBP homologous protein (CHOP). AD brains activate the ISR and we hypothesize mitochondrial defects may contribute to ISR activation. Here we review current recognized contributions of the mitochondria to AD, with an emphasis on their potential contribution to brain stress responses.

Keywords: Alzheimer's disease; eIF2α; metabolism; mitochondria; proteostasis; stress response

1. Introduction

Sporadic Alzheimer's disease (AD) brains possess profound mitochondrial defects, including changes in number, morphology, and enzyme activity [1–3]. Mitochondrial dysfunction in AD is not restricted to the nervous system. Systemic mitochondrial defects occur in AD patients compared to controls [4,5]. Metabolic defects occur alongside mitochondrial abnormalities in AD, providing early markers of disease progression [6]. Mitochondrial dysfunction may contribute to hallmark AD pathology and stimulate stress response pathways.

2. AD Brain Hypometabolism

Brain glucose uptake studies provided some of the earliest evidence for AD metabolic defects. Changes in cerebral glucose utilization occur during AD, demonstrated by numerous studies using [18F]-2-fluoro-2-deoxy-D-glucose (FDG) coupled with positron emission tomography (PET) [7–10]. In these studies, researchers administer radiolabeled FDG to patients intravenously. Cells take FDG up through glucose importers and subsequently phosphorylate FDG via hexokinase. Unlike glucose, FDG cannot be processed further by glycolytic enzymes and accumulates within the cell. Cells taking up more radiolabeled FDG display a stronger PET signal [11]. AD patients consistently display reduced cerebral PET signals following [18F] FDG administration suggesting reductions in glucose uptake and neuronal activity [12–14].

Classically, AD brains display decreased temporo-parietal glucose uptake in both hemispheres [15]. Studies have attempted to correlate numerous AD pathological changes with cognitive decline. FDG PET studies show cerebral glucose utilization correlates reasonably well with cognitive decline. Amyloid plaques, on the other hand, correlate poorly with cognitive decline, while neurofibrillary tangles (NFTs) show better correlation [16]. Although a definitive diagnosis of Alzheimer's disease requires the presence of amyloid plaques and NFTs, brain hypometabolism may provide a sensitive and early marker of neurodegeneration [17].

Decreased cerebral glucose utilization occurs early in AD and could prove useful diagnostically [6,18,19]. Studies performing FDG PET analysis on patients with "very early Alzheimer's disease" found changes in glucose uptake. The study divided participants into a very early Alzheimer's disease group and an age-matched control group based on mini-mental state exam performance. The very early Alzheimer's disease group displayed brain region specific decreases in glucose utilization relative to age-matched controls. Early glucose uptake deficits presented most prominently in the posterior cingulate cortex (PCC) and cinguloparietal transition regions. Reports describe neurodegeneration in these regions in neuropathologically confirmed AD cases. Reduced glucose utilization in the very early Alzheimer's disease brain did not correlate with AD pathology. Neuropathological examination of very early Alzheimer's disease brains found NFT accumulation in medial and inferior temporal cortex but not in the PCC. It is interesting to note that metabolic deficiencies occur in the absence of AD pathology, suggesting NFTs and plaques do not need to be present for reduced glucose utilization to occur [13]. Longitudinal FDG PET studies, followed up with neuropathological diagnosis, demonstrate further AD specific changes in brain metabolism. This study improved upon prior work by confirming eventual AD diagnosis. The results support brain glucose utilization as a potential tool in AD diagnosis [20]. Several studies show metabolic defects can be detected long before the onset of cognitive decline [21,22]. Meta-analysis of studies evaluating FDG-PET for AD diagnosis shows FDG-PET performs better in diagnosing AD than current diagnostic methodologies [23]. Brain hypometabolism's early appearance and correlation with dementia in AD patients suggests altered metabolism is intimately linked with disease progression.

A clear association between brain hypometabolism and dementia exists but researchers do not understand why AD brains display reduced glucose utilization. Glucose transporter studies in AD brains provide one potential explanation for decreased glucose utilization. Glucose transporters move glucose across cell membranes and into the cytoplasm. Neurons import glucose mainly through GLUT3, while astrocytes import glucose mainly through GLUT1 [24]. Both GLUT1 and GLUT3 protein levels decrease in AD brain and these changes in GLUT1 and GLUT3 persist after correcting for cell death. For this reason, Simpson et al. argue glucose transporter loss contributes to neurodegeneration [25]. Further studies found reduced glucose transporter levels at the blood-brain barrier in AD brains, another likely contributor to decreased glucose uptake [26]. Decreasing glucose transporter levels speak to broad metabolic defects in AD brain. Mitochondrial dysfunction likely contributes to these broad and general AD metabolic defects.

3. AD Mitochondrial Defects

Altered metabolism in AD coincides with numerous mitochondrial changes. AD platelet cytochrome oxidase (COX) activity studies provided early evidence for mitochondrial dysfunction. Parker et al. showed altered COX activity in AD platelets, later extending their findings to AD brain tissue [4,27]. At this time, they postulated mitochondrial DNA (mtDNA) alterations may trigger AD COX deficiencies [28]. An additional study characterized AD mitochondrial complex I and II–III activities, finding no consistent activity changes in various brain areas. However, the study confirms decreased COX activity in multiple AD cortical brain regions [29]. Mitochondrial tricarboxylic acid (TCA) cycle enzymes also display altered activity in AD brain. Post-mortem AD brain activity assays reveal increases and decreases in TCA enzyme activities. Among mitochondrial enzyme activities, clinical decline correlates most closely with changes in pyruvate dehydrogenase complex activity [30]. Defects in numerous mitochondrial enzymes exist in AD brain, likely contributing to metabolic abnormalities.

Further changes in mitochondrial enzymes exist in AD. Post-mortem AD brain tissue analysis finds that COX subunits decrease during disease progression. In one study, the authors analyzed COXIV (nuclear-encoded) and COXII (mitochondrial-encoded) subunit levels in cerebellar Purkinje neurons, an area relatively preserved in AD subjects compared to age-matched controls. The study found decreased COXIV and COXII protein levels in AD Purkinje neurons relative to age matched controls, as well as COX subunit reductions in aged controls relative to young controls. Based upon this finding, the authors argue COX deficiency occurs during normal aging and accelerated COX deficiency contributes to AD progression [31]. Cottrell et al. [32] discovered increased COX deficient neurons in the AD hippocampus. The study examined COX and mitochondrial complex II (succinate dehydrogenase) levels in individual cells via immunohistochemistry (IHC). Neurons containing drastically reduced COX levels with normal succinate dehydrogenase levels were classified as COX deficient [32]. The specific reduction in COX levels relative to succinate dehydrogenase suggests mitochondrial mass is maintained while COX is preferentially depleted. Subsequent studies correlated AD pathology with COX deficiency. Correlational studies revealed COX deficient neurons contain decreased NFTs relative to surrounding COX positive neurons. The study found no correlation between COX levels and plaque burden [33].

While the AD hippocampus contains many COX deficient neurons, studies also observe AD neurons displaying increases in COX and mtDNA. In AD neurons with increased COX and mtDNA, lysosomal structures tend to accumulate mitochondrial components. These findings suggest increases in COX and mtDNA do not reflect increased intact mitochondria. Instead, mtDNA and COX accumulation likely signals deficient mitochondrial degradation [34]. AD neurons upregulate lysosomal components early in the disease process. AD neurons increase lysosomal protease (cathepsin D) mRNA and protein with concomitant lysosomal accumulation [35]. Furthermore, diseased neurons accumulate autophagosomes at a high level, suggesting either an increased autophagic rate, decreased autophagosome maturation, or both [36]. Disruptions to autophagy and lysosomal degradation likely contribute to AD mitochondrial defects. Defective mitochondria generally undergo selective degradation through an autophagosome dependent process known as mitophagy. To begin the process of mitophagy, autophagic vacuoles surround and envelope mitochondria. Autophagic vacuoles containing mitochondrial components then acidify, maturing to lysosomes, and degrading their contents [37]. Mitophagy maintains a healthy mitochondrial pool, so disruptions in this process compounds other mitochondrial defects [38]. Mitophagy is altered in AD, as studies observe increased mitochondria-lysosome associations.

AD mitochondria also display alterations in morphology. AD brain electron microscopy (EM) studies reveal changes in mitochondrial physical structure. Numerous AD brain regions display increased variability in mitochondrial shape and disrupted cristae, as well as decreased mitochondrial surface area [39]. Mitochondrial morphology relies on fission and fusion processes. Mitochondrial fission and fusion defects occur in AD and likely contribute to morphological changes. Zhang et al. [40] performed three-dimensional (3D) reconstruction of serial AD hippocampal EM sections. 3D EM revealed a novel AD mitochondrial morphology termed "mitochondria on a string" (MOAS). Earlier methodologies could not detect this morphological feature, likely classifying MOAS as fragmented mitochondria. MOAS likely form when fission machinery malfunctions. The authors propose AD bioenergetic defects inhibit fission machinery, triggering mitochondrial morphology changes [40]. Additional studies suggest AD disrupts fusion and fission. Wang et al. describe altered mitochondrial localization in AD pyramidal neurons along with altered fusion and fission proteins [41]. Experiments also show that amyloid beta can cause fusion and fission defects, and inhibiting mitochondrial fission proves beneficial in AD mouse models. Mutant amyloid precursor protein overexpression in primary mouse hippocampal neurons altered fusion and fission genes and disrupted mitochondrial structure [42]. Additionally, amyloid beta treatment in neuronal cells caused dynamin related protein 1

(Drp1) phosphorylation and increased mitochondrial fission. A mitochondrial fission inhibitor reduced reactive oxygen species (ROS) and reduced mitochondrial dysfunction caused by amyloid beta treatment [43,44]. Studies of mitochondrial fission in AD models also suggest that mitochondrial fission favors cell death. In fact, amyloid beta oligomers trigger mitochondrial fragmentation and subsequent cell death via the loss of a mitochondrial fusion factor [45]. Inhibiting mitochondrial fission in an AD mouse model decreased brain pathology and improved memory, as well as synaptic connections, suggesting mitochondrial fission inhibitors may have therapeutic potential [46]. Mitochondrial morphological changes in AD speak to widespread mitochondrial dysfunction.

Whether metabolic and mitochondrial defects represent a cause or consequence of AD remains controversial. Numerous groups propose mitochondrial dysfunction initiates AD pathological cascades and therapeutics should target mitochondrial dysfunction [1,47–49]. The cause for mitochondrial dysfunction in AD remains unclear, however. Many theories explaining AD mitochondrial dysfunction exist. Groups viewing mitochondrial dysfunction as a primary event in AD progression point to mtDNA as a potential disease driver [28,50]. Groups viewing mitochondrial dysfunction as a disease consequence propose AD pathology, namely amyloid protein and tau tangles, initiates mitochondrial dysfunction [51–53]. Still other groups propose defective mitochondrial quality control and oxidative damage contributes to mitochondrial dysfunction [34]. Each of these mechanisms may in fact contribute to AD mitochondrial dysfunction and, increasingly, mitochondrial function is viewed favorably as a therapeutic target.

4. Role of Mitochondrial DNA in AD

Somatic mtDNA mutations may contribute to AD and mtDNA inheritance may influence AD risk. Mitochondrial function relies on coordinated expression of genes from the nuclear and mitochondrial genomes. Inherited mtDNA polymorphisms cause a range of disorders known as primary mitochondrial diseases. Many of these diseases primarily affect cognition, demonstrating that neurons possess high sensitivity to mitochondrial defects. One of the most widely recognized mtDNA deletions, a 4997 bp deletion called the "common deletion", increases in the brain during normal aging. Common deletion rates increase most drastically in regions with high metabolic activity, causing some to speculate that mtDNA somatic deletions dispose individuals to neurological disease [54]. AD mtDNA alterations surpass those observed in age-matched controls. High common deletion rates occur early in the AD cortex. As AD patients reach age 80, however, common deletion rates typically decline. The opposite trend exists in age-matched control cortex, with low common deletion rates early and increasing rates as individuals age [55,56].

AD brains also display increased mtDNA oxidative damage. Interestingly, mtDNA oxidative damage occurs most heavily in the parietal lobe, which displays early and consistent hypometabolism [57]. Increased oxidative damage correlates with mitochondrial dysfunction. For this reason, researchers speculated that AD oxidative damage favors mtDNA mutations. Indeed, mtDNA control region mutations increase in AD frontal cortex. Increased control region mutations associate with decreased mtDNA transcription and replication [58]. Subsequent analyses utilizing next generation sequencing (NGS) discovered increased AD hippocampal mtDNA point mutations. However, the authors conclude AD point mutations likely stem from mtDNA replication errors rather than oxidative damage [59].

AD inheritance pattern studies implicate mtDNA inheritance as an AD risk factor. In a group of families with one AD affected parent and two affected siblings, Edland et al. discovered increased AD rates among individuals with a maternal AD history [60]. These findings suggest AD favors a maternal inheritance pattern. Maternal AD history also increases risk for brain hypometabolism, potentially increasing AD risk [61]. Groups propose mtDNA inheritance explains AD's subtle but identifiable maternal inheritance predominance [48]. mtDNA largely passes from mother to child, and therefore AD's bias towards maternal inheritance is consistent with mtDNA influencing AD risk.

Cytoplasmic hybrid (cybrid) studies provide further evidence mtDNA contributes to AD mitochondrial abnormalities. Cybrid generation occurs by repopulating cells lacking mtDNA (ρ0) with exogenous mtDNA. Exogenous mtDNA often comes from patient platelets, allowing creation of cybrids containing AD patient mtDNA. Cybrids, therefore, effectively model AD mitochondrial function on a stable nuclear background. AD cybrids recapitulate numerous AD features. Initial AD cybrid studies demonstrate COX activity deficits that recapitulate those of AD patient mitochondria. AD cybrid COX deficits provide strong evidence that mtDNA contributes to AD mitochondrial defects [62,63]. Further studies suggest mtDNA deletions contribute to AD hippocampal COX deficiency. As referenced earlier, COX deficient neurons increase in the AD hippocampus. COX deficient AD neurons contain increased mtDNA deletions, suggesting mtDNA deletions contribute to COX deficiency [64]. Additional AD cybrid studies describe enlarged, swollen mitochondria with reductions in membrane potential and increases in ROS and antioxidant enzymes [63,65]. AD cybrid studies also suggest mitochondrial dysfunction can drive changes in AD neuropathology.

AD cybrids display amyloid changes reminiscent of those observed in AD and possess increased sensitivity to amyloid beta fragments. AD cybrids release amyloid beta at greater rates than controls. Furthermore, AD cybrids contain increased intracellular amyloid beta. Elevations in amyloid beta coincide with increased cytochrome c release and caspase-3 activity, suggesting cell death pathway activation may contribute to elevated amyloid beta [66]. AD cybrids treated with amyloid beta display enhanced cell death pathway activity compared to control. Mitochondrial membrane potential, cytochrome c release and caspase 3 activity all change to a greater extent in amyloid beta treated AD cybrids [67]. AD mitochondrial function predisposes cells to increased amyloid beta production and cell death.

5. Mitochondrial Interaction with AD Pathology

Further studies demonstrate mitochondrial function influences AD pathology. Treating fibroblasts from control subjects with a mitochondrial membrane potential uncoupler (CCCP) triggers tau phosphorylation at sites altered in AD [68]. Complex I inhibitors also initiate AD-like tau alterations. Chronic rotenone treatment in rat brain triggers tau hyperphosphorylation and aggregation [69]. Studies often utilize triple transgenic mice to model AD. Triple transgenic mice express mutated forms of APP, tau and presenilin 1, causing them to develop amyloid plaques and tau tangles. Studies in female triple transgenic mice observe mitochondrial dysfunction prior to amyloid plaque formation. Female triple transgenic mice eventually experience increased mitochondrial amyloid beta levels which may exacerbate mitochondrial dysfunction. However, female triple transgenic mice experience decreased COX activity and increased glycolytic rates prior to amyloidosis [70]. Overexpressing a form of mutant APP in mice also causes mitochondrial gene upregulation in the hippocampus long before amyloid plaque deposition. Most of these upregulated genes contribute to oxidative phosphorylation (OXPHOS) [71].

Studies question whether mitochondrial dysfunction triggers Alzheimer's pathology. Fukui et al. deleted the *COX10* gene, which encodes a necessary COX assembly factor, in triple transgenic mice. *COX10* deletion inhibits COX assembly, causing loss of function. *COX10* deficient triple transgenic mice produce fewer amyloid plaques and amyloid beta than triple transgenic mice with functional COX [72]. This finding suggests loss of COX function reduces amyloid plaque production. However, it should be noted that loss of COX function via *COX10* deletion likely stimulates different responses than those elicited by defective functioning of intact COX. Additional studies are needed to more fully examine mitochondrial dysfunction's effects on AD pathology.

A reciprocal relationship exists between AD pathology and mitochondrial function. Amyloid beta treatment in cell culture causes mitochondrial dysfunction, including decreases in membrane potential, electron transport chain activity and oxygen consumption [73]. Amyloid beta inhibits COX activity in isolated mitochondria [74]. In AD brains APP accumulates in mitochondrial translocases, potentially inhibiting their function [53]. Further work describes AD mitochondrial amyloid beta accumulation and interaction with an alcohol dehydrogenase within the mitochondrial matrix [75,76]. Tau also interacts with mitochondria and their biology. Tau overexpression in cell culture changes mitochondrial localization, likely by disrupting mitochondrial transport along microtubules. Post mortem AD brain studies observe decreased synaptic mitochondria suggesting AD disturbs neuronal mitochondrial transport [77]. Pathological tau may contribute to microtubule disruption and subsequent mitochondrial localization changes in AD. Hyperphosphorylated tau associates with voltage dependent anion channel 1 (VDAC1) on the outer mitochondrial membrane. AD increases hyperphosphorylated tau bound to VDAC1, another potential contributor to mitochondrial dysfunction [78].

Tau truncation also occurs in AD, potentially contributing to mitochondrial dysfunction. AD NFTs contain truncated tau and these truncated tau species may be toxic [79,80]. Overexpressing a specific N-terminal tau fragment (NH2-26-44) causes primary neurons to die. N-terminal tau fragment treatment inhibits adenine nucleotide transporter (ANT) function, causing mitochondrial dysfunction [81]. Further studies need to determine whether this N-terminal tau fragment increases during AD progression. Overexpressing another tau fragment (Asp-421 cleaved tau), known to increase during AD, causes mitochondrial fragmentation and increased oxidative stress in cell culture [82]. Tau fragment generation likely occurs through caspase cleavage during apoptosis. Additional AD-associated protein fragments disrupt mitochondrial function.

Apolipoprotein E allele ε4 (apoE4) increases risk for AD. Relative to other apoE isoforms, apoE4 accumulates in endosomal compartments and stimulates cholesterol efflux less efficiently [83]. Furthermore, apoE4 appears susceptible to c-terminal protease cleavage. C-terminal apoE fragments occur in AD brain and truncated apoE colocalizes with NFTs. Overexpressing apoE4 fragments (apoE4 Δ272–299) in cell culture stimulates NFT formation [84]. ApoE associates with mitochondrial proteins, with apoE4 fragments binding mitochondrial proteins more strongly than apoE2 and apoE3. Overexpressing apoE4 fragments decreases mitochondrial complex III and COX activity [85], suggesting apoE4 increases AD risk partly through mitochondrial effects.

6. Mitochondrial Contributions to Proteostasis

Emerging evidence suggests mitochondria contribute to cellular proteostasis (Figure 1). In yeast, mitochondria degrade misfolded cytosolic proteins through resident proteases. Ruan et al. [86] show aggregated protein degradation in yeast relies on mitochondrial import machinery and proteases. When the authors blocked mitochondrial protein import and deleted mitochondrial proteases, protein aggregates became more stable. Defective cytosolic chaperones caused misfolded proteins to accumulate in mitochondria. Together, these observations highlight mitochondrial contributions to yeast proteostasis. The authors refer to mitochondrial protein degradation as "Mitochondria as Guardians in the Cytosol" (MAGIC) [86]. Whether MAGIC contributes substantially to proteostasis in human cells remains unclear. If MAGIC occurs in human cells, defective mitochondrial proteastasis could contribute to AD plaque and tangle formation. Another study shows mitochondrial degradation via mitophagy reduces amyloid burden in mAPP transgenic mice. mAPP mice lacking PTEN-induced putative kinase (PINK1) accumulate amyloid pathology earlier than mAPP mice expressing PINK1. PINK1 accumulation in mitochondrial membranes stimulates mitophagy. PINK1 knockout, therefore, seems to increase amyloid pathology in mAPP mice by disrupting mitophagy. Alternatively, PINK1 overexpression in mAPP mice enhances mitophagy and reduces amyloid beta plaques [87]. Mitophagy induction likely reduces mAPP mouse plaque burden by degrading amyloid beta filled mitochondria. In line with these findings, another report highlights mitophagy's role in clearing protein aggregates. Findings suggest mitochondrial fission facilitates selective mitophagy of regions containing protein aggregates [88]. These studies suggest mitochondria act as disposal sites for aggregated proteins. Pathological protein aggregates may signal defective mitochondrial proteostasis or mitophagy.

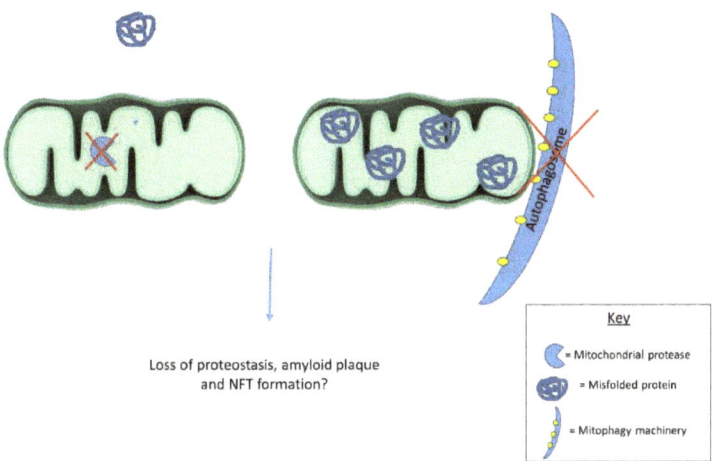

Figure 1. Mitochondrial proteases and mitophagy contribute to cellular proteostasis. Dysfunctional proteases and/or mitophagy could contribute to protein misfolding in disease states. Future work should examine mitochondrial contributions to cellular proteostasis in human cells as this process has largely been described in yeast.

Mitochondria possess intrinsic mechanisms for responding to unfolded proteins. Mitochondrial protein misfolding triggers a compensatory mechanism termed the mitochondrial unfolded response (mtUPR). Mutant ornithine transcarbamylase (OTC) overexpression leads misfolded OTC to accumulate within mitochondria, stimulating the mtUPR [89]. The mtUPR induces mitochondrial proteases and chaperones to restore proteostasis. Studies in Caenorhabditis elegans (C. elegans) provide most of the evidence for a mtUPR. mtDNA depletion by ethidium bromide, doxycycline treatment and mitochondrial ribosomal protein knockdown all trigger the mtUPR in C. elegans. Also, disrupting mitochondrial protein complexes and knocking down mitochondrial proteases and chaperones activates the mtUPR [90]. In C. elegans, Activating Transcription Factor associated with Stress-1 (ATFS-1) mediates mtUPR activation. ATFS-1 controls the mtUPR based on its subcellular localization. Functional mitochondria import and degrade ATFS-1. When mitochondrial dysfunction occurs, ATFS-1 accumulates in the nucleus due to a nuclear targeting sequence. Nuclear ATFS-1 activates mitochondrial protease and chaperone transcription. The mtUPR and ATFS-1 gained notoriety following discoveries of lifespan extension in C. elegans upon electron transport chain (ETC) gene knockdown [91]. Groups posited mtUPR activation mediates the lifespan extension gained from ETC gene knockdown. However, C. elegans lifespan studies suggest mtUPR activation and ATFS-1 activity do not facilitate the observed lifespan extension following mitochondrial insult [92].

In mammalian cells, activating transcription factor 5 (ATF5) may regulate an mtUPR similar to how ATFS-1 functions in C. elegans. However, distinct mitochondrial stress response pathways appear to predominate in mammalian cells. Studies examining diverse mitochondrial stressors suggest ATF5 activation occurs under specific circumstances. Paraquat treatment and mutant OTC overexpression stimulate mitochondrial chaperone and protease transcription in an ATF5-dependent manner. While ATF5 appears responsive to paraquat and mutant OTC, studies reveal activating transcription factor 4 (ATF4) responds to numerous mitochondrial stressors. Quiros et al. [93] introduced mammalian cells to distinct mitochondrial stressors, including membrane depolarization, translation inhibition, OXPHOS inhibition and protein import suppression. Mitochondrial stressors failed to induce either mtUPR or ATF5 activation, instead stimulating ATF4 dependent stress response pathways. Numerous studies implicate ATF4 in the mitochondrial stress response. ATF4 orchestrates diverse metabolic changes to help cells cope with mitochondrial dysfunction. Abrogating ATF4 decreases cellular proliferation, especially following mitochondrial stress [93]. In line with these findings, another study finds mitochondrial OXPHOS inhibitors stimulate stress response genes via ATF4 induction [94]. ATF4 clearly responds to diverse mitochondrial stressors, leading researchers to examine how mitochondrial dysfunction activates ATF4. While this review will focus on ATF4 related stress responses, mitochondrial dysfunction stimulates diverse compensatory mechanisms.

7. Mitochondrial Dysfunction Triggers Numerous Retrograde Responses, Including the Integrated Stress Response (mtISR)

Mitochondrial dysfunction triggers numerous changes in nuclear gene expression [95,96]. Referred to as retrograde responses, mitochondrial stress responses preserve cell viability by modulating metabolic pathways and mitochondrial function. Studies in *Saccharomyces cerevisiae* have elucidated mitochondrial-nuclear communication pathways in great detail [97]. While fewer studies on mammalian retrograde responses exist, certain pathways consistently respond to mitochondrial stressors. Multiple mitochondrial stressors perturb cytosolic calcium (Ca^{2+}) and ROS levels, activating nuclear factor kappa B (NFκB). Although NFκB activation is canonically associated with immune system function, diverse cellular stressors, including mitochondrial dysfunction, activate NFκB [97]. Mitochondrial dysfunction activates NFκB in a manner distinct from cytokine mediated NFκB activation. Furthermore, NFκB may regulate c-Myc transcription, a transcription factor consistently upregulated by mitochondrial dysfunction [98,99]. c-Myc forms a Myc-Max heterodimer homologous to yeast retrograde response mediators [97]. NFκB and Myc activity increase in aged tissues and decrease during cell senescence [100]. Numerous cell signaling pathways participate in tightly orchestrated, context dependent retrograde responses. In yeast, retrograde responses facilitate replicative lifespan extension. Some groups speculate retrograde responses act similarly in mammalian cells to compensate for age-related mitochondrial deficits [101,102].

Many studies show that mitochondrial stress activates ATF4 signaling, suggesting ATF4 plays a role in retrograde signaling. ATF4 activation occurs through a pathway known as the integrated stress response (ISR). The ISR begins with eukaryotic initiation factor 2 alpha (eIF2α) phosphorylation [103]. Four kinases, heme-regulated inhibitor (HRI), protein kinase R (PKR), PKR-like endoplasmic reticulum kinase, (PERK) and general control non-depressible 2 (GCN2), phosphorylate eIF2α. Heme depletion, viral infection, endoplasmic reticulum stress, and amino acid starvation activate each kinase, respectively [104]. Studies implicate eIF2α phosphorylation in long term potentiation and long term memory through downstream effects on cyclic AMP responsive element binding protein (CREB), providing a potential link between the ISR and cognitive decline [105]. eIF2α phosphorylation triggers diverse cellular effects.

One of eIF2α phosphorylation's most important effects is to pause general protein translation, assisting in cellular stress recovery. However, eIF2α phosphorylation paradoxically increases protein translation from mRNAs possessing alternative open reading frames (ORFs). Numerous stress responsive factors contain alternative ORFs. Therefore, eIF2α phosphorylation reduces cell protein loads while preferentially increasing stress response factors [106]. ATF4 translation increases following eIF2α phosphorylation due to ATF4's alternative ORFs [107]. Increased ATF4 translation stimulates downstream target transcription, including C/EBP homologous protein (CHOP) [108]. CHOP induction favors cell cycle arrest and, upon chronic activation, apoptosis [109,110]. In summary, the ISR responds to numerous stressors by reducing general protein translation while upregulating stress responsive factors. Major ISR mediators include eIF2α, ATF4, and CHOP.

Several reports show mitochondrial dysfunction stimulates the ISR in mammalian cells. Rotenone treated oligodendroglia increase eIF2α phosphorylation as well as ATF4 and CHOP protein [111]. Earlier studies show CHOP mediates a mitochondrial specific stress response [112]. mtDNA depletion and doxycycline treatment in cell culture activate CHOP expression in an ATF4-dependent manner without concomitant mtUPR activation [113]. Multiple studies indicate mitochondrial protease inhibition specifically induces the ISR. Knocking out a mitochondrial serine protease, HtrA2, triggers the ISR in mouse brain [114]. Similarly, LON protease (LONP1) deficient cell lines exhibit mitochondrial protein aggregation along with ISR activation. LONP1 functions as an important mitochondrial matrix protease. Mitochondrial protein aggregates stemming from LONP1 depletion only modestly induce the mtUPR. The authors conclude LONP1 depletion prominently activates the ISR, while slightly increasing some mitochondrial proteases and chaperones [115].

Further studies show mitochondrial stress induces the ISR in muscle and brain tissue. Deletor mice possessing a dominant Twinkle (helicase involved in mtDNA replication) mutation model mitochondrial myopathy. Deletor mice rapidly accumulate mtDNA mutations leading to OXPHOS deficiency. OXPHOS deficiency in deletor mice triggers ISR components resulting in altered one carbon metabolism, serine synthesis, and glutathione production pathways (transulfuration) [116]. Quiros et al. note similar metabolic changes following mitochondrial dysfunction and ISR activation [93]. Furthermore, mammalian target of rapamycin complex 1 (mTORC1) inhibition in deletor mice rescues metabolic alterations by reducing ISR activity. In this model, mitochondrial dysfunction activates mTORC1 which subsequently activates the ISR [116].

Inducible Drp1 knockout in mouse neurons also stimulates the ISR. Drp1 knockout disrupts mitochondrial fission causing mitochondrial dysfunction and ISR activation. Drp1 knockout neurons increase fibroblast growth factor 21 (Fgf21) plasma protein and mRNA levels. A cytokine associated with mitochondrial myopathies, Fgf21 release increases upon mitochondrial dysfunction. Neuronal Drp1 knockout mouse studies show that brain mitochondrial dysfunction triggers Fgf21 release in an ISR dependent manner [117]. Some consider Fgf21 a mitokine, transmitting mitochondrial stress signals between organs [118]. ISR stimulation of Fgf21 expression further demonstrates a link between mitochondrial dysfunction and the ISR. However, the ISR responds to numerous cellular stressors. A mitochondrial stress-induced ISR refers to a unique ISR subgroup, a mitochondrial ISR (mtISR). Ample evidence of the mtISR exists, however, future research should examine signaling cascades stimulating the mtISR.

As referenced earlier, numerous signals could activate ISRs. Heme depletion, viral infection, endoplasmic reticulum stress, and amino acid starvation all stimulate eIF2α kinases [104]. Determining specific signals responsible for the mtISR may prove difficult. Few studies associate mtISRs with specific eIF2α kinases. One study finds doxycycline treatment increases eIF2α phosphorylation through GCN2, the amino acid starvation sensitive kinase [113]. However, Quiros et al. knocked down all four eIF2α kinases following mitochondrial depolarization and saw no reductions in eIF2α phosphorylation. The authors concluded multiple kinases increase eIF2α phosphorylation during the mtISR [93]. Determining whether a mtISR occurs in sporadic diseases such as AD remains difficult since we do not know specific mtISR signatures. Numerous stressors occur in AD brain which may feed into the ISR. For example, endoplasmic reticulum (ER) stress occurs in AD and is known to strongly induce the ISR. Changes in ER calcium levels and protein glycosylation as well as misfolded protein accumulation trigger ER stress leading to eIF2α phosphorylation and increased ATF4 and CHOP [119,120]. ER stress activates another unique ISR subgroup, an ER stress-induced ISR (erISR).

Mitochondria associate with the ER and assist in calcium maintenance, leading investigators to speculate whether mitochondrial dysfunction triggers ISR by causing ER stress. To determine whether mitochondrial dysfunction stimulates ER stress, studies examined classical ER stress markers not involved in the ISR. Studies found that mitochondrial dysfunction triggers the mtISR independently of general ER stress [93]. Although the mtISR does not appear to involve ER stress, these phenomena are not mutually exclusive (Figure 2). AD neurons exhibit increased ER stress markers concomitant with eIF2α phosphorylation. IHC studies show the ER stress markers, p-PERK and p-IRE1, increase in AD hippocampal neurons along with p-eIF2α. p-eIF2α and p-PERK antibodies stain similar granular structures in AD pyramidal hippocampal neurons, suggesting concomitant ER stress and ISR. While correlative, these findings suggest erISR occurs in AD [121]. Further work should attempt to determine whether a mtISR occurs in AD. To our knowledge, no articles discuss potential mitochondrial contributions to the observed AD ISR activation.

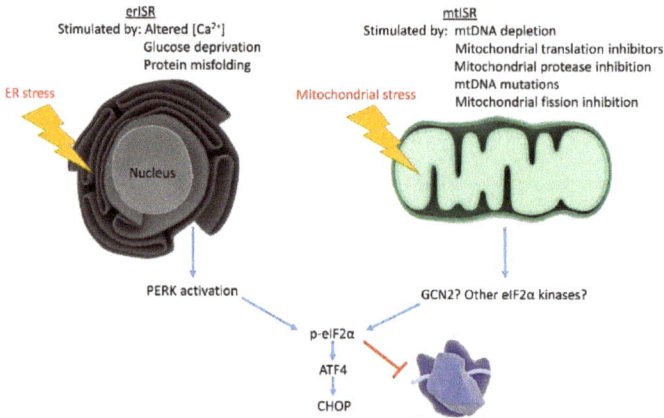

Figure 2. Mitochondrial stress and ER stress stimulate the ISR, which reduces general protein translation while upregulating stress responsive factors. ER stress activates the ISR through PERK, while mitochondrial stress may activate the ISR through multiple eIF2α kinases.

8. AD Activates the ISR: Could Mitochondrial Dysfunction Contribute?

Declining mitochondrial function during aging could theoretically stimulate retrograde responses [122]. AD brains display mitochondrial dysfunction beyond those observed with normal aging. Groups hypothesize that compensatory responses fail in AD once mitochondrial dysfunction passes a threshold, thus favoring disease progression [47]. In this hypothesized paradigm, beneficial compensatory responses decline during disease progression while potentially maladaptive stress responses predominate. Under certain conditions, ISR activation may represent a maladaptive response [123]. Maladaptive retrograde responses seem counterintuitive, however, inhibiting retrograde responses proves beneficial in multiple scenarios. For example, chronic ISR activation appears to favor cell death and inhibiting ISR activity proves beneficial in traumatic brain injury models [124]. Similarly, NFκB activation can favor apoptosis in cases of severe stress [125]. Researchers speculate that many retrograde responses may prove beneficial in the short term but become detrimental upon chronic activation [126].

Few studies focus on retrograde responses in the AD brain. In fact, many mammalian retrograde response studies use cancer cells which may respond differently to mitochondrial stress than postmitotic neurons. However, several lines of evidence suggest retrograde responses occur in neurons. *Drosophila melanogaster* models of neuronal mitochondrial dysfunction identify hypoxia inducible factor 1α, forkhead box O (FOXO) and ATF4 as key retrograde responders [127]. Human primary mitochondrial diseases often present with neurological deficits and cell lines carrying associated mtDNA mutations display retrograde responses [128]. Differentiated dopaminergic neurons treated with a complex I inhibitor upregulated ATF4 signaling pathways according to transcriptional profiling [129]. The diversity in retrograde responses makes it difficult to describe a canonical retrograde response in disease states, however, AD brains display changes in numerous factors implicated in mammalian retrograde responses.

Postmortem AD brains generally display NFκB activation and cyclic AMP response element binding protein (CREB) alterations, both of which can be affected by mitochondrial dysfunction and subsequent changes in Ca^{2+} concentration [130]. AD brains also possess decreased HIF1α and cortical SIRT1 levels [131,132]. The AD parietal lobe also displays activated Akt/mTOR as well as increases in downstream targets [133]. All of these factors can participate in retrograde responses, however, they all respond to diverse stimuli. AD mitochondrial dysfunction could activate these response or, alternatively, deficits in these responses could make cells more vulnerable to mitochondrial dysfunction.

Currently, there is not enough evidence to support strong conclusions regarding the nature and function of retrograde responses in AD.

Another hypothetical consequence of AD mitochondrial dysfunction is ISR activation. Post-mortem AD brains display ISR activation. Given the well documented mitochondrial dysfunction in AD and evidence suggesting mitochondrial dysfunction stimulates the mtISR, it seems possible a mtISR occurs in AD brain. Numerous studies describe increased eIF2α phosphorylation in AD brains, particularly in hippocampal neurons [134,135] ATF4 protein levels increase in AD frontal cortex and increased ATF4 correlates fairly well with increased p-eIF2α [135]. IHC studies reveal increases in ATF4 positive cells in AD entorhinal cortex and subiculum, but decreases in the hippocampus [136]. CHOP protein also increases in AD cortex [137]. AD brains display defects in ribosome function and protein translation, although the ISR's role in these deficits remains unclear [138]. Whether the ISR activation in AD stems largely from ER stress or mitochondrial dysfunction remains unknown.

Several groups propose amyloid beta and tau alterations trigger the ISR. In embryonic rat hippocampal cultures, amyloid beta oligomer treatment induces axonal ATF4 and CHOP synthesis. ATF4 siRNA desensitizes rat hippocampal cultures to amyloid beta's negative effects, suggesting ATF4 potentiates amyloid beta toxicity [136]. However, aged Tg2576 mice, which accumulate amyloid plaques, do not display CHOP induction [137]. Colocalization experiments with ER stress markers and tau antibodies reveal a correlation between p-PERK staining and pretangle neurons containing hyperphosphorylated tau. Interestingly, neurons decorated with NFTs rarely display p-PERK staining, suggesting ER stress markers appear early in disease progression [139]. ER stress and mitochondrial dysfunction independently activate the ISR. Both ER stress and mitochondrial dysfunction occur in AD. Therefore, it seems reasonable that the mtISR may occur in AD alongside the erISR. While evidence suggests the possibility of a mtISR in AD, more work is needed to support this hypothetical relationship. If AD mitochondrial dysfunction triggers disease relevant retrograde responses, then therapeutic approaches should consider whether the compensatory mechanisms represent beneficial or maladaptive responses.

9. Conclusions

Mitochondrial dysfunction occurs in AD. Mitochondrial and metabolic abnormalities present early in disease progression. Systemic AD metabolic changes may prove useful diagnostically, and mitochondrial dysfunction seems to be a reasonable therapeutic target. Mitochondria-specific stress responses help cells cope with mitochondrial dysfunction. Certain mitochondrial stress response components are activated in AD, although to what extent these stress responses contribute to retarding or promoting AD progression remains unclear.

Author Contributions: I.W. and R.H.S. wrote this manuscript.

Funding: This research was funded by the National Institute on Aging, grant number P30AG035982 (R.H.S.), the University of Kansas Madison and Lila Self Graduate Fellowship (I.W.), and a Mabel Woodyard Fellowship from the University of Kansas Medical Center Institute for Neurological Discoveries (I.W.).

Acknowledgments: Thanks to the Library of Science and Medical Illustrations for illustrations that were incorporated into the figures. Free access to the Library of Science and Medical Illustrations can be found at http://www.somersault1824.com/science-illustrations/.

Conflicts of Interest: The authors declare no conflict of interest.

References

1. Onyango, I.G.; Dennis, J.; Khan, S.M. Mitochondrial Dysfunction in Alzheimer's Disease and the Rationale for Bioenergetics Based Therapies. *Aging Dis.* **2016**, *7*, 201–214. [CrossRef]
2. Lin, M.T.; Beal, M.F. Mitochondrial dysfunction and oxidative stress in neurodegenerative diseases. *Nature* **2006**, *443*, 787. [CrossRef] [PubMed]
3. Wang, X.; Wang, W.; Li, L.; Perry, G.; Lee, H.G.; Zhu, X. Oxidative stress and mitochondrial dysfunction in Alzheimer's disease. *Biochim. Biophys. Acta* **2014**, *1842*, 1240–1247. [CrossRef]

4. Parker, W.D., Jr.; Filley, C.M.; Parks, J.K. Cytochrome oxidase deficiency in Alzheimer's disease. *Neurology* **1990**, *40*, 1302–1303. [CrossRef]
5. Coskun, P.E.; Wyrembak, J.; Derbereva, O.; Melkonian, G.; Doran, E.; Lott, I.T.; Head, E.; Cotman, C.W.; Wallace, D.C. Systemic mitochondrial dysfunction and the etiology of Alzheimer's disease and down syndrome dementia. *J. Alzheimers. Dis.* **2010**, *20*, S293–S310. [CrossRef]
6. Mosconi, L. Brain glucose metabolism in the early and specific diagnosis of Alzheimer's disease. *Eur. J. Nucl. Med. Mol. Imaging* **2005**, *32*, 486–510. [CrossRef] [PubMed]
7. Rapoport, S.I. Positron emission tomography in Alzheimer's disease in relation to disease pathogenesis: A critical review. *Cerebrovasc. Brain Metab. Rev.* **1991**, *3*, 297–335.
8. Drzezga, A.; Lautenschlager, N.; Siebner, H.; Riemenschneider, M.; Willoch, F.; Minoshima, S.; Schwaiger, M.; Kurz, A. Cerebral metabolic changes accompanying conversion of mild cognitive impairment into Alzheimer's disease: A PET follow-up study. *Eur. J. Nucl. Med. Mol. Imaging* **2003**, *30*, 1104–1113.
9. Friedland, R.P.; Budinger, T.F.; Ganz, E.; Yano, Y.; Mathis, C.A.; Koss, B.; Ober, B.A.; Huesman, R.H.; Derenzo, S.E. Regional cerebral metabolic alterations in dementia of the Alzheimer type: positron emission tomography with [18F]fluorodeoxyglucose. *J. Comput. Assist. Tomogr.* **1983**, *7*, 590–598. [CrossRef] [PubMed]
10. Small, G.W.; Kuhl, D.E.; Riege, W.H.; Fujikawa, D.G.; Ashford, J.W.; Metter, E.J.; Mazziotta, J.C. Cerebral glucose metabolic patterns in Alzheimer's disease. Effect of gender and age at dementia onset. *Arch. Gen. Psychiatry* **1989**, *46*, 527–532. [CrossRef] [PubMed]
11. Mosconi, L. Glucose metabolism in normal aging and Alzheimer's disease: Methodological and physiological considerations for PET studies. *Clin. Transl. Imaging* **2013**, *1*, 217–233. [CrossRef] [PubMed]
12. Grady, C.L.; Haxby, J.V.; Schlageter, N.L.; Berg, G.; Rapoport, S.I. Stability of metabolic and neuropsychological asymmetries in dementia of the Alzheimer type. *Neurology* **1986**, *36*, 1390–1392. [CrossRef] [PubMed]
13. Minoshima, S.; Giordani, B.; Berent, S.; Frey, K.A.; Foster, N.L.; Kuhl, D.E. Metabolic reduction in the posterior cingulate cortex in very early Alzheimer's disease. *Ann. Neurol.* **1997**, *42*, 85–94. [CrossRef] [PubMed]
14. Smith, G.S.; de Leon, M.J.; George, A.E.; Kluger, A.; Volkow, N.D.; McRae, T.; Golomb, J.; Ferris, S.H.; Reisberg, B.; Ciaravino, J.; et al. Topography of cross-sectional and longitudinal glucose metabolic deficits in Alzheimer's disease: Pathophysiologic implications. *Arch. Neurol.* **1992**, *49*, 1142–1150. [CrossRef]
15. Hoffman, J.M.; Welsh-Bohmer, K.A.; Hanson, M.; Crain, B.; Hulette, C.; Earl, N.; Coleman, R.E. FDG PET imaging in patients with pathologically verified dementia. *J. Nucl. Med.* **2000**, *41*, 1920–1928.
16. Giannakopoulos, P.; Herrmann, F.R.; Bussiere, T.; Bouras, C.; Kovari, E.; Perl, D.P.; Morrison, J.H.; Gold, G.; Hof, P.R. Tangle and neuron numbers, but not amyloid load, predict cognitive status in Alzheimer's disease. *Neurology* **2003**, *60*, 1495–1500. [CrossRef]
17. Salmon, E.; Sadzot, B.; Maquet, P.; Degueldre, C.; Lemaire, C.; Rigo, P.; Comar, D.; Franck, G. Differential diagnosis of Alzheimer's disease with PET. *J. Nucl. Med.* **1994**, *35*, 391–398.
18. Marcus, C.; Mena, E.; Subramaniam, R.M. Brain PET in the diagnosis of Alzheimer's disease. *Clin. Nucl. Med.* **2014**, *39*, e413–e426. [CrossRef]
19. Ishii, K. Clinical application of positron emission tomography for diagnosis of dementia. *Ann. Nucl. Med.* **2002**, *16*, 515–525. [CrossRef]
20. Silverman, D.H.; Small, G.W.; Chang, C.Y.; Lu, C.S.; Kung De Aburto, M.A.; Chen, W.; Czernin, J.; Rapoport, S.I.; Pietrini, P.; Alexander, G.E.; et al. Positron emission tomography in evaluation of dementia: Regional brain metabolism and long-term outcome. *Jama* **2001**, *286*, 2120–2127. [CrossRef]
21. de Leon, M.J.; Convit, A.; Wolf, O.T.; Tarshish, C.Y.; DeSanti, S.; Rusinek, H.; Tsui, W.; Kandil, E.; Scherer, A.J.; Roche, A.; et al. Prediction of cognitive decline in normal elderly subjects with 2-[(18)F]fluoro-2-deoxy-D-glucose/poitron-emission tomography (FDG/PET). *Proc. Natl. Acad. Sci. USA* **2001**, *98*, 10966–10971. [CrossRef]
22. Mosconi, L.; De Santi, S.; Li, J.; Tsui, W.H.; Li, Y.; Boppana, M.; Laska, E.; Rusinek, H.; de Leon, M.J. Hippocampal hypometabolism predicts cognitive decline from normal aging. *Neurobiol. Aging* **2008**, *29*, 676–692. [CrossRef]
23. Mosconi, L.; Tsui, W.H.; Herholz, K.; Pupi, A.; Drzezga, A.; Lucignani, G.; Reiman, E.M.; Holthoff, V.; Kalbe, E.; Sorbi, S.; et al. Multicenter standardized 18F-FDG PET diagnosis of mild cognitive impairment, Alzheimer's disease, and other dementias. *J. Nucl. Med.* **2008**, *49*, 390–398. [CrossRef]
24. Vannucci, S.J.; Maher, F.; Simpson, I.A. Glucose transporter proteins in brain: Delivery of glucose to neurons and glia. *Glia* **1997**, *21*, 2–21. [CrossRef]

25. Simpson, I.A.; Chundu, K.R.; Davies-Hill, T.; Honer, W.G.; Davies, P. Decreased concentrations of GLUT1 and GLUT3 glucose transporters in the brains of patients with Alzheimer's disease. *Ann. Neurol.* **1994**, *35*, 546–551. [CrossRef]
26. Kalaria, R.N.; Harik, S.I. Reduced glucose transporter at the blood-brain barrier and in cerebral cortex in Alzheimer disease. *J. Neurochem.* **1989**, *53*, 1083–1088. [CrossRef]
27. Parker, W.D., Jr.; Parks, J.K. Cytochrome c oxidase in Alzheimer's disease brain: Purification and characterization. *Neurology* **1995**, *45 (Pt 1)*, 482–486. [CrossRef]
28. Parker, W.D., Jr. Cytochrome oxidase deficiency in Alzheimer's disease. *Ann. N. Y. Acad. Sci.* **1991**, *640*, 59–64. [CrossRef]
29. Mutisya, E.M.; Bowling, A.C.; Beal, M.F. Cortical Cytochrome Oxidase Activity Is Reduced in Alzheimer's Disease. *J. Neurochem.* **1994**, *63*, 2179–2184. [CrossRef]
30. Bubber, P.; Haroutunian, V.; Fisch, G.; Blass, J.P.; Gibson, G.E. Mitochondrial abnormalities in Alzheimer brain: mechanistic implications. *Ann. Neurol.* **2005**, *57*, 695–703. [CrossRef]
31. Ojaimi, J.; Masters, C.L.; McLean, C.; Opeskin, K.; McKelvie, P.; Byrne, E. Irregular distribution of cytochrome c oxidase protein subunits in aging and Alzheimer's disease. *Ann. Neurol.* **1999**, *46*, 656–660. [CrossRef]
32. Cottrell, D.A.; Blakely, E.L.; Johnson, M.A.; Ince, PG.; Turnbull, D.M. Mitochondrial enzyme-deficient hippocampal neurons and choroidal cells in AD. *Neurology* **2001**, *57*, 260–264. [CrossRef]
33. Cottrell, D.A.; Borthwick, G.M.; Johnson, M.A.; Ince, P.G.; Turnbull, D.M. The role of cytochrome c oxidase deficient hippocampal neurones in Alzheimer's disease. *Neuropathol. Appl. Neurobiol.* **2002**, *28*, 390–396. [CrossRef]
34. Hirai, K.; Aliev, G.; Nunomura, A.; Fujioka, H.; Russell, R.L.; Atwood, C.S.; Johnson, A.B.; Kress, Y.; Vinters, H.V.; Tabaton, M.; et al. Mitochondrial abnormalities in Alzheimer's disease. *J. Neurosci.* **2001**, *21*, 3017–3023. [CrossRef]
35. Cataldo, A.M.; Barnett, J.L.; Berman, S.A.; Li, J.; Quarless, S.; Bursztajn, S.; Lippa, C.; Nixon, R.A. Gene expression and cellular content of cathepsin D in Alzheimer's disease brain: evidence for early up-regulation of the endosomal-lysosomal system. *Neuron* **1995**, *14*, 671–680. [CrossRef]
36. Nixon, R.A.; Wegiel, J.; Kumar, A.; Yu, W.H.; Peterhoff, C.; Cataldo, A.; Cuervo, A.M. Extensive involvement of autophagy in Alzheimer disease: An immuno-electron microscopy study. *J. Neuropathol. Exp. Neurol.* **2005**, *64*, 113–122. [CrossRef]
37. Ding, W.-X.; Yin, X.-M. Mitophagy: Mechanisms, pathophysiological roles, and analysis. *Biol. Chem.* **2012**, *393*, 547–564. [CrossRef]
38. Shi, R.; Guberman, M.; Kirshenbaum, L.A. Mitochondrial quality control: The role of mitophagy in aging. *Trends Cardiovasc. Med.* **2018**, *28*, 246–260. [CrossRef]
39. Baloyannis, S.J.; Costa, V.; Michmizos, D. Mitochondrial alterations Alzheimer's disease. *Am. J. Alzheimer's Dis. Other Dement.* **2004**, *19*, 89–93. [CrossRef]
40. Zhang, L.; Trushin, S.; Christensen, T.A.; Bachmeier, B.V.; Gateno, B.; Schroeder, A.; Yao, J.; Itoh, K.; Sesaki, H.; Poon, W.W.; et al. Altered brain energetics induces mitochondrial fission arrest in Alzheimer's Disease. *Sci. Rep.* **2016**, *6*, 18725. [CrossRef]
41. Wang, X.; Su, B.; Lee, H.G.; Li, X.; Perry, G.; Smith, M.A.; Zhu, X. Impaired balance of mitochondrial fission and fusion in Alzheimer's disease. *J. Neurosci.* **2009**, *29*, 9090–9103. [CrossRef]
42. Reddy, P.H.; Yin, X.; Manczak, M.; Kumar, S.; Pradeepkiran, J.A.; Vijayan, M.; Reddy, A.P. Mutant APP and amyloid beta-induced defective autophagy, mitophagy, mitochondrial structural and functional changes and synaptic damage in hippocampal neurons from Alzheimer's disease. *Hum. Mol. Genet.* **2018**, *27*, 2502–2516. [CrossRef]
43. Kim, D.I.; Lee, K.H.; Gabr, A.A.; Choi, G.E.; Kim, J.S.; Ko, S.H.; Han, H.J. Abeta-Induced Drp1 phosphorylation through Akt activation promotes excessive mitochondrial fission leading to neuronal apoptosis. *Biochim. Biophys. Acta* **2016**, *1863*, 2820–2834. [CrossRef]
44. Manczak, M.; Kandimalla, R.; Yin, X.; Reddy, P.H. Mitochondrial division inhibitor 1 reduces dynamin-related protein 1 and mitochondrial fission activity. *Hum. Mol. Genet.* **2019**, *28*, 177–199. [CrossRef]
45. Park, J.; Choi, H.; Min, J.S.; Kim, B.; Lee, S.R.; Yun, J.W.; Choi, M.S.; Chang, K.T.; Lee, D.S. Loss of mitofusin 2 links beta-amyloid-mediated mitochondrial fragmentation and Cdk5-induced oxidative stress in neuron cells. *J. Neurochem.* **2015**, *132*, 687–702. [CrossRef] [PubMed]

46. Wang, W.; Yin, J.; Ma, X.; Zhao, F.; Siedlak, S.L.; Wang, Z.; Torres, S.; Fujioka, H.; Xu, Y.; Perry, G.; et al. Inhibition of mitochondrial fragmentation protects against Alzheimer's disease in rodent model. *Hum. Mol. Genet.* **2017**, *26*, 4118–4131. [CrossRef] [PubMed]
47. Swerdlow, R.H.; Burns, J.M.; Khan, S.M. The Alzheimer's disease mitochondrial cascade hypothesis. *J. Alzheimer's Dis.* **2010**, *20* (Suppl. 2), S265–S279. [CrossRef]
48. Moreira, P.I.; Carvalho, C.; Zhu, X.; Smith, M.A.; Perry, G. Mitochondrial dysfunction is a trigger of Alzheimer's disease pathophysiology. *Biochim. Biophys. Acta* **2010**, *1802*, 2–10. [CrossRef]
49. Beal, M.F.; Hyman, B.T.; Koroshetz, W. Do defecs in mitochondrial energy metabolism underlie the pathology of neurodegenerative diseases? *Trends Neurosci.* **1993**, *16*, 125–131. [CrossRef]
50. Swerdlow, R.H.; Khan, S.M. A "mitochondrial cascade hypothesis" for sporadic Alzheimer's disease. *Med. Hypotheses* **2004**, *63*, 8–20. [CrossRef] [PubMed]
51. Reddy, P.H. Abnormal tau, mitochondrial dysfunction, impaired axonal transport of mitochondria, and synaptic deprivation in Alzheimer's disease. *Brain Res.* **2011**, *1415*, 136–148. [CrossRef]
52. Caspersen, C.; Wang, N.; Yao, J.; Sosunov, A.; Chen, X.; Lustbader, J.W.; Xu, H.W.; Stern, D.; McKhann, G.; Yan, S.D. Mitochondrial Abeta: a potential focal point for neuronal metabolic dysfunction in Alzheimer's disease. *Faseb J.* **2005**, *19*, 2040–2041. [CrossRef]
53. Devi, L.; Prabhu, B.M.; Galati, D.F.; Avadhani, N.G.; Anandatheerthavarada, H.K. Accumulation of amyloid precursor protein in the mitochondrial import channels of human Alzheimer's disease brain is associated with mitochondrial dysfunction. *J. Neurosci.* **2006**, *26*, 9057–9068. [CrossRef] [PubMed]
54. Corral-Debrinski, M.; Horton, T.; Lott, M.T.; Shoffner, J.M.; Beal, M.F.; Wallace, D.C. Mitochondrial DNA deletions in human brain: regional variability and increase with advanced age. *Nat. Genet.* **1992**, *2*, 324–329. [CrossRef] [PubMed]
55. Corral-Debrinski, M.; Horton, T.; Lott, M.T.; Shoffner, J.M.; McKee, A.C.; Beal, M.F.; Graham, B.H.; Wallace, D.C. Marked changes in mitochondrial DNA deletion levels in Alzheimer brains. *Genomics* **1994**, *23*, 471–476. [CrossRef]
56. Hamblet, N.S.; Castora, F.J. Elevated levels of the Kearns–Sayre syndrome mitochondrial DNA deletion in temporal cortex of Alzheimer's patients. *Mutat. Res./Fundam. Mol. Mech. Mutagenes.* **1997**, *379*, 253–262. [CrossRef]
57. Mecocci, P.; MacGarvey, U.; Beal, M.F. Oxidative damage to mitochondrial DNA is increased in Alzheimer's disease. *Ann. Neurol.* **1994**, *36*, 747–751. [CrossRef]
58. Coskun, P.E.; Beal, M.F.; Wallace, D.C. Alzheimer's brains harbor somatic mtDNA control-region mutations that suppress mitochondrial transcription and replication. *Proc. Natl. Acad. Sci. USA* **2004**, *101*, 10726–10731. [CrossRef]
59. Hoekstra, J.G.; Hipp, M.J.; Montine, T.J.; Kennedy, S.R. Mitochondrial DNA mutations increase in early stage Alzheimer disease and are inconsistent with oxidative damage. *Ann. Neurol.* **2016**, *80*, 301–306. [CrossRef] [PubMed]
60. Edland, S.D.; Silverman, J.M.; Peskind, E.R.; Tsuang, D.; Wijsman, E.; Morris, J.C. Increased risk of dementia in mothers of Alzheimer's disease cases. *Neurology* **1996**, *47*, 254. [CrossRef]
61. Mosconi, L.; Mistur, R.; Switalski, R.; Brys, M.; Glodzik, L.; Rich, K.; Pirraglia, E.; Tsui, W.; De Santi, S.; de Leon, M.J. Declining brain glucose metabolism in normal individuals with a maternal history of Alzheimer disease. *Neurology* **2009**, *72*, 513–520. [CrossRef] [PubMed]
62. Ghosh, S.S.; Swerdlow, R.H.; Miller, S.W.; Sheeman, B.; Parker, W.D., Jr.; Davis, R.E. Use of cytoplasmic hybrid cell lines for elucidating the role of mitochondrial dysfunction in Alzheimer's disease and Parkinson's disease. *Ann. NY Acad. Sci.* **1999**, *893*, 176–191. [CrossRef]
63. Swerdlow, R.H.; Parks, J.K.; Cassarino, D.S.; Maguire, D.J.; Maguire, R.S.; Bennett, J.P., Jr.; Davis, R.E.; Parker, W.D., Jr. Cybrids in Alzheimer's disease: a cellular model of the disease? *Neurology* **1997**, *49*, 918–925. [CrossRef] [PubMed]
64. Krishnan, K.J.; Ratnaike, T.E.; De Gruyter, H.L.; Jaros, E.; Turnbull, D.M. Mitochondrial DNA deletions cause the biochemical defect observed in Alzheimer's disease. *Neurobiol. Aging* **2012**, *33*, 2210–2214. [CrossRef] [PubMed]
65. Trimmer, P.A.; Swerdlow, R.H.; Parks, J.K.; Keeney, P.; Bennett, J.P., Jr.; Miller, S.W.; Davis, R.E.; Parker, W.D., Jr. Abnormal mitochondrial morphology in sporadic Parkinson's and Alzheimer's disease cybrid cell lines. *Exp. Neurol.* **2000**, *162*, 37–50. [CrossRef] [PubMed]

66. Khan, S.M.; Cassarino, D.S.; Abramova, N.N.; Keeney, P.M.; Borland, M.K.; Trimmer, P.A.; Krebs, C.T.; Bennett, J.C.; Parks, J.K.; Swerdlow, R.H.; et al. Alzheimer's disease cybrids replicate beta-amyloid abnormalities through cell death pathways. *Ann. Neurol.* **2000**, *48*, 148–155. [CrossRef]
67. Cardoso, S.M.; Santana, I.; Swerdlow, R.H.; Oliveira, C.R. Mitochondria dysfunction of Alzheimer's disease cybrids enhances Abeta toxicity. *J. Neurochem.* **2004**, *89*, 1417–1426. [CrossRef] [PubMed]
68. Blass, J.P.; Baker, A.C.; Ko, L.; Black, R.S. Induction of Alzheimer antigens by an uncoupler of oxidative phosphorylation. *Arch. Neurol.* **1990**, *47*, 864–869. [CrossRef]
69. Hoglinger, G.U.; Lannuzel, A.; Khondiker, M.E.; Michel, P.P.; Duyckaerts, C.; Feger, J.; Champy, P.; Prigent, A.; Medja, F.; Lombes, A.; et al. The mitochondrial complex I inhibitor rotenone triggers a cerebral tauopathy. *J. Neurochem.* **2005**, *95*, 930–939. [CrossRef]
70. Yao, J.; Irwin, R.W.; Zhao, L.; Nilsen, J.; Hamilton, R.T.; Brinton, R.D. Mitochondrial bioenergetic deficit precedes Alzheimer's pathology in female mouse model of Alzheimer's disease. *Proc. Natl. Acad. Sci. USA* **2009**, *106*, 14670–14675. [CrossRef]
71. Reddy, P.H.; McWeeney, S.; Park, B.S.; Manczak, M.; Gutala, R.V.; Partovi, D.; Jung, Y.; Yau, V.; Searles, R.; Mori, M.; et al. Gene expression profiles of transcripts in amyloid precursor protein transgenic mice: up-regulation of mitochondrial metabolism and apoptotic genes is an early cellular change in Alzheimer's disease. *Hum. Mol. Genet.* **2004**, *13*, 1225–1240. [CrossRef]
72. Fukui, H.; Diaz, F.; Garcia, S.; Moraes, C.T. Cytochrome c oxidase deficiency in neurons decreases both oxidative stress and amyloid formation in a mouse model of Alzheimer's disease. *Proc. Natl. Acad. Sci. USA* **2007**, *104*, 14163–14168. [CrossRef]
73. Pereira, C.; Santos, M.S.; Oliveira, C. Mitochondrial function impairment induced by amyloid beta-peptide on PC12 cells. *Neuroreport* **1998**, *9*, 1749–1755. [CrossRef]
74. Casley, C.S.; Canevari, L.; Land, J.M.; Clark, J.B.; Sharpe, M.A. Beta-amyloid inhibits integrated mitochondrial respiration and key enzyme activities. *J. Neurochem.* **2002**, *80*, 91–100. [CrossRef]
75. Manczak, M.; Anekonda, T.S.; Henson, E.; Park, B.S.; Quinn, J.; Reddy, P.H. Mitochondria are a direct site of A beta accumulation in Alzheimer's disease neurons: implications for free radical generation and oxidative damage in disease progression. *Hum. Mol. Genet.* **2006**, *15*, 1437–1449. [CrossRef]
76. Lustbader, J.W.; Cirilli, M.; Lin, C.; Xu, H.W.; Takuma, K.; Wang, N.; Caspersen, C.; Chen, X.; Pollak, S.; Chaney, M.; et al. ABAD directly links Abeta to mitochondrial toxicity in Alzheimer's disease. *Science* **2004**, *304*, 448–452. [CrossRef]
77. Pickett, E.K.; Rose, J.; McCrory, C.; McKenzie, C.A.; King, D.; Smith, C.; Gillingwater, T.H.; Henstridge, C.M.; Spires-Jones, T.L. Region-specific depletion of synaptic mitochondria in the brains of patients with Alzheimer's disease. *Acta Neuropathol.* **2018**, *136*, 747–757. [CrossRef]
78. Manczak, M.; Reddy, P.H. Abnormal interaction of VDAC1 with amyloid beta and phosphorylated tau causes mitochondrial dysfunction in Alzheimer's disease. *Hum. Mol. Genet.* **2012**, *21*, 5131–5146. [CrossRef]
79. Derisbourg, M.; Leghay, C.; Chiappetta, G.; Fernandez-Gomez, F.J.; Laurent, C.; Demeyer, D.; Carrier, S.; Buee-Scherrer, V.; Blum, D.; Vinh, J.; et al. Role of the Tau N-terminal region in microtubule stabilization revealed by new endogenous truncated forms. *Sci. Rep.* **2015**, *5*, 9659. [CrossRef]
80. Guillozet-Bongaarts, A.L.; Garcia-Sierra, F.; Reynolds, M.R.; Horowitz, P.M.; Fu, Y.; Wang, T.; Cahill, M.E.; Bigio, E.H.; Berry, R.W.; Binder, L.I. Tau truncation during neurofibrillary tangle evolution in Alzheimer's disease. *Neurobiol. Aging* **2005**, *26*, 1015–1022. [CrossRef]
81. Atlante, A.; Amadoro, G.; Bobba, A.; de Bari, L.; Corsetti, V.; Pappalardo, G.; Marra, E.; Calissano, P.; Passarella, S. A peptide containing residues 26-44 of tau protein impairs mitochondrial oxidative phosphorylation acting at the level of the adenine nucleotide translocator. *Biochim. Biophys. Acta* **2008**, *1777*, 1289–1300. [CrossRef]
82. Quintanilla, R.A.; Matthews-Roberson, T.A.; Dolan, P.J.; Johnson, G.V. Caspase-cleaved tau expression induces mitochondrial dysfunction in immortalized cortical neurons: implications for the pathogenesis of Alzheimer disease. *J. Biol. Chem.* **2009**, *284*, 18754–18766. [CrossRef]
83. Heeren, J.; Grewal, T.; Laatsch, A.; Becker, N.; Rinninger, F.; Rye, K.A.; Beisiegel, U. Impaired recycling of apolipoprotein E4 is associated with intracellular cholesterol accumulation. *J. Biol. Chem.* **2004**, *279*, 55483–55492. [CrossRef]

84. Huang, Y.; Liu, X.Q.; Wyss-Coray, T.; Brecht, W.J.; Sanan, D.A.; Mahley, R.W. Apolipoprotein E fragments present in Alzheimer's disease brains induce neurofibrillary tangle-like intracellular inclusions in neurons. *Proc. Natl. Acad. Sci. USA* **2001**, *98*, 8838–8843. [CrossRef]
85. Nakamura, T.; Watanabe, A.; Fujino, T.; Hosono, T.; Michikawa, M. Apolipoprotein E4 (1-272) fragment is associated with mitochondrial proteins and affects mitochondrial function in neuronal cells. *Mol. Neurodegener.* **2009**, *4*, 35. [CrossRef]
86. Ruan, L.; Zhou, C.; Jin, E.; Kucharavy, A.; Zhang, Y.; Wen, Z.; Florens, L.; Li, R. Cytosolic proteostasis through importing of misfolded proteins into mitochondria. *Nature* **2017**, *543*, 443–446. [CrossRef]
87. Du, F.; Yu, Q.; Yan, S.; Hu, G.; Lue, L.F.; Walker, D.G.; Wu, L.; Yan, S.F.; Tieu, K.; Yan, S.S. PINK1 signalling rescues amyloid pathology and mitochondrial dysfunction in Alzheimer's disease. *Brain* **2017**, *140*, 3233–3251. [CrossRef]
88. Burman, J.L.; Pickles, S.; Wang, C.; Sekine, S.; Vargas, J.N.S.; Zhang, Z.; Youle, A.M.; Nezich, C.L.; Wu, X.; Hammer, J.A.; et al. Mitochondrial fission facilitates the selective mitophagy of protein aggregates. *J. Cell Biol.* **2017**, *216*, 3231. [CrossRef]
89. Horibe, T.; Hoogenraad, N.J. The chop gene contains an element for the positive regulation of the mitochondrial unfolded protein response. *PLoS ONE* **2007**, *2*, e835. [CrossRef]
90. Yoneda, T.; Benedetti, C.; Urano, F.; Clark, S.G.; Harding, H.P.; Ron, D. Compartment-specific perturbation of protein handling activates genes encoding mitochondrial chaperones. *J. Cell Sci.* **2004**, *117*, 4055–4066. [CrossRef]
91. Houtkooper, R.H.; Mouchiroud, L.; Ryu, D.; Moullan, N.; Katsyuba, E.; Knott, G.; Williams, R.W.; Auwerx, J. Mitonuclear protein imbalance as a conserved longevity mechanism. *Nature* **2013**, *497*, 451–457. [CrossRef] [PubMed]
92. Bennett, C.F.; Vander Wende, H.; Simko, M.; Klum, S.; Barfield, S.; Choi, H.; Pineda, V.V.; Kaeberlein, M. Activation of the mitochondrial unfolded protein response does not predict longevity in Caenorhabditis elegans. *Nat. Commun.* **2014**, *5*, 3483. [CrossRef]
93. Quirós, P.M.; Prado, M.A.; Zamboni, N.; D'Amico, D.; Williams, R.W.; Finley, D.; Gygi, S.P.; Auwerx, J. Multi-omics analysis identifies ATF4 as a key regulator of the mitochondrial stress response in mammals. *J. Cell Biol.* **2017**, *216*, 2027. [CrossRef] [PubMed]
94. Garaeva, A.A.; Kovaleva, I.E.; Chumakov, P.M.; Evstafieva, A.G. Mitochondrial dysfunction induces SESN2 gene expression through Activating Transcription Factor 4. *Cell Cycle* **2016**, *15*, 64–71. [CrossRef] [PubMed]
95. Epstein, C.B.; Waddle, J.A.; Hale, W., IV; Davé, V.; Thornton, J.; Macatee, T.L.; Garner, H.R.; Butow, R.A. Genome-wide responses to mitochondrial dysfunction. *Mol. Biol. Cell* **2001**, *12*, 297–308. [CrossRef]
96. Butow, R.A.; Avadhani, N.G. Mitochondrial Signaling: The Retrograde Response. *Mol. Cell* **2004**, *14*, 1–15. [CrossRef]
97. Jazwinski, S.M.; Kriete, A. The Yeast Retrograde Response as a Model of Intracellular Signaling of Mitochondrial Dysfunction. *Front. Physiol.* **2012**, *3*, 139. [CrossRef]
98. Biswas, G.; Anandatheerthavarada, H.K.; Zaidi, M.; Avadhani, N.G. Mitochondria to nucleus stress signaling: a distinctive mechanism of NFkappaB/Rel activation through calcineurin-mediated inactivation of IkappaBbeta. *J. Cell Biol.* **2003**, *161*, 507–519. [CrossRef]
99. Srinivasan, V.; Kriete, A.; Sacan, A.; Jazwinski, S.M. Comparing the yeast retrograde response and NF-kappaB stress responses: implications for aging. *Aging Cell* **2010**, *9*, 933–941. [CrossRef]
100. Tilstra, J.S.; Clauson, C.L.; Niedernhofer, L.J.; Robbins, P.D. NF-kappaB in Aging and Disease. *Aging Dis.* **2011**, *2*, 449–465.
101. Kirchman, P.A.; Kim, S.; Lai, C.Y.; Jazwinski, S.M. Interorganelle signaling is a determinant of longevity in Saccharomyces cerevisiae. *Genetics* **1999**, *152*, 179–190. [PubMed]
102. Traven, A.; Wong, J.M.; Xu, D.; Sopta, M.; Ingles, C.J. Interorganellar communication. Altered nuclear gene expression profiles in a yeast mitochondrial DNA mutant. *J. Biol. Chem.* **2001**, *276*, 4020–4027. [CrossRef] [PubMed]
103. Harding, H.P.; Zhang, Y.; Zeng, H.; Novoa, I.; Lu, P.D.; Calfon, M.; Sadri, N.; Yun, C.; Popko, B.; Paules, R.; et al. An integrated stress response regulates amino acid metabolism and resistance to oxidative stress. *Mol. Cell* **2003**, *11*, 619–633. [CrossRef]
104. Taniuchi, S.; Miyake, M.; Tsugawa, K.; Oyadomari, M.; Oyadomari, S. Integrated stress response of vertebrates is regulated by four eIF2alpha kinases. *Sci. Rep.* **2016**, *6*, 32886. [CrossRef]

105. Costa-Mattioli, M.; Gobert, D.; Stern, E.; Gamache, K.; Colina, R.; Cuello, C.; Sossin, W.; Kaufman, R.; Pelletier, J.; Rosenblum, K.; et al. eIF2alpha phosphorylation bidirectionally regulates the switch from short- to long-term synaptic plasticity and memory. *Cell* **2007**, *129*, 195–206. [CrossRef] [PubMed]
106. Harding, H.P.; Novoa, I.; Zhang, Y.; Zeng, H.; Wek, R.; Schapira, M.; Ron, D. Regulated translation initiation controls stress-induced gene expression in mammalian cells. *Mol. Cell* **2000**, *6*, 1099–1108. [CrossRef]
107. Lu, P.D.; Harding, H.P.; Ron, D. Translation reinitiation at alternative open reading frames regulates gene expression in an integrated stress response. *J. Cell Biol.* **2004**, *167*, 27–33. [CrossRef]
108. Averous, J.; Bruhat, A.; Jousse, C.; Carraro, V.; Thiel, G.; Fafournoux, P. Induction of CHOP expression by amino acid limitation requires both ATF4 expression and ATF2 phosphorylation. *J. Biol. Chem.* **2004**, *279*, 5288–5297. [CrossRef]
109. Jauhiainen, A.; Thomsen, C.; Strombom, L.; Grundevik, P.; Andersson, C.; Danielsson, A.; Andersson, M.K.; Nerman, O.; Rorkvist, L.; Stahlberg, A.; et al. Distinct cytoplasmic and nuclear functions of the stress induced protein DDIT3/CHOP/GADD153. *PLoS One* **2012**, *7*, e33208. [CrossRef]
110. Li, Y.; Guo, Y.; Tang, J.; Jiang, J.; Chen, Z. New insights into the roles of CHOP-induced apoptosis in ER stress. *Acta Biochim. Biophys. Sin. (Shanghai)* **2014**, *46*, 629–640. [CrossRef]
111. Silva, J.M.; Wong, A.; Carelli, V.; Cortopassi, G.A. Inhibition of mitochondrial function induces an integrated stress response in oligodendroglia. *Neurobiol. Dis.* **2009**, *34*, 357–365. [CrossRef]
112. Zhao, Q.; Wang, J.; Levichkin, I.V.; Stasinopoulos, S.; Ryan, M.T.; Hoogenraad, N.J. A mitochondrial specific stress response in mammalian cells. *EMBO J.* **2002**, *21*, 4411–4419. [CrossRef]
113. Michel, S.; Canonne, M.; Arnould, T.; Renard, P. Inhibition of mitochondrial genome expression triggers the activation of CHOP-10 by a cell signaling dependent on the integrated stress response but not the mitochondrial unfolded protein response. *Mitochondrion* **2015**, *21*, 58–68. [CrossRef]
114. Moisoi, N.; Klupsch, K.; Fedele, V.; East, P.; Sharma, S.; Renton, A.; Plun-Favreau, H.; Edwards, R.E.; Teismann, P.; Esposti, M.D.; et al. Mitochondrial dysfunction triggered by loss of HtrA2 results in the activation of a brain-specific transcriptional stress response. *Cell Death Differ.* **2009**, *16*, 449–464. [CrossRef]
115. Zurita Rendón, O.; Shoubridge, E.A. LONP1 Is Required for Maturation of a Subset of Mitochondrial Proteins, and Its Loss Elicits an Integrated Stress Response. *Mol. Cell. Biol.* **2018**, *38*. [CrossRef]
116. Khan, N.A.; Nikkanen, J.; Yatsuga, S.; Jackson, C.; Wang, L.; Pradhan, S.; Kivela, R.; Pessia, A.; Velagapudi, V.; Suomalainen, A. mTORC1 Regulates Mitochondrial Integrated Stress Response and Mitochondrial Myopathy Progression. *Cell Metab.* **2017**, *26*, 419–428.e415. [CrossRef]
117. Restelli, L.M.; Oettinghaus, B.; Halliday, M.; Agca, C.; Licci, M.; Sironi, L.; Savoia, C.; Hench, J.; Tolnay, M.; Neutzner, A.; et al. Neuronal Mitochondrial Dysfunction Activates the Integrated Stress Response to Induce Fibroblast Growth Factor 21. *Cell Rep.* **2018**, *24*, 1407–1414. [CrossRef]
118. Kim, K.H.; Jeong, Y.T.; Oh, H.; Kim, S.H.; Cho, J.M.; Kim, Y.N.; Kim, S.S.; Kim, D.H.; Hur, K.Y.; Kim, H.K.; et al. Autophagy deficiency leads to protection from obesity and insulin resistance by inducing Fgf21 as a mitokine. *Nat. Med.* **2013**, *19*, 83–92. [CrossRef]
119. Sano, R.; Reed, J.C. ER stress-induced cell death mechanisms. *Biochim. Biophys. Acta (BBA) Mol. Cell Res.* **2013**, *1833*, 3460–3470. [CrossRef]
120. Bi, M.; Naczki, C.; Koritzinsky, M.; Fels, D.; Blais, J.; Hu, N.; Harding, H.; Novoa, I.; Varia, M.; Raleigh, J.; et al. ER stress-regulated translation increases tolerance to extreme hypoxia and promotes tumor growth. *Embo J.* **2005**, *24*, 3470–3481. [CrossRef]
121. Scheper, W.; Hoozemans, J.J.M. The unfolded protein response in neurodegenerative diseases: A neuropathological perspective. *Acta Neuropathol.* **2015**, *130*, 315–331. [CrossRef]
122. De Benedictis, G.; Carrieri, G.; Garasto, S.; Rose, G.; Varcasia, O.; Bonafe, M.; Franceschi, C.; Jazwinski, S.M. Does a retrograde response in human aging and longevity exist? *Exp. Gerontol.* **2000**, *35*, 795–801. [CrossRef]
123. Pakos-Zebrucka, K.; Koryga, I.; Mnich, K.; Ljujic, M.; Samali, A.; Gorman, A.M. The integrated stress response. *EMBO Rep.* **2016**, *17*, 1374–1395. [CrossRef]
124. Chou, A.; Krukowski, K.; Jopson, T.; Zhu, P.J.; Costa-Mattioli, M.; Walter, P.; Rosi, S. Inhibition of the integrated stress response reverses cognitive deficits after traumatic brain injury. *Proc. Natl. Acad. Sci. USA* **2017**, *114*, E6420–e6426. [CrossRef]
125. Kaltschmidt, B.; Kaltschmidt, C.; Hofmann, T.G.; Hehner, S.P.; Droge, W.; Schmitz, M.L. The pro- or anti-apoptotic function of NF-kappaB is determined by the nature of the apoptotic stimulus. *Eur. J. Biochem.* **2000**, *267*, 3828–3835. [CrossRef]

126. Duncan, O.F.; Bateman, J.M. Mitochondrial retrograde signaling in the Drosophila nervous system and beyond. *Fly* **2016**, *10*, 19–24. [CrossRef]
127. Hunt, R.J.; Bateman, J.M. Mitochondrial retrograde signaling in the nervous system. *FEBS Lett.* **2018**, *592*, 663–678. [CrossRef]
128. Picard, M.; Zhang, J.; Hancock, S.; Derbeneva, O.; Golhar, R.; Golik, P.; O'Hearn, S.; Levy, S.; Potluri, P.; Lvova, M.; et al. Progressive increase in mtDNA 3243A>G heteroplasmy causes abrupt transcriptional reprogramming. *Proc. Natl. Acad. Sci. USA* **2014**, *111*, E4033–E4042. [CrossRef]
129. Krug, A.K.; Gutbier, S.; Zhao, L.; Pöltl, D.; Kullmann, C.; Ivanova, V.; Förster, S.; Jagtap, S.; Meiser, J.; Leparc, G.; et al. Transcriptional and metabolic adaptation of human neurons to the mitochondrial toxicant MPP(+). *Cell Death Dis.* **2014**, *5*, e1222.
130. Snow, W.M.; Albensi, B.C. Neuronal Gene Targets of NF-κB and Their Dysregulation in Alzheimer's Disease. *Front. Mol. Neurosci.* **2016**, *9*, 118. [CrossRef]
131. Liu, Y.; Liu, F.; Iqbal, K.; Grundke-Iqbal, I.; Gong, C.-X. Decreased glucose transporters correlate to abnormal hyperphosphorylation of tau in Alzheimer disease. *FEBS Lett.* **2008**, *582*, 359–364. [CrossRef] [PubMed]
132. Julien, C.; Tremblay, C.; Emond, V.; Lebbadi, M.; Salem, N., Jr.; Bennett, D.A.; Calon, F. Sirtuin 1 reduction parallels the accumulation of tau in Alzheimer disease. *J. Neuropathol. Exp. Neurol.* **2009**, *68*, 48–58. [CrossRef] [PubMed]
133. Tramutola, A.; Triplett, J.C.; Di Domenico, F.; Niedowicz, D.M.; Murphy, M.P.; Coccia, R.; Perluigi, M.; Butterfield, D.A. Alteration of mTOR signaling occurs early in the progression of Alzheimer disease (AD): analysis of brain from subjects with pre-clinical AD, amnestic mild cognitive impairment and late-stage AD. *J. Neurochem.* **2015**, *133*, 739–749. [CrossRef] [PubMed]
134. Chang, R.C.; Wong, A.K.; Ng, H.K.; Hugon, J. Phosphorylation of eukaryotic initiation factor-2alpha (eIF2alpha) is associated with neuronal degeneration in Alzheimer's disease. *Neuroreport* **2002**, *13*, 2429–2432. [CrossRef]
135. Lewerenz, J.; Maher, P. Basal levels of eIF2alpha phosphorylation determine cellular antioxidant status by regulating ATF4 and xCT expression. *J. Biol. Chem.* **2009**, *284*, 1106–1115. [CrossRef] [PubMed]
136. Baleriola, J.; Walker, C.A.; Jean, Y.Y.; Crary, J.F.; Troy, C.M.; Nagy, P.L.; Hengst, U. Axonally synthesized ATF4 transmits a neurodegenerative signal across brain regions. *Cell* **2014**, *158*, 1159–1172. [CrossRef]
137. Lee, J.H.; Won, S.M.; Suh, J.; Son, S.J.; Moon, G.J.; Park, U.J.; Gwag, B.J. Induction of the unfolded protein response and cell death pathway in Alzheimer's disease, but not in aged Tg2576 mice. *Exp. Mol. Med.* **2010**, *42*, 386–394. [CrossRef]
138. Ding, Q.; Markesbery, W.R.; Chen, Q.; Li, F.; Keller, J.N. Ribosome dysfunction is an early event in Alzheimer's disease. *J. Neurosci.* **2005**, *25*, 9171–9175. [CrossRef] [PubMed]
139. Hoozemans, J.J.; van Haastert, E.S.; Nijholt, D.A.; Rozemuller, A.J.; Eikelenboom, P.; Scheper, W. The unfolded protein response is activated in pretangle neurons in Alzheimer's disease hippocampus. *Am. J. Pathol.* **2009**, *174*, 1241–1251. [CrossRef] [PubMed]

© 2019 by the authors. Licensee MDPI, Basel, Switzerland. This article is an open access article distributed under the terms and conditions of the Creative Commons Attribution (CC BY) license (http://creativecommons.org/licenses/by/4.0/).

Review

Mitochondria and Aging—The Role of Exercise as a Countermeasure

Mats I Nilsson [1,2] and Mark A Tarnopolsky [1,2,*]

1. Department of Pediatrics and Medicine, McMaster University Medical Center, Hamilton, ON L8S 4L8, Canada; mats.nilsson@exerkine.com or mnilsson7714@gmail.com
2. Exerkine Corporation, McMaster University Medical Center, Hamilton, ON L8N 3Z5, Canada
* Correspondence: tarnopol@mcmaster.ca; Tel.: +1-905-521-2100 (ext. 75226)

Received: 27 February 2019; Accepted: 12 April 2019; Published: 11 May 2019

Abstract: Mitochondria orchestrate the life and death of most eukaryotic cells by virtue of their ability to supply adenosine triphosphate from aerobic respiration for growth, development, and maintenance of the 'physiologic reserve'. Although their double-membrane structure and primary role as 'powerhouses of the cell' have essentially remained the same for ~2 billion years, they have evolved to regulate other cell functions that contribute to the aging process, such as reactive oxygen species generation, inflammation, senescence, and apoptosis. Biological aging is characterized by buildup of intracellular debris (e.g., oxidative damage, protein aggregates, and lipofuscin), which fuels a 'vicious cycle' of cell/DNA danger response activation (CDR and DDR, respectively), chronic inflammation ('inflammaging'), and progressive cell deterioration. Therapeutic options that coordinately mitigate age-related declines in mitochondria and organelles involved in quality control, repair, and recycling are therefore highly desirable. Rejuvenation by exercise is a non-pharmacological approach that targets all the major hallmarks of aging and extends both health- and lifespan in modern humans.

Keywords: aging; exercise; mitochondria; aerobic; ROS; inflammation; senescence; lysosome; autophagy; mitophagy

1. Introduction

Mitochondria are the energy-producing organelles of nearly all eukaryotic cells, which arose ~1.5–2 billion years ago when a phototrophic α-proteobacterium was endocytosed by an ancestral eukaryote [1]. This endosymbiotic relationship is thought to have conferred significant evolutionary advantages to the anaerobic host at a time when Earth was becoming more oxygenated [2]. The increased energy availability allowed for expansion of the eukaryotic genome, enhanced protein expression, and more complex signaling pathways and cellular traits [3], allowing for the rise of complex life [4].

In terms of human evolution, the first marked increase in hominin brain size emerged ~2 million years ago concurrent with increased exploration (ranging, scavenging, and hunting), a dietary shift to higher quality/nutrient-dense food (meat), and technological sophistication [5,6]. Striding bipedalism, such as long-distance walking and running, is a unique human trait contingent upon aerobic prowess (e.g., lungs, heart, and muscles), and allowed for divergence from their apelike forbears to become successful hunters [7,8]. It is now widely accepted that the ability to deliver and utilize oxygen by the cardiorespiratory system and skeletal muscles, respectively (e.g., maximal aerobic capacity; VO_2max), is a strong determinant of health and longevity in modern humans [9]. For example, runners have ~45–70% and ~30–50% reduced risk of mortality from cardiovascular disease (CVD) and cancer, respectively, and live 3–10% (2–8 years) longer than non-runners [10,11].

Considering the profound role of mitochondria in the evolution of aerobic life, it is not surprising that they hold a central position in cellular homeostasis and drive many aspects of the biological aging process. Aging is characterized by a progressive impairment of all body organs, including those that regulate VO$_2$max and locomotion (e.g., cardiorespiratory, nervous, and musculoskeletal systems), resulting in a ~10% decline in aerobic capacity per decade in both males and females after ~30 years of age [12,13]. Post-mitotic cells are particularly susceptible to the 'wear and tear' of aging, as exemplified by the progressive build-up of intracellular debris over a lifetime [14]. Concomitant oxidative damage, protein aggregation, and lipofuscinogenesis are interrelated features of the aging process, neurodegenerative disease, and lysosomal storage disorders [14–16]. Collectively, these danger-associated molecular patterns (DAMPs) fuel a 'vicious cycle' of cell/DNA danger response activation (CDR and DDR, respectively), senescence, and systemic inflammation ('inflammaging') [17]. Organelles that regulate reactive oxygen species (ROS) production (mitochondria), protein quality control/repair (unfolded protein response: endoplasmic reticulum (UPR^{ER}) and mitochondria (UPR^{MT})), and recycling (autophagosomes, lysosomes, and proteasomes) therefore constitute the cell's major defense systems against aging. Arguably, no other organelle is more important than the mitochondria in this context because they provide the bulk of the energy needed to sustain the 'physiologic reserve' and regulate other vital functions for cell survival, including ROS production, inflammation, senescence, and apoptosis (Figure 1) [18–20]. Currently, physical activity (PA) and caloric restriction represent the only non-pharmacologic means to enhance health-span and life expectancy by their ability to coordinately rejuvenate the systems that drive the biological aging process [21,22]; however, exercise is the only factor confirmed to lower morbidity and all-cause mortality in epidemiological studies.

Herein, we highlight the integrative nature of cell aging, review the evidence for age-associated mitochondrial dysfunction, and discuss how habitual PA attenuates the biological aging process, specifically aerobic (AET) and resistance (RET) exercise training. Because mitochondria have been extensively studied in heart and muscle, the main organs that limit VO$_2$max, we primarily focus on myocellular aging in this review.

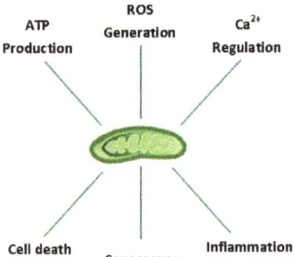

Figure 1. Major eukaryotic cell functions regulated by mitochondria. ATP: adenosine triphosphate; ROS: reactive oxygen species; Ca^{2+}: calcium ion.

2. Integrated Systems Hypothesis of Aging

Although over 300 hypotheses of aging have been proposed to date, with the vast majority focused on observable age effects ('wear and tear') or presumed root causes of age-associated pathology ('primary damage') [23], the common denominator across species has not yet been identified. The works of Schrödinger [24], Bortz [25], and Hayflick [26,27] collectively point to biological aging being a stochastic process occurring after reproductive maturity that is driven by entropy and results in progressive accumulation of random, irreparable losses in molecular fidelity. In line with the second law of thermodynamics, entropy is the tendency of a system to spontaneously disperse energy and evolve toward thermodynamic equilibrium, which is exemplified by a molecule's altered energy state following breakage of its intra- and/or inter-molecular bonds. According to Hayflick [27], entropic changes are circumvented by the cell's high-fidelity repair and replacement mechanisms until reproductive maturity, at which the rate of damage accumulation exceeds the rate of self-renewal.

One of the most prominent hypotheses in the category of 'primary damage' is Denham Harman's free radical theory of aging [28], originally conceptualized in 1954 but still garnering interest in the research community [29]. The original hypothesis posits that oxygen free radicals are the driving factors of aging, but has evolved to include all forms of reactive oxygen species (ROS) and mitochondria as the main source of ROS (oxidative stress and mitochondrial theories of aging, respectively) [28,30,31]. Subsequent observations of buildup of indigestible material in the lysosomes of post-mitotic cells (e.g., ferritin, mitochondrial fragments, and lipofuscin/'age-pigment') connected the aging process with an impairment in autolysosomal clearance (the mitochondrial–lysosomal axis theory of aging) [14]. Franceschi et al. further expanded on this integrated hypothesis by arguing that chronic, systemic inflammation at old age (inflammaging) is fueled by intracellular DAMPs originating from this 'garbage catastrophe' [32].

In summary, aging may be driven by entropy and manifests as a progressive accumulation of molecules with altered energy states, rendering them inactive or malfunctioning, prone to posttranslational modifications (e.g., oxidation, acetylation, methylation, glycation, etc.), cross-linking, and aggregation, and ultimately resistant to normal recycling mechanisms (for example, advanced glycation end products (AGEs) and lipofuscin). While mitochondria still remain central to the biological aging process in this view (by virtue of altered ROS and energy production), other organelles involved in recycling and quality control/repair also age, thus contributing to the 'vicious cycle' of debris accumulation, DAMP-activation of CDR/DDR, inflammation, and induction of cell-death (Figure 2).

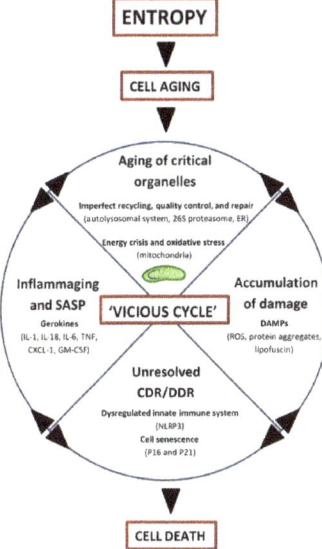

Figure 2. Integrated Systems Hypothesis of Aging (see text for details). DAMPs: danger-associated molecular patterns; ER: endoplasmic reticulum; CDR: cell danger response; DDR: DNA danger response; Inflammaging: chronic, low-grade inflammation with aging; SASP: senescence-associated secretory phenotype; Gerokines: cytokines, chemokines, growth factors, and proteases increased with aging; Vicious cycle: self-reinforcing feedback loop with detrimental outcome(s); NLRP3: inflammasome; P16: tumor suppressor protein P16^{INK4A}/CDKN2A; P21: tumor suppressor protein P21^{Cip1}/CDKN1A; IL: interleukin; TNF: tumor necrosis factor; CXCL-1: chemokine (C-X-C motif) ligand 1 (also KC and GROα); GM-CSF: granulocyte-macrophage colony-stimulating factor.

3. Mitochondrial Respiration: Yin and Yang of Aerobic Life

From an aging perspective, thermodynamically unfavorable/endergonic processes, such as biosynthesis and repair, are essential to counteract inevitable energy dispersal by entropy. A consistent supply of energy in the form of adenosine triphosphate (ATP) is integral to maintain tissue order and function; a task that is governed by the mitochondria [33]. Mammalian mitochondria generate >80% of cellular ATP under normal conditions and are composed of ~1158 proteins encoded by the nuclear genome and to a lesser extent by mitochondrial DNA (mtDNA) (MitoCarta2.0 [34]). The 37 gene products transcribed by mtDNA (e.g., 2 ribosomal ribonucleic acids (rRNAs), 22 transfer ribonucleic acids (tRNAs), 13 protein sub-units (7 complex I, 1 complex III, 3 complex IV, and 2 complex V)) are synthesized within the organelle itself, while the vast majority of mitochondrial proteins are encoded by the nuclear genome, synthesized in the cytoplasm, and imported by the mitochondrial translocation machinery. Mitochondrial biogenesis, and by extension energy supply, cell homeostasis, and human longevity, rely on the synchronous and concerted action of these processes.

Aerobic energy production by mitochondria, referred to as oxidative phosphorylation (OXPHOS), consumes the vast majority of cellular oxygen and is driven by a series of redox reactions/electron transfers in the inner mitochondrial membrane (mitochondrial respiratory chain (MRC)) [35]. In this process, electrons are successively transferred from electron donors (reducing agents) generated by macronutrient oxidation (glucose, fatty acids, and amino acids) to successively more electronegative electron-acceptors (oxidizing agents) in order to establish a proton gradient to drive ATP synthesis. Ironically, molecular oxygen is not only essential for ATP synthesis, but also represents a major source of reactive oxygen species (ROS) in mammalian cells, making oxidative metabolism a double-edged sword requiring careful cellular coordination.

As a natural by-product of respiration, ~0.2–2% of molecular oxygen undergoes a one-electron reduction into superoxide radicals in complexes I and III ($O_2^{-\bullet}$), which may be further converted into membrane-permeable singlet oxygen ($^1\Delta gO_2$ and $^2\Sigma gO_2$) or hydrogen peroxide (H_2O_2) [36–39]. Although mainly generated by complexes I and III, H_2O_2 and $O_2^{-\bullet}$ are also produced by the monoamine oxidases and NADPH oxidases in mitochondria [39,40]. Transition metals in iron–sulfur clusters in the MRC and lysosomes may react with H_2O_2 to generate hydroxyl radicals ($^\bullet OH$; via Fenton-type reactions), which are short-lived but indiscriminate oxidants that are highly dangerous to biological organisms [20]. Superoxide may also become protonated into perhydroxyl radicals ($HO_2^\bullet$) and have been proposed to play a central role in mediating the toxic side effects of aerobic respiration because of their high reactivity and membrane permeability [41]. Other potentially damaging molecules are also produced by the mitochondria, such as nitric oxide ($NO^\bullet$) and peroxynitrite ($ONOO^-$), but are technically considered reactive nitrogen species (RNS).

Chronic overproduction of ROS can lead to oxidative damage, cell toxicity, and apoptosis, and is linked to neurodegenerative diseases, cancer, and aging [42,43]. Paradoxically, ROS are also integral for regulation of cell signaling pathways, gene expression, and exercise adaptations [20]. Consequently, the complete amelioration of pro-oxidants is not advantageous for cell viability or health [43,44]. ROS levels are thereby exquisitely fine-tuned by the cell's principal enzymatic (EA) and non-enzymatic (NEA) antioxidant defense systems (EA: superoxide dismutases 1 and 2 (Cu/Zn-SOD and Mn-SOD, respectively), catalase, glutathione reductase, and glutathione peroxidases (GPx 1-4); NEA: reduced vs. oxidized glutathione (GSH: GSSG ratio), vitamin E, and vitamin C). Perturbations in the 'redox state' of the cell, generally defined as an imbalance between (pro-oxidants)/(anti-oxidants), predisposes towards oxidative damage [20]. Biological targets include lipids and proteins of cell membranes and nucleic acids of either genome being the most vulnerable in post-mitotic tissues. Oxidative modifications may lead to inactivation, fragmentation, and degradation of proteins, decomposition of membrane lipids, and significant RNA/DNA damage, including strand breaks, cross-links, and mutations, which predispose for senescence and cell-death [43].

4. Mitochondrial Aging

4.1. Oxidative Stress, mtDNA Mutagenesis, Apoptosis, and Respiration

Although direct evidence from human trials is lacking, mitochondrial $O_2^{-\bullet}$ and H_2O_2 production increases with advancing age and is inversely correlated to lifespan in multiple mammalian species and flies [45–48]. Excessive mitochondrial ROS production (and/or reduced antioxidant capacity) is associated with oxidative damage, MRC dysfunction, loss of mitochondrial membrane potential ($\Delta\Psi_m$), and induction of cell-death pathways in post-mitotic tissues of both prematurely (progeroid) and physiologically aged animal models. For example, mtDNA polymerase gamma-deficient mice (PolG; ↑ mtDNA mutagenesis) exhibit an accelerated aging phenotype (shorter lifespan, muscle atrophy, cardiomyopathy, anemia, thin dermis, gray fur, and kyphosis), deficits in OXPHOS function and ATP synthesis, and increased ROS-induced damage to mitochondrial proteins and nucleic acids [49]. Consistent with observations made in old Fisher 344 Brown Norway rats [48], reduced $\Delta\Psi_m$ in PolG mice is associated with the release of pro-apoptotic factors and induction of apoptosis, which likely contributes to organ dysfunction and muscle wasting in this model [50–52]. As cogently summarized by others [53,54], ROS imbalance, Ca^{2+} dysregulation, and/or loss of $\Delta\Psi_m$ may mediate mitochondrial outer membrane permeabilization and activation of intrinsic apoptotic pathways by opening of the mitochondrial permeability transition pore (mPTP) and the Bax/Bcl2-controlled mitochondrial apoptosis channel. ROS also contribute to telomere shortening and nuclear DNA instability (mainly in stem cells [55]), and genotoxic damage is a known activator of p53-mediated mPTP opening and apoptosis [56], which is the basis of the telomere-p53-mitochondrion model of aging [57]. In other words, several intrinsic (mitochondrial, ER, and lysosomal) and extrinsic (death receptor-induction by TNF-α and FasL) pathways may cooperate in myonuclear and satellite cell apoptosis, while mitochondria-driven cell death is believed to play the most important role in sarcopenia of aging [53,54,58,59].

Biological aging in humans is characterized by a progressive accumulation of oxidative damage and mutations to the mitochondrial genome from the third decade of life onward in several post-mitotic tissues (for example, muscle, heart, and brain) [31,60–63]. Concurrent with (or as a result of) increased ROS-induced damage and/or mtDNA mutagenesis, aging mitochondria display morphological abnormalities [30,64], lower MRC and OXPHOS activities [65,66], and impaired ATP synthesis [67,68]. Age-associated mitochondrial dysfunction, as assessed in vivo or at the whole tissue level [69], is attributable to intrinsic mitochondrial deficiency and a reduction in organellar number [68,70,71]. Due to the close proximity of mtDNA to the source of ROS, lack of protection by histones, and limited capacity for DNA repair [72,73], mtDNA is more susceptible to oxidative damage than nuclear DNA (nDNA), resulting in a nearly 20-fold higher mutation rate [74], including deletions [75–80], tandem duplications [81], and single base modifications [82]. In a series of landmark publications by the groups of Aiken and Turnbull, it was shown that clonal expansion of mtDNA mutations were linked to energy-deficient, cytochrome c oxidase-negative (COX^-) areas within skeletal muscle that contained atrophied and broken myofibers with high apoptotic susceptibility [75,76,83–87]. In one elegant study, Bua et al. found that a significant number of vastus lateralis (VL) muscle fibers displayed a 'ragged-blue phenotype' (e.g., succinate dehydrogenase-hyperactive (SDH^{++}) and COX^-) in older humans (>90 years), and that >80% of the total mtDNA pool was mutated in affected fibers [76]. Other findings suggest that random deletions may be present in up to 70% of mtDNA molecules in VL muscle of 'the oldest old', primarily affecting MRC complexes that contain mtDNA-encoded subunits [88].

Collectively, animal and human studies indicate that MRC dysfunction and OXPHOS deficits are common features of biological aging across multiple species (e.g., flies, mice, rats, dogs, monkeys, and humans) [89], and that a loss of $\Delta\Psi_m$, redox imbalance, and mtDNA mutagenesis confer a significant challenge to a plethora of organ systems and cell functions in mammals (Figure 3). It is well-known that the major growth-regulatory processes in skeletal muscle (GRPs; synthesis, degradation, satellite cell function, and apoptosis) are sensitive to perturbations in ROS, Ca^{2+}, ATP, and immunological homeostasis. Progressive dysfunction of organelles that regulate the aforementioned signaling molecules in skeletal muscle may therefore underlie the age-associated induction of intrinsic and extrinsic apoptotic pathways [53,54], reduction in proliferation and differentiation potentials of satellite cells [90], and desensitization to the anabolic and anti-proteolytic effects of insulin receptor (IR) stimulation [91–94]. Although these observations are consistent with the mitochondrial theory of aging [28,30,95], the question as to whether mitochondrial dysfunction drives the aging process and sarcopenia remains to be answered.

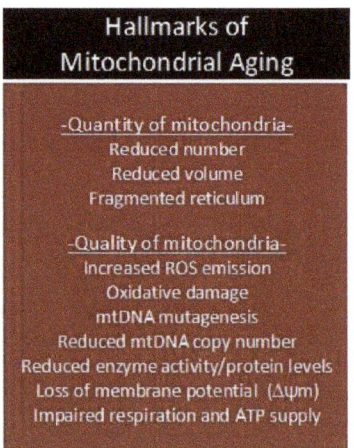

Figure 3. Age-associated deterioration in 'quantity and quality' of mitochondria. ROS: reactive oxygen species; mtDNA: mitochondrial DNA; $\Delta\Psi_m$: mitochondrial membrane potential; ATP: adenosine triphosphate

4.2. Garbage Catastrophe—The Role of Mitochondria

Recycling of biologic waste provides the cell with new building-blocks and substrates for energy metabolism; an integral housekeeping process predominately executed by the proteasome and lysosomes. Clearance of damaged organelles and macromolecules is critically important to maintain tissue homeostasis, particularly in post-mitotic cells that are unable to undergo waste dilution by cell division. Mitochondrial proteostasis is governed by an integrated network of pathways that include the organelles specialized in recycling and protein quality control (e.g., 26S proteasome, autolysosomal system, and PERK-mediated UPR^{ER}) and mitochondria-specific QC mechanisms (fusion/fission, mitophagy, various proteases, and the GCN2-mediated UPR^{MT}) [96,97]. Failure to maintain cellular clearance causes clumping of oxidatively damaged and misfolded proteins, formation of insoluble aggregates, and cell death by apoptosis or necrosis. The importance of efficient waste disposal is demonstrated by the fact that its disruption leads to neurodegenerative disease and lysosomal storage disorders; conditions linked to accelerated aging of neurons and muscle cells. Ablation of genes coding for lysosomal hydrolases or proteins that regulate intracellular waste delivery to lysosomes (e.g., autophagy) is associated with autophagic blockage, mitochondrial dysfunction, and tissue deterioration. In the case of acid α-glucosidase deficiency (Pompe disease), failure to clear lysosomal glycogen leads to cardiorespiratory insufficiency, muscle wasting, and premature death [98]. Conversely, pharmacological or genetic manipulations that prolong lifespan in model organisms typically activate cellular clearance pathways, and their inhibition may negate the life-extending effects, as in the case of caloric restriction [99].

A unifying feature in the pathogenesis of mammalian aging and accelerated aging conditions is the progressive deposition of cytotoxic debris impervious to lysosomal and proteasomal degradation. Age-related functional declines in autophagy, including macroautophagy and microautophagy (and likely aggrephagy), are linked to impaired mitochondrial turnover, protein aggregation, and accumulation of lipofuscin [14,16,99–103]. Lipofuscin, or 'aging pigment', is a degradation-resistant, redox-active biomolecule composed of oxidized proteins (30–70%), lipids (20–50%), and transition metals (iron, copper etc.) and increases with advancing age in lysosomes of post-mitotic cells [100,104–106]. In humans, lipofuscin has been demonstrated in heart, liver, kidney, and skin, but is believed to play the most fundamental role in the aging process of neurons and muscle cells [100]. Motor neurons in the anterior horn of the spinal cord, which innervate muscles necessary for voluntary movement of the limbs and trunk, appear particularly susceptible to lipofuscin deposition [104]. In addition, lipofuscin in skeletal muscle has been proposed to be a more robust marker of age-induced pathology compared to oxidative stress/damage [107].

Due to degradation of iron-containing macromolecules (ferritin, myoglobin, cytochrome c (e.g., mitochondrial complexes)), lysosomes, and by extension lipofuscin, contain significant amounts of low-mass, reactive iron (Fe^{2+}), which catalyze Fenton-type reactions ($H_2O_2 \rightarrow {}^{\bullet}OH$)[108]. Hydroxyl radicals are extremely harmful and induce ubiquitous damage to biologic material, including peroxidation and permeabilization of the lysosomal membrane (LMP). Aspartic and cysteine cathepsins are released from LMP and cleave targets in the apoptotic pathway (Bid, Bcl-2 family members, caspase 8, and XIAP), culminating in activation of apoptosis, amplification of the apoptotic response, and/or necrosis [109]. Other known LMP inducers are DNA damage, lysosomotropic agents, calpain 1, and extrinsic stimuli such as death receptor ligands and signaling enhancers (TNF-α, FasL, IFN-γ) [109]. In addition to being independent activators of apoptosis, significant crosstalk occurs between lysosomes, mitochondria, and the ER in response to cellular stress via H_2O_2, cathepsins, and Ca^{2+}. According to the mitochondrial–lysosomal axis theory of aging, mitochondrial ROS serves as an accelerant of lipofuscinogenesis, which impairs lysosomal degradative capacity and recycling of damaged mitochondria, further perpetuating redox imbalance, cytotoxicity, and debris aggregation [14]. Collectively, mitochondria and lysosomes generate the vast majority of 'accelerating agents' for oxidation, aggregation, and lipofuscinogenesis (ROS and Fe^{2+}, respectively), play major roles in the induction of cell death, and likely contribute significantly to sarcopenia and the biological aging process.

4.3. Inflammaging—The Role of Mitochondria

Inflammation is a basic biological response to prevent, limit, and repair damage by invading pathogens or endogenous biomolecules. Cell stress and infectious agents trigger transmembrane (Toll-like (TLR) and C-type lectin) and cytosolic (NOD-like (NLR), RIG-I-like (RLR), and PYHIN protein family) signaling receptors in immune and non-immune cells, which activate intracellular and humoral components of the innate and acquired immune systems [110]. While the transient inflammatory response is beneficial (removal of pathogens, mitigation of injury, and clearance of dying cells), persistent inflammation is associated with tissue dysfunction and pathology (obesity, type 2 diabetes, atherosclerosis, asthma, and neurodegenerative diseases) [111]. Chronic low-grade inflammation (inflammaging) is a hallmark of biological aging and is characterized by a 2 to 4-fold increase in circulating cytokines, chemokines, growth factors, and proteases, collectively termed 'gerokines', which may be broadly classified into pro- (TNF-α, IL-1α/β, IL-8, IFNγ, VEGF, etc.) and anti-inflammatory (IL-2, IL-4, IL-10, IL-13, TGF-β, etc.) factors [17,32]. Inflammaging is attributed to DAMP-activation of the innate immune response, cell senescence (e.g., SASP; senescence-associated secretory phenotype), and immunosenescence, and has been linked to an elevation in all-cause mortality and sarcopenia [112–118].

In 2002, Jürg Tschopp discovered a molecular platform that mediates the induction of the innate immune response in myeloid (monocytes, macrophages, dendritic cells, and neutrophils) and nonmyeloid cells (nerve, muscle, heart, endothelial etc.) [119,120]. The multi-protein complex, referred to as the inflammasome, is a cytosolic receptor that senses pathogen- and damage-associated molecular patterns (PAMPs and DAMPs, respectively), activates caspase-1, and causes IL-1β/IL-18 maturation. Subsequent secretion of IL-1β/IL-18 recruits immune cells to the site of damage, which leads to further release of cytokines and chemokines (TNF-α, IL-1β etc.), cell death, and phagocytosis of apoptotic bodies. NLRP3, the most widely studied inflammasome, requires a priming step by NF-κB and a danger/pathogen signal to become fully activated [121]. Tschopp's group demonstrated that ROS overgeneration by mitochondria, induced by inhibition of mitophagy (via 3-MA and Beclin1/ATG5 knockdown) or complex I and III inactivation (via rotenone and anti-mycin, respectively), activated the NLRP3 inflammasome and promoted IL-1β secretion [122,123]. Other DAMPs, including mitochondrial- (mtDNA, cardiolipin, mitofusins, and the mitochondrial antiviral-signaling protein (MAVS)), lysosomal- (cathepsins), and ER-derived (Ca^{2+}), directly regulate NLRP3 activity or are integral in the recruitment and docking of the inflammasome to the mitochondria [110,124]. Oxidized mtDNA released during apoptosis is a known inducer of NLRP3 [125], increases gradually after the fifth decade of life, and is positively correlated to systemic proinflammatory cytokine levels [115]. Failure to recycle damaged mitochondria causes ROS overgeneration, mtDNA damage, and exacerbation of the inflammatory response (as shown by Tschopp et al.). Taken together, these data suggest that age-associated danger signals generated from mitochondria, lysosomes, and the ER contribute to inflammaging and sarcopenia [113].

5. The Anti-Aging Benefits of Physical Activity

In light of the hormetic effects of low levels and/or pulses of oxidative stress [20,126], aging intervention strategies should be aimed at dampening (but not ameliorating) persistent ROS overgeneration and removing oxidative damage and protein aggregates as expediently as possible, which would limit the formation of waste products refractory to normal enzyme catalysis and inflammation. Attenuation of the major hallmarks of aging will not halt entropy per se, but will delay downstream pathology, extend health-span, and add longevity. Accrual of an excess physiologic reserve before the reproductive peak in humans (20–30 years), and maintenance of this reserve capacity by efficient repair and recycling in adulthood (40 years onwards), are synonymous with life extension, but may necessitate a combined approach of pharmacotherapy, rejuvenative biotechnology, and lifestyle modification. The feasibility of using non-exercise strategies for life extension has been discussed elsewhere [127].

5.1. Acute Exercise is Hormesis

Acute contractile activity is a hormetic stress stimulus that temporarily alters intracellular danger signals (ROS, Ca^{2+}, pH, and hypoxia), lowers cellular energy state (NAD^+/NADH and AMP/ATP), and promotes release of hormones and circulatory factors ('exerkines'), which synchronously activate signaling pathways that stimulate mitochondrial biogenesis (CaMK II, PGC-1α, SIRT1, and AMPK), antioxidant defense (Nrf2-Keap1, NF-κB, and MAPK), waste recycling (autophagy (ULK1-Beclin1) and 26S proteasome (FOXO3a)), and the immune response (IL-1β, IL-18, IL-6, IL-10, IL-1ra, sTNF-R, etc.) [20,21,128–139]. Anabolic GRPs, mainly mediated by Akt-mTOR signaling (e.g., protein synthesis), are activated following exercise concomitant with energy repletion, and may stay elevated for 1–2 days in older adults [140]. Consistent with the concept of hormesis [44,141], repeated exposure to a single-stress stimulus such as exercise improves stress resistance and immunity, rejuvenates mitochondria (increased biogenesis, recycling, and damage removal), and increases the organ functional reserve [17].

5.2. Mitochondrial Rejuvenation

The long-term benefits of PA are multi-systemic (muscular, nervous, vascular, endocrine, and immune systems) and culminate in reduced all-cause mortality and enhanced longevity (e.g., ~3–10% in average life expectancy) [10,11]. AET is considered the gold standard to improve mitochondrial biogenesis, insulin sensitivity, and cardiorespiratory fitness across all age groups. In older adults, AET partially reverses mitochondrial dysfunction by augmenting mtDNA copy number, mitochondrial transcript and protein expression, oxidative enzyme function, ATP synthesis, and total mitochondrial volume [142–144]. Short et al. demonstrated that the capacity for mitochondrial biogenesis (e.g., PGC-1α, NRF1, and TFAM), mitochondrial gene expression (COX IV and ND4), and Kreb's cycle/MRC enzyme activities (CS and COX) may be enhanced by AET regardless of age [143]. Indeed, 12 weeks of progressive moderate-intensity AET (50–70% VO2 max) increased total mitochondrial content (mtDNA and cardiolipin), MRC function (NADH oxidase and succinate oxidase), and HOMA-IR (Homeostatic Model Assessment of Insulin Resistance) in older adults [142]. In the latter study, both pools of mitochondria (e.g., subsarcolemmal and intermyofibrillar) were responsive to contractile activity and the majority of mitochondrial variables were improved by >50%. Although findings by Broskey et al. suggest that AET-induced mitochondrial benefits are largely ascribed to higher mitochondrial volume density [144], other studies (both cross-sectional and longitudinal) indicate that the intrinsic quality of individual mitochondria may also be enhanced by AET [145–147].

RET is generally considered to have minimal effects on mitochondrial biogenesis, but our group and others have clearly shown that strength training rejuvenates the mitochondrial transcriptome profile, enhances MRC and antioxidant enzyme activities, and reduces oxidative damage in skeletal muscle of older adults [148–153]. A study by Jubrias et al. demonstrated that gains in mitochondrial volume density may even be greater following RET vs. AET in elderly (30% vs. 10%, respectively) [146], and one report suggests that mitochondrial adaptations are similar regardless of exercise mode [154]. Porter and Rasmussen found that the intrinsic quality of mitochondria is improved by regular strength training, with a potential shift in the relative contribution of complex I and complex II to maximal electron transfer [155]. Data from our studies suggest that RET-induced mitochondrial benefits are partially mediated by activation of satellite cells, which fuse with the mature myofiber and bring in wildtype mtDNA to 'dilute down' the mutant mtDNA pool [153]. The concept of mtDNA shifting following muscle overload and subtle myofiber injury by concentric and eccentric contractile activity, respectively, was first introduced by Taivassalo et al. and tested in mitochondrial disease patients [156,157]. We have now expanded this concept and demonstrated that progressive RET (50–75% of one-repetition maximum (1-RM)) lowers mtDNA deletions and increases lean mass, muscle strength, and function in older adults [152,153].

5.3. Intracellular Garbage Clearance

Aging is associated with ROS overgeneration that overwhelms antioxidant defense systems and leads to oxidative modifications of proteins, lipids, and nDNA/mtDNA [158]. Oxidative damage and other DAMPs fuel a vicious cycle that culminates in blunted recycling, debris accumulation, and inflammation. Oxidative stress reduction is thereby a key aspect of anti-aging therapies and a substantial amount data supports the antioxidant role of PA [20,158]. Regular training induces a shift from fast to intermediate muscle fiber types (e.g., glycolytic → more oxidative), which strengthens antioxidant defense and protects against buildup of damage [20]. Importantly, these observations are not limited to the muscles, as PA has been shown to have multi-systemic benefits on diverse tissues (skin and brain, for example), including reductions in nDNA/mtDNA adducts, AGE cross-links, and amyloid plaques [159–161].

Although the effects of contractile activity on quality control, repair, and recycling mechanisms remain largely unknown, several research teams have contributed to an increased understanding of how exercise modulates autolysosomal and 26S proteasomal pathways. Collectively, animal and human studies suggest that aerobic exercise reduces the cellular energy state, which stimulates lysosomal biogenesis, macroautophagy, proteasomal activity, and mitochondrial recycling via activation of TFEB (CLEAR network), ULK1-Beclin1, and FOXO transcription factors [132,162–164]. The magnitude of the response appears to be dependent upon the duration of exercise and nutrient status (e.g., fed or fasted state) [132]. In a study by Pagano et al, it was shown that aerobic exercise sequentially activates AMPK (preceding Akt-mTOR inhibition), ULK1, and FOXO3, leading to an increased LC3BII/LC3BI ratio, expression of E3 ubiquitin ligases (MuRF1 and MAFbx), and enhanced mitochondrial turnover (Mul1 and DRP1)[131]. Additionally, the integral role of Bcl2-mediated activation of Beclin1 for exercise-induced autophagy has previously been demonstrated in multiple tissues by Beth Levine's group (muscle, heart, liver, pancreas, adipose tissue, and brain) [129,130]. Importantly, basal autophagic flux is improved in skeletal muscle following AET, concomitant with a shift from glycolytic to oxidative myofiber phenotype [165]. Data from our laboratory suggest that AET also augments autophagic debris removal in LSDs such as Pompe disease [15]. Conversely, RET may preferentially modulate the ubiquitin-proteasome pathway [132], but more research is needed to determine the effects of exercise mode and training intensity on recycling/repair processes.

5.4. Boosting the Immune System

High-intensity and unaccustomed exercise may cause tissue damage, elevated levels of pro- and anti-inflammatory factors, and delayed onset muscle soreness (DOMS) [17,166,167]. Typically, the acute inflammatory response is followed by a healing phase, structural remodeling, and muscle adaptation, which mitigates DOMS from subsequent exercise sessions. Over time, exercise leads to physiological adaptations into a more stress-resistant, homeostatic level, which protects against age-related chronic diseases such as systemic inflammation and cancer [17].

Although the anti-inflammatory benefits of PA are traditionally attributed to a reduction in visceral fat mass and/or induction of an anti-inflammatory environment with each bout of exercise [137,139], it is plausible that mitochondrial rejuvenation in multiple cell populations concomitantly enhances immunity via enhanced control of the inflammasome (e.g., ↓ DAMP-mediated activation). Considering that skeletal muscle is an endocrine organ that makes up ~35–50% of total body mass, immune benefits in older adults may be partially mediated by preservation of muscle mass. Results from randomized controlled trials indicate that AET and combined AET/RET boost the vaccination response, reduce circulatory levels of pro-inflammatory cytokines, and augment proliferative capacity and/or function of multiple cell types in the innate and adaptive immune systems [118]. Our group recently demonstrated that lifelong AET potently dampens inflammaging, including master regulators of cytokine cascade and tumorigenesis (IL1-α/β, TNF-α, and IL-6), which partially preserved muscle mass, protected against multi-systemic cancers, and enhanced health-span of naturally-aged mice [17]. Lastly, the anti-inflammatory effects of strength training in isolation are understudied, but there is sufficient evidence of improved immune function following long-term RET in frail elderly [168,169].

6. Conclusions

Concomitant oxidative damage, protein aggregation, lipofuscinogenesis, and inflammation are unifying features of the normal aging process, neurodegenerative disease, and lysosomal storage disorders. Activation of clearance pathways extends lifespan in multiple species, collectively suggesting that the ability to neutralize cytotoxins, recycle debris, and repair stress-induced damage is integral for survival. Although the 'ground zero' of aging may be entropy, we propose that the rate of aging is predominately dictated by the organelles/processes that govern the most critical needs of the cell, such as energy production (mitochondria), recycling (autophagosome, lysosome, and 26S proteasome), and quality control (UPR^{ER} and UPT^{MT}). Given their importance in eukaryotic evolution, cell homeostasis, and growth, mitochondria may be considered the 'hubs of aerobic life', and are therefore assigned a central role in the Integrated Systems Hypothesis of Aging.

An impressive body of knowledge over the last 50 years unequivocally proves that regular exercise lowers all-cause and cardiovascular mortality risks, enhances health and longevity, and that an inactive lifestyle is inherently unsafe. Both major types of exercise, aerobic and resistance training, bestow multi-systemic benefits and protect against the major hallmarks of aging, including mitochondrial dysfunction, recycling deficiency, impaired quality control, and systemic inflammation, thus providing a compelling argument in support of exercise as a front-line modality to decelerate the aging process.

Author Contributions: M.I.N. and M.A.T. conceptualized, wrote, reviewed, and edited the original draft/final manuscript.

Funding: This work was funded by CIHR 143325 Mark A Tarnopolsky: "Characterization of novel exersomes that mediate the systemic pro-metabolic effects of exercise in aging and neuromuscular disorders" and Exerkine Corporation.

Conflicts of Interest: Exerkine Corporation is a biotechnology company that develops and commercializes therapies based on supplements, exercise-derived factors ('exerkines'), and extracellular vesicles to treat genetic disorders, chronic diseases, and aging. MAT is the founder, CEO, and CSO of Exerkine Corporation, which provided support in the form of salary to MIN. MAT and MIN are also shareholders in the company. The funders had no additional roles in the decision to publish or preparation of the manuscript.

References

1. Kurland, C.G.; Andersson, S.G.E. Origin and Evolution of the Mitochondrial Proteome. *Microbiol. Mol. Biol. Rev.* **2000**, *64*, 786–820. [CrossRef]
2. Holland, H. The oxygenation of atmosphere and oceans. *Philos. Trans. R. Soc. Lond. B Biol. Sci.* **2006**, *361*, 903–915. [CrossRef]
3. Lane, N. Bioenergetic Constraints on the Evolution of Complex Life. *Cold Harb. Perspect. Boil.* **2014**, *6*, a015982. [CrossRef] [PubMed]
4. Stamati, K.; Mudera, V.; Cheema, U. Evolution of oxygen utilization in multicellular organisms and implications for cell signalling in tissue engineering. *J. Tissue Eng.* **2011**, *2*. [CrossRef] [PubMed]
5. Burini, R.C.; Leonard, W.R. The evolutionary roles of nutrition selection and dietary quality in the human brain size and encephalization. *Nutrire* **2018**, *43*, 19. [CrossRef]
6. Wood, B.; Collard, M. The Human Genus. *Science* **1999**, *284*, 65–71. [CrossRef]
7. Bramble, D.M.; Lieberman, D.E. Endurance running and the evolution of Homo. *Nature* **2004**, *432*, 345. [CrossRef] [PubMed]
8. Pontzer, H. Economy and Endurance in Human Evolution. *Curr. Biol.* **2017**, *27*, R613–R621. [CrossRef]
9. Kodama, S.; Saito, K.; Tanaka, S.; Maki, M.; Yachi, Y.; Asumi, M.; Sugawara, A.; Totsuka, K.; Shimano, H.; Ohashi, Y.; et al. Cardiorepiratory fitness as a quantitative predictor of all-cause mortality and cardiovascular events in healthy men and women: A meta-analysis. *JAMA* **2009**, *301*, 2024–2035. [CrossRef] [PubMed]
10. Reimers, C.D.; Knapp, G.; Reimers, A.K. Does physical activity increase life expectancy? A review of the literature. *J. Aging Res.* **2012**, *9*, 2012. [CrossRef] [PubMed]
11. Lee, D.; Brellenthin, A.; Thompson, P.; Sui, X.; Lee, I.; Lavie, C. Running as a key lifestyle medicine for longevity. *Prog. Cardiovasc. Dis.* **2017**, *60*, 45–55. [CrossRef]
12. Pimentel, A.E.; Gentile, C.L.; Tanaka, H.; Seals, D.R.; Gates, P.E. Greater rate of decline in maximal aerobic capacity with age in endurance-trained than in sedentary men. *J. Appl. Physiol.* **2003**, *94*, 2406–2413. [CrossRef] [PubMed]
13. Tanaka, H.; Desouza, C.A.; Jones, P.P.; Stevenson, E.T.; Davy, K.P.; Seals, D.R. Greater rate of decline in maximal aerobic capacity with age in physically active vs. sedentary healthy women. *J. Appl. Physiol.* **1997**, *83*, 1947–1953. [CrossRef]
14. Terman, A.; Kurz, T.; Navratil, M.; Arriaga, E.A.; Brunk, U.T. Mitochondrial turnover and aging of long-lived postmitotic cells: The Mitochondrial-Lysosomal Axis Theory of Aging. *Antioxid. Redox Signal.* **2010**, *12*, 503–535. [CrossRef] [PubMed]
15. Nilsson, M.; MacNeil, L.; Kitaoka, Y.; Suri, R.; Young, S.; Kaczor, J.; Nates, N.; Ansari, M.; Wong, T.; Ahktar, M.; et al. Combined aerobic exercise and enzyme replacement therapy rejuvenates the mitochondrial-lysosomal axis and alleviates autophagic blockage in Pompe disease. *Free Radic. Biol. Med.* **2015**, *87*, 98–112. [CrossRef]
16. Hyttinena, J.M.; Amadiob, M.; Viiri, J.; Pascaleb, A.; Salminenc, A.; Kaarnirantaa, K. Clearance of misfolded and aggregated proteins by aggrephagy and implications for aggregation diseases. *Ageing Res. Rev.* **2014**, *18*, 16–28. [CrossRef]
17. Nilsson, M.I.; Bourgeois, J.M.; Nederveen, J.P.; Leite, M.R.; Hettinga, B.P.; Bujak, A.L.; May, L.; Lin, E.; Crozier, M.; Rusiecki, D.R.; et al. Lifelong aerobic exercise protects against inflammaging and cancer. *PLoS ONE* **2019**, *14*, e0210863. [CrossRef]
18. Mehta, M.M.; Weinberg, S.E.; Chandel, N.S. Mitochondrial control of immunity: Beyond ATP. *Nat. Rev. Immunol.* **2017**, *17*, 608. [CrossRef]
19. Marzetti, E.; Calvani, R.; Cesari, M.; Buford, T.W.; Lorenzi, M.; Behnke, B.J.; Leeuwenburgh, C. Mitochondrial dysfunction and sarcopenia of aging: From signaling pathways to clinical trials. *Int. J. Biochem. Cell Biol.* **2013**, *45*, 2288–2301. [CrossRef] [PubMed]
20. Powers, S.K.; Jackson, M.J. Exercise-induced oxidative stress: Cellular mechanisms and impact on muscle force production. *Physiol. Rev.* **2008**, *88*, 1243–1276. [CrossRef]

21. Garatachea, N.; Pareja-Galeano, H.; Sanchis-Gomar, F.; Santos-Lozano, A.; Fiuza-Luces, C.; Moran, M.; Emanuele, E.; Joyner, M.J.; Lucia, A. Exercise attenuates the major hallmarks of aging. *Rejuvenation Res.* **2014**, *18*, 57–89. [CrossRef] [PubMed]
22. Fiuza-Luces, C.; Garatachea, N.; Berger, N.; Lucia, A. Exercise is the real polypill. *Physiology* **2013**, *28*, 330–358. [CrossRef] [PubMed]
23. Medvedev, Z.A. An attempt at a rational classification of theories of ageing. *Biol. Rev.* **1990**, *65*, 375–398.
24. Schrodinger, E. What is life. In *The Physical Aspect of the Living Cell*; McMillan Co.: New York, NY, USA, 1947.
25. Bortz, W.M. Aging as entropy. *Exp. Gerontol.* **1986**, *21*, 321–328. [CrossRef]
26. Hayflick, L. Aging: The reality: Anti-aging is an oxymoron. *J. Gerontol. Ser. A Biol. Sci. Med Sci.* **2004**, *59*, B573–B578. [CrossRef]
27. Hayflick, L. Entropy explains aging, genetic determinism explains longevity, and undefined terminology explains misunderstanding both. *PLoS Genet.* **2007**, *3*, e220. [CrossRef]
28. Harman, D. The biologic clock: The mitochondria. *J. Am. Geriatr. Soc.* **1972**, *20*, 145–147. [CrossRef] [PubMed]
29. Sanz, A.; Pamplona, R.; Barja, G. Is the mitochondrial free radical theory of aging intact? *Antioxid. Redox Signal.* **2006**, *8*, 582–599. [CrossRef]
30. Miquel, J.; Economos, A.C.; Fleming, J.; Johnson, J.E.J. Mitochondrial role in cell aging. *Exp. Gerontol.* **1980**, *15*, 575–591. [CrossRef]
31. Bokov, A.; Chaudhuri, A.; Richardson, A. The role of oxidative damage and stress in aging. *Mech. Ageing Dev.* **2004**, *125*, 811–826. [CrossRef]
32. Franceschi, C.; Garagnani, P.; Vitale, G.; Capri, M.; Salvioli, S. Inflammaging and 'Garb-Aging'. *Trends Endocrinol. Metab.* **2017**, *28*, 199–209. [CrossRef]
33. Gray, M. Origin and evolution of organelle genomes. *Curr. Opin. Genet. Dev.* **1993**, *3*, 884–890. [CrossRef]
34. Calvo, S.; Clauser, K.; Mootha, V. MitoCarta2.0: An updated inventory of mammalian mitochondrial proteins. *Nucleic Acids Res.* **2016**, *44*, D1251–D1257. [CrossRef] [PubMed]
35. Rolfe, D.F.; Brown, G.C. Cellular energy utilization and molecular origin of standard metabolic rate in mammals. *Physiol. Rev.* **1997**, *77*, 731–758. [CrossRef]
36. Harper, M.; Bevilacqua, L.; Hagopian, K.; Weindruch, R.; Ramsey, J. Ageing, oxidative stress, and mitochondrial uncoupling. *Acta Physiol. Scand.* **2004**, *182*, 321–331.
37. Muller, F.L.; Liu, Y.; Van Remmen, H. Complex III releases superoxide to both sides of the inner mitochondrial membrane. *J. Biol. Chem.* **2004**, *279*, 49064–49073. [CrossRef] [PubMed]
38. Chance, B.; Sies, H.; Boveris, A. Hydroperoxide metabolism in mammalian organs. *Physiol. Rev.* **1979**, *59*, 527–605. [CrossRef] [PubMed]
39. Wong, H.-S.; Benoit, B.; Brand, M.D. Mitochondrial and cytosolic sources of hydrogen peroxide in resting C2C12 myoblasts. *Free Radic. Biol. Med.* **2019**, *130*, 140–150. [CrossRef]
40. Maggiorani, D.; Manzella, N.; Edmondson, D.E.; Mattevi, A.; Parini, A.; Binda, C.; Mialet-Perez, J. Monoamine Oxidases, Oxidative Stress, and Altered Mitochondrial Dynamics in Cardiac Ageing. *Oxidative Med. Cell. Longev.* **2017**, *2017*, 3017947. [CrossRef] [PubMed]
41. De Grey, A.D. $HO_2 \bullet$: The Forgotten Radical. *DNA Cell Biol.* **2002**, *21*, 251–257. [CrossRef] [PubMed]
42. Mattson, M.P. Hormesis defined. *Ageing Res. Rev.* **2008**, *7*, 1–7. [CrossRef]
43. Kohen, R.; Nyska, A. Invited review: Oxidation of biological systems: Oxidative stress phenomena, antioxidants, redox reactions, and methods for their quantification. *Toxicol. Pathol.* **2002**, *30*, 620–650. [CrossRef] [PubMed]
44. Ristow, M.; Zarse, K. How increased oxidative stress promotes longevity and metabolic health: The concept of mitochondrial hormesis (mitohormesis). *Exp. Gerontol.* **2010**, *45*, 410–418. [CrossRef]
45. Sohal, R.; Ku, H.; Agarwal, S.; Forster, M.; Lal, H. Oxidative damage, mitochondrial oxidant generation, and antioxidant defenses during aging and in response to food restriction in the mouse. *Mech. Ageing Dev.* **1994**, *74*, 121–133. [CrossRef]
46. Sohal, R.; Sohal, B.; Orr, W. Mitochondrial superoxide and hydrogen peroxide generation, protein oxidative damage, and longevity in different species of flies. *Free Radic. Biol. Med.* **1995**, *19*, 499–504. [CrossRef]
47. Ku, H.; Brunk, U.; Sohal, R. Relationship between mitochondrial superoxide and hydrogen peroxide production and longevity of mammalian species. *Free Radic. Biol. Med.* **1993**, *15*, 621–627. [CrossRef]

48. Chabi, B.; Ljubicic, V.; Menzies, K.J.; Huang, J.H.; Saleem, A.; Hood, D.A. Mitochondrial function and apoptotic susceptibility in aging skeletal muscle. *Aging Cell* **2008**, *7*, 2–12. [CrossRef]
49. Dai, D.F.; Chen, T.; Wanagat, J.; Laflamme, M.; Marcinek, D.J.; Emond, M.J.; Ngo, C.P.; Prolla, T.A.; Rabinovitch, P.S. Age-dependent cardiomyopathy in mitochondrial mutator mice is attenuated by overexpression of catalase targeted to mitochondria. *Aging Cell* **2010**, *9*, 536–544. [CrossRef]
50. Hiona, A.; Sanz, A.; Kujoth, G.C.; Pamplona, R.; Seo, A.Y.; Hofer, T.; Someya, S.; Miyakawa, T.; Nakayama, C.; Samhan-Arias, A.K.; et al. Mitochondrial DNA mutations induce mitochondrial dysfunction, apoptosis and sarcopenia in skeletal muscle of mitochondrial DNA mutator mice. *PLoS ONE* **2010**, *5*, e11468. [CrossRef] [PubMed]
51. Kujoth, G.C.; Hiona, A.; Pugh, T.D.; Someya, S.; Panzer, K.; Wohlgemuth, S.E.; Hofer, T.; Seo, A.Y.; Sullivan, R.; Jobling, W.A.; et al. Mitochondrial DNA mutations, oxidative stress, and apoptosis in mammalian aging. *Science* **2005**, *309*, 481–484. [CrossRef]
52. Trifunovic, A.; Hansson, A.; Wredenberg, A.; Rovio, A.T.; Dufour, E.; Khvorostov, I.; Spelbrink, J.N.; Wibom, R.; Jacobs, H.T.; Larsson, N.G. Somatic mtDNA mutations cause aging phenotypes without affecting reactive oxygen species production. *Proc. Natl. Acad. Sci. USA* **2005**, *102*, 17993–17998. [CrossRef]
53. Marzetti, E.; Privitera, G.; Simili, V.; Wohlgemuth, S.E.; Aulisa, L.; Pahor, M.; Leeuwenburgh, C. Multiple Pathways to the Same End: Mechanisms of Myonuclear Apoptosis in Sarcopenia of Aging. *Sci. J.* **2010**, *10*, 340–349. [CrossRef] [PubMed]
54. Alway, S.E.; Siu, P.M. Nuclear apoptosis contributes to sarcopenia. *Exerc. Sport Sci. Rev.* **2008**, *36*, 51–57. [CrossRef] [PubMed]
55. Von Zglinicki, T. Oxidative stress shortens telomeres. *Trends Biochem. Sci.* **2002**, *27*, 339–344. [CrossRef]
56. Mihara, M.; Erster, S.; Zaika, A.; Petrenko, O.; Chittenden, T.; Pancoska, P.; Moll, U.M. p53 has a direct apoptogenic role at the mitochondria. *Mol. Cell* **2003**, *11*, 577–590. [CrossRef]
57. Sahin, E.; DePinho, R.A. Axis of ageing: Telomeres, p53 and mitochondria. *Nat. Rev. Mol. Cell Biol.* **2013**, *13*, 397–404. [CrossRef]
58. Boya, P.; Kroemer, G. Lysosomal membrane permeabilization in cell death. *Oncogene* **2008**, *27*, 6434–6451. [CrossRef] [PubMed]
59. Szegezdi, E.; Logue, S.E.; Gorman, A.M.; Samali, A. Mediators of endoplasmic reticulum stress-induced apoptosis. *EMBO Rep.* **2006**, *7*, 880–885. [CrossRef] [PubMed]
60. Hayakawa, M.; Hattori, K.; Sugiyama, S.; Ozawa, T. Age-associated oxygen damage and mutations in mitochondrial DNA in human hearts. *Biochem. Biophys. Res. Commun.* **1992**, *189*, 979–985. [CrossRef]
61. Hayakawa, M.; Torii, K.; Sugiyama, S.; Tanaka, M.; Ozawa, T. Age-associated accumulation of 8-hydroxydeoxyguanosine in mitochondrial DNA of human diaphragm. *Biochem. Biophys. Res. Commun.* **1991**, *179*, 1023–1029. [CrossRef]
62. Mecocci, P.; Fan, G.; Fulle, S.; MacGarvey, U.; Shinobu, L.; Polidori, M.C.; Cherubini, A.; Vecchiet, J.; Senin, U.; Beal, M.F. Age-dependent increases in oxidative damage to DNA, lipids, and proteins in human skeletal muscle. *Free Radic. Biol. Med.* **1999**, *26*, 303–308. [CrossRef]
63. Parise, G.; Kaczor, J.J.; Mahoney, D.J.; Phillips, S.M.; Tarnopolsky, M.A. Oxidative stress and the mitochondrial theory of aging in human skeletal muscle. *Exp. Gerontol.* **2004**, *39*, 1391–1400.
64. Beregi, E.; Regius, O. Comparative morphological study of age related mitochondrial changes of the lymphocytes and skeletal muscle cells. *Acta Morphol. Hung.* **1987**, *35*, 219–224.
65. Greco, M.; Villani, G.; Mazzucchelli, F.; Bresolin, N.; Papa, S.; Attardi, G. Marked aging-related decline in efficiency of oxidative phosphorylation in human skin fibroblasts. *FASEB J. Off. Publ. Fed. Am. Soc. Exp. Biol.* **2003**, *17*, 1706–1708. [CrossRef]
66. Papa, S. Mitochondrial oxidative phosphorylation changes in the life span. Molecular aspects and physiopathological implications. *Biochim. Biophys. Acta (BBA) Bioenerg.* **1996**, *1276*, 87–105. [CrossRef]
67. Short, K.; Bigelow, M.; Kahl, J.; Singh, R.; Coenen-Schimke, J.; Raghavakaimal, S. Decline in skeletal muscle mitochondrial function with aging in humans. *Proc. Natl. Acad. Sci. USA* **2005**, *102*, 5618–5623. [CrossRef] [PubMed]
68. Johnson, M.L.; Robinson, M.M.; Nair, K.S. Skeletal muscle aging and the mitochondrion. *Trends Endocrinol. Metab.* **2013**, *24*, 247–256. [CrossRef]
69. Petersen, K.; Befroy, D.; Dufour, S.; Dziura, J.; Ariyan, C.; Rothman, D. Mitochondrial dysfunction in the elderly: Possible role in insulin resistance. *Science* **2003**, *300*, 1140–1142. [CrossRef] [PubMed]

70. Crane, J.; Devries, M.; Safdar, A.; Hamadeh, M.; Tarnopolsky, M. The effect of aging on human skeletal muscle mitochondrial and intramyocellular lipid ultrastructure. *J. Gerontol. A Biol. Sci. Med. Sci.* **2010**, *65*, 119–128. [CrossRef]
71. Lanza, I.; Nair, K.S. Mitochondrial function as a determinant of life span. *Pflug. Arch. Eur. J. Physiol.* **2010**, *459*, 277–289. [CrossRef]
72. Richter, C.; Park, J.W.; Ames, B.N. Normal oxidative damage to mitochondrial and nuclear DNA is extensive. *Proc. Natl. Acad. Sci. USA* **1988**, *85*, 6465–6467. [CrossRef] [PubMed]
73. Yakes, F.M.; Van Houten, B. Mitochondrial DNA damage is more extensive and persists longer than nuclear DNA damage in human cells following oxidative stress. *Proc. Natl. Acad. Sci. USA* **1997**, *94*, 514–519. [CrossRef] [PubMed]
74. Johnson, F.; Sinclair, D.; Guarente, L. Molecular biology of aging. *Cell* **1999**, *96*, 291–302. [CrossRef]
75. Aiken, J.; Bua, E.; Cao, Z.; Lopez, M.; Wanagat, J.; McKenzie, D.; McKiernan, S. Mitochondrial DNA deletion mutations and sarcopenia. *Ann. N. Y. Acad. Sci.* **2002**, *959*, 412–423. [CrossRef] [PubMed]
76. Bua, E.; Johnson, J.; Herbst, A.; Delong, B.; McKenzie, D.; Salamat, S.; Aiken, J.M. Mitochondrial DNA deletion mutations accumulate intracellularly to detrimental levels in aged human skeletal muscle fibers. *Am. J. Hum. Genet.* **2006**, *79*, 469–480. [CrossRef] [PubMed]
77. Corral-Debrinski, M.; Horton, T.; Lott, M.T.; Shoffner, J.M.; Flint Beal, M.; Wallace, D.C. Mitochondrial DNA deletions in human brain: Regional variability and increase with advanced age. *Nat. Genet.* **1992**, *2*, 324–329. [CrossRef] [PubMed]
78. Katayama, M.; Tanaka, M.; Yamamoto, H.; Ohbayashi, T.; Nimura, Y.; Ozawa, T. Deleted mitochondrial DNA in the skeletal muscle of aged individuals. *Biochem. Int.* **1991**, *25*, 47–56. [PubMed]
79. Lee, H.-C.; Pang, C.-Y.; Hsu, H.-S.; Wei, Y.-H. Differential accumulations of 4977 bp deletion in mitochondrial DNA of various tissues in human ageing. *Biochim. Biophys. Acta (BBA) Mol. Basis Dis.* **1994**, *1226*, 37–43. [CrossRef]
80. Melov, S.; Shoffner, J.M.; Kaufman, A.; Wallace, D.C. Marked increase in the number and variety of mitochondrial DNA rearrangements in aging human skeletal muscle. *Nucleic Acids Res.* **1995**, *23*, 4122–4126. [CrossRef]
81. Lee, H.-C.; Pang, C.-Y.; Hsu, H.-S.; Weia, Y.-H. Ageing-associated tandem duplications in the D-loop of mitochondrial DNA of human muscle. *FEBS Lett.* **1994**, *354*, 79–83. [CrossRef]
82. Zhang, C.F.; Linnane, A.W.; Nagley, P. Occurrence of a particular base substitution (3243 A to G) in mitochondrial DNA of tissues of ageing humans. *Biochem. Biophys. Res. Commun.* **1993**, *195*, 1104–1110. [CrossRef] [PubMed]
83. Cao, Z.; Wanagat, J.; McKiernan, S.H.; Aiken, J.M. Mitochondrial DNA deletion mutations are concomitant with ragged red regions of individual, aged muscle fibers: Analysis by laser-capture microdissection. *Nucleic Acids Res.* **2001**, *29*, 4502–4508. [CrossRef]
84. Pak, J.W.; Herbst, A.; Bua, E.; Gokey, N.; McKenzie, D.; Aiken, J.M. Mitochondrial DNA mutations as a fundamental mechanism in physiological declines associated with aging. *Aging Cell* **2003**, *2*, 1–7. [CrossRef]
85. Wanagat, J.; Cao, Z.; Pathare, P.; Aiken, J.M. Mitochondrial DNA deletion mutations colocalize with segmental electron transport system abnormalities, muscle fiber atrophy, fiber splitting, and oxidative damage in sarcopenia. *FASEB J. Off. Publ. Fed. Am. Soc. Exp. Biol.* **2001**, *15*, 322–332. [CrossRef] [PubMed]
86. Brierley, E.J.; Johnson, M.A.; Lightowlers, R.N.; James, O.F.; Turnbull, D.M. Role of mitochondrial DNA mutations in human aging: Implications for the central nervous system and muscle. *Ann. Neurol.* **1998**, *43*, 217–223. [CrossRef] [PubMed]
87. Krishnan, K.J.; Reeve, A.K.; Samuels, D.C.; Chinnery, P.F.; Blackwood, J.K.; Taylor, R.W.; Wanrooij, S.; Spelbrink, J.N.; Lightowlers, R.N.; Turnbull, D.M. What causes mitochondrial DNA deletions in human cells? *Nat. Genet.* **2008**, *40*, 275–279. [CrossRef] [PubMed]
88. Chabi, B.; de Camaret, B.M.; Chevrollier, A.; Boisgard, S.; Stepien, G. Random mtDNA deletions and functional consequence in aged human skeletal muscle. *Biochem. Biophys. Res. Commun.* **2005**, *332*, 542–549. [CrossRef] [PubMed]
89. Scatena, R.; Bottoni, P.; Giardina, B. *Advances in Mitochondrial Medicine*; Springer Publishing: Dordrecht, The Netherlands, 2012.
90. Alway, S.; Myers, M.; Mohamed, J. Regulation of satellite cell function in sarcopenia. *Front. Aging Neurosci.* **2014**, *6*, 246. [CrossRef]

91. Dickinson, J.M.; Volpi, E.; Rasmussen, B.B. Exercise and nutrition to target protein synthesis impairments in aging skeletal muscle. *Exerc. Sport Sci. Rev.* **2013**, *41*, 216–223. [CrossRef] [PubMed]
92. Breen, L.; Phillips, S.M. Skeletal muscle protein anabolism in elderly: Intervention to counteract the "anabolic resistance" of ageing. *Nutr. Metab.* **2011**, *8*, 68. [CrossRef] [PubMed]
93. Kumar, V.; Selby, A.; Rankin, D.; Patel, R.; Atherton, P.; Hildebrandt, W.; Williams, J.; Smith, K.; Seynnes, O.; Hiscock, N.; et al. Age-related differences in the dose–response relationship of muscle protein synthesis to resistance exercise in young and old men. *J. Physiol.* **2009**, *587*, 211–217. [CrossRef]
94. Churchward-Venne, T.A.; Breen, L.; Phillips, S.M. Alterations in human muscle protein metabolism with aging: Protein and exercise as countermeasures to offset sarcopenia. *BioFactors* **2013**, *40*, 199–205. [CrossRef]
95. Linnane, A.; Ozawa, T.; Marzuki, S.; Tanaka, M. Mitochondrial DNA mutations as an important contributor to aging and degenerative diseases. *Lancet* **1989**, *333*, 642–645. [CrossRef]
96. Fischer, F.; Hamann, A.; Osiewacz, H.D. Mitochondrial quality control: An integrated network of pathways. *Trends Biochem. Sci.* **2012**, *37*, 284–292. [CrossRef]
97. Jensen, M.B.; Jasper, H. Mitochondrial Proteostasis in the Control of Aging and Longevity. *Cell Metab.* **2014**, *20*, 214–225. [CrossRef]
98. Lim, J.-A.; Li, L.; Raben, N. Pompe disease: From pathophysiology to therapy and back again. *Front. Aging Neurosci.* **2014**, *6*, 177. [CrossRef]
99. Rubinsztein, D.; Marino, G.; Kroemer, G. Autophagy and aging. *Cell* **2011**, *146*, 682–695. [CrossRef]
100. Höhn, A.; König, J.; Grune, T. Protein oxidation in aging and the removal of oxidized proteins. *J. Proteom.* **2013**, *92*, 132–159. [CrossRef] [PubMed]
101. Rajawat, Y.S.; Hilioti, Z.; Bossis, I. Aging: Central role for autophagy and the lysosomal degradative system. *Ageing Res. Rev.* **2009**, *8*, 199–213. [CrossRef]
102. Cuervo, A.M.; Bergamini, E.; Brunk, U.T.; Droge, W.; Ffrench, M.; Terman, A. Autophagy and aging: The importance of maintaining "clean" cells. *Autophagy* **2005**, *1*, 131–140. [CrossRef]
103. Zhang, C.; Cuervo, A.M. Restoration of chaperone-mediated autophagy in aging liver improves cellular maintenance and hepatic function. *Nat. Med.* **2008**, *14*, 959–965. [CrossRef]
104. Gray, D.A.; Woulfe, J. Lipofuscin and aging: A matter of toxic waste. *Sci. Aging Knowl. Knowl. Environ.* **2005**, *2005*, 1–5. [CrossRef] [PubMed]
105. Melis, J.P.; Jonker, M.J.; Vijg, J.; Hoeijmakers, J.H.; Breit, T.M.; Van Steeg, H. Aging on a different scale chronological versus pathology-related aging. *Aging* **2013**, *5*, 782–788. [CrossRef] [PubMed]
106. Höhn, A.; Grune, T. Lipofuscin: Formation, effects and role of macroautophagy. *Redox Biol.* **2013**, *1*, 140–144. [CrossRef] [PubMed]
107. Hütter, E.; Skovbro, M.; Lener, B.; Prats, C.; Rabøl, R.; Dela, F.; Jansen-Dürr, P. Oxidative stress and mitochondrial impairment can be separated from lipofuscin accumulation in aged human skeletal muscle. *Aging Cell* **2007**, *6*, 245–256. [CrossRef]
108. Terman, A.; Kurz, T. Lysosomal iron, iron chelation, and cell death. *Antioxid. Redox Signal.* **2013**, *18*, 888–898. [CrossRef]
109. Mrschtik, M.; Ryan, K.M. Lysosomal proteins in cell death and autophagy. *FEBS J.* **2015**, *282*, 1858–1870. [CrossRef] [PubMed]
110. Hornung, V.; Latz, E. Critical functions of priming and lysosomal damage for NLRP3 activation. *Eur. J. Immunol.* **2010**, *40*, 620–623. [CrossRef]
111. Medzhitov, R. Origin and physiological roles of inflammation. *Nature* **2008**, *454*, 428–435. [CrossRef]
112. Álvarez-Rodríguez, L.; López-Hoyos, M.; Muñoz-Cacho, P.; Martínez-Taboada, V.M. Aging is associated with circulating cytokine dysregulation. *Cell. Immunol.* **2012**, *273*, 124–132. [CrossRef] [PubMed]
113. Franceschi, C.; Bonafe, M.; Valensin, S.; Olivieri, F.; De Luca, M.; Ottaviani, E.; De Benedictis, G. Inflamm-aging: An evolutionary perspective on immunosenescence. *Ann. N. Y. Acad. Sci.* **2000**, *908*, 244–254. [CrossRef]
114. Krabbe, K.S.; Pedersen, M.; Bruunsgaard, H. Inflammatory mediators in the elderly. *Exp. Gerontol.* **2004**, *39*, 687–699. [CrossRef]
115. Pinti, M.; Cevenini, E.; Nasi, M.; De Biasi, S.; Salvioli, S.; Monti, D.; Benatti, S.; Gibellini, L.; Cotichini, R.; Stazi, M.A.; et al. Circulating mitochondrial DNA increases with age and is a familiar trait: Implications for "inflamm-aging". *Eur. J. Immunol.* **2014**, *44*, 1552–1562. [CrossRef]
116. Singh, T.; Newman, A.B. Inflammatory markers in population studies of aging. *Ageing Res. Rev.* **2010**, *10*, 319–329. [CrossRef]

117. Jo, E.; Lee, S.-R.; Park, B.-S.; Kim, J.-S. Potential mechanisms underlying the role of chronic inflammation in age-related muscle wasting. *Aging Clin. Exp. Res.* **2012**, *24*, 412–422.
118. Simpson, R.J.; Lowder, T.W.; Spielmann, G.; Bigley, A.B.; LaVoy, E.C.; Kunz, H. Exercise and the aging immune system. *Ageing Res. Rev.* **2012**, *11*, 404–420. [CrossRef]
119. Martinon, F.; Burns, K.; Tschopp, J. The inflammasome: A molecular platform triggering activation of inflammatory caspases and processing of pro IL-β. *Mol. Cell* **2002**, *10*, 417–426. [CrossRef]
120. Yazdi, A.; Drexler, S.; Tschopp, J. The role of the inflammasome in nonmyeloid cells. *J. Clin. Immunol.* **2010**, *30*, 623–627. [CrossRef] [PubMed]
121. Martinon, F.; Mayor, A.; Tschopp, J. The inflammasomes: Guardians of the body. *Annu. Rev. Immunol.* **2009**, *27*, 229–265. [CrossRef]
122. Tschopp, J. Mitochondria: Sovereign of inflammation? *Eur. J. Immunol.* **2011**, *41*, 1196–1202. [CrossRef]
123. Zhou, R.; Yazdi, A.S.; Menu, P.; Tschopp, J. A role for mitochondria in NLRP3 inflammasome activation. *Nature* **2011**, *469*, 221–225. [CrossRef] [PubMed]
124. Gurung, P.; Lukens, J.R.; Kanneganti, T.-D. Mitochondria: Diversity in the regulation of the NLRP3 inflammasome. *Trends Mol. Med.* **2014**, *21*, 193–201. [CrossRef] [PubMed]
125. Shimada, K.; Crother, T.R.; Karlin, J.; Dagvadorj, J.; Chiba, N.; Chen, S.; Ramanujan, V.K.; Wolf, A.J.; Vergnes, L.; Ojcius, D.M.; et al. Oxidized mitochondrial DNA activates the NLRP3 inflammasome during apoptosis. *Immunity* **2012**, *36*, 401–414. [CrossRef]
126. Goto, S.; Naito, H.; Kaneko, T.; Chung, H.Y.; Radak, Z. Hormetic effects of regular exercise in aging: Correlation with oxidative stress. *Appl. Physiol. Nutr. Metab.* **2007**, *32*, 948–953. [CrossRef] [PubMed]
127. Zealley, B.; De Grey, A.D. Strategies for engineered negligible senescence. *Gerontology* **2013**, *59*, 183–189. [CrossRef] [PubMed]
128. Piantadosi, C.A.; Suliman, H.B. Redox regulation of mitochondrial biogenesis. *Free Radic. Biol. Med.* **2012**, *53*, 2043–2053. [CrossRef] [PubMed]
129. He, C.; Bassik, M.C.; Moresi, V.; Sun, K.; Wei, Y.; Zou, Z.; An, Z.; Loh, J.; Fisher, J.; Sun, Q.; et al. Exercise-induced BCL2-regulated autophagy is required for muscle glucose homeostasis. *Nature* **2012**, *481*, 511–515. [CrossRef]
130. He, C.; Sumpter, J.R.; Levine, B. Exercise induces autophagy in peripheral tissues and in the brain. *Autophagy* **2012**, *8*, 1548–1551. [CrossRef]
131. Pagano, A.F.; Py, G.; Bernardi, H.; Candau, R.B.; Sanchez, A.M.J. Autophagy and protein turnover signaling in slow-twitch muscle during exercise. *Med. Sci. Sports Exerc.* **2014**, *46*, 1314–1325. [CrossRef]
132. Sanchez, A.M.; Bernardi, H.; Py, G.; Candau, R. Autophagy is essential to support skeletal muscle plasticity in response to endurance exercise. *Am. J. Physiol. Regul. Integr. Comp. Physiol.* **2014**, *307*, R956–R969. [CrossRef]
133. Powers, S.K.; Nelson, W.B.; Hudson, M.B. Exercise-induced oxidative stress in humans: Cause and consequences. *Free Radic. Biol. Med.* **2011**, *51*, 942–950. [CrossRef]
134. Ji, L.L.; Gomez-Cabrera, M.-C.; Vina, J. Role of nuclear factor kB and mitogen-activated protein kinase signaling in exercise-induced antioxidant enzyme adaptation. *Appl. Physiol. Nutr. Metab.* **2007**, *32*, 930–935. [CrossRef]
135. Gounder, S.S.; Kannan, S.; Devadoss, D.; Miller, C.J.; Whitehead, K.S.; Odelberg, S.J.; Firpo, M.A.; Paine, R., III; Hoidal, J.R.; Abel, E.D.; et al. Impaired transcriptional activity of Nrf2 in age-related myocardial oxidative stress is reversible by moderate exercise training. *PLoS ONE* **2012**, *7*, e45697. [CrossRef]
136. MacNeil, L.G.; Safdar, A.; Baker, S.K.; Melov, S.; Tarnopolsky, M.A. Eccentric exercise affects Nrf2-mediated oxidative stress response in skeletal muscle by increasing nuclear Nrf2 content. *Med. Sci. Sports Exerc.* **2011**, *43*, 383. [CrossRef]
137. Gleeson, M.; Bishop, N.C.; Stensel, D.J.; Lindley, M.R.; Mastana, S.S.; Nimmo, M.A. The anti-inflammatory effects of exercise: Mechanisms and implications for the prevention and treatment of disease. *Nat. Rev. Immunol.* **2011**, *11*, 607–615. [CrossRef]
138. Petersen, A.; Pedersen, B. The anti-inflammatory effect of exercise. *J. Appl. Physiol.* **2005**, *98*, 1154–1162. [CrossRef] [PubMed]
139. Pedersen, B.; Febbraio, M. Muscle as an endocrine organ: Focus on muscle-derived interleukin-6. *Physiol. Rev.* **2008**, *88*, 1379–1406. [CrossRef]

140. Bell, K.E.; Seguin, C.; Parise, G.; Baker, S.K.; Phillips, S.M. Day-to-day changes in muscle protein synthesis in recovery from resistance, aerobic, and high-intensity interval exercise in older men. *J. Gerontol. Ser. A* **2015**, *70*, 1024–1029. [CrossRef]
141. Tapia, P.C. Sublethal mitochondrial stress with an attendant stoichiometric augmentation of reactive oxygen species may precipitate many of the beneficial alterations in cellular physiology produced by caloric restriction, intermittent fasting, exercise and dietary phytonutrients: Mitohormesis for health and vitality. *Med. Hypotheses* **2006**, *66*, 832–843.
142. Menshikova, E.V.; Ritov, V.B.; Fairfull, L.; Ferrell, R.E.; Kelley, D.E.; Goodpaster, B.H. Effects of exercise on mitochondrial content and function in aging human skeletal muscle. *J. Gerontol. Ser. A Biol. Sci. Med Sci.* **2006**, *61*, 534–540. [CrossRef]
143. Short, K.; Vittone, J.; Bigelow, M. Impact of aerobic exercise training on age-related changes in insulin sensitivity and muscle oxidative capacity. *Diabetes Care* **2003**, *52*, 1888–1896. [CrossRef]
144. Broskey, N.T.; Greggio, C.; Boss, A.; Boutant, M.; Dwyer, A.; Schlueter, L.; Hans, D.; Gremion, G.; Kreis, R.; Boesch, C.; et al. Skeletal muscle mitochondria in the elderly: Effects of physical fitness and exercise training. *J. Clin. Endocrinol. Metab.* **2015**, *99*, 1852–1861. [CrossRef]
145. Conley, K.; Jubrias, S.; Cress, M.; Esselman, P. Elevated energy coupling and aerobic capacity improves exercise performance in endurance-trained elderly subjects. *Exp. Physiol.* **2013**, *98*, 899–907.
146. Jubrias, S.A.; Esselman, P.C.; Price, L.B.; Cress, M.E.; Conley, K.E. Large energetic adaptations of elderly muscle to resistance and endurance training. *J. Appl. Physiol.* **2001**, *90*, 1663–1670. [CrossRef] [PubMed]
147. Jacobs, R.A.; Lundby, C. Mitochondria express enhanced quality as well as quantity in association with aerobic fitness across recreationally active individuals up to elite athletes. *J. Appl. Physiol.* **2013**, *114*, 344–350. [CrossRef]
148. Parise, G.; Brose, A.N.; Tarnopolsky, M.A. Resistance exercise training decreases oxidative damage to DNA and increases cytochrome oxidase activity in older adults. *Exp. Gerontol.* **2005**, *40*, 173–180. [CrossRef] [PubMed]
149. Parise, G.; Phillips, S.M.; Kaczor, J.J.; Tarnopolsky, M.A. Antioxidant enzyme activity is up-regulated after unilateral resistance exercise training in older adults. *Free Radic. Biol. Med.* **2005**, *39*, 289–295. [CrossRef]
150. Ji, L.L.; Zhang, Y. Antioxidant and anti-inflammatory effects of exercise: Role of redox signaling. *Free Radic. Res.* **2014**, *48*, 3–11. [CrossRef] [PubMed]
151. Melov, S.; Tarnopolsky, M.A.; Beckman, K.; Felkey, K.; Hubbard, A. Resistance exercise reverses aging in human skeletal muscle. *PLoS ONE* **2007**, *2*, e465. [CrossRef]
152. Tarnopolsky, M.; Zimmer, A.; Paikin, J.; Safdar, A.; Aboud, A.; Pearce, E.; Roy, B.; Doherty, T. Creatine monohydrate and conjugated linoleic acid improve strength and body composition following resistance exercise in older adults. *PLoS ONE* **2007**, *2*, e991. [CrossRef] [PubMed]
153. Tarnopolsky, M.A. Mitochondrial DNA shifting in older adults following resistance exercise training. *Appl. Physiol. Nutr. Metab.* **2009**, *34*, 348–354. [CrossRef]
154. Pesta, D.; Hoppel, F.; Macek, C.; Messner, H.; Faulhaber, M.; Kobel, C.; Parson, W.; Burtscher, M.; Schocke, M.; Gnaiger, E. Similar qualitative and quantitative changes of mitochondrial respiration following strength and endurance training in normoxia and hypoxia in sedentary humans. *Am. J. Physiol. Regul. Integr. Comp. Physiol.* **2011**, *301*, R1078–R1087. [CrossRef] [PubMed]
155. Porter, C.; Reidy, P.; Bhattarai, N.; Sidossis, L.; Rasmussen, B. Resistance exercise training alters mitochondrial function in human skeletal muscle. *Med. Sci. Sports Exerc.* **2014**, *47*, 1922. [CrossRef] [PubMed]
156. Taivassalo, T.; Fu, K.; Johns, T.; Arnold, D.; Karpati, G.; Shoubridge, E.A. Gene shifting: A novel therapy for mitochondrial myopathy. *Hum. Mol. Genet.* **1999**, *8*, 1047–1052. [CrossRef] [PubMed]
157. Murphy, J.L.; Blakely, E.L.; Schaefer, A.M.; He, L.; Wyrick, P.; Haller, R.G.; Taylor, R.W.; Turnbull, D.M.; Taivassalo, T. Resistance training in patients with single, large-scale deletions of mitochondrial DNA. *Brain* **2008**, *131*, 2832–2840. [CrossRef]
158. Ji, L.L. Exercise at old age: Does it increase or alleviate oxidative stress? *Ann. N. Y. Acad. Sci.* **2001**, *928*, 236–247. [CrossRef]
159. Izzotti, A. Genomic biomarkers and clinical outcomes of physical activity. *Ann. N. Y. Acad. Sci.* **2011**, *1229*, 103–114. [CrossRef]

160. Couppe, C.; Svensson, R.B.; Grosset, J.-F.; Kovanen, V.; Nielsen, R.; Olsen, M.; Larsen, J.; Praet, S.E.; Skovgaard, D.; Hansen, M.; et al. Life-long endurance running is associated with reduced glycation and mechanical stress in connective tissue. *Age* **2013**, *36*, 9665. [CrossRef]
161. Head, D.; Bugg, J.M.; Goate, A.M.; Fagan, A.M.; Mintun, M.A.; Benzinger, T.; Holtzman, D.M.; Morris, J.C. Exercise engagement as a moderator of the effects of apoe genotype on amyloid deposition. *Arch. Neurol.* **2012**, *69*, 636–643.
162. Tam, B.T.; Siu, P.M. Autophagic cellular responses to physical exercise in skeletal muscle. *Sports Med.* **2014**, *44*, 625–640. [CrossRef]
163. Erlich, A.T.; Brownlee, D.M.; Beyfuss, K.; Hood, D.A. Exercise induces TFEB expression and activity in skeletal muscle in a PGC-1α-dependent manner. *Am. J. Physiol. Cell Physiol.* **2018**, *314*, C62–C72. [CrossRef]
164. Kim, Y.; Hood, D.A. Regulation of the autophagy system during chronic contractile activity-induced muscle adaptations. *Physiol. Rep.* **2017**, *5*, e13307. [CrossRef]
165. Lira, V.A.; Okutsu, M.; Zhang, M.; Greene, N.P.; Laker, R.C.; Breen, D.S.; Hoehn, K.L.; Yan, Z. Autophagy is required for exercise training-induced skeletal muscle adaptation and improvement of physical performance. *FASEB J. Off. Publ. Fed. Am. Soc. Exp. Biol.* **2013**, *27*, 4184–4193. [CrossRef]
166. Pedersen, B.K.; Ostrowski, T.; Rohde, K.; Bruunsgaard, H. The cytokine response to strenuous exercise. *Can. J. Physiol. Pharm.* **1998**, *76*, 505–511. [CrossRef]
167. Proske, U.; Morgan, D.L. Muscle damage from eccentric exercise: Mechanism, mechanical signs, adaptation and clinical applications. *J. Physiol.* **2001**, *537*, 333–345. [CrossRef]
168. Greiwe, J.S.; Cheng, B.O.; Rubin, D.C.; Yarasheski, K.E.; Semenkovich, C.F. Resistance exercise decreases skeletal muscle tumor necrosis factor alpha in frail elderly humans. *FASEB J. Off. Publ. Fed. Am. Soc. Exp. Biol.* **2001**, *15*, 475–482.
169. McFarlin, B.K.; Flynn, M.G.; Phillips, M.D.; Stewart, L.K.; Timmerman, K.L. Chronic resistance exercise training improves natural killer cell activity in older women. *J. Gerontol. A Biol. Sci. Med. Sci.* **2005**, *60*, 1315–1318. [CrossRef]

© 2019 by the authors. Licensee MDPI, Basel, Switzerland. This article is an open access article distributed under the terms and conditions of the Creative Commons Attribution (CC BY) license (http://creativecommons.org/licenses/by/4.0/).

MDPI
St. Alban-Anlage 66
4052 Basel
Switzerland
Tel. +41 61 683 77 34
Fax +41 61 302 89 18
www.mdpi.com

Biology Editorial Office
E-mail: biology@mdpi.com
www.mdpi.com/journal/biology

www.ingramcontent.com/pod-product-compliance
Lightning Source LLC
LaVergne TN
LVHW071940080526
838202LV00064B/6646